W0036743

QCD@WORK

Related Titles from AIP Conference Proceedings

594 Hadrons and Nuclei: First International Symposium
Edited by Il-Tong Cheon, Taekeun Choi, Seung-Woo Hong, and Su Houng Lee, November 2001, 0-7354-0037-7

562 Particles and Fields: Ninth Mexican School
Edited by Gerardo Herrera Corral and Lukas Nellen, April 2001, 1-56396-998-X

549 Intersections of Particle and Nuclear Physics: 7th Conference, CIPANP2000
Edited by Zohreh Parsa and William J. Marciano, December 2000, 1-56396-978-5

539 Symmetries in Subatomic Physics: 3rd International Symposium
Edited by X.-H. Guo, A. W. Thomas, and A. G. Williams, October 2000, 1-56396-964-5

536 Instrumentation in Elementary Particle Physics: VIII ICFA School
Edited by Sehban Kartal, September 2000, 1-56396-960-2

531 Particles and Fields: Seventh Mexican Workshop
Edited by Alejandro Ayala, Guillermo Contreras, and Gerardo Herrera, July 2000, 1-56396-954-8

508 Hadron Physics: Effective Theories of Low Energy QCD
Edited by A. H. Blin, B. Hiller, M. C. Ruivo, C. A. Sousa, and E. van Beveren, March 2000, 1-56396-927-0

494 New Directions in Quantum Chromodynamics
Edited by Chueng-Ryong Ji and Dong-Pil Min, November 1999, 1-56396-908-4

482 RHIC Physics and Beyond: Kay Kay Gee Day
Edited by Berndt Müller and Robert Pisarski, July 1999, 1-56396-878-9

448 Workshop on Space Charge Physics in High Intensity Hadron Rings
Edited by A. U. Luccio and W. T. Weng, October 1998, 1-56396-824-X

432 Hadron Spectroscopy: Seventh International Conference
Edited by Suh-Urk Chung and Hans J. Willutzki, June 1998, 1-56396-765-0

To learn more about these titles, or the AIP Conference Proceedings Series, please visit the webpage **http://proceedings.aip.org**

QCD@WORK

International Workshop on Quantum Chromodynamics: Theory and Experiment

Martina Franca, Bari, Italy 16–20 June 2001

EDITORS

Pietro Colangelo
I.N.F.N., Sezione di Bari, Italy

Giuseppe Nardulli
University of Bari, Italy

Melville, New York, 2001

AIP CONFERENCE PROCEEDINGS ■ VOLUME 602

Editors:

Pietro Colangelo
Istituto Nazionale di Fisica Nucleare
Sezione di Bari
via Amendola n. 173
I-70126 Bari
ITALY
E-mail: Pietro.Colangelo@ba.infn.it

Giuseppe Nardulli
Dipartimento di Fisica
Università di Bari
via Amendola n. 173
I-70126 Bari
ITALY
E-mail: Giuseppe.Nardulli@ba.infn.it

L.C. Catalog Card No. 2001097638
ISBN 0-7354-0046-6
ISSN 0094-243X
Printed in the United States of America

CONTENTS

Preface ix

LOW ENERGY AND NONPERTURBATIVE QCD

Pion Physics at Low Energy and High Accuracy 3
H. Leutwyler
Large-N_c QCD and Low Energy Interactions 14
E. de Rafael
A New Result on the Direct CP Violation in Two Pion Decays of the Neutral Kaon 28
D. Madigojine for the NA 48 Collaboration
Unitarization of the Complete Meson-Meson Scattering at One Loop in Chiral Perturbation Theory 34
J. R. Peláez and A. Gómez Nicola
Simple Check of the Vacuum Structure in full QCD Lattice Simulations 40
S. Dürr

CHALLENGES IN PERTURBATIVE QCD

Problems and Challenges in Perturbative QCD 51
P. Nason
Theoretical Aspects of HERA Physics 60
S. Forte
Deep Inelastic Scattering Experiments with Unpolarised Beams 69
A. De Roeck
HERMES Past and Future: 1995–2000 and 2001–2005 83
E. C. Aschenauer for the HERMES Collaboration
QCD Analysis of Polarized Deep Inelastic Scattering Data 94
J. Blümlein and *H. Böttcher*
Experimental Status of α_s Measurements at LEP 100
R. W. L. Jones
Multi-Jet Event Shapes in QCD Hard Processes 108
A. Banfi, *G. Marchesini*, G. Smye, and G. Zanderighi
Measurements Using Multijet Final States from Hadronic Decays of the Z Boson 114
H. Jeremie for the OPAL Collaboration
Measurements of Color Transparency 121
S. Heppelmann for the E850 (EVA) Collaboration
Soft Gluons and the Energy Dependence of Total Cross-Sections 127
R. M. Godbole, A. Grau, *G. Pancheri*, and Y. N. Srivastava

*Italicized name indicates the author who presented the paper.

The Diagonal Ghost Equation Ward Identity for Yang-Mills Theories in the Maximal Abelian Gauge 139
A. R. Fazio
SUSY Scaling Violations and UHECR 145
C. Coriano' and A. E. Faraggi

HEAVY QUARKS

Charmless 2-Body B Decays: A Way to α and γ 153
F. Ferroni
CP Violation at the B Factories 165
L. Lanceri
Aspects of QCD Factorization 168
M. Neubert
Two Body B Decays, Factorization, and Λ_{QCD}/m_b Corrections 180
M. Ciuchini, E. Franco, *G. Martinelli*, M. Pierini, and L. Silvestrini
QCD Sum Rules for Heavy Flavour Physics 194
A. Khodjamirian
Charming Penguin in Nonleptonic B Decays 206
T. N. Pham
Direct CP Violation in Radiative B Decays 212
T. Hurth and T. Mannel
New Results on Charm Photoproduction at Fermilab 220
S. P. Ratti, J. Link, M. Reyes, P. M. Yager, J. Anjos, I. Bediaga, C. Gobel, J. Magnin, A. Massafferri, J. M. de Miranda, I. M. Pepe, A. C. dos Reis, S. Carrillo, E. Casimiro, A. Sánchez-Hernández, C. Uribe, F. Vasquez, L. Cinquini, J. P. Cumalat, B. O'Reilly, J. E. Ramirez, E. W. Vaandering, J. N. Butler, H. W. K. Cheung, I. Gaines, P. H. Garbincius, L. A. Garren, E. Gottschalk, P. H. Kasper, A. E. Kreymer, R. Kuschke, S. Bianco, F. L. Fabbri, S. Sarwar, A. Zallo, C. Cawlfield, D. Y. Kim, A. Rahimi, J. Wiss, R. Gardner, A. Kryemadhi, Y. S. Chung, J. S. Kang, B. R. Ko, J. W. Kwak, K. B. Lee, H. Park, G. Alimonti, M. Boschini, B. Caccianiga, P. D'Angelo, M. DiCorato, P. Dini, M. Giammarchi, P. Inzani, F. Leveraro, S. Malvezzi, D. Menasce, M. Mezzadri, L. Milazzo, L. Moroni, D. Pedrini, C. Pontoglio, F. Prelz, M. Rovere, S. Sala, T. F. Davenport III, L. Agostino, V. Arena, G. Boca, G. Bonomi, G. Gianini, G. Liguori, M. M. Merlo, D. Pantea, C. Riccardi, I. Segoni, P. Vitulo, H. Hernandez, A. M. Lopez, H. Mendez, L. Mendez, E. Montiel, D. Olaya, A. Paris, J. Quinones, C. Rivera, W. Xiong, Y. Zhang, J. R. Wilson, K. Cho, T. Handler, R. Mitchell, D. Engh, M. Hosack, W. E. Johns, M. Nehring, P. D. Sheldon, K. Stenson, M. Webster, and M. Sheaff
Dalitz Decays of Charmed Mesons 229
A. Palano
The $D^0 \to K^0 \overline{K}^0$ Decay Beyond Factorization 242
J. O. Eeg, *S. Fajfer*, and J. Zupan

*Italicized name indicates the author who presented the paper.

Final State Interaction for Non-Leptonic Exclusive Charm Decays 248
F. Buccella
A Quark Loop Model for Heavy Mesons 253
A. Deandrea
Mass Effects in the Emission of Gluons from Heavy Quarks at High Energies 259
J. Fuster, M. J. Costa, and P. Tortosa
Large Order Behavior in Perturbation Theory of the Pole Mass and the Singlet Static Potential 265
A. Pineda
Renormalization of HQET at Three Loops 271
A. G. Grozin

HOT AND DENSE QCD

From SPS to RHIC: Breaking the Barrier to the Quark-Gluon Plasma 281
U. Heinz
Charmonia States Suppression and Transverse Momentum Distribution in Pb-Pb Collisions at the CERN SPS 293
A. B. Kurepin for the NA50 Collaboration
Results on Hyperon Production from CERN NA57 Experiment 299
V. Manzari for the NA57 Collaboration
The First Year at RHIC 307
T. S. Ullrich
Identified Charged Particle Production in Au+Au Interactions at STAR-RHIC 319
D. Cozza for the STAR Collaboration and the STAR-RICH Collaboration
Thermodynamics of 2 and 3 Flavour QCD 323
F. Karsch
Strangeness Production in a Statistical Effective Model of Hadronisation 333
F. Becattini and G. Pettini
Crystalline Color Superconductivity 339
K. Rajagopal
Color Superconductivity: Symmetries and Effective Lagrangians 352
F. Sannino
Effective Lagrangians for QCD at High Density 358
R. Casalbuoni

Participants 369
Author Index 371

*Italicized name indicates the author who presented the paper.

Preface

This volume is a record of QCD@Work, the International Workshop on QCD: Theory and Experiment, which was held at Martina Franca, near Bari (Italy) on June 16-20, 2001. It contains versions of the talks given at the meeting and of the posters presented there.

The aim of the workshop was to gather physicists involved in experiments and in theory for discussing the most recent and outstanding results in hadronic physics and quantum chromodynamics. It was possible to discuss the first results from the Brookhaven RHIC machine, together with the latest data from CERN, that are already providing us with deeper insights in the problem of the quark-gluon plasma and in the fascinating issue of the QCD phase diagram. On this subject, the discovery of the Colour Flavour Locking phase at high chemical potential and low temperature envisages the possibility of producing interesting predictions on the phenomenon of colour superconductivity, with observable phenomenological consequences.

Results on both polarized and unpolarized structure functions provide more and more stringent tests of the theoretical predictions based on perturbative QCD; other very interesting data result from experiments on color transparency and the precise measurements of the fundamental parameters of the Standard Model. Fascinating results are also obtained by multijet physics and the study of total cross sections.

The new measurements from B-factories already challenge our theoretical understanding of the interplay between weak and strong interactions. In this regard, significant amounts of data on non leptonic heavy meson decays from the factories has already been produced last year. These data, as well as the results on CP violation, will be among the most productive output of the experiments in the next few years and will seriously challenge the present theoretical approaches, be they based on systematic approximations such as QCD factorization, QCD Sum Rules and Lattice QCD, or on more phenomenological hypotheses.

Finally, the aspects of low energy QCD generally investigated by theoretical tools such as Chiral Perturbation Theory and large N_c QCD, have been recently discussed and offer new and stimulating challenges to the experiments. A similar effort is also continuously produced in the non perturbative regime of QCD, most notably by various Lattice QCD collaborations.

In order to discuss all these aspects, about 65 physicists met for the International Workshop on QCD at Martina Franca, a wonderful resort in the area of Valle d'Itria (Apulia); all the sessions were held in the Sala dell'Arcadia of Palazzo Ducale.

We had the privilege to list Tom Ypsilantis among the advisors of the workshop. He participated in the organization of the workshop at its early stage, before his premature departure. Tom was a very talented physicist and a very good friend; certainly he would have actively contributed to the meeting had he still been with us at the time of the workshop. This volume is dedicated to his memory.

The workshop was sponsored by Istituto Nazionale di Fisica Nucleare (INFN), by the Ministry of University and Scientific Research (MURST) under the project (PRIN) "Theoretical Physics of Fundamental Interactions", by the Administrative Council and by the Physics Department of the University of Bari. We gratefully acknowledge the financial support from all these Institutions. We also acknowledge the patronage of the Municipality of Martina Franca, which provided us with organization support and the availability of the magnificent Palazzo Ducale, and the advice of the Archimedes Center for Scientific Culture located in Palazzo Ducale. Finally, we are grateful to the Twin's Travels Agency and to the secretary of the workshop, Mrs. Fausta Cannillo, for their invaluable help in the organization of the event.

Pietro Colangelo
Giuseppe Nardulli
INFN and University of Bari

International Advisory Committee
G. Altarelli (CERN)
R. Gatto (University of Geneva, Chairman)
U. Heinz (Ohio State University)
N. Paver (University of Trieste)
S. Stone (University of Syracuse)
T. Ypsilantis (deceased)

Local Organizing Committee
P. Colangelo (INFN, Bari)
R.A. Fini (INFN, Bari)
E. Nappi (INFN, Bari)
G. Nardulli (University of Bari, Chairman)

LOW ENERGY AND NONPERTURBATIVE QCD

Pion Physics at Low Energy and High Accuracy

H. Leutwyler

Institute for Theoretical Physics, University of Bern, Sidlerstr. 5, CH-3012 Switzerland

Abstract. The role of the quark condensate for the low energy structure of QCD is discussed in some detail. In particular, the dependence of M_π on m_u and m_d and the low energy theorems for the $\pi\pi$ scattering amplitude are reviewed. The new data on K_{e_4} decay beautifully confirm the standard picture, according to which the quark condensate is the leading order parameter of the spontaneously broken chiral symmetry.

1. STANDARD MODEL FOR $\mathbf{E \ll M_W}$

At energies that are small compared to $\{M_W, M_Z, M_H\} = O(100\,\text{GeV})$, the weak interaction freezes out, because these energies do not suffice to excite the relevant degrees of freedom. As a consequence, the gauge group of the Standard Model, SU(3)×SU(2)×U(1), breaks down to the subgroup SU(3)×U(1) – only the photons, the gluons, the quarks and the charged leptons are active at low energies. Since the neutrini neither carry colour nor charge, they decouple.

The effective Lagrangian relevant at low energies is the one of QCD + QED. The strength of the interaction is characterized by the two coupling constants g and e. In contrast to the Standard Model, the SU(3)×U(1) Lagrangian does contain mass terms: the quark and lepton mass matrices m_q, m_ℓ. The field basis may be chosen such that m_q and m_ℓ are diagonal and positive.

The two gauge fields behave in a qualitatively different manner: while the photons do not carry electric charge, the gluons do carry colour. This difference is responsible for the fact that the strong interaction becomes strong at low energies, while the electromagnetic interaction becomes weak there, in fact remarkably weak: the photons and leptons essentially decouple from the quarks and gluons. For the QCD part, on the other hand, perturbation theory is useful only at high energies. In the low energy domain, the strong interaction is so strong that it confines the quarks and gluons. For the same reason, a term in the Lagrangian of the form $\sim \theta G_{\mu\nu}\tilde{G}^{\mu\nu}$ (where $G_{\mu\nu}$ is the gluon field strength) cannot a priori be dismissed, despite the fact that it represents a total derivative: it generates an electric dipole moment in the neutron, for instance. Conversely, the experimental fact that the dipole moment is smaller than 10^{-25} ecm implies that (in the basis where the quark mass matrix is diagonal, real and positive) the vacuum angle θ must be tiny, so that the Lagrangian is invariant under the discrete symmetries P, C and T, to a very high degree of accuracy.

The resulting effective low energy theory is mathematically even more satisfactory than the Standard Model as such – it does not involve scalar degrees of freedom and has fewer free parameters. Remarkably, this simple theory must describe the structure

CP602, *QCD@Work: International Workshop on Quantum Chromodynamics*
edited by P. Colangelo and G. Nardulli

of cold matter to a very high degree of precision, once the parameters in the Lagrangian are known. It in particular explains the size of the atoms in terms of the scale

$$a_B = \frac{4\pi}{e^2 m_e} ,$$

which only contains the two parameters e and m_e – these are indeed known to an incredible precision. Unfortunately, our ability to solve the QCD part of the theory is rather limited – in particular, we are still far from being able to demonstrate on the basis of the QCD Lagrangian that the strong interaction actually confines colour. Likewise, our knowledge of the magnitude of the light quark masses leaves to be desired – we need to know these more accurately in order to test ideas that might lead to an understanding of the mass pattern, such as the relations with the lepton masses that emerge from attempts at unifying the electroweak and strong forces.

2. SYMMETRIES OF MASSLESS QCD

In the following, I focus on the QCD part and switch the electromagnetic interaction off. It so happens that the interactions of u,d,s with the Higgs fields are weak, so that the masses m_u, m_d, m_s are small. Let me first set these parameters equal to zero and, moreover, send the masses of the heavy quarks, m_c, m_b, m_t to infinity. In this limit, the theory becomes a theoreticians paradise: the Lagrangian contains a single parameter, g. In fact, since the value of g depends on the running scale used, the theory does not contain any dimensionless parameter that would need to be adjusted to observation. In principle, this theory fully specifies all dimensionless observables as pure numbers, while dimensionful quantities like masses or cross sections can unambiguously be predicted in terms of the scale Λ_{QCD} or the mass of the proton. The resulting theory – QCD with three massless flavours – is among the most beautiful quantum field theories we have. I find it breathtaking that, at low energies, nature reduces to this beauty, as soon as the dressing with the electromagnetic interaction is removed and the Higgs condensate is replaced by one that does not hinder the light quarks, but is impenetrable for W and Z waves as well as for heavy quarks.

The Lagrangian of the massless theory, $\mathcal{L}^0_{QCD}$, has a high degree of symmetry that originates in the fact that the interaction among the quarks and gluons is flavour-independent and conserves helicity: $\mathcal{L}^0_{QCD}$ is invariant under independent flavour rotations of the three right- and left-handed quark fields. These form the group $G = \text{SU}(3)_R \times \text{SU}(3)_L$. The corresponding 16 currents $V^\mu_i \bar{q}\gamma^\mu \frac{1}{2}\lambda_i q$ and $A^\mu_i = \bar{q}\gamma^\mu\gamma_5 \frac{1}{2}\lambda_i q$ are conserved, so that their charges commute with the Hamiltonian:

$$[Q^V_i, H^0_{QCD}] = [Q^A_i, H^0_{QCD}] = 0 , \qquad i = 1, \ldots, 8 .$$

Vafa and Witten [1] have shown that the state of lowest energy is necessarily invariant under the vector charges: $Q^V_i|0\rangle = 0$. For the axial charges, however, there are the two possibilities characterized in table 1.

The observed spectrum does not contain parity doublets. In the case of the lightest meson, the pion, for instance, the lowest state with the same spin and flavour quantum

TABLE 1. Alternative realizations of the symmetry group $G = \mathrm{SU}(3)_R \times \mathrm{SU}(3)_L$.

$Q_i^A \|0\rangle = 0$	$Q_i^A \|0\rangle \neq 0$
Wigner-Weyl realization of G ground state is symmetric $\langle 0\|\bar{q}_R q_L\|0\rangle = 0$ ordinary symmetry spectrum contains parity partners degenerate multiplets of G	Nambu-Goldstone realization of G ground state is asymmetric $\langle 0\|\bar{q}_R q_L\|0\rangle \neq 0$ spontaneously broken symmetry spectrum contains Goldstone bosons degenerate multiplets of $\mathrm{SU}(3)_V \subset G$

numbers, but opposite parity is the $a_0(980)$. So, experiment rules out the first possibility: for dynamical reasons that yet remain to be understood, the state of lowest energy is an asymmetric state. Since the axial charges commute with the Hamiltonian, there must be eigenstates with the same energy as the ground state:

$$H^0_{QCD}\, Q_i^A\, |0\rangle = Q_i^A\, H^0_{QCD}\, |0\rangle = 0\,.$$

The spectrum must contain 8 states $Q_1^A|0\rangle, \ldots, Q_8^A|0\rangle$ with $E = \vec{P} = 0$, describing massless particles, the Goldstone bosons of the spontaneously broken symmetry. Moreover, these must carry spin 0, negative parity and form an octet of SU(3).

3. QUARK MASSES AS SYMMETRY BREAKING PARAMETERS

Indeed, the 8 lightest hadrons, $\pi^+, \pi^0, \pi^-, K^+, K^0, \bar{K}^0, K^-, \eta$, do have these quantum numbers, but massless they are not. This has to do with the deplorable fact that we are not living in paradise: the masses m_u, m_d, m_s are different from zero and thus allow the left-handed quarks to communicate with the right-handed ones. The full Hamiltonian is of the form

$$H_{QCD} = H^0_{QCD} + H^1_{QCD}\,, \qquad H^1_{QCD} = \int d^3x\, \bar{q}_R m\, q_L + \bar{q}_L m^\dagger q_R\,, \qquad m = \begin{pmatrix} m_u & & \\ & m_d & \\ & & m_s \end{pmatrix}.$$

The quark masses may be viewed as symmetry breaking parameters: the QCD-Hamiltonian is only approximately symmetric under independent rotations of the right- and left-handed quark fields, to the extent that these parameters are small. Chiral symmetry is thus broken in two ways:

- spontaneously $\qquad \langle 0|\bar{q}_R q_L|0\rangle \neq 0$
- explicitly $\qquad m_u, m_d, m_s \neq 0$

The consequences of the fact that the explicit symmetry breaking is small may be worked out by means of an effective field theory, "chiral perturbation theory" [2, 3, 4]. In this context, the heavy quarks do not play an important role – as the corresponding fields are singlets under $\mathrm{SU}(3)_R \times \mathrm{SU}(3)_L$, we may include their contributions in the symmetric part of the Hamiltonian, irrespective of the size of their mass.

Since the masses of the two lightest quarks are particularly small, the Hamiltonian of QCD is almost exactly invariant under the subgroup $SU(2)_R \times SU(2)_L$. The ground state spontaneously breaks that symmetry to the subgroup $SU(2)_V$ – the good old isospin symmetry discovered in the thirties of the last century [5]. The pions represent the corresponding Goldstone bosons [6], while the kaons and the η remain massive if the limit $m_u, m_d \to 0$ is taken at fixed m_s. In the following, I consider this framework and, moreover, ignore isospin breaking, setting $m_u = m_d = \hat{m}$.

If $SU(2)_R \times SU(2)_L$ was an exact symmetry, the pions would be strictly massless. According to Gell-Mann, Oakes and Renner [7], the square of the pion mass is proportional to the product of the quark masses and the quark condensate:

$$M_\pi^2 \simeq \frac{1}{F_\pi^2} \times (m_u + m_d) \times |\langle 0|\, \bar{u}u\, |0\rangle| \,. \tag{1}$$

The factor of proportionality is given by the pion decay constant F_π. The term $m_u + m_d$ measures the explicit breaking of chiral symmetry, while the quark condensate,

$$\langle 0|\, \bar{u}u\, |0\rangle = \langle 0|\, \bar{u}_R u_L\, |0\rangle + \text{c.c.} = \langle 0|\, \bar{d}d\, |0\rangle \,,$$

is a measure of the spontaneous symmetry breaking: it may be viewed as an order parameter and plays a role analogous to the spontaneous magnetization of a magnet.

4. ROLE OF THE QUARK CONDENSATE

The approximate validity of the relation (1) was put to question by Stern and collaborators [8], who pointed out that there is no experimental evidence for the quark condensate to be different from zero. Indeed, the dynamics of the ground state of QCD is not understood – it could resemble the one of an antiferromagnet, where, for dynamical reasons, the most natural candidate for an order parameter, the magnetization, happens to vanish. There are a number of theoretical reasons indicating that this scenario is unlikely:

(i) The fact that the pseudoscalar meson octet satisfies the Gell-Mann-Okubo formula remarkably well would then be accidental.

(ii) The value obtained for the quark condensate on the basis of QCD sum rules, in particular for the baryonic correlation functions [9], confirms the standard picture.

(iii) The lattice values [10] for the ratio $m_s/\hat{m}$ agree very well with the result of the standard chiral perturbation theory analysis [11], corroborating this picture further.

Quite irrespective, however, of whether or not the scenario advocated by Stern et al. is theoretically appealing, the issue can be subject to experimental test. In fact, significant progress has recently been achieved in this direction [12, 13]. The remainder of the talk concerns this matter.

The Gell-Mann-Oakes-Renner formula is not exact. The expansion of M_π^2 in powers of m_u, m_d contains an infinite sequence of contributions. The expansion starts with a term linear in the quark masses:

$$M_\pi^2 = M^2 - \frac{\bar{\ell}_3}{32\pi^2 F^2}\, M^4 + O(M^6)\,, \qquad M^2 \equiv (m_u + m_d)\, B \,. \tag{2}$$

The coefficient B of the linear term is given by the value of $|\langle 0|\,\bar{u}u\,|0\rangle|/F_\pi^2$ in the limit $m_u, m_d \to 0$, and F is the value of F_π in that limit. The Gell-Mann-Oakes-Renner formula is obtained by dropping the higher order contributions. These are dominated by the term of order M^4, which involves one of the coupling constants occurring in the effective Lagrangian at order p^4. More precisely, the formula involves the value of the running coupling constant ℓ_3 at scale $\mu = M$, which logarithmically depends on M. Expressed in terms of the corresponding intrinsic scale Λ_3, we have

$$\bar{\ell}_3 = \ln \frac{\Lambda_3^2}{M^2} \,. \tag{3}$$

The symmetry does not determine the numerical value of this scale. The crude estimates underlying the standard version of chiral perturbation theory [3] yield numbers in the range

$$0.2\,\text{GeV} < \Lambda_3 < 2\,\text{GeV} \,. \tag{4}$$

The term of order M^4 is then very small compared to the one of order M^2, so that the Gell-Mann-Oakes-Renner formula is obeyed very well. Stern and collaborators investigate the more general framework, referred to as "generalized chiral perturbation theory", where arbitrarily large values of $\bar{\ell}_3$ are considered. The quartic term in eq. (2) can then take values comparable to the "leading", quadratic one. If so, the dependence of M_π^2 on the quark masses would fail to be approximately linear, even for values of m_u and m_d that are small compared to the intrinsic scale of QCD. A different bookkeeping for the terms occurring in the chiral perturbation series is then needed [8] – the standard chiral power counting is adequate only if $\bar{\ell}_3$ is not too large.

5. QUARK MASS DEPENDENCE OF M_π AND F_π

The behaviour of the ratio M_π^2/M^2 as a function of $\hat{m}$ is indicated in fig. 1, taken from ref. [14]. The fact that the information about the value of Λ_3 is very meagre shows up through very large uncertainties. In particular, with $\Lambda_3 \simeq 0.5\,\text{GeV}$, the ratio M_π^2/M^2 would remain close to 1, on the entire interval shown. Note that outside the range (4), the dependence of M_π^2 on the quark masses would necessarily exhibit strong curvature.

The figure illustrates the fact that brute force is not the only way the very small values of m_u and m_d observed in nature can be reached through numerical simulations on a lattice. It suffices to equip the strange quark with the physical value of m_s and to measure the dependence of the pion mass on m_u, m_d in the region where M_π is comparable to M_K. A fit to the data based on eq.(2) should provide an extrapolation to the physical quark masses that is under good control[1]. Moreover, the fit would allow a determination of the scale Λ_3 on the lattice. This is of considerable interest, because that scale also shows up in other contexts, in the $\pi\pi$ scattering lengths, for example. For recent work in this direction, I refer to [17, 18].

[1] The logarithmic singularities occurring at next-to-next-to-leading order are also known [15] – for a detailed discussion, I refer to [16].

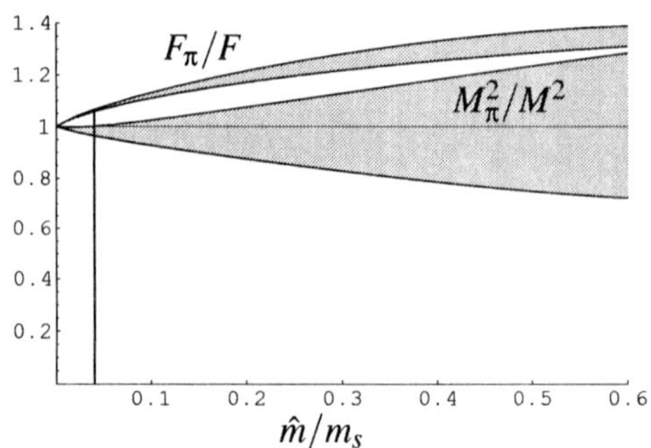

FIGURE 1. Dependence of the ratios F_π/F and M_π^2/M^2 on $\hat{m} = \frac{1}{2}(m_u + m_d)$. The strange quark mass is held fixed at the physical value. The vertical line corresponds to the physical value of $\hat{m}$.

For the pion decay constant, the expansion analogous to eq. (2) reads

$$F_\pi = F\left\{1 + \frac{\bar{\ell}_4 M^2}{16\pi^2 F^2} + O(M^4)\right\}, \qquad \bar{\ell}_4 = \ln \frac{\Lambda_4^2}{M^2}. \tag{5}$$

In this case, the relevant effective coupling constant is known rather well: chiral symmetry implies that it also determines the slope of the scalar form factor of the pion,

$$F_s(t) = \langle \pi(p')|\, \bar{u}u + \bar{d}d \,|\pi(p)\rangle = F_s(0)\left\{1 + \tfrac{1}{6}\langle r^2\rangle_s t + O(t^2)\right\}.$$

As shown in ref. [3], the expansion of $\langle r^2\rangle_s$ in powers of m_u, m_d starts with

$$\langle r^2\rangle_s = \frac{6}{(4\pi F)^2}\left\{\bar{\ell}_4 - \frac{13}{12} + O(M^2)\right\}. \tag{6}$$

Analyticity relates the scalar form factor to the $I = 0$ S–wave phase shift of $\pi\pi$ scattering [19]. Evaluating the relevant dispersion relation with the remarkably accurate information about the phase shift that follows from the Roy equations [16], one finds $\langle r^2\rangle_s = 0.61 \pm 0.04\,\mathrm{fm}^2$. Expressed in terms of the scale Λ_4, this amounts to

$$\Lambda_4 = 1.26 \pm 0.14\,\mathrm{GeV}. \tag{7}$$

Fig. 1 shows that this information determines the quark mass dependence of the decay constant to within rather narrow limits. The change in F_π occurring if $\hat{m}$ is increased from the physical value to $\frac{1}{2}m_s$ is of the expected size, comparable to the difference between F_K and F_π. The curvature makes it evident that a linear extrapolation from values of order $\hat{m} \sim \frac{1}{2}m_s$ down to the physical region is meaningless.

6. $\pi\pi$ SCATTERING

The experimental test of the hypothesis that the quark condensate represents the leading order parameter relies on the fact that $\langle 0|\, \bar{q}q \,|0\rangle$ not only manifests itself in the dependence of the pion mass on m_u and m_d, but also in the low energy properties of the $\pi\pi$ scattering amplitude.

At low energies, the scattering amplitude is dominated by the contributions from the S– and P–waves, because the angular momentum barrier suppresses the higher partial waves. Bose statistics implies that configurations with two pions and $\ell = 0$ are symmetric in flavour space and thus carry either isospin $I = 0$ or $I = 2$, so that there are two distinct S–waves. For $\ell = 1$, on the other hand, the configuration must be antisymmetric in flavour space, so that there is a single P–wave, $I = 1$. If the relative momentum tends to zero, only the S–waves contribute, through the corresponding scattering lengths a_0^0 and a_0^2 (the lower index refers to angular momentum, the upper one to isospin).

As shown by Roy [20], analyticity, unitarity and crossing symmetry subject the partial waves to a set of coupled integral equations. These equations involve two subtraction constants, which may be identified with the two S–wave scattering lengths a_0^0, a_0^2. If these two constants are given, the Roy equations allow us to calculate the scattering amplitude in terms of the imaginary parts above 800 MeV and the available experimental information suffices to evaluate the relevant dispersion integrals, to within small uncertainties [21]. In this sense, a_0^0, a_0^2 represent the essential parameters in low energy $\pi\pi$ scattering.

As a general consequence of the hidden symmetry, Goldstone bosons of zero momentum cannot interact with one another. Hence the scattering lengths a_0^0 and a_0^2 must vanish in the symmetry limit, $m_u, m_d \to 0$. These quantities thus also measure the explicit symmetry breaking generated by the quark masses, like M_π^2. In fact, Weinberg's low energy theorem [22] states that, to leading order of the expansion in powers of m_u and m_d, the scattering lengths are proportional to M_π^2, the factor of proportionality being fixed by the pion decay constant:[2]

$$a_0^0 = \frac{7M_\pi^2}{32\pi F_\pi^2} + O(\hat{m}^2)\,, \qquad a_0^2 = -\frac{M_\pi^2}{16\pi F_\pi^2} + O(\hat{m}^2)\,. \tag{8}$$

Chiral symmetry thus provides the missing element: in view of the Roy equations, Weinberg's low energy theorem fully determines the low energy behaviour of the $\pi\pi$ scattering amplitude. The prediction (8) corresponds to the dot on the left of fig. 2.

The prediction is of limited accuracy, because it only holds to leading order of the expansion in powers of the quark masses. In the meantime, the chiral perturbation series of the scattering amplitude has been worked out to two loops [23]. At first nonleading order of the expansion in powers of momenta and quark masses, the scattering amplitude can be expressed in terms of F_π, M_π and the coupling constants $\ell_1, \ldots, \ell_4$ that occur in the derivative expansion of the effective Lagrangian at order p^4. The terms ℓ_1 and ℓ_2 manifest themselves in the energy dependence of the scattering amplitude and can thus be determined phenomenologically. As discussed in section 5, the coupling constant ℓ_4 is known rather accurately from the dispersive analysis of the scalar form factor. The crucial term is ℓ_3 – the range considered for this coupling constant makes the difference between standard and generalized chiral perturbation theory. In the standard framework,

---

[2] The standard definition of the scattering length corresponds to a_0/M_π. It is not suitable in the present context, because it differs from the invariant scattering amplitude at threshold by a kinematic factor that diverges in the chiral limit.

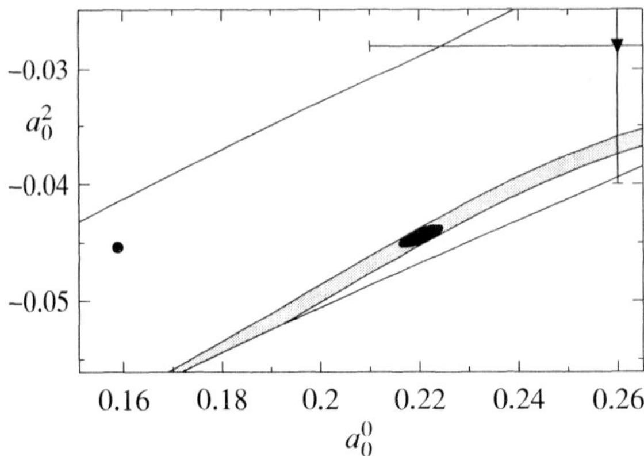

FIGURE 2. S–wave scattering lengths. The Roy equations only admit solutions in the "universal band", spanned by the two tilted lines. The dot indicates Weinberg's leading order result, while the small ellipse includes the higher order corrections, evaluated in the framework of standard chiral perturbation theory. In the generalized scenario, there is no prediction for a_0^0, but there is a correlation between a_0^0 and a_0^2, shown as a narrow strip. The triangle with error bars indicates the phenomenological range permitted by the old data, $a_0^0 = 0.26 \pm 0.05$, $a_0^2 = -0.028 \pm 0.012$ [27].

where the relevant scale is in the range (4), one finds that the leading order result is shifted into the small ellipse shown in fig. 2, which corresponds to [24, 25]:

$$a_0^0 = 0.220 \pm 0.005\,, \qquad a_0^2 = -0.0444 \pm 0.0010\,. \tag{9}$$

The numerical value quoted includes the higher order corrections (in the standard framework, the contributions from the corresponding coupling constants are tiny).

The corrections from the higher order terms in the Gell-Mann-Oakes-Renner relation can only be large if the estimate (4) for Λ_3 is totally wrong. As pointed out long ago [26], there is a low energy theorem that holds to first nonleading order and relates the S–wave scattering lengths to the scalar radius:

$$2a_0^0 - 5a_0^2 = \frac{3M_\pi^2}{4\pi F_\pi^2}\left\{1 + \frac{1}{3}M_\pi^2\langle r^2\rangle_s + \frac{41\,M_\pi^2}{192\,\pi^2 F_\pi^2}\right\} + O(\hat{m}^3)\,. \tag{10}$$

In this particular combination of scattering lengths, the term ℓ_3 drops out. The theorem thus correlates the two scattering lengths, independently of the numerical value of Λ_3. The correlation holds both in standard and generalized chiral perturbation theory. The corrections occurring in eq. (10) at order $\hat{m}^3$ have also been worked out. These are responsible for the fact that the narrow strip, which represents the correlation in fig. 2, is slightly curved.

7. IMPACT OF THE NEW K DECAY DATA

The final state interaction theorem implies that the phases of the form factors relevant for the decay $K \to \pi\pi e\nu$ are determined by those of the $I = 0$ S–wave and of the P–wave of elastic $\pi\pi$ scattering, respectively. Conversely, the analysis of the final state distribution observed in this decay yields a measurement of the phase difference $\delta(s) \equiv \delta_0^0(s) - \delta_1^1(s)$, in the region $4M_\pi^2 < s < M_K^2$. As discussed above, the Roy equations determine the behaviour of the phase shifts in terms of the two S–wave scattering lengths. Moreover, in view of the correlation between the two scattering lengths, a_0^2 is determined by a_0^0, so that the phase difference $\delta(s)$ can be calculated as a function of a_0^0 and q, where q is the c.m. momentum in units of M_π, $s = 4M_\pi^2(1+q^2)$. In the region of interest ($q < 1$, $0.18 < a_0^0 < 0.26$), the prediction reads

$$\delta_0^0 - \delta_1^1 = \frac{q}{\sqrt{1+q^2}}\,(a_0^0 + q^2 b + q^4 c + q^6 d) \pm e \tag{11}$$

$$b = 0.2527 + 0.151\,\Delta a_0^0 + 1.14\,(\Delta a_0^0)^2 + 35.5\,(\Delta a_0^0)^3\,,$$

$$c = 0.0063 - 0.145\,\Delta a_0^0\,, \qquad d = -0.0096\,,$$

with $\Delta a_0^0 = a_0^0 - 0.22$. The uncertainty in this relation mainly stems from the experimental input used in the Roy equations and is not sensitive to a_0^0:

$$e = 0.0035\,q^3 + 0.0015\,q^5\,. \tag{12}$$

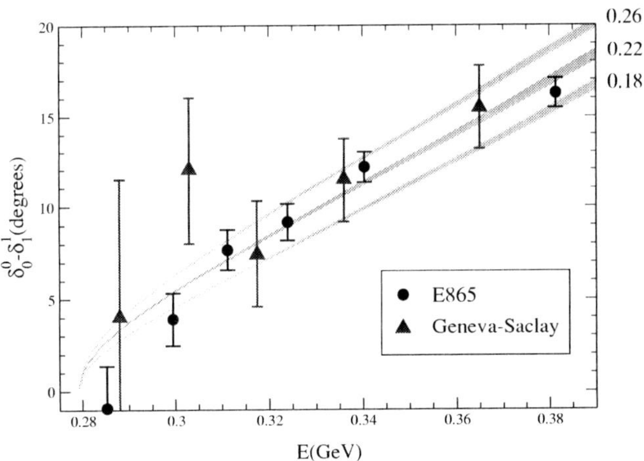

FIGURE 3. Phase relevant for the decay $K \to \pi\pi e\nu$. The three bands correspond to the three indicated values of the S–wave scattering length a_0^0. The uncertainties are dominated by those from the experimental input used in the Roy equations. The triangles are the data points of ref. [28], while the full circles represent the E865 results [13].

The prediction (11) is illustrated in fig. 3, where the energy dependence of the phase difference is shown for $a_0^0 = 0.18$, 0.22 and 0.26. The width of the corresponding bands indicates the uncertainties, which according to (12) grow in proportion to q^3 – in the range shown, they amount to less than a third of a degree.

The figure shows that the data of ref. [28] barely distinguish between the three values of a_0^0 shown. The results of the E865 experiment at Brookhaven [13] are significantly

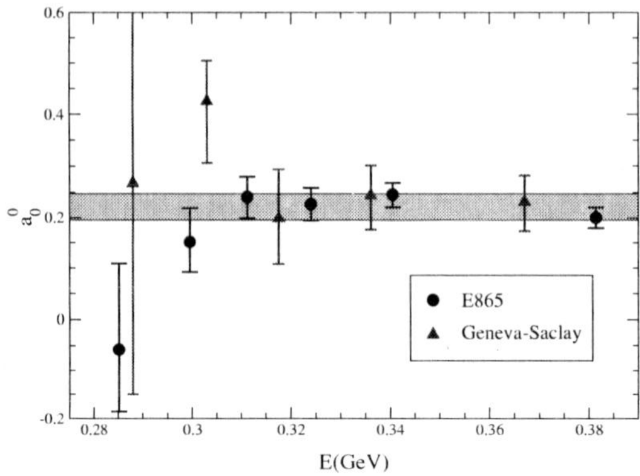

FIGURE 4. K_{e_4} data on the scattering length a_0^0. The triangles are the data points of ref. [28], while the full circles represent the E865 results [13]. The horizontal band indicates the statistical average of the 11 values for a_0^0 shown in the figure.

more precise, however. The best fit to these data is obtained for $a_0^0 = 0.218$, with $\chi^2 = 5.7$ for 5 degrees of freedom. This beautifully confirms the value in eq. (9), obtained on the basis of standard chiral perturbation theory. There is a marginal problem only with the bin of lowest energy: the corresponding scattering lengths are outside the region where the Roy equations admit solutions. In view of the experimental uncertainties attached to that point, this discrepancy is without significance: the difference between the central experimental value and the prediction amounts to 1.5 standard deviations. Note also that the old data are perfectly consistent with the new ones: the overall fit yields $a_0^0 = 0.221$ with $\chi^2 = 8.3$ for 10 degrees of freedom.

The relation (11) can be inverted, so that each one of the values found for the phase difference yields a measurement of the scattering length a_0^0. The result is shown in fig. 4. The experimental errors are remarkably small. It is not unproblematic, however, to treat the data collected in the different bins as statistically independent: in the presence of correlations, this procedure underestimates the actual uncertainties. Also, since the phase difference rapidly rises with the energy, the binning procedure may introduce further uncertainties. To account for this, the final result given in ref. [12],

$$a_0^0 = 0.221 \pm 0.026 \,, \tag{13}$$

corresponds to the 95% confidence limit – in effect, this amounts to stretching the statistical error bar by a factor of two.

We may translate the result into an estimate for the magnitude of the coupling constant $\bar{\ell}_3$: the range (13) corresponds to $|\bar{\ell}_3| \lesssim 16$. Although this is a coarse estimate, it implies that the Gell-Mann-Oakes-Renner relation does represent a decent approximation: more than 94% of the pion mass stems from the first term in the quark mass expansion (2), i.e. from the term that originates in the quark condensate. This demonstrates that there is no need for a reordering of the chiral perturbation series based on $SU(2)_R \times SU(2)_L$. In that context, the generalized scenario has served its purpose and can now be dismissed.

A beautiful experiment is under way at CERN [29], which exploits the fact that $\pi^+\pi^-$ atoms decay into a pair of neutral pions, through the strong transition $\pi^+\pi^- \rightarrow \pi^0\pi^0$. Since the momentum transfer nearly vanishes, only the scattering lengths are relevant: at leading order in isospin breaking, the transition amplitude is proportional to $a_0^0 - a_0^2$. The corrections at next–to–leading order are now also known [30]. Hence a measurement of the lifetime of a $\pi^+\pi^-$ atom amounts to a measurement of this combination of scattering lengths. At the planned accuracy of 10% for the lifetime, the experiment will yield a measurement of the scattering lengths to 5%, thereby subjecting chiral perturbation theory to a very sensitive test.

REFERENCES

1. C. Vafa and E. Witten, *Nucl. Phys.* B **234** (1984) 173.
2. S. Weinberg, *Physica* A **96** (1979) 327.
3. J. Gasser and H. Leutwyler, *Annals Phys.* **158** (1984) 142.
4. For an overview of recent work based on this method, see for instance: *Proc. 3rd Workshop on Chiral Dynamics: Theory and Experiment, Newport News, Virginia, July 2000*, eds. A. Bernstein, J. Goity and U. Meissner, *World Scientific*, in press.
5. W. Heisenberg, *Z. Phys.* **77** (1932) 1; N. Kemmer, *Proc. Cambridge Phil. Soc.* **34** (1938) 354.
6. Y. Nambu, *Phys. Rev. Lett.* **4** (1960) 380; *Phys. Rev.* **117** (1960) 648.
7. M. Gell-Mann, R. J. Oakes and B. Renner, *Phys. Rev.* **175** (1968) 2195.
8. M. Knecht, B. Moussallam, J. Stern and N. H. Fuchs, *Nucl. Phys.* B **457** (1995) 513 [hep-ph/9507319]; *ibid.* B **471** (1996) 445 [hep-ph/9512404].
9. B. L. Ioffe, *Nucl. Phys.* B **188** (1981) 317.
10. V. Lubicz, *Nucl. Phys.* B *(Proc. Suppl.)* **94** (2001) 116 [hep-lat/0012003].
11. H. Leutwyler, *Phys. Lett.* B **378** (1996) 313 [hep-ph/9602366].
12. G. Colangelo, J. Gasser and H. Leutwyler, *Phys. Rev. Letters* **86** (2001) 5008 [hep-ph/0103063].
13. S. Pislak et al. [BNL-E865 Collaboration], hep-ex/0106071.
14. H. Leutwyler, *Nucl. Phys.* B *(Proc. Suppl.)* **94** (2001) 108 [hep-ph/0011049].
15. G. Colangelo, *Phys. Lett.* B **350** (1995) 85; *ibid.* B **361** (1995) 234 (E) [hep-ph/9502285].
16. G. Colangelo, J. Gasser and H. Leutwyler, *Nucl. Phys.* B **603** (2001) 125 [hep-ph/0103088].
17. J. Heitger, R. Sommer and H. Wittig [ALPHA Collaboration], *Nucl. Phys.* B **588** (2000) 377 [hep-lat/0006026].
18. S. Dürr, hep-lat/0103011.
19. J. F. Donoghue, J. Gasser and H. Leutwyler, *Nucl. Phys.* B **343** (1990) 341.
20. S. M. Roy, *Phys. Lett.* B **36** (1971) 353.
21. B. Ananthanarayan, G. Colangelo, J. Gasser and H. Leutwyler, hep-ph/0005297, *Phys. Rep.*, in print.
22. S. Weinberg, *Phys. Rev. Lett.* **17** (1966) 616.
23. J. Bijnens, G. Colangelo, G. Ecker, J. Gasser and M. E. Sainio, *Phys. Lett.* B **374** (1996) 210 [hep-ph/9511397]; *Nucl. Phys.* B **508** (1997) 263; *ibid.* B **517** (1998) 639 (E) [hep-ph/9707291].
24. G. Amoros, J. Bijnens and P. Talavera, *Nucl. Phys.* B **585** (2000) 293, *ibid.* B **598** (2000) 293 (E) [hep-ph/0003258].
25. G. Colangelo, J. Gasser and H. Leutwyler, *Phys. Lett.* B **488** (2000) 261 [hep-ph/0007112].
26. J. Gasser and H. Leutwyler, *Phys. Lett.* B **125** (1983) 325.
27. C. D. Froggatt and J. L. Petersen, *Nucl. Phys.* B **129** (1977) 89; M. M. Nagels et al., *Nucl. Phys.* B **147** (1979) 189.
28. L. Rosselet et al., *Phys. Rev.* D **15** (1977) 574.
29. B. Adeva et al., *CERN proposal CERN/SPSLC* 95-1 (1995); http://dirac.web.cern.ch/DIRAC/.
30. J. Gasser, V. E. Lyubovitskij and A. Rusetsky, *Phys. Lett.* B **471** (1999) 244 [hep-ph/9910438]; H. Sazdjian, *Phys. Lett.* B **490** (2000) 203 [hep-ph/0004226] and references therein.

Large–N_c QCD and Low Energy Interactions

Eduardo de Rafael

Centre de Physique Théorique
CNRS–Luminy, case 907, F-13288 Marseille cedex 9, France

Abstract. This talk reviews recent progress in formulating the dynamics of the electroweak interactions of hadrons at low energies, within the framework of the $1/N_c$–expansion in QCD. The emphasis is put on the basic issues of the approach.

INTRODUCTION

In the Standard Model, the electroweak interactions of hadrons at very low energies can be described by an effective Lagrangian which only has as active degrees of freedom the flavour $SU(3)$ octet of the low–lying pseudoscalar particles. The underlying theory, however, is the gauge theory $SU(3)_C \times SU(2)_L \times U(1)_{Y_w}$ which has as dynamical degrees of freedom quarks and gauge fields. Going from these degrees of freedom at high energies to an effective description in terms of mesons at low energies is, in principle, a problem which should be understood in terms of the evolution of the renormalization group from short–distances to long–distances. Unfortunately, it is difficult to carry out explicitly this evolution because at energies, typically of a few GeV, non–perturbative dynamics like spontaneous chiral symmetry breaking and color confinement sets in.

It has been possible, however, to integrate out the heavy degrees of freedom of the Standard Model gauge theory, in the presence of the strong interactions, perturbatively, thanks to the asymptotic freedom property of QCD at short–distances. This procedure results in an effective Lagrangian which consists of the usual QCD Lagrangian with the light quark flavours u, d, and s still active, plus a sum of effective four–quark operators of the light quarks, modulated by c–number coefficients (the Wilson coefficients,) which are functions of the masses of the heavy particles which have been integrated out, and of the renormalization scale. We are still left with the evolution from this effective field theory, appropriate at intermediate scales higher than a few GeV, to an effective Lagrangian description in terms of the low–lying pseudoscalar particles which are the Goldstone modes associated to the spontaneous chiral symmetry breaking of the Standard Gauge Theory in the light quark sector. In this talk, I shall review recent progress which has been made in approaching this last step, when the problem is formulated within the framework of QCD in the limit of a large number of colours N_c. The emphasis is put on basic issues. Details of the applications reviewed here can be found in the original publications.

The suggestion to keep N_c as a free parameter was first made by G. 't Hooft [1] as a possible way to approach the study of non–perturbative aspects of QCD. The limit $N_c \to \infty$ is taken with the product $\alpha_s N_c$ kept fixed and it is highly non–trivial. In spite

CP602, *QCD@Work: International Workshop on Quantum Chromodynamics*
edited by P. Colangelo and G. Nardulli

of the efforts of many illustrious theorists who have worked on the subject, QCD in the large–N_c limit still remains unsolved; but many interesting properties have been proved, which suggest that, indeed, the theory in this limit has the bulk of the non–perturbative properties of the full QCD. In particular, it has been shown that, if confinement persists in this limit, there is spontaneous chiral symmetry breaking [2].

The spectrum of the theory in the large–N_c limit consists of an infinite number of narrow stable meson states which are flavour nonets [3]. This spectrum looks *a priori* rather different to the real world. The vector and axial–vector spectral functions measured in $e^+e^- \to$ hadrons and in the hadronic τ–decay show indeed a richer structure than just a sum of narrow states. There are, however, many instances where one is only interested in observables which are given by weighted integrals of some hadronic spectral functions. In these cases, it may be enough to know a few *global* properties of the hadronic spectrum, and to have a good interpolation. Typical examples of that are, as we shall see, the coupling constants of the effective chiral Lagrangian of QCD at low energies, as well as the coupling constants of the effective chiral Lagrangian of the electroweak interactions of pseudoscalar particles in the Standard Model. Some of these couplings are needed in order to understand K–Physics quantitatively. In these examples the *hadronic world* predicted by large–N_c QCD provides already a good approximation to the real hadronic spectrum. It is in this sense that I shall show that large–N_c QCD is a very useful phenomenological approach for understanding non–perturbative QCD physics at low energies.

There are a number of good articles and lecture notes on large–N_c QCD in the literature [1]. Here I shall limit myself to make a couple of comments on prejudices one often encounters concerning the QCD large–N_c limit.

- The first prejudice has to do with the "extrapolation" from $N_c = 3$ to $N_c = \infty$. In fact, N_c is really used as a label to select specific topologies among Feynman diagrams. The topology which corresponds to the highest power in the N_c–label is the one which selects *planar* diagrams only; and the claim is that it is this class that already provides a good approximation to the full theory.
- The second prejudice has to do with the fact that some physical quantities are absent in the large–N_c limit; the η'–mass e.g. is zero in that limit. That does not mean that the $1/N_c$–expansion fails, as sometimes it is argued, but rather that some observables only appear at subleading topologies, (*planar* diagrams with *one handle* in this case,) much the same as in QED, there is no light–by–light scattering at the Born approximation and one has to go to *one loop* diagrams to evaluate its leading behaviour.

THE CHIRAL LAGRANGIAN AT LOW ENERGIES

The strong and electroweak interactions of the Goldstone modes at very low energies are described by an effective Lagrangian which has terms with an increasing number of

---

[1] See e.g., the book in ref. [4] and the lectures in [5]

derivatives (and quark masses if explicit chiral symmetry breaking is taken into account.) These terms are modulated by couplings which encode the dynamics of the underlying theory. The evaluation of these couplings from the underlying theory is the question we are interested in. Typical terms of the chiral Lagrangian are

$$\mathcal{L}_{\text{eff}} = \underbrace{\frac{1}{4}F_0^2 \operatorname{tr}\left(D_\mu U D^\mu U^\dagger\right)}_{\pi\pi\to\pi\pi,\quad K\to\pi e\nu} + \underbrace{L_{10}\operatorname{tr}\left(U^\dagger F_{R\mu\nu} U F_L^{\mu\nu}\right)}_{\pi\to e\nu\gamma} + \cdots$$

$$\underbrace{e^2 C \operatorname{tr}\left(Q_R U Q_L U^\dagger\right)}_{-e^2 C\frac{2}{F_0^2}(\pi^+\pi^- + K^+K^-)+\cdots} + \cdots - \underbrace{\frac{G_F}{\sqrt{2}} V_{ud} V_{us}^* \, g_8 F_0^2 \left(D_\mu U D^\mu U^\dagger\right)_{23}}_{K\to\pi\pi,\quad K\to\pi\pi\pi,\quad +\cdots} + \cdots, \tag{1}$$

where U is a 3×3 unitary matrix in flavour space which collects the Goldstone fields and which under chiral rotations transforms as $U \to V_R U V_L^\dagger$. The first line shows typical terms of the strong interactions in the presence of external currents [6],[7, 8]; the second line shows typical terms which appear when photons, $W's$ and $Z's$ have been integrated out in the presence of the strong interactions. We show under the braces the typical physical processes to which each term contributes. Each term is modulated by a constant: F_0^2, L_{10},... C...g_8... which encode the underlying dynamics responsible for the appearance of the corresponding effective term. Knowing g_8 for example, means that we can calculate from first principles the dominant $\Delta I = 1/2$ transitions for K–decays to leading order in the chiral expansion.

There are two crucial observations concerning the relation of these low energy constants to the underlying theory, that I want to discuss.

- The low–energy constants of the Strong Lagrangian, like F_0^2 and L_{10}, are the coefficients of the *Taylor expansion* of appropriate QCD Green's Functions. For example, with $\Pi_{LR}(Q^2)$ the correlation function of a left–current with a right–current in the chiral limit, (where the light quark masses are neglected,) i.e.,

$$2i\int d^4x\, e^{iq\cdot x}\langle 0|T\left(\bar{u}_L\gamma^\mu d_L(x)\bar{u}_R\gamma^\nu d_R(0)^\dagger\right)|0\rangle = (q^\mu q^\nu - g^{\mu\nu}q^2)\Pi_{LR}(Q^2), \tag{2}$$

  the Taylor expansion

$$-Q^2\Pi_{LR}(Q^2)|_{Q^2\to 0} = F_0^2 - 4L_{10}\,Q^2 + O\left(Q^4\right), \tag{3}$$

  defines the constants F_0^2 and L_{10}.
- By contrast, the low–energy constants of the ElectroWeak Lagrangian, like e.g. C and g_8, are *integrals* of appropriate QCD Green's Functions. For example, including the effect of weak neutral currents [9],

$$C = \frac{3}{32\pi^2}\int_0^\infty dQ^2\left(1 - \frac{Q^2}{Q^2+M_Z^2}\right)\left(-Q^2\Pi_{LR}(Q^2)\right). \tag{4}$$

  Their evaluation appears to be, a priori, quite a formidable task because they require the knowledge of Green's functions at all values of the euclidean momenta;

i.e. they require a precise *matching* of the *short–distance* and the *long–distance* contributions of the underlying Green's functions.

The observations above are completely generic to the Standard Model independently of the $1/N_c$–expansion. The large–N_c approximation helps, however, because it restricts the *analytic structure* of the Green's functions in general, and $\Pi_{LR}(Q^2)$ in particular, to be *meromorphic functions*: they only have poles as singularities; e.g., in large–N_c QCD,

$$\Pi_{LR}(Q^2) = \sum_V \frac{f_V^2 M_V^2}{Q^2 + M_V^2} - \sum_A \frac{f_A^2 M_A^2}{Q^2 + M_A^2} - \frac{F_0^2}{Q^2}, \tag{5}$$

where the sums are, in principle, extended to an infinite number of states.

There are two types of important restrictions on Green's functions like $\Pi_{LR}(Q^2)$. One type follows from the fact that, as already stated above, the Taylor expansion at low euclidean momenta must match the low energy constants of the strong chiral Lagrangian. This results in a series of *long–distance sum rules* like e.g.

$$\sum_V f_V^2 - \sum_A f_A^2 = -4L_{10}. \tag{6}$$

Another type of constraints follows from the *short–distance properties* of the underlying Green's functions. The behaviour at large euclidean momenta of the Green's functions which govern the low energy constants of the chiral Lagrangian in Eq. (1) can be obtained from the operator product expansion (OPE) of local currents in QCD. In the large–N_c limit, this results in a series of algebraic sum rules [10] which restrict the coupling constants and masses of the hadronic poles. In the case of the LR–correlation function in Eq. (5) one has,

$$\text{No } \tfrac{1}{Q^2} \text{ term in OPE} \Rightarrow \sum f_V^2 M_V^2 - \sum f_A^2 M_A^2 - F_0^2 = 0\,, \quad \text{1st Weinberg sum rule.} \tag{7}$$

$$\text{No } \tfrac{1}{Q^4} \text{ term in OPE} \Rightarrow \sum f_V^2 M_V^4 - \sum f_A^2 M_A^4 = 0\,, \quad \text{2nd Weinberg sum rule.} \tag{8}$$

$$\text{Matching } \tfrac{1}{Q^6} \text{ terms in the OPE} \Rightarrow \sum f_V^2 M_V^6 - \sum f_A^2 M_A^6 = \langle O^{(6)} \rangle\,, \quad \text{ref. [10]}. \tag{9}$$

where [11], in large–N_c QCD,

$$\langle O^{(6)} \rangle = \left[-4\pi\alpha_s + O(\alpha_s^2)\right] \langle \bar{\psi}\psi \rangle^2. \tag{10}$$

The Minimal Hadronic Ansatz Approximation to Large–N_c QCD

In most cases of interest, the Green's functions which govern the low–energy constants of the chiral Lagrangian are order parameters of spontaneous chiral symmetry breaking; i.e. they vanish, in the chiral limit, order by order in the perturbative vacuum of QCD. That implies that they have a power fall–out in $1/Q^2$ at large–Q^2; like e.g., the function $\Pi_{LR}(Q^2)$, which as explicitly shown in Eq. (9), falls as $1/Q^6$. That also implies that within a finite radius in the complex Q^2–plane, these Green's functions in large–N_c

QCD, only have a *finite number of poles*. The natural question which arises is: WHAT IS THE MINIMAL NUMBER OF POLES REQUIRED TO SATISFY THE OPE CONSTRAINTS? The answer to that follows from a well known theorem in analysis [12] which we illustrate with the example of the Green's function

$$-Q^2\Pi_{LR}(Q^2) \equiv \Delta[z] \quad \text{with} \quad z = \frac{Q^2}{M_{V_1}^2}, \tag{11}$$

where for convenience we normalize Q^2 to the mass of the lowest vector state. The function $\Delta[z]$ has the property that

$$N - P = \frac{1}{2\pi i}\oint \frac{\Delta'[z]}{\Delta[z]} dz, \tag{12}$$

where N is the number of zeros and P is the number of poles inside the integration contour (a zero and/or a pole of order m is counted m times.) For a contour of radius sufficiently large so that the OPE applies, we simply have that $N - P = -p_{\mathrm{OPE}}$ where p_{OPE} denotes the *leading power* fall–out in $1/z$ predicted by the OPE. Since $N \geq 0$, it follows that $P \geq p_{\mathrm{OPE}}$. In our case [2] $p_{\mathrm{OPE}} = 2 \Rightarrow P \geq 2$ and the *minimal hadronic ansatz* (MHA) compatible with the OPE requires two poles: one vector state and one axial–vector state. The MHA approximation to the large–N_c expression in Eq. (5) is then the simple function

$$-Q^2\Pi_{LR}(Q^2) = F_0^2 \frac{M_V^2 M_A^2}{(Q^2 + M_V^2)(Q^2 + M_A^2)}. \tag{13}$$

Inserting this function in Eq. (4) gives a prediction to the low–energy constant C which governs the electromagnetic $\pi^+ - \pi^0 \equiv \Delta m_\pi$ mass difference, with the result [3]

$$\Delta m_\pi = (4.9 \pm 0.4)\,\mathrm{MeV}, \qquad \text{MHA to Large–}N_c\text{ QCD}, \tag{14}$$

to be compared with the experimental value

$$\Delta m_\pi = (4.5936 \pm 0.0005)\,\mathrm{MeV}, \qquad \text{Particle Data Book [15]}. \tag{15}$$

The shape of the function in Eq. (13), normalized to its value at $Q^2 = 0$, is shown in Fig. 1 below, (the continuous red curve.) It provides a good interpolation between the low–Q^2 regime where χPT applies and the high–Q^2 regime where the OPE applies. Also shown in the same plot is the experimental curve, (the green dotted curve) obtained from the ALEPH collaboration data [16]. Except for the intermediate energy region, where the MHA overestimates slightly the experimental curve, the overall agreement is quite remarkable.

---

[2] Notice that with the definition of $\Delta[z]$ in Eq. (11) the pion pole is removed.

[3] This is the result for $M_V = (748 \pm 29)\,\mathrm{MeV}$ and $g_A = \frac{M_V^2}{M_A^2} = 0.50 \pm 0.06$. These values follow from an overall fit to predictions of the low energy constants [13, 14].

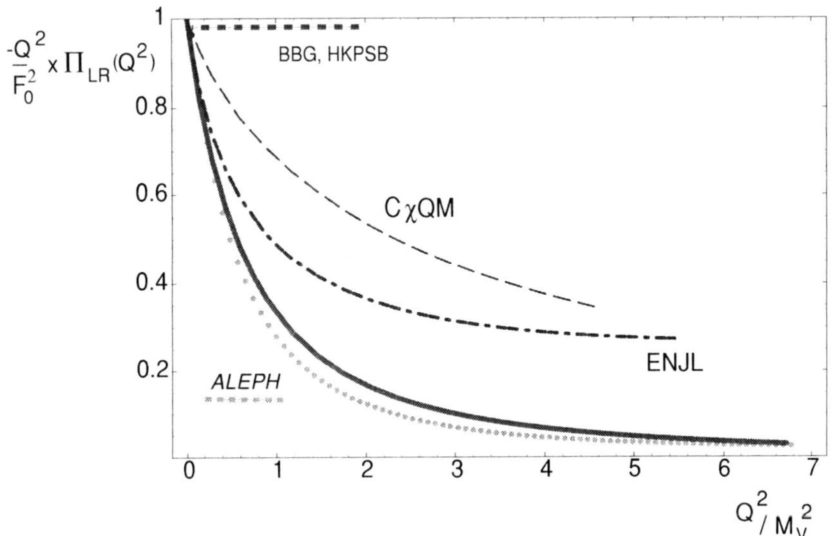

Fig. 1 *MHA (continuous curve,) versus ALEPH (dots) and other predictions*

The MHA approximation to large–N_c QCD is a starting point to do well defined approximations in nonperturbative QCD. The approximation has been tested with the ALEPH data [17]. In principle it is improvable: inserting more terms in the OPE provides extra sum rules which can be used to fix the extra hadronic parameters. We have made tests with models of large–N_c QCD [18, 19]. It has also been shown [20] that in the case of Π_{LR}, inserting an extra ρ'–like vector state, improves the overall picture; the Δm_π prediction in particular.

At this stage, it is perhaps illustrative to compare the large–N_c approach we have discussed so far with other analytic approaches which exist in the literature. The Π_{LR} correlation function provides us with an excellent theoretical laboratory to do that. The different shapes of this correlation function predicted by other analytic approaches are also collected in Fig. 1. Let us comment on them individually.

- The suggestion to use large–N_c QCD combined with lowest order χPT loops, was first proposed by Bardeen, Buras and Gérard in a series of seminal papers [4]. The same approach has been applied by the Dortmund group [24], in particular to the evaluation of ε'/ε. In this approach the *hadronic ansatz* to the Green's functions consists of Goldstone poles only and their integrals, (which become of course UV–divergent, often quadratically divergent since the correct QCD short–distance behaviour is not implemented,) are cut–off sharply. In this case, the predicted hadronic shape of the LR–correlation function normalized to its value at $Q^2 = 0$ is constant, as shown by the BBG, HKPSB line (black dotted) in Fig. 1.
- The Trieste group evaluate the relevant Green's functions using the constituent chiral quark model (CχQM) proposed in refs. [25] and [26, 27]. They have obtained

[4] See refs. [21, 22, 23] and references therein.

a long list of predictions [28], in particular ε'/ε. The model gives an educated first guess of the low–Q^2 behaviour of the Green's functions, as one can judge from the CχQM–curve (green dashed) in Fig. 1, but it fails to reproduce the short–distance QCD–behaviour. Another objection to this approach is that the "natural matching scale" to the short–distance behaviour in this model should be $\sim 4M_Q^2$, (M_Q the constituent quark mass,) too low to be trusted.

- The extended Nambu–Jona-Lasinio (ENJL) model was developed as an improvement on the CχQM, since in a certain way it incorporates the vector–like fluctuations of the underlying QCD theory, which are known to be phenomenologically important [5]. The model is, indeed, rather successful in predicting the low–energy $O(p^4)$ constants of the chiral Lagrangian. It has, indeed, a better low–energy behaviour than the CχQM, as the ENJL–curve (blue dot–dashed) in Fig. 1 shows; but it fails to reproduce the short–distance behaviour of the OPE in QCD. Arguments to do the matching to short–distance QCD have been forcefully elaborated in refs. [30], which also have made a lot of predictions; a large value for ε'/ε in particular.
- The problem with the ENJL model as a plausible model of large–N_c QCD, is that the on–shell production of unconfined constituent quark $Q\bar{Q}$ pairs that it predicts violates the large–N_c QCD counting rules. In fact, as shown in ref. [13], when the unconfining pieces in the ENJL spectral functions are removed by adding an appropriate series of local counterterms, the resulting theory is entirely equivalent to an effective chiral meson theory with three narrow states V, A and S; very similar to the phenomenological *Resonance Dominance* Lagrangians proposed in refs. [31, 32]. These *Resonance Dominance* Lagrangians can be viewed as particular models of large–N_c QCD. They predict the same Green's functions as the MHA approximation to large–N_c QCD discussed above, in <u>some</u> particular cases but not in general [6].

In view of the difficulties that these analytic approaches have in reproducing the shape of the simplest Green's function one can think of, it is difficult to attribute more than a qualitative significance to their "predictions"; ε'/ε in particular, which requires the interplay of several other Green's functions much more complex than $\Pi_{LR}(Q^2)$.

METHODOLOGY AND APPLICATIONS

The approach that we propose in order to compute a specific coupling of the chiral electroweak Lagrangian consists of the following steps:

1. *Identify the underlying QCD Green's functions.*
   In most cases of interest, the Green's functions in question are two–point functions with zero momentum insertions of vector, axial vector, scalar and pseudoscalar currents. The higher the power in the chiral expansion, the higher will be the

[5] For a review, see e.g. ref. [29] where earlier references can be found.

[6] See e.g. the three–point functions discussed in ref. [33].

number of insertions. This step is totally general and does not invoke any large–N_c approximation.

2. *Work out the short–distance behaviour and the long–distance behaviour of the relevant Green's functions.*
The long–distance behaviour is governed by the Goldstone singularities and can be obtained from χPT. The short–distance behaviour is governed by the OPE of the currents through which the hard momenta flows. Again, this step is well defined independently of the large–N_c expansion; in practice, however, the calculations simplify a lot when restricted to the appropriate order in the $1/N_c$–expansion one is interested in.
3. *Large–N_c ansatz for the underlying Green's functions.*
As already mentioned, the large–N_c ansatz involves only sums of poles; the *minimal hadronic ansatz* consists in limiting these sums to the minimum number required to satisfy the leading power fall–out at short–distances, as well as the appropriate χPT long–distance constraints.

All the three steps can be done analytically which helps to show the crucial points of the underlying dynamics. The method is, in principle, improvable [7] by adding constraints from the next–to–leading short–distance inverse power behaviour and/or higher orders in the chiral expansion.

We have tested this approach with the calculation of a few low–energy observables:

- *The electroweak Δm_π mass difference* [9] which we have already discussed.
- *The hadronic vacuum polarization contribution to the anomalous magnetic moment of the muon a_μ* [34]. The MHA in this case requires one vector–state pole and a pQCD continuum. The absence of dimension two operators in QCD in the chiral limit, constrains the threshold of the continuum. The result, which includes an estimate of the systematic error of the approach, is

$$a_\mu|_{HVP} = (5.7 \pm 1.7) \times 10^{-8}, \tag{16}$$

to be compared with an average of recent phenomenological determinations [8]

$$a_\mu|_{HVP} = (6.949 \pm 0.064) \times 10^{-8}. \tag{17}$$

- *The $\pi^0 \to e^+e^-$ and $\eta \to \mu^+\mu^-$ decay rates* [36]. These processes are governed by a $\langle PVV \rangle$ three–point function, with the Q^2–momentum flowing through the two V insertions. The MHA in this case requires a vector–pole and a double vector–pole. The predictions of the branching ratios

$$R(P \to l^+l^-) = \frac{\Gamma(P \to l^+l^-)}{\Gamma(P \to \gamma\gamma)}, \tag{18}$$

compared to the experimental determinations are shown in Table 1 below.

---

[7] Unlike the other analytic methods discussed above.

[8] See e.g. Prades's talk at KAON2001 [35] and references therein.

TABLE 1. *Summary of branching ratios results*

Branching Ratio (18)	**Large–N_c Approach**	**Experiment** [15]
$R(\pi^0 \to e^+e^-) \times 10^8$	6.2 ± 0.3	6.28 ± 0.55
$R(\eta \to \mu^+\mu^-) \times 10^5$	1.4 ± 0.2	1.47 ± 0.20
$R(\eta \to e^+e^-) \times 10^8$	1.15 ± 0.05	$< 1.8 \times 10^4$

These successful predictions have encouraged us to start a project of a systematic analysis of non–leptonic K–decays within this large–N_c approach. We have first used the example of the neutral current contribution to the Δm_π mass difference, (see Eq. (4),) as a theoretical laboratory to show explicitly the cancellation between the renormalization scale in the Wilson coefficient of four–quark operators and in the hadronic matrix elements evaluated in our approach. We have later shown that this cancellation can also be made renormalization scheme independent [37, 38].

So far we have completed two calculations of K–matrix elements within this large–N_c approach, which we next discuss.

The B_K–Factor of $K^0 - \bar{K}^0$ Mixing

The factor in question is conventionally defined by the matrix element of the four–quark operator $Q_{\Delta S=2}(x) = (\bar{s}_L\gamma^\mu d_L)(\bar{s}_L\gamma_\mu d_L)(x)$:

$$\langle \bar{K}^0 | Q_{\Delta S=2}(0) | K^0 \rangle = \frac{4}{3} f_K^2 M_K^2 B_K(\mu)\,. \tag{19}$$

To lowest order in the chiral expansion the operator $Q_{\Delta S=2}(x)$ bosonizes into a term of $O(p^2)$

$$Q_{\Delta S=2}(x) \Rightarrow -\frac{F_0^4}{4} g_{\Delta S=2}(\mu) \left[(D^\mu U^\dagger)U\right]_{23} \left[(D_\mu U^\dagger)U\right]_{23}, \tag{20}$$

with $g_{\Delta S=2}(\mu)$ a low energy constant, to be determined, which is a function of the renormalization scale μ of the Wilson coefficient $C(\mu)$ which modulates the operator $Q_{\Delta S=2}(x)$ in the four–quark effective Lagrangian. A convenient choice of the underlying Green's function here is the four point function $W_{LRLR}(Q^2)$ of two left–currents which carry the Q^2–momentum one has to integrate over, and two right–currents with zero momentum insertions. The coupling constant $g_{\Delta S=2}(\mu)$, which has to be evaluated in the same renormalization scheme as the Wilson coefficient $C(\mu)$ has been calculated, is then given by an integral [37]

$$g_{\Delta S=2}(\mu) = 1 - \frac{1}{32\pi^2 F_0^2} \int_0^\infty dQ^2 \left(\frac{4\pi\mu^2}{Q^2}\right)^{\varepsilon/2} W_{LRLR}(Q^2)\,, \tag{21}$$

conceptually similar to the one which determines the electroweak constant C in Eq. (4). The *hadronic ansatz*, in the $1/N_c$–expansion, of the Green's function $W_{LRLR}(Q^2)$, which fulfills the leading short–distance constraint and the long–distance constraints which fix $W_{LRLR}(0)$ and $W'_{LRLR}(0)$, requires one vector–pole, a double vector–pole and a triple vector–pole [9]. The invariant $\hat{B}_K$ defined as

$$\hat{B}_K = \frac{3}{4}C(\mu) \times g_{\Delta S=2}(\mu)\,, \tag{22}$$

can then be evaluated, with no free parameters, with the result [37]

$$\hat{B}_K = 0.38 \pm 0.11\,. \tag{23}$$

When comparing this result to other determinations, specially in lattice QCD, it should be realized that the unfactorized contribution in Eq. (21) is the one in the chiral limit. It is possible, in principle, to calculate chiral corrections within the same large–N_c approach, but this has not yet been done.

The result in Eq. (23) is compatible with the old current algebra prediction [39] which, to lowest order in chiral perturbation theory, relates the B_K factor to the $K^+ \to \pi^+\pi^0$ decay rate. In fact, our calculation of the B_K factor can be viewed as a successful prediction of the $K^+ \to \pi^+\pi^0$ decay rate!

As discussed in ref. [40] the bosonization of the four–quark operator $Q_{\Delta S=2}$ and the bosonization of the operator $Q_2 - Q_1$ which generates $\Delta I = 1/2$ transitions are related to each other in the combined chiral limit and next–to–leading order in the $1/N_c$–expansion. Lowering the value of $\hat{B}_K$ from the leading large–N_c prediction $\hat{B}_K = 3/4$ to the result in Eq. (23) is correlated with an increase of the coupling constant g_8 in the lowest order effective chiral Lagrangian, (see Eq. (1),) which generates $\Delta I = 1/2$ transitions, and provides a first step towards a quantitative understanding of the dynamical origin of the $\Delta I = 1/2$ rule.

ElectroWeak Four–Quark Operators

These are the four–quark operators generated by the so called electroweak Penguin like diagrams [10]

$$\mathcal{L} \Rightarrow \cdots C_7(\mu)Q_7 + C_8(\mu)Q_8\,, \tag{24}$$

with

$$Q_7 = 6(\bar{s}_L\gamma^\mu d_L) \sum_{q=u,d,s} e_q(\bar{q}_R\gamma_\mu q_R) \quad \text{and} \quad Q_8 = -12 \sum_{q=u,d,s} e_q(\bar{s}_L q_R)(\bar{q}_R d_L)\,, \tag{25}$$

and $C_7(\mu), C_8(\mu)$ their corresponding Wilson coefficients. They generate a term of $O(p^0)$ in the effective chiral Lagrangian[42]; therefore, the matrix elements of these operators,

---

[9] This goes beyond the strict MHA which, in this case, only requires a vector–pole. It is the *extra* information of knowing $W_{LRLR}(0)$ and $W'_{LRLR}(0)$ in χPT which forces the presence of the double and triple poles.

[10] See e.g., Buras lectures[41]

although suppressed by an e^2 factor, are chirally enhanced. Furthermore, the Wilson coefficient C_8 has a large imaginary part, which makes the matrix elements of the Q_8 operator to be particularly important in the evaluation of ε'/ε.

Within the large–N_c framework, the bosonization of these operators produce matrix elements with the following counting

$$\langle Q_7\rangle|_{O(p^0)} = \underline{O(N_c)} + O(N_c^0) \quad \text{and} \quad \langle Q_8\rangle|_{O(p^0)} = \underline{O(N_c^2)} + \underbrace{O(N_c^0)}_{\text{Zweig suppressed}} \tag{26}$$

A first estimate of the underlined contributions was made in ref. [43] [11]. The bosonization of the Q_7 operator to $O(p^0)$ in the chiral expansion and to $O(N_c)$ is very similar to the calculation of the Z–contribution to the coupling constant C in Eq. (4). An evaluation which also takes into account the renormalization scheme dependence has been recently made in ref. [38].

The contribution of $O(N_c^0)$ to $\langle Q_8\rangle|_{O(p^0)}$ is Zweig suppressed. It involves the sector of scalar (pseudoscalar) Green's functions where it is hinted from various phenomenological sources that the restriction to just the leading large–N_c contribution may not always be a good approximation. Fortunately, as first pointed out by the authors of ref. [47], independently of large–N_c considerations, the bosonization of the Q_8 operator to $O(p^0)$ in the chiral expansion can be related to the four–quark condensate $\langle O_2\rangle \equiv \langle 0|(\bar{s}_L s_R)(\bar{d}_R d_L)|0\rangle$ by current algebra Ward identities, the same four–quark condensate which also appears in the OPE of the $\Pi_{LR}(Q^2)$ function discussed above. More precisely

$$\lim_{Q^2\to\infty} \left(-Q^2\Pi_{LR}(Q^2)\right) Q^4 = 4\pi^2\frac{\alpha_s}{\pi}\left(4\langle O_2\rangle + \frac{2}{N_c}\langle O_1\rangle\right) + O\left(\frac{\alpha_s}{\pi}\right)^2, \tag{27}$$

where $\langle O_1\rangle \equiv \langle 0|(\bar{s}_L\gamma^\mu d_L)(\bar{d}_R\gamma_\mu s_R)|0\rangle$ is the vev which governs $\langle Q_7\rangle|_{O(p^0)}$. In fact, in the $1/N_c$–expansion [38]

$$\langle O_1\rangle = \left(-\frac{1}{2} i g_{\mu\nu}\int\frac{d^4q}{(2\pi)^4}\Pi_{LR}^{\mu\nu}(q)\right)^{\text{ren.}}_{\overline{\text{MS}}} \tag{28}$$

$$= -\frac{3}{32\pi^2}\left[\sum_A f_A^2 M_A^6\log\frac{\Lambda^2}{M_A^2} - \sum_V f_V^2 M_V^6\log\frac{\Lambda^2}{M_V^2}\right], \tag{29}$$

with (NDR means naive dimensional renormalization scheme; HV means 't Hooft–Veltman scheme,)

$$\Lambda^2 = \mu^2\exp(1/3+\kappa); \qquad \kappa = -1/2 \quad \text{in NDR}, \quad \text{and} \quad \kappa = +3/2 \quad \text{in HV}. \tag{30}$$

[11] The inclusion of final state interaction effects based on the leading large–N_c determination of $\langle Q_8\rangle$, (and $\langle Q_6\rangle$,) in connection with a phenomenological determination of ε'/ε, has been recently discussed in ref. [44].

The crucial observation is that large–N_c QCD gives a rather good description of the $\Pi_{LR}(Q^2)$–function, as we have seen earlier; in particular it implies that, (see Eq. (9),)

$$\lim_{Q^2\to\infty} \left(-Q^2\Pi_{LR}(Q^2)\right) Q^4 = \sum_V f_V^2 M_V^6 - \sum_A f_A^2 M_A^6 . \tag{31}$$

Solving these equations in the MHA approximation, results in a determination of the matrix elements of $\langle Q_8\rangle|_{O(p^0)}$ which does not require the separate knowledge of the Zweig suppressed $O(N_c^0)$ term in Eq. (26).

The numerical results we get for the matrix elements

$$M_{7,8} \equiv \langle(\pi\pi)_{I=2}|Q_{7,8}|K^0\rangle_{\,2\text{GeV}} \tag{32}$$

at the renormalization scale $\mu = 2$ GeV in the two schemes NDR and HV and in units of GeV^3 are shown in Table 2 below, (the first line.)

TABLE 2. *Matrix elements results, (see Eq. (32))*

METHOD	M_7(NDR)	M_7(HV)	M_8(NDR)	M_8(HV)
Large–N_c Approach				
Ref. [38]	0.11 ± 0.03	0.67 ± 0.20	3.5 ± 1.1	3.5 ± 1.1
Lattice QCD				
Ref. [46]	0.11 ± 0.04	0.18 ± 0.06	0.51 ± 0.10	0.62 ± 0.12
Dispersive Approach				
Ref. [47]	0.22 ± 0.05		1.3 ± 0.3	
Ref. [48]	0.35 ± 0.10		2.7 ± 0.6	
Ref. [49]	0.18 ± 0.12	0.50 ± 0.06	2.13 ± 0.85	2.44 ± 0.86
Ref. [51]	0.16 ± 0.10	0.49 ± 0.07	2.22 ± 0.67	2.46 ± 0.70

Also shown in the same table are other evaluations of matrix elements with which we can compare scheme dependences explicitly [12]. Several remarks are in order

- Our evaluations of M_8 do not include the terms of $O(\alpha_s^2)$ in Eq. (27) because, as pointed out in Ref. [38], the available results in the literature [45] were not calculated in the right basis of four–quark operators needed here.
- We find that our results for M_7 are in very good agreement with the lattice results in the NDR scheme, but not in the HV scheme. This disagreement is, very likely, correlated with the strong discrepancy we have with the lattice result for M_8(NDR).
- The recent revised dispersive approach results [49, 51], which now include the effect of higher terms in the OPE, are in agreement, within errors, with the large–N_c approach results. In fact, the agreement improves further if the new $O(\alpha_s^2)$ corrections, which have now been calculated in the right basis [51], are also incorporated in the large–N_c approach.

[12] There is another "dispersive determination" in the literature [50] since the HEP-2001 conference, but it is controversial as yet; this is why we do not include it in the Table.

- Both the revised dispersive approach results and the large–N_c approach results for M_8 are higher than the lattice results. The discrepancy may originate in the fact that, for reasonable values of $\langle\bar{\psi}\psi\rangle$, most of the contribution to M_8 appears to come from an OZI–violating Green's function which is something inaccessible in the quenched approximation at which the lattice results, so far, have been obtained.

Conclusion and Outlook

We hope to have shown with these examples that the large–N_c approach that we have discussed, provides a very useful framework to formulate calculations of genuinely non–perturbative nature, like the low–energy constants of the effective chiral Lagrangian, both in the strong and the electroweak sector.

Other calculations in progress, by various groups of people and in order of advancement, are

- *The electroweak hadronic contributions to* $g_\mu - 2$.
- *The matrix elements of the strong Penguin operator* Q_6, relevant for ε'/ε.
- *The light–by–light hadronic contributions to* $g_\mu - 2$; in particular the one generated by the convolution of two $\langle PVV\rangle$ three–point functions.
- *The chiral corrections to the* B_K*–factor.*

We hope to have the results in the near future.

ACKNOWLEDGEMENTS

My knowledge on the subject reported here owes much to enjoyable discussions and work with Marc Knecht, Santi Peris, Michel Perrottet, and Toni Pich, as well as, more recently, with Thomas Hambye, Laurent Lellouch, Andreas Nyffeler and Boris Phily. I wish to thank them all here.

REFERENCES

1. G. 't Hooft, Nucl. Phys. **B72** (1974) 461; **B73** (1974) 461.
2. S. Coleman and E. Witten, Phys. Rev. Lett., **45** (1980) 100.
3. E. Witten, Nucl. Phys. **B160** (1979) 57.
4. *The large N Expansion in Quantum Field Theory and Statistical Physics*, Editors E. Brezin and S.R. Wadia, World Scientific, 1993.
5. A.V. Manohar, *Large–N_c QCD*, in les Houches Session LXVIII, North Holland, 1999.
6. S. Weinberg, Physica **A96** (1979) 327.
7. J. Gasser and H. Leutwyler, Ann. of Phys.(N.Y.) **158** (1984) 142.
8. J. Gasser and H. Leutwyler, Nucl. Phys. **B250** (1985) 465.
9. M. Knecht, S. Peris and E. de Rafael, Phys. Lett. **B443** (1998) 255.
10. M. Knecht and E. de Rafael, Phys. Lett. **B424** (1998) 355.
11. M.A. Shifman, A.I. Vainshtein and V.I. Zakharov, Nucl. Phys. **B147** (1979) 385, *ibid* 447.
12. E.C. Titchmarsh, *The Theory of Functions*, 2nd edition, OUP 1939.

13. S. Peris, M. Perrottet and E. de Rafael, JHEP **05** (1998) 011.
14. M. Golterman and S. Peris, Phys. Rev. **D61** (2000) 034018.
15. *Review of Particle Physics*, Eur. Phys. J. **C15** (2000) 1.
16. ALEPH Collaboration, R. Barate *et al*, Z. Phys. **C76** (1997) 15; *ibid* Eur. Phys. J. **C4** (1998) 409.
17. S. Peris, B. Phily and E. de Rafael, Phys. Rev. Lett. **86** (2001) 14.
18. S. Peris, B. Phily and E. de Rafael, *talk at the EURODAPHNE–Marseille meeting Feb. 2001* to be published.
19. M. Golterman and S. Peris, JHEP **01** (2001) 028.
20. B. Phily, PhD–Thesis, University of Marseille, Luminy.
21. A. Buras, *The $1/N$ approach to nonleptonic weak interactions*, in CP violation, ed. C. Jarlskog, World Scientific, Singapore, 1998.
22. W. Bardeen, *Weak decay amplitudes in large N_c QCD*, in Proc. of Ringberg Workshop, Nucl. Phys.B (Proc. Suppl.) **7A** (1989) 149.
23. J.-M. Gérard, Acta Phys. Pol. **B21** (1990) 257.
24. T. Hambye, G.O. Köhler, E.A. Paschos, P.H. Soldan and W.A. Bardeen, Phys. Rev. **D58** (1998) 014017, and references therein.
25. A. Manohar and H. Georgi, Nucl. Phys. **B234** (1984) 189.
26. D. Espriu, E. de Rafael and J. Taron, Nucl. Phys. **B345** (1990) 22.
27. A. Pich and E. de Rafael, Nucl. Phys. **B358** (1991) 311.
28. S. Bertolini, M. Fabbrichesi and J.O. Egg, Rev. Mod. Phys. **72** (2000) 65 and references therein.
29. J. Bijnens, Phys. Rep. **265(6)** (1996) 369.
30. J. Bijnens and J. Prades, JHEP **9901** (1999) 023 ; *ibid* **0001** (2000) 002; *ibid* **0006** (2000) 035.
31. G. Ecker, J. Gasser, A. Pich and E. de Rafael, Nucl. Phys. **B321** (1989) 311.
32. G. Ecker, J. Gasser, H. Leutwyler, A. Pich and E. de Rafael, Phys. Lett. **B321** (1989) 425.
33. M. Knecht and A. Nyffeler, hep-ph/0106034.
34. M. Perrottet and E. de Rafael, *unpublished.*
35. J. Prades, hep-ph/0108192.
36. M. Knecht, S. Peris, M. Perrottet and E. de Rafael, Phys. Rev. Lett. **83** (1999) 5230.
37. S. Peris and E. de Rafael, Phys. Lett. **B490** (2000) 213, *erratum* hep-ph/0006146 v3.
38. M. Knecht, S. Peris and E. de Rafael, Phys. Lett. **B508** (2001) 117.
39. J.F. Donoghue, E. Golowich and B.R. Holstein, Phys. Lett. **B119** (1982) 412.
40. A. Pich and E. de Rafael, Phys. Lett. **B374** (1996) 186.
41. A.J. Buras, *Weak Hamiltonian, CP Violation and Rare Decays*, in les Houches Session LXVIII, North Holland, 1999.
42. J. Bijnens and M. Wise, Phys. Lett. **B137** (1984) 245.
43. M. Knecht, S. Peris and E. de Rafael, Phys. Lett. **B457** (1999) 227.
44. E. Pallante, A. Pich and I. Scimemi, hep-ph/0105011 and references therein.
45. L.V. Lanin, V.P. Spiridonov and K. G. Chetyrkin, Sov. J. Nucl. Phys. **44** (1986) 896.
46. A. Donini *et al*, Phys. Lett. **B470** (1999) 233.
47. J.F. Donoghue and E. Golowich, Phys. Lett. **B478** (2000) 172.
48. S. Narison, Nucl. Phys. **B593** (2001) 3; hep-ph/0012019; *and private communication.*
49. E. Golowich, Talk at the EPS-HEP-2001 Conference, Budapest.
50. J. Bijnens, E. Gámiz and J. Prades, hep-ph/0108240, submitted to JHEP.
51. V. Cirigliano, J. Donoghue, E. Golowich and K. Maltman, hep-ph/0109113 v1.

A new result on the direct CP violation in two pion decays of the neutral kaon

D.Madigojine [1]

Joint Institute for Nuclear Research, 141980, Dubna, Russia

Abstract. The direct CP violation parameter $Re(\varepsilon'/\varepsilon)$ has been measured from the neutral kaon decays into two pions with the NA48 detector at the CERN SPS. The result, based on the data collected during 1998 and 1999 years, is $Re(\varepsilon'/\varepsilon) = (15.0 \pm 2.7) \times 10^{-4}$.

INTRODUCTION

The parameters of indirect (ε) and direct (ε') CP violation are defined from the amplitude relations:

$$\eta_{\pm} = \frac{A(K_L \to \pi^+\pi^-)}{A(K_S \to \pi^+\pi^-)} = \varepsilon + \varepsilon'$$

$$\eta_{00} = \frac{A(K_L \to \pi^0\pi^0)}{A(K_S \to \pi^0\pi^0)} = \varepsilon - 2\varepsilon'$$

It is convenient to measure the double ratio:

$$R = \frac{\Gamma(K_L \to \pi^0\pi^0)}{\Gamma(K_S \to \pi^0\pi^0)} / \frac{\Gamma(K_L \to \pi^+\pi^-)}{\Gamma(K_S \to \pi^+\pi^-)} \approx 1 - 6 \times Re(\varepsilon'/\varepsilon) \tag{1}$$

The first evidence for direct CP violation was published in 1988 [1]. In 1993, two experiments ([2] and [3]) have published their results without a consistent answer on the existence of the direct CP violation. Recently, the KTeV collaboration measured an effect of $Re(\varepsilon'/\varepsilon) = (28.0 \pm 4.1) \times 10^{-4}$ [4] and NA48 published a result from a sample of the total statistics: $Re(\varepsilon'/\varepsilon) = (18.5 \pm 7.3) \times 10^{-4}$ [5]. These results confirmed the existence of a direct CP violation component.

A new NA48 measurement of $Re(\varepsilon'/\varepsilon)$with high precision, using data samples recorded in 1998 and 1999, is presented in this talk.

---

[1] **On behalf of the NA48 Collaboration: Cagliari, Cambridge, CERN, Dubna, Edinburgh, Ferrara, Firenze, Mainz, Orsay, Perugia, Pisa, Saclay, Siegen, Torino, Warsaw, Wien.**

CP602, *QCD@Work: International Workshop on Quantum Chromodynamics*
edited by P. Colangelo and G. Nardulli

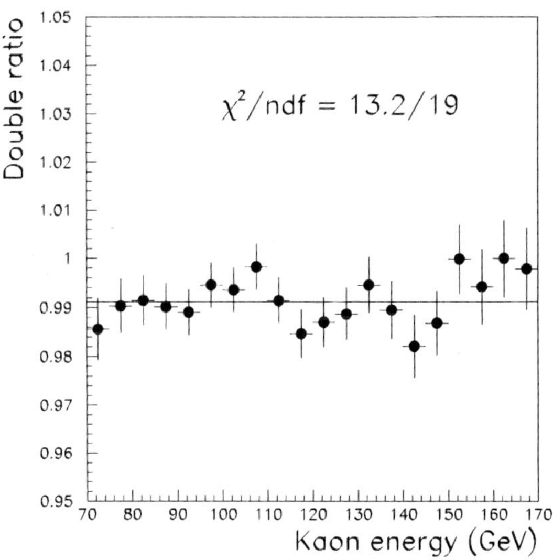

FIGURE 2. Stability of the corrected double ratio as a function of the kaon energy

performed by several independent groups. Fully consistent results have been obtained in all the analyses.

Converting the result using Eq. 1 and adding two uncertainties in quadrature, one can write the following value of the direct CP violation parameter:

$$Re(\varepsilon'/\varepsilon) = (15.0 \pm 2.7) \times 10^{-4}. \tag{4}$$

It is in a good agreement with the NA48 result based on 1997 data [5].

REFERENCES

1. H.Brukhardt et al., *Phys. Lett.*, **B 206**, 169 (1988).
2. G.Barr et al., *Phys. Lett.*, **B 317**, 233 (1993).
3. L.K.Gibbons et al., *Phys. Rev. Lett.*, **70**, 1203 (1993).
4. A.Alavi-Harati et al., *Phys. Rev. Lett.*, **83**, 22 (1999).
5. V.Fanti et al., *Phys. Lett.*, **B 465**, 335 (1999).
6. C.Biino et al., CERN-SL-98-033(EA) (1999).
7. N.Doble, L.Gatignon, P.Grafstrom, *Nucl.Instr. and Methods*, **B 119**, 181 (1996).
8. P.Grafstrom et al., *Nucl. Instr. and Methods*, **A 344**, 487 (1994).
9. D.Bederede et al., *Nucl. Instr. and Methods*, **A 367**, 88 (1995).
10. G.D.Barr et al., *Nucl. Instr. and Methods*, **A 370**, 413 (1993).
11. R.Moore et al., *Nucl. Instr. and Methods*, **B 119**, 149 (1996).

Unitarization of the complete meson-meson scattering at one loop in Chiral Perturbation Theory.

José R. Peláez and A. Gómez Nicola

Departamento de Física Teórica, Universidad Complutense, 28040 Madrid, SPAIN

Abstract. We report on our one-loop calculation of all the two meson scattering amplitudes within SU(3) Chiral Perturbation Theory, i.e. with pions, kaons and etas. Once the amplitudes are unitarized with the coupled channel Inverse Amplitude Method, they satisfy simultaneously the correct low-energy chiral constraints and unitarity. We obtain a remarkable description of meson-meson scattering data up to 1.2 GeV including the scattering lengths and seven light resonances.

Introduction

Chiral Perturbation Theory (ChPT) [1] provides a powerful tool to describe the interactions of the lightest mesons. These particles correspond to the Goldstone bosons associated to the spontaneous breaking of the $SU(3)_L \times SU(3)_R$ chiral symmetry down to $SU(3)_{L+R}$. This would be the symmetry breaking pattern of QCD if the three lightest quarks were massless. Of course, quarks are not massless, but since their masses are very small compared to the typical hadronic scales, $O(1$ GeV), their explicit symmetry breaking effect only yields a small mass contribution for the lightest mesons, which become pseudo-Goldstone bosons. Thus, the three pions are the pseudo-Goldstone bosons of the $SU(2)$ spontaneous breaking when only the u and d quarks are considered. Similarly, when s is also included, the eight $SU(3)$ pseudo-Goldstone bosons can be identified with the meson octet formed by the pions, kaons and the eta. The low energy interactions of pions, kaons and the eta have to be described with an effective Lagrangian respecting the above described chiral symmetry breaking pattern. Within ChPT, only pseudo-Goldstone bosons are included in the Lagrangian, thus providing a low energy description. The possible terms compatible with the symmetry breaking pattern are organized in a derivative and mass expansion (generically p). For instance, amplitudes are obtained as an expansion in powers of the external momenta and the quark masses over a typical chiral scale of $O(1$ GeV). One remarkable feature of the ChPT scheme is that all loop divergences appearing at a given order in the expansion can be absorbed by a finite number of (low energy) constants of the counterterms that appear in the Lagrangian to the same order. Therefore, order by order, the theory is finite and depends on a few parameters, providing a predictive framework. Thus, once the low-energy constants are determined from just a few experiments, predictions can be made for other processes. This approach is very successful, but only at low energies (usually, less than 500 MeV). For that reason, there is a growing interest in developing methods to extend the ChPT applicability range. Among them, the explicit introduction of heavier resonances in the Lagrangian [2], resummation of diagrams in a Lippmann-Schwinger or

CP602, *QCD@Work: International Workshop on Quantum Chromodynamics*
edited by P. Colangelo and G. Nardulli

Bethe-Salpeter approach [3], or unitarization and dispersive techniques like the Inverse Amplitude Method (IAM) [4, 5] applied to one-loop amplitudes. A version of the latter, generalized to coupled channels provided a remarkable description of meson-meson scattering up to 1.2 GeV, generating dynamically seven light resonances [6]. In principle, these methods respect the good low energy properties of ChPT, since they are built from the perturbative results. However, not all the meson-meson scattering processes had been calculated at one loop in ChPT. The amplitudes available so far are $\pi\pi \to \pi\pi$ [7], $K\pi \to K\pi$ [7], $\eta\pi \to \eta\pi$ [7] and the two independent $K^+K^- \to K^+K^-$, $K^+K^- \to K^0\bar{K}^0$ [8]. Therefore, the IAM has only been applied rigorously to the $\pi\pi$, $K\bar{K}$ final states, whereas for the complete low-energy meson-meson scattering, additional approximations had to be done [6], meaning in particular that it was not possible to compare with the low energy parameters of standard ChPT in dimensional regularization or to describe simultaneously the low and high energy regimes. Here we report on our recent work [9] where we have completed the calculation of the meson-meson scattering in one-loop ChPT. There are three completely new amplitudes: $K\eta \to K\eta$, $\eta\eta \to \eta\eta$ and $K\pi \to K\eta$. In addition, we have recalculated independently the other five amplitudes and all of them will be given together in a unified notation, ensuring exact perturbative unitarity and also correcting some misprints in the literature. Once all the amplitudes are available, we have done a coupled channel IAM fit to describe the whole meson-meson scattering data below 1.2 GeV. Our results allow for a direct comparison with the standard low-energy chiral parameters. Indeed, we find a very good agreement with previous determinations from low-energy data using standard ChPT. The main differences of our work with [6] are that we consider the full one-loop calculation of the amplitudes, which ensures their finiteness and scale independence in dimensional regularization, we take into account the new processes mentioned above and we are able to describe simultaneously the low energy and the resonance regions.

The amplitudes

The lowest order, $O(p^2)$, meson-meson scattering amplitudes (the low energy theorems) are obtained just from the tree level diagrams of the lowest order Lagrangian. In contrast, the calculation of the $O(p^4)$ contribution involves the evaluation of the following Feynman diagrams: First, the one-loop diagrams in Fig.1, which are divergent. In particular those in Fig.1e, provide the wave function, mass and decay constant renormalizations, and that in Fig.1a gives the imaginary part to ensure perturbative unitarity. Second, the tree level graphs with the second order Lagrangian, which depend on the chiral parameters L_i, that will absorb the previous divergences through renormalization. In Table I, we list the L_i values from recent determinations. Note that the parameters have been renormalized in the usual $\overline{MS}-1$ scheme of ChPT, using dimensional regularization. Thus, the renormalized parameters have a scale dependence $L_i^r(\mu)$ (except L_3 and L_7), and they are given at $\mu = M_\rho$. After renormalization, the amplitudes are finite and scale independent. The details and results of the calculation will be published elsewhere [9]. We will just recall that, in order to compare with experiment, the amplitudes are projected into partial waves t_{IJ} of definite isospin I and angular momentum J. Therefore, in the chiral expansion we will have, omitting the I,J subindices, $t \simeq t_2 + t_4 + ...$, where t_2 and t_4 the $O(p^2)$ and $O(p^4)$ contributions, respectively.

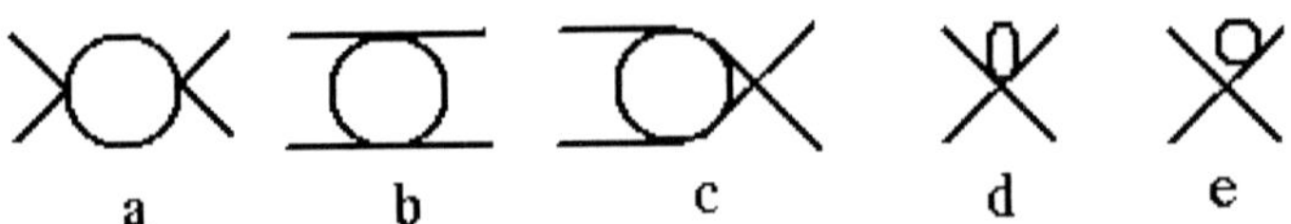

FIGURE 1. Generic one-loop Feynman diagrams that have to be evaluated in meson-meson scattering.

TABLE 1. Different sets of chiral parameters ($\times 10^3$). The first two columns come from recent analysis of K_{l4} decays at different orders [10] (L_4 and L_6 are set to zero). In the third column L_1, L_2, L_3 come from [11] and the rest from [1]. The last one corresponds to the values from the IAM including the uncertainty due to different systematic error used on different fits.

Chiral Parameter	$O(p^6)$ K_{l4} decays	$O(p^4)$ K_{l4} decays	ChPT	IAM fits
$L_1^r(M_\rho)$	0.53 ± 0.25	0.46	0.4 ± 0.3	0.56 ± 0.10
$L_2^r(M_\rho)$	0.71 ± 0.27	1.49	1.35 ± 0.3	1.21 ± 0.10
L_3	-2.72 ± 1.12	-3.18	-3.5 ± 1.1	-2.79 ± 0.14
$L_4^r(M_\rho)$	0	0	-0.3 ± 0.5	-0.36 ± 0.17
$L_5^r(M_\rho)$	0.91 ± 0.15	1.46	1.4 ± 0.5	1.4 ± 0.5
$L_6^r(M_\rho)$	0	0	-0.2 ± 0.3	0.07 ± 0.08
L_7	-0.32 ± 0.15	-0.49	-0.4 ± 0.2	-0.44 ± 0.15
$L_8^r(M_\rho)$	0.62 ± 0.2	1.00	0.9 ± 0.3	0.78 ± 0.18

Unitarity

The S matrix unitarity relation $SS^\dagger = 1$ translates into simple relations for the elements of the T matrix t_{ij} if they are projected into partial waves, where i, j denote the different states physically available. For instance, if there is only one possible state, "1", the partial wave t_{11} satisfies

$$\mathrm{Im}\, t_{11} = \sigma_1 |t_{11}|^2 \quad \Rightarrow \quad \mathrm{Im} \frac{1}{t_{11}} = -\sigma_1 \quad \Rightarrow \quad t_{11} = \frac{1}{\mathrm{Re}\, t_{11} - i\sigma_1} \tag{1}$$

where $\sigma_i = 2q_i/\sqrt{s}$ and q_i is the C.M. momentum of the state i. Written in this way it can be readily noted that *we only need to know the real part of the Inverse Amplitude*. The imaginary part is fixed by unitarity. In principle, this relation *only holds above threshold* up to the energy where another state, "2", is physically accessible. Above that point, the unitarity relation for the partial waves can be written in matrix form as:

$$\mathrm{Im}\, T = T \Sigma T^* \quad \Rightarrow \quad \mathrm{Im}\, T^{-1} = -\Sigma \quad \Rightarrow \quad T = (\mathrm{Re}\, T - i\Sigma)^{-1} \tag{2}$$

with

$$T = \begin{pmatrix} t_{11} & t_{12} \\ t_{12} & t_{22} \end{pmatrix} \quad , \quad \Sigma = \begin{pmatrix} \sigma_1 & 0 \\ 0 & \sigma_2 \end{pmatrix}, \tag{3}$$

which allows for an straightforward generalization to the case of n accessible states. Once more, unitarity means that we would only need to calculate the real part of the inverse amplitude matrix.

Note that the above unitarity relations are non-linear. This implies that they will never be satisfied exactly with polynomials like the amplitudes obtained from ChPT.

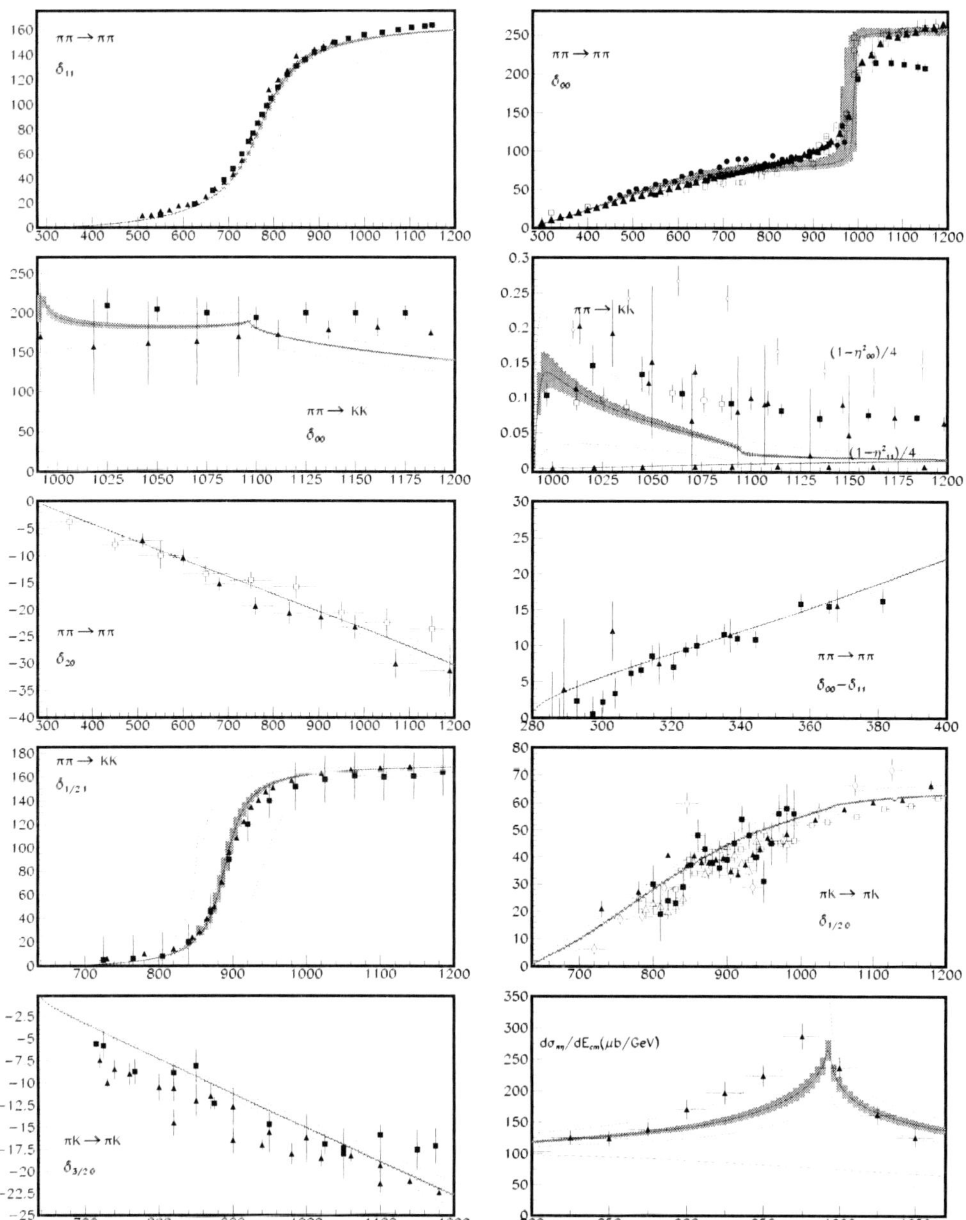

FIGURE 2. Result of the coupled channel IAM fit to meson-meson scattering data (see [9] for references). The shaded area covers the uncertainty due to MINUIT errors. The area between the dotted lines corresponds to the uncertainty in the L_i^r due to the use of different systematic errors on the fits. The dashed line in the last plot is the continuous background underneath the resonant contribution.

Nevertheless, unitarity holds perturbatively, i.e,

$$\mathrm{Im}\, T_2 = 0 + O(p^4), \qquad \mathrm{Im}\, T_4 = T_2 \Sigma T_2^* + O(p^6), \tag{4}$$

Unitarization: The inverse Amplitude Method

One of the simplest methods to unitarize the chiral amplitudes is to introduce the $\mathrm{Re}\, T$ in eq.(2), calculated as a ChPT expansion

$$T^{-1} \simeq T_2^{-1}(1 - T_4 T_2^{-1} + \ldots), \tag{5}$$

$$\mathrm{Re}\, T^{-1} \simeq T_2^{-1}(1 - (\mathrm{Re}\, T_4) T_2^{-1} + \ldots), \tag{6}$$

Taking into account the perturbative unitarity conditions, eq.(4), we find

$$T \simeq T_2 (T_2 - T_4)^{-1} T_2, \tag{7}$$

which is the coupled channel Inverse Amplitude Method, which will use to unitarize simultaneously all the one-loop ChPT meson-meson scattering amplitudes. This method is able to generate seven resonant states. The novelty of our approach is that, since we have the complete $O(p^4)$ ChPT amplitudes, we can simultaneously recover the very same ChPT amplitudes up to $O(p^4)$, and thus have a good low energy limit.

Results and conclusion

We can now use previous determinations of the chiral parameters with the IAM and even the correct resonant behavior resonances. Once more, we can use the L_i^r because we have the complete amplitudes renormalized in the $\bar{MS}-1$ scheme. We have nevertheless carried out a fit (using MINUIT [12]) of the presently available data on meson-meson scattering. Since there are incompatibilities between different experiments, customarily a 1%, 3% and 5% systematic error has been added, which introduces an additional source of error. We give in Table 1 the resulting chiral parameters from the fit, whose errors correspond to those of MINUIT combined with those from the systematic uncertainty. Note that they are compatible with previous determinations. In Fig.2 we show the results of the IAM fit to these data, which is given in terms of phase shifts, inelasticities, and mass distributions of different processes (see [9] for details). The gray error bands cover the uncertainties in the L_i due to MINUIT, and are calculated by a Monte-Carlo gaussian sampling of the parameters. The area between the dotted lines has been calculated similarly but with the errors in the chiral parameters due to the different choice of systematic error. It can be noticed that all the resonant features are reproduced. However, thanks to the new amplitudes we are also able to obtain simultaneously values for the threshold parameters (they have not been fitted) which are listed in table 2. Note the good agreement with the experimental values when they exist.

Acknowledgments. Work partially supported from the Spanish CICYT projects AEN97-1693, FPA2000-0956, PB98-0782 and BFM2000-1326.

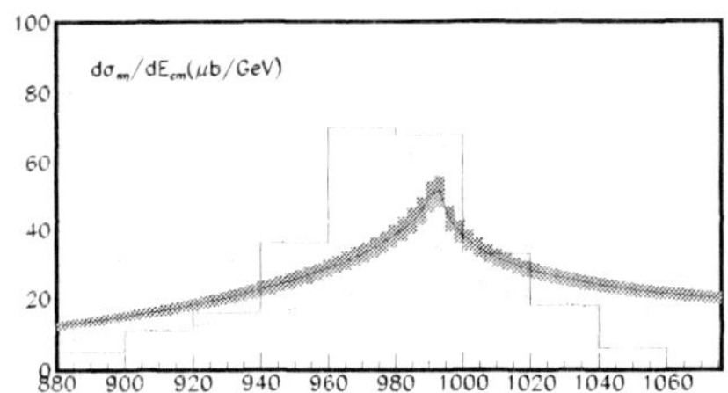

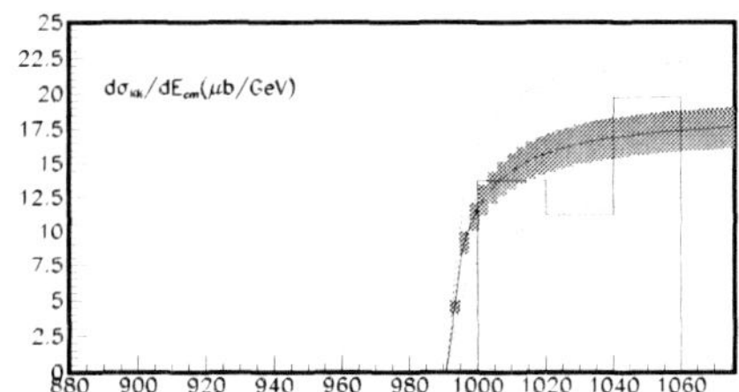

FIGURE 3. Effective mass distributions of the two mesons in the final state of $K^-p \to \Sigma^+(1385)\pi\eta$ and $K^-p \to \Sigma^+(1385)K\bar{K}$. This plots are not part of the IAM fit. For data references see [9].

TABLE 2. Scattering lengths a_{IJ} and slope parameters b_{IJ} for different meson-meson scattering channels. For experimental references see [9]. Let us remark that our one-loop IAM results are very similar to those of two-loop ChPT.

Threshold parameter	Experiment	IAM fit	ChPT $O(p^4)$ [2, 5, 7]	ChPT $O(p^6)$ [13]
a_{00}	0.26 ±0.05	$0.231^{+0.003}_{-0.006}$	0.20	0.219±0.005
b_{00}	0.25 ±0.03	0.30± 0.01	0.26	0.279±0.011
a_{20}	-0.028±0.012	$-0.0411^{+0.0009}_{-0.001}$	-0.042	-0.042±0.01
b_{20}	-0.082±0.008	-0.074±0.001	-0.070	-0.0756±0.0021
a_{11}	0.038±0.002	0.0377±0.0007	0.037	0.0378±0.0021
$a_{1/20}$	0.13...0.24	$0.11^{+0.06}_{-0.09}$	0.17	
$a_{3/20}$	-0.13...-0.05	$-0.049^{+0.002}_{-0.003}$	-0.5	
$a_{1/21}$	0.017...0.018	0.016±0.002	0.014	
a_{10}		$0.15^{+0.07}_{-0.11}$	0.0072	

REFERENCES

1. S. Weinberg, Physica A96, (1979) 327. J. Gasser and H. Leutwyler, Ann. Phys. 158, (1984) 142. J. Gasser and H. Leutwyler, Nucl. Phys. B250, (1985) 465,517,539.
2. V. Bernard, N. Kaiser and U.G. Meißner, Nucl. Phys. B364 (1991), 283. J.A. Oller, E. Oset, Phys.Rev.D60:074023,1999. M.Jamin, J.A. Oller, A.Pich, Nucl.Phys.B587 (2000), 331-362.
3. J.A. Oller, and E. Oset Nucl. Phys. A620 (1997), 438. J. Nieves, E. Ruiz Arriola. Phys.Rev.D63 (2001) 076001.
4. T. N. Truong, Phys. Rev. Lett. 661, (1988) 2526 ;Phys. Rev. Lett. 67, (1991) 2260; A. Dobado, M.J.Herrero and T.N. Truong, Phys. Lett. B235, (1990) 134;
5. A. Dobado and J.R. Peláez, Phys. Rev. D47, (1993) 4883; Phys. Rev. D56, (1997) 3057.
6. J. A. Oller, E. Oset and J. R. Peláez, Phys. Rev. Lett. 80, (1998) 3452; Phys. Rev. D59, (1999) 074001; Erratum-ibid. D60, (1999) 099906.
7. V. Bernard, N. Kaiser, U.G. Meissner, Phys. Rev. D43 (1991), 2757; Nucl. Phys. B357 (1991), 129; Phys. Rev. D44 (1991), 3698.
8. F. Guerrero and J. A. Oller, Nucl. Phys. B537, (1999) 459.
9. A. Gómez Nicola and J. R. Peláez, in preparation.
10. G. Amorós, J. Bijnens and P. Talavera, Nucl. Phys. B602(2001),87.
11. J. Bijnens, G. Colangelo and J. Gasser, Nucl. Phys. B427, (1994) 427.
12. F. James, Minuit Reference Manual D506 (1994).
13. G. Amorós, J. Bijnens and P. Talavera, Nucl.Phys.B585:293-352,2000, Erratum-ibid.B598:665-666,2001.

Simple check of the vacuum structure in full QCD lattice simulations

S. Dürr

Paul Scherrer Institut, 5232 Villigen PSI, Switzerland

Abstract. Given the increasing availability of lattice data for (unquenched) QCD with $N_f = 2$, it is worth while to check whether the generated vacuum significantly deviates from the quenched one. I discuss a specific attempt to do this on the basis of topological susceptibility data gained at various sea-quark masses, since for this observable detailed predictions are available. The upshot is that either discretization effects in dynamical simulations are still untolerably large or the vacuum structure in 2-flavour QCD substantially deviates from that in the theory with 3 (or 2+1) light quarks.

INTRODUCTION

In a historical perspective, the path towards phenomenological predictions of QCD by means of lattice techniques involves three steps: Pure Yang-Mills theory (where there is just glueball physics), quenched QCD (where the vacuum is the one in the SU(3) theory, but observables may involve so-called *current*-quarks) and full QCD (where the fermion determinant with the dynamical *sea*-quarks accounts for the quark loops in the vacuum). Today, the lattice community makes the final push towards full QCD, despite the fact that state-of-the-art simulations are modestly announced as "partially quenched" which means that the sea- and current-quark masses in the (euclidean) generating functional

$$Z[\bar{\eta},\eta] = \int DA\ e^{-S_G} \prod_{N_f} \det(\not{D}+m_{\rm sea})\ \exp\Big(\int \bar{\eta}\frac{1}{\not{D}+m_{\rm cur}}\eta\Big) \tag{1}$$

are (in general) unequal and in most cases significantly heavier than the physical u- and d-quarks, so that phenomenological statements require a twofold extrapolation.

Since the finite sea-quark mass constitutes the key ingredient in this ultimate step (note that the determinant turns into a constant for $m_{\rm sea} \to \infty$, hence (1) reduces to the quenched generating functional in that limit), an obvious task is to check whether these "partially quenched" or "full" QCD simulations exhibit the change in the vacuum structure expected to occur if the fermions are "active" (i.e. if the back-reaction of the "dynamical" fermions on the gauge background is taken into account). The prime observable used to distinguish the respective vacua is the topological susceptibility

$$\chi(m_{\rm sea}) = \frac{\langle q^2 \rangle}{V}\ , \tag{2}$$

with q the (global) topological charge, because detailed theoretical predictions show that χ behaves rather different in the quenched ($m_{\rm sea} \to \infty$) and chiral ($m_{\rm sea} \to 0$) limits,

CP602, *QCD@Work: International Workshop on Quantum Chromodynamics*
edited by P. Colangelo and G. Nardulli

respectively. Even though in the lattice-regulated theory (and with certain definitions of the topological charge operator) q may be somewhat ambiguous on the level of a single configuration, the moment of the q-distribution which enters (2) can be measured with controlled error-bars, and as a purely gluonic object the resulting $\chi = \chi(m_{\text{sea}})$ encodes nothing but the vacuum structure of the theory.

Below, I give a quick survey of recent lattice determinations of χ at various sea-quark masses in $N_f = 2$ QCD, I discuss the available continuum knowledge of the functional form $\chi = \chi(m_{\text{sea}})$, and I present a non-standard lattice determination of the quenched topological susceptibility χ_∞ and the chiral condensate Σ based on it. The outcome is that either certain observables in todays phenomenological studies with light dynamical quarks suffer from large discretization effects or – the more speculative view – that the low-energy structure of QCD with $N_f = 2$ is substantially different from that with $N_f = 3$.

LATTICE DATA

I start with a quick survey of recent lattice data for the topological susceptibility in QCD with 2 dynamical flavours; the selection reflects nothing but my personal awareness.

CP-PACS: The CP-PACS collaboration has simulated full QCD on several grids at various (β, κ)-values, using an RG-improved gauge action and an $O(a)$-improved fermion action with mean-field values for the associate c_{SW} coefficients. Below, I concentrate on the data generated on the $24^3 \times 48$ lattice at $\beta = 2.1$ with LW-cooling [1].

UKQCD: The UKQCD collaboration has simulated full QCD on a $16^3 \times 32$ grid at various (β, κ)-values, using the standard (Wilson) gauge action and an $O(a)$-improved fermion action with the non-perturbative values for the associate c_{SW} coefficients [2].

SESAM/TXL: The SESAM/TXL collaboration has simulated full QCD on two grids ($16^3 \times 32$ and $24^3 \times 40$) at several (β, κ)-values, combining the unimproved (Wilson) gauge action with unimproved (Wilson) fermions (i.e. setting $c_{SW} = 0$) [3].

Thin link staggered: Trusting a continuum identity for the relationship between the 2- and the 4-flavour functional determinant, the staggered fermion action may be used to simulate QCD with $N_f = 2$. There are data by the Pisa group [4] and by A. Hasenfratz based on configurations by the MILC collaboration and the Columbia/BNL project [5].

Fig. 1 displays the data, along with continuum constraints to be discussed next.

CONTINUUM KNOWLEDGE

As mentioned in the introduction, the data for the topological susceptibility χ versus the sea-quark mass $m \equiv m_{\text{sea}}$ prove useful to test the vacuum structure, because continuum QCD provides us with detailed predictions: There are analytic upper bounds for $\chi(m)$ at both asymptotically small and large sea-quark masses and there is a "semi-analytic" formula for $\chi(m)$ valid at intermediary quark masses (where the bulk of the lattice data reside). The only caveat is that these bounds hold true in the continuum limit, but so far no continuum extrapolation for $\chi(m)$ in $N_f = 2$ QCD is available yet. Before stating the complications due to this, the continuum functional forms shall be discussed.

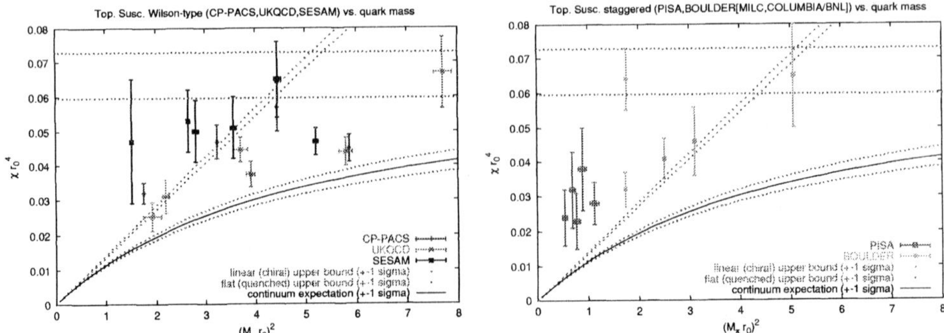

FIGURE 1. Topological susceptibility versus quark mass (each in dimensionless units) in $N_f = 2$ QCD with Wilson-type (left) or staggered (right) sea-quarks [1, 2, 3, 4, 5]. For comparison, 1σ bands indicating the constraints in the deep chiral regime (based on $F_\pi = 93 \pm 1$ MeV) and in the heavy sea-quark (quenched) limit (from $\chi_\infty = (200 \pm 5\,\text{MeV})^4$) are shown as well as the associate continuum band (5) (full line).

Asymptotically small sea-quark masses: For $m \ll \Lambda_{\text{QCD}}$ and to leading order in the chiral expansion the axial WT-identity yields (see Refs. cited in [6] for details)

$$\chi(m) = \frac{\Sigma m}{N_f}(1+O(m)) = \frac{M_\pi^2 F_\pi^2}{2N_f}(1+O(m)) \equiv \chi_0(1+O(m)), \tag{3}$$

where, in the second equality, the Gell-Mann–Oakes–Renner relation has been assumed.

Asymptotically large sea-quark masses: For $m \gg \Lambda_{\text{QCD}}$ the topological susceptibility gradually approaches its quenched counterpart (see Refs. cited in [6] for details)

$$\chi(m) = \chi_\infty(1+O(1/m)) = (200 \pm 5\,\text{MeV})^4, \tag{4}$$

Intermediate quark masses: For other quark masses (i.e. for $(M_\pi r_0)^2 \in [1.5, 15]$ or so, with the Sommer scale $r_0 = 0.5$ fm throughout) neither one of the asymptotic predictions is applicable. Fortunately, there is the "reduced" interpolation formula

$$\chi(m) = 1/(1/\chi_0 + 1/\chi_\infty) \tag{5}$$

with χ_0 defined in (3), which is, of course, not exact but represents an "educated guess"; it follows either from the chiral Lagrangian together with pure entropy considerations (which makes it very robust) [6] or from large-N_c arguments [7].

NAIVE EVALUATION OF Σ_2 AND χ_∞

Disregarding possible lattice artefacts, one may fit the available data to the continuum curve (5) and extract Σ and χ_∞ from the fit parameters [6]. It is worth emphasizing that this evaluation of Σ is *distinct from the usual fermionic determination* which is via the trace of the Green's function of the Dirac operator at various *current*-quark masses and

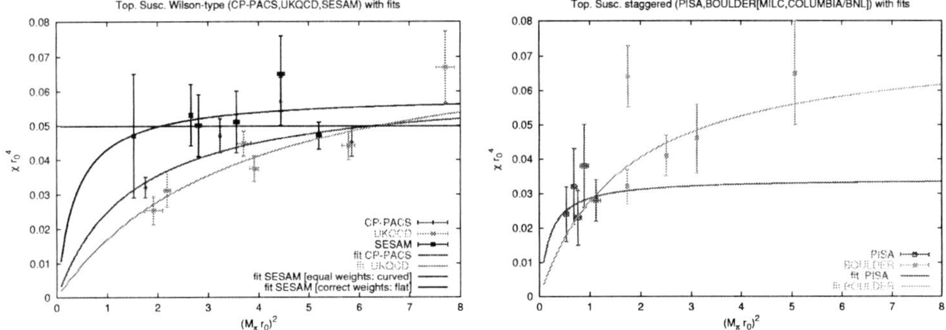

FIGURE 2. Topological susceptibility versus quark mass in $N_f = 2$ QCD with Wilson-type (left) or staggered (right) sea-quarks [1, 2, 3, 4, 5] together with naive fits of the susceptibility curve (5) to the data, neglecting possible discretization effects. The associate values for Σ_2 and χ_∞ are tabulated in Table 1.

extrapolating (after proper renormalization) to the physical (or chiral) point. Regardless how convincing this sounds, the results as tabulated in Table 1 look rather devastating: While our value for the quenched topological susceptibility $\chi_\infty \simeq (200 \pm 10\,\text{MeV})^4$ nicely agrees with previous direct determinations in the SU(3)-theory, the suggested value for the (full) chiral condensate in the chiral limit $\Sigma \simeq (450 \pm 100\,\text{MeV})^3$ *dramatically exceeds* (by more than a factor 2) the "phenomenological" value $\Sigma \simeq (288\,\text{MeV})^3$ (which follows from the GOR-relation with $m_{u,d}(\overline{\text{MS}}, \mu = 2\,\text{GeV}) \simeq 3.5\,\text{MeV}$ [8]).

Looking back at Fig. 1, one may argue that this hardly comes as a surprise, since both the "Wilson-type" and the "staggered" data sets are much more likely to violate the linear upper bound in the deep chiral regime than the flat ceiling in the heavy-(sea)-quark limit. Besides, Fig. 1 tells us how important it is to compare the data to the right prediction: Knowing nothing but the chiral constraint (3), one might be tempted to say that the data are in nice agreement with the leading order chiral prediction. The outcome of our analysis shows that there is absolutely no point in comparing lattice data gained at $(M_\pi r_0)^2 \geq 1.5$ to the leading order chiral behaviour (3), because the "true" prediction (5) is *substantially lower*: If lattice data at $(M_\pi r_0)^2 = 2.5$ are found to be in "nice agreement" with the chiral prediction (3) based on the phenomenological value $\Sigma \simeq (288\,\text{MeV})^3$, then it means that they are $\sim$50% in excess of what they should be.

The bottom line is that todays full QCD simulations (with both Wilson-type and staggered sea-quarks) – if taken at face value – do show unquenching effects in their vacuum structure but, in general, far less than expected at their respective sea-quark masses. Unpleasant as it is, we are invited to think about the reasons for this finding.

TWO ALTERNATIVES

There are two main reasons why Σ as determined via fitting full QCD topological susceptibility data to (5) could substantially exceed the standard phenomenological value $\Sigma \simeq (288\,\text{MeV})^3$ while the simultaneously determined χ_∞ takes a regular value.

TABLE 1. Naive determination of Σ and χ_∞ from full QCD vacuum data with (5), using $r_0 = 0.5$ fm and the GOR-relation to convert to physical units.

	CP-PACS	UKQCD	SESAM *	PISA	BOULDER
$(F_\pi r_0)^2/4$	0.0417	0.0216	0.1600	0.1728	0.0441
F_π[MeV] †	161.	116.	316.	328.	166.
$\Sigma^{1/3}$[MeV]	415.	334.	650.	667.	423.
$\chi_\infty r_0^4$	0.0616	0.0781	0.0587	0.0343	0.0748
$\chi_\infty^{1/4}$[MeV]	197.	209.	194.	170.	206.

* All data get equal weights, otherwise the best direct fit is almost flat (cf. Fig. 2).
† Using the convention in which $F_\pi \simeq 93$ MeV in nature; note that – except for the one in the UKQCD column – all entries are substantially larger than that value.

Large lattice artefacts: The simple reason is that discretization effects could be large, since for the case of the topological susceptibility finite-volume effects are analytically shown to be well under control in most of todays simulations [6]. With the ascent of fermion actions which satisfy the Ginsparg-Wilson relation it makes sense to disentangle chirality violation effects from ordinary scaling violation effects, and in a recent paper A. Hasenfratz has shown some evidence [5] that one should primarily suspect the former type of discretization effects to give rise to the excessive Σ values listed in Table 1.

Proximity of the "conformal window": The more "exotic" view is that the fitted value for $\chi_0(m) \simeq \Sigma m/N_f$ is appropriate for $N_f = 2$ QCD, while the standard phenomenological evaluation – even if it involves (non-strange) pions only – concerns QCD with 3 (2+1) light flavours. In this scenario the way χ depends on m ($m \ll \Lambda_{\text{QCD}}$) has an N_f-dependence beyond the one indicated in (3), i.e. the low-energy constant Σ depends (strongly) on N_f, e.g. $\Sigma_2 \simeq (450\,\text{MeV})^3$ while $\Sigma_3 \simeq (288\,\text{MeV})^3$. The latter gap could then be interpreted as a hint that the "conformal window" (where, for appropriate N_f, one has $\Sigma_{N_f} \ll \Lambda_{\text{QCD}}^3$ and chiral symmetry is primarily broken through higher dimensional condensates) might be "close" – see [6] for a discussion and some references.

REFINED EVALUATION OF Σ_2 AND χ_∞

In the following, I concentrate on the first alternative and show that – in the absence of data suitable for a continuum extrapolation – basic knowledge regarding the dominant discretization effects allows for a more sophisticated evaluation of the parameters in (5).

The key observation on which this analysis relies is that the leading lattice artefacts in both elements of (5) – the chiral piece (3) and the quenched piece (4) – are known: On the chiral side the dominant effect is chirality violation (for a discussion see [5, 2]), i.e. $\hat{F}_\pi \hat{r}_0 = F_\pi r_0 (1 + \text{const}\,(a/r_0)^p)$ with p reflecting the fermion formulation. On the quenched side scaling violations are known to result in $\hat{\chi}_\infty \hat{r}_0^4 = \chi_\infty r_0^4 - 0.208(a/r_0)^2$ (with a known coefficient !) [2, 1]. Combining all the ingredients, one ends up with

$$\hat{\chi} \hat{r}_0^4 = 1/\{\, 2N_f/[(M_\pi r_0)^2 (F_\pi r_0)^2 (1 + \text{const}\,(a/r_0)^p)^q] + 1/[\chi_\infty r_0^4 - 0.208(a/r_0)^2] \,\}\,, \quad (6)$$

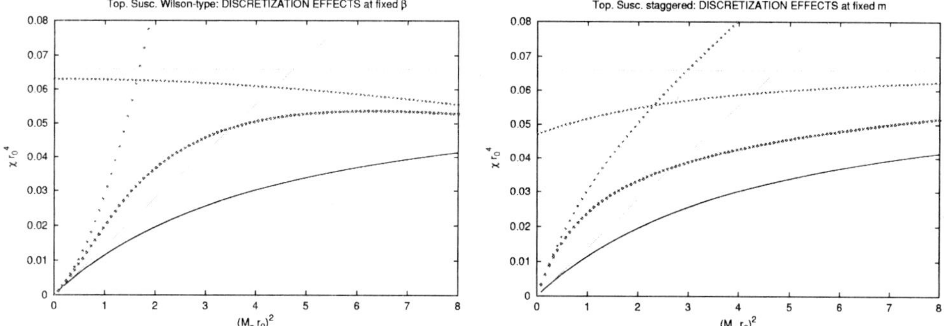

FIGURE 3. Schematic representation how discretization effects typically affect topological susceptibility data in full QCD with Wilson-type (left) and staggered (right) sea-quarks, if simulations are run at fixed β and fixed m, respectively. In either case discretization effects tend to enhance the effective F_π or Σ (in particular for large lattice spacing a) and they reduce the associate χ_∞ by an amount $\propto a^2$. The lattice susceptibility (6) is the reduced mean of the modified functions (fat dots) and may, for $2 < (M_\pi r_0)^2 < 8$, show little variation with m and lie substantially above the expected continuum curve (black line).

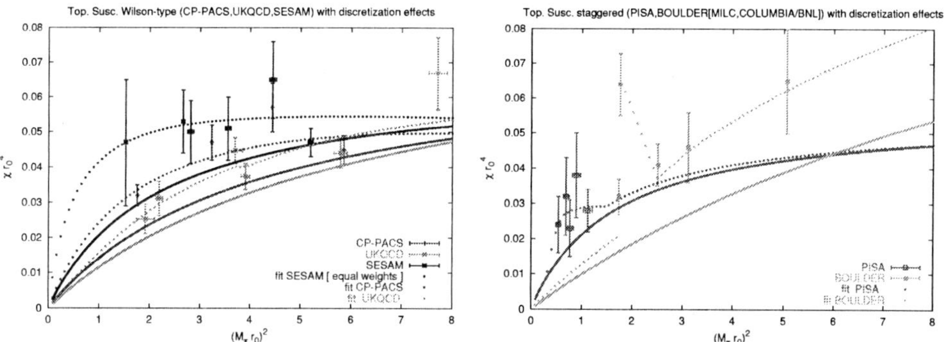

FIGURE 4. Topological susceptibility versus quark mass in $N_f = 2$ QCD with Wilson-type (left) or staggered (right) sea-quarks [1, 2, 3, 4, 5] together with fits of the functional form (6) to the data (dotted lines). This time, the values for Σ_2 and χ_∞ suggest "reasonable" continuum curves (full lines, c.f. Table 2).

where q may be chosen between 1 and 2, since $(1+O(a^p))^2 = 1+O(a^p)$; I use $q=2$. A qualitative picture how these modifications affect the measured topological susceptibility is drawn in Fig. 3. In this respect it is important to know that in a series of full QCD simulations at fixed β the lattice spacing will *shrink* if the (sea-)quark mass gets reduced (which is often the case in studies with Wilson-type sea-quarks; the only exception is the one by UKQCD, where β gets relaxed when κ is increased in such a way that a stays approximately constant), whereas if one works in a staggered formulation at fixed quark mass $\hat{m}$ the lattice will get *coarser* as one approaches the chiral limit – eventually the measure resembles more that of a theory with a single pseudo-Goldstone rather than one with N_f^2-1 pion type degrees of freedom as in the continuum [5].

In order to make use of this knowledge (i.e. to utilize (5) to fit the data) one has to

TABLE 2. Refined determination of Σ and χ_∞ from full QCD vacuum data with (6), using $r_0=0.5\,\text{fm}$ and the GOR-relation to convert to physical units. For a cautionary statement regarding the fitting procedure see text.

	CP-PACS	UKQCD	SESAM	PISA	BOULDER
$(F_\pi r_0)^2/4$	0.0174	0.0134	0.0299	0.0340	0.0106
F_π[MeV]	104.	91.	137.	146.	81.
$\Sigma^{1/3}$[MeV]	311.	284.	372.	388.	263.
$\chi_\infty r_0^4$	0.0737	0.0845	0.0661	0.0562	0.1466
$\chi_\infty^{1/4}$[MeV]	206.	213.	200.	192.	244.

decide on the parameters (p, const) showing up in (6). The former choice is relatively easy – I take $p_{\text{CP-PACS}}=1.5, p_{\text{UKQCD}}=2, p_{\text{SESAM}}=1, p_{\text{PISA}}=p_{\text{BOULDER}}=2$ to account for the formulation and the different strategies regarding c_{SW}. The latter choice – which value "const" shall be given – is more delicate: Ideally, one would like to determine it directly from the data. However, it turns out that the quality of the data at hand is not sufficient to allow for an additional (i.e. third) fitting parameter. A reasonable option would be to determine it from conventional F_π measurements on the individual ensembles. A simpler option is to use (6) twice – in a first round "const" is given a likely value by fitting it while $((F_\pi r_0)^2/4, \chi_\infty r_0^4)$ is held fixed at (0.014, 0.066); in a second round the latter get adjusted while "const" is kept fixed at the previously determined value. Obviously, with this simpler option the final outcome for $((F_\pi r_0)^2/4, \chi_\infty r_0^4)$ reflects, to some extent, the corresponding initial values. The simplest option is just to set "const" to a generic value like 1. My person choice is to take the arithmetic average of the "const" values suggested by the last two options and to use that value to fit for $(F_\pi r_0)^2/4$ and $\chi_\infty r_0^4$.

The result of this exercise is shown in Table 2 and Fig. 4, where dotted lines represent the lattice curve (6) with $(F_\pi r_0)^2/4$ and $\chi_\infty r_0^4$ adjusted such as to make it go through the data points while full lines indicate the associate continuum curve (5). The data in Table 2 should be taken with a grain of salt since, as explained above, there is a remnant trace of the initial values in the final fitting parameters $(F_\pi r_0)^2/4$ and $\chi_\infty r_0^4$. Nontheless, the result is interesting because it supports the standard view that QCD with $N_f=2$ is not in the "conformal window" and that its low-energy structure agrees with that suggested by phenomenological investigations in "real" (2+1 flavour) QCD [9], i.e. at least for $N_f=2$ chiral symmetry is predominantly broken through a *distinctively nonzero condensate* (for references to an alternative scenario see [6]) and Σ_2 as suggested by Table 2 is *compatible with the value from the GOR-relation*, $\Sigma_{2+1}\simeq(288\,\text{MeV})^3$.

From a lattice perspective it reassuring to see that through a simple ansatz for the dominant discretization effects todays state-of-the-art simulations (which, from a naive perspective, seemed to support an almost flat topological susceptibility curve and hence to reproduce – in spite of all unquenching efforts – a more or less quenched vacuum structure) may, in fact, be shown to give *supportive evidence in favour of a decreased topological susceptibility near the chiral limit* and hence "bear" the knowledge of the difference between the quenched and the unquenched vacuum structure in them. The ultimate goal is, of course, to make discretization effects sufficiently small so that the expected continuum pattern of the vacuum structure gets visible in the raw data already.

REFERENCES

1. A. Ali Khan *et al.* [CP-PACS collab.], hep-lat/0106010 (2001).
2. A. Hart and M. Teper [UKQCD collab.], hep-lat/0108006 (2001).
3. G. Bali *et al.* [SESAM/TXL collab.], *Phys. Rev. D*, **64**, 054502 (2001).
4. B. Allés *et al.* [PISA group], *Phys. Lett. B*, **483**, 139 (2000).
5. A. Hasenfratz, hep-lat/0104015 (2001).
6. S. Dürr, hep-lat/0103011_v2, to appear in Nucl. Phys. B (2001).
7. H. Leutwyler and A. Smilga, *Phys. Rev. D*, **46**, 5607 (1992).
8. A. Ali Khan *et al.* [CP-PACS collab.], *Phys. Rev. Lett.*, **85**, 4674 (2000).
9. G. Colangelo, J. Gasser and H. Leutwyler, *Phys. Rev. Lett.*, **86**, 5008 (2001).

CHALLENGES IN PERTURBATIVE QCD

Problems and Challenges in Perturbative QCD

P. Nason

INFN, sez. di Milano, Milan, Italy

Abstract. I discuss problems in perturbative QCD where progress is likely to take place in the near future. I will illustrate recent progress in next-to-next-to leading order calculation, current efforts in extending the validity of parton shower models, and some open problems in soft gluon resummation.

INTRODUCTION

At present, perturbative QCD constitutes an advanced, well tested framework for the description of hadronic processes. In particular, QCD studies at LEP have given a convincing evidence that the perturbative expansion, including NLO (next-to-leading order) terms is at work. Furthermore, a considerable theoretical effort has gone into computation of high energy QCD processes at the NLO level, so that we can say that at present NLO calculations have become the standard.

NLO calculations have been used mostly for QCD studies and tests, and in some cases (i.e. top and higgs production) for the computation of the cross section of an undiscovered particle. For more practical applications of everyday experimental physics, shower montecarlo programs, which are capable to give a detailed (although approximate) description of the final state, are the standard tool to perform background studies and efficiency corrections.

In extreme kinematic regions of phase space, like production near threshold, the high energy limit, and very high transverse momentum production of massive objects, fixed order perturbative QCD is inadequate, because of large logarithmic enhancements arising in all orders of perturbation theory. Resummation techniques have been developed to deal with these cases.

In the near future, hadron collider physics will have a prominent role in the high energy physics program. Perturbative QCD will be therefore an essential theoretical tool, and will have to develop appropriately to match the future challenges. In this talk I will discuss what are the directions in which progress is needed and is likely to be achieved. I will discuss the ongoing effort in raising the precision of QCD calculations, by going to higher perturbative orders, the open problems in the improvement of shower Montecarlo programs, and some open problem in soft gluon resummations.

BEYOND THE NLO LEVEL

Although today's typical QCD calculations do include terms up to the NLO level, there are a few classes of problems where higher order terms have been computed. These

CP602, *QCD@Work: International Workshop on Quantum Chromodynamics*
edited by P. Colangelo and G. Nardulli

are problems in which the result depends upon a single physical scale. Among the most important results, the total cross section for $e^+e^- \to$hadrons, computed to order α_s^3 [1, 2], The QCD β function, computed at the 4-th loop level ($O(\alpha_s^5)$) [3], up to $N = 12$ singlet, $N = 14$ non-singlet crossing even, and $N = 13$ crossing odd moments of the $O(\alpha_s^3)$ splitting functions, together with the $O(\alpha_s^3)$ coefficient functions for DIS [4] [5] [6]. In all these cases, the problem can be reduced to the computation of a massless propagator type graph, which can be computed with techniques developed in refs. [7, 8] (the MINCER program). These results have an important impact on α_s determinations.

At LEP, α_s has been measured most accurately from the Z^0 hadronic width (for a 100 GeV Higgs mass), yielding $\alpha_s(M_Z) = 0.1224 \pm 0.0038$ (from R_l) and 0.1180 ± 0.0030 (from σ_l^0)[1]), and from τ hadronic decays yielding $\alpha_s(M_Z) = 0.1181 \pm 0.0007(\text{exp.}) \pm 0.003(\text{theo.})$ where the average is taken from ref. [9] For τ decays, both ALEPH [10] and OPAL [11] have performed measurements of the vector and axial vector spectral mass distributions, allowing several consistency checks. Consistency of the two determination, both performed at the NLLO level, is very encouraging, since they are theoretically very similar and are performed at very different scales. Determinations of comparable accuracy have recently become available in deep-inelastic scattering (DIS). The authors of refs. [12, 13, 14] analyzed the revised CCFR data in NNLO, and obtained

$$\alpha_s(M_Z) = 0.118 \pm 0.002(\text{stat.}) \pm 0.005(\text{sys}) \pm 0.003(\text{theo.}) \tag{1}$$

The authors of ref. [15], considered both μ/e and ν DIS. From a NNLO analysis of F_2 in μ and e DIS from 2.5 to 230 GeV^2 they obtained $\alpha_s(M_Z) = 0.1166 \pm 0.0013$, and from νN scattering $\alpha_s(M_Z) = 0.1153 \pm 0.0063$ Besides the knowledge of NNLO corrections, inclusive hadronic decays and DIS determination of α_s have the advantage of small power corrections ($1/Q^4$ in $e^+e^- \to$hadrons, $1/Q^2$ in DIS), which allow precise determinations even at relatively low scales. The consistency of all these very accurate determinations is quite remarkable. In fact, in the review of α_s measurements of S. Bethke [9], they end up dominating the final result.

Aside from these cases, the standard QCD calculations for collider physics are done at NLO, NNLO results being available only in the simplest case of Drell-Yan pair production [16]. Among the most important results, jet production in e^+e^- annihilation [17], and for hadron collisions the Drell-Yan process [18, 19], direct photon production [20], heavy flavour production [21, 22, 23], jet production [24, 25, 26] and higgs production [27].

It is standard practice to regularize real and virtual corrections in $d = 4 - 2\varepsilon$ dimensions, and singularities in $1/\varepsilon$ and $1/\varepsilon^2$ appear in both real and virtual corrections, as a consequence of the collinear and soft divergences in the calculation. These singularities cancel in inclusive quantities. Remnants of the divergences remain however in the final results, in the form of distributions in the kinematical variables. Because of this, the practical implementation of the results is often problematic. As an example, practical implementations of jet production calculations appeared four years after the computation of the matrix elements. The presence of distributions in the final result makes

[1] From the Lep Electroweak Working Group homepage, summer 2001 updates

it difficult to handle them to experimentalists, to be used as black box generators of events. Practical implementations of these results are obtained either by slicing away a small region around the singularities (Slicing method), or subtracting under the integral sign when integrating the cross section against a physical observable (Subtraction method). The subtraction method, first introduced in ref. [17], has the advantage that it does not need an unphysical cut-off, and it does not lead to cancellations between large numbers when the cut-off is small. In the program EVENT [28] this method has been implemented as a black box parton montecarlo generator, i.e. a generator of weighted parton events, that gives the correct $O(\alpha_S^2)$ results when used to compute I.R. safe shape variables. The same technique has also been introduced in the context of hadronic collisions in vector boson pair production [29] and in jet physics [26]. Inclusive, I.R. safe quantities are correctly computed using this procedure. However, the net finite result is the sum of a positive infinite result and a negative infinite one. In practice, this causes difficulties in the convergence of the calculations, so that much work is needed to get good convergence properties. The slicing method has unavoidably the same problem, and furthermore, it has the problem that for small slicing parameter one has to control a large cancellation between two integrals.

Do we need to go to NNLO?

NLO radiative corrections are typically quite large. NNLO corrections are therefore certainly important, but they may turn out to be of the same order as NLO corrections, thus spoiling our effort to increase precision. An interesting example to look at is jet physics in e^+e^- collisions. The typical size of the corrections to shape variable distributions is 40%. We thus know two terms of the perturbative expansion, the second term being 40% of the previous one. On this basis, it is difficult to assess the convergence of the perturbative expansion. The data, however, seems to imply that there is convergence, as can be seen from fig. 1 [30]. The figure reports several determinations of $\alpha_S(M_Z)$ for each histogram bin of a shape variable, performed using OPAL data. No attempt is made to correct for hadronization effects. The remarkable increase of consistency from LO to NLO shows that the perturbative expansion really seems to be at work. Inclusion of NNLO may further improve the consistency, and teach us something about the real size of hadronization effects.

In collider physics, in many instances we find marginal consistency of theoretical prediction with data [31]. In all these cases, better knowledge of theoretical predictions would help.

Status of NNLO calculations

The focus is on jet production in hadronic collision, that is to say, on the parton-parton scattering process (jet production in e^+e^- annihilation is next in complexity). One needs

- the square of the $2 \to 4$ tree amplitude
- the interference of the $2 \to 3$ tree level and one loop amplitude

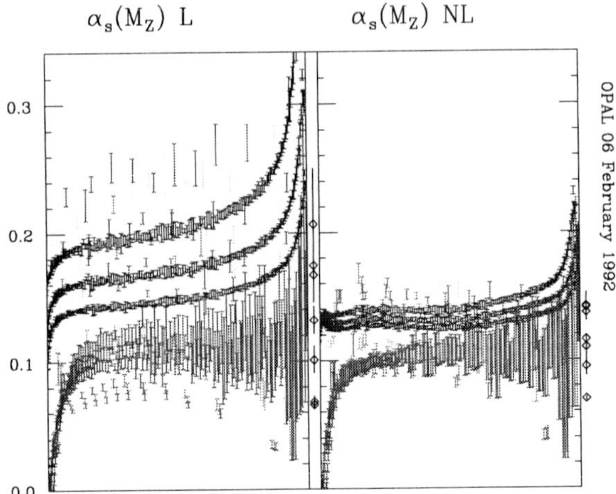

FIGURE 1. Determination of $\alpha_S(M_Z)$ at leading order (left) and next-to-leading order, for each bin of the histogram for the shape variables thrust, oblateness, c parameter, heavy jet mass, jet mass difference, energy-energy correlation and asymmetry in the energy-energy correlation. For each determination the normalization scale μ is chosen equal to M_Z, $M_Z/2$ and $M_Z/4$.

- the square of the 2→2 one loop amplitude
- the interference of the 2→2 two loop and tree level amplitude
- for a consistent phenomenological treatment, the $O(\alpha_S^3)$ Altarelli-Parisi splitting functions are also needed

For the last item, an approximate expressions, based upon constraints coming from the large and small x behaviour, and the moments known at the NNLO level has been obtained [32]. A calculation of the full NNLO splitting function is under way [33]

Techniques for the computation of tree level amplitude for the 2→4 process have been available for a long time. The problem is the collinear and soft limits of such amplitudes, that must be regularized in order to implement the cancellation and factorization of soft and collinear divergences. In dimensional regularization, these singularities will appear as poles in $1/\varepsilon$ up to the fourth power. Therefore, the structure of these singularities must be understood with the required accuracy in ε, in order to get a correct result after cancellation. Indeed, all these results are available now:

- Double soft limit [34] when two final state particles become soft. Can yield singularities up to $1/\varepsilon^4$ after final state integration of the soft particles
- Double collinear and soft-collinear [35, 36, 37]: a subset of 3 final state partons become collinear, or 2 become collinear and one is soft (up to $1/\varepsilon^3$ singularities)

The computation of 2→3 amplitude at one loop level are also known [38, 39] Again, the collinear and soft limit of this amplitude is needed in order to meet the accuracy required by NNLO calculations:

- Collinear limit of one loop amplitudes [40, 41, 42]

• Soft limit of one loop amplitudes [43, 44, 45]

The two loop 2→2 contribution has been recently computed [46, 47, 48]. Recursion relations among Feynman integrals are used to reduce the problem to the computation of a small number of master integrals. The hardest of those (double box, planar and crossed) have been solved only recently [49, 50] The structure of the $1/\varepsilon$ singularities of these amplitude can be checked against a general factorization formula [51].

All ingredient needed for a full NLO computation of jet cross sections in hadronic collisions are in place. The implementation of the various terms into a useful result still appears as a formidable task, as we have learned from the difficulties encountered in the one-loop case. However, in view of the enormous progress achieved in the last few years, it appears now that a result will become available in useful time.

NLO AND SHOWER MONTECARLO

As discussed in the previous section, NLO corrections lead to singular distributions in the kinematic variables. This problem arise because of the presence of infrared singularities of collinear and soft nature. Soft and final state singularities cancel in the sum of real and virtual contributions, and initial state collinear singularities are subtracted and absorbed into the parton densities. It remains the problema, however, that in particular regions of phase space one can get a negative result for a physical cross section. In these regions, only an all-order resummation of the perturbative expansion can lead to positive results. Thus, any implementation of a QCD perturbative calculations at fixed order cannot be blindly used to describe the final state. There are always regions where the calculation cannot be trusted, and all order resummation of enhanced contribution is needed to get sensible answers. While analytical techniques are available to perform these resummation with next-to-leading logarithmic accuracy for particular observables, shower Montecarlo program address the same problem at the leading logarithmic level, but with the advantage of generating explicitly the multiparton final state associated with the resummation of an infinite set of graphs. The multiparton final state can be completed by some phenomenological model of hadron formation. Thus, Shower Montecarlo programs offer the advantage of providing a model of the final state that can be comfortably used by experimentalists as a black box generator of physical events. It does not, however, include NLO corrections. There are basically three problems that arise in the implementation of NLO results:

i The negative weights.

ii Implementations of NLO results have also the problem of being weighted event generators, since it is not easy to unweight a generator that has negative weights. Experimentalists prefer implementation of theoretical results as generators of unweighted events, because of the high cost of detector simulation.

iii It would be desirable to interface the NLO computation with a shower Montecarlo program, in order to generate the detailed structure of the final state. This is possible with leading order calculation, but very difficult with NLO, because of possible overcounting.

What was commonly done up to now is to use NLO calculation to fix the normalization of certain inclusive distributions, but use the LL Montecarlo to simulate the full event. Attempts to overcome the negative weights problem [52] have been implemented, that use the inclusive character of the experimental measurement to implement real and virtual cancellation. In the case of W pair production, followed by semileptonic decay of the W, the neutrino is not detected, and jets are detected only above a certain energy. Integrating the real and virtual contributions in the unmeasured kinematical variables yields finite results. This approach is however not always applicable, and furthermore, it does not easily allow to interface the NLO result with a Montecarlo program.

A more interesting solution would be to improve shower Montecarlo programs in such a way that they can handle NLO matrix elements. Extending a Montecarlo program to NLL accuracy seems to be a formidable task, requiring the inclusion of NLO branching processes, multiple branching, and NLO corrections at the interaction vertex. A more viable program would be to include NLO corrections at the interaction vertex, including exact matrix element for the emission of the hardest parton, making sure that it correctly describes infrared safe variables at the NLO level.

Webber [53] proposed a method to generate hard multijet configuration which are at least correct at leading order, and incorporate jet fragmentation at NNLO accuracy. One borrows a clustering algorithm from e^+e^- jet physics (k_T clustering). Multijet cross sections have been computed in next-to-leading logarithmic order in the clustering parameter. For example, the two jet rate is essentially a Sudakov form factor for no emission of resolvable jets with a y parameter above the cut. The generation of jet clusters is performed according to the exact n-parton matrix elements, supplemented with the appropriate Sudakov factor and running couplings. The remaining structure of the event is generated applying the showering algorithm to each cluster, modified in such a way that no overcounting is done in the showering, and the y dependence of the cross section cancels in the appropriate way at the end.

In ref. [54] a method was proposed for correcting the parton shower when hard emissions are present. Essentially, events are reweighted with the exact matrix element when there is a hard emission. To avoid inconsistencies, reweighting is performed at each branching, if that branching is the hardest so far. This method has been applied to top decay and vector boson production (Corcella and Seymour).

Collins [55] has proposed a method for correcting the shower at the NLO level. This is done by adding the NLO contribution, after subtraction of the approximate emission probability that is used in the shower model, to avoid overcounting. The method has been applied so far to the qg contribution to the W production cross section.

Although no satisfactory answer is available yet, we may expect progress in this field in the near future.

SOFT GLUON EFFECTS

When approaching the end of phase space, enhanced contributions arise to all order in perturbation theory, due to the uncompensated cancellation of enhanced soft gluon emission corrections. The case study for this phenomenon has been the Drell-Yan cross section, where formulae for the NLL resummation of these enhanced contributions have been obtained long ago [56, 57]. Several problems in the use and interpretation of these formulae have arisen in the past, having to do in particular with factorial growth in the perturbative expansion (see [58] and references therein). Extension of the resummation formalism to heavy flavour and direct photon production have appeared in refs. [59, 60] [61, 62, 58, 63, 64].

Here I would like to illustrate some conceptual problems that arise when the threshold limit is entangled with the large x limit of the parton densities and the fragmentation functions [65]. In particular, large x_T direct photon production (traditionally one of the most important tool to measure the gluon structure function at large x values) is an interesting example that shows a few puzzling feature. It is usually assumed that the fragmentation contribution (i.e. the contribution coming from a final state parton fragmenting into a photon) is suppressed at large x_T (the scaled photon transverse energy) in this process. A closer look shows that, in processes of interest, this may not be the case.

We use the notation of ref. [62]. One proceeds by taking moments in x_T, and looks at large N,

$$\sigma_{\gamma,N}(E_T)=\sum_{a,b} f_{a/H_1}^{(N+1)} f_{b/H_2}^{(N+1)}\left\{\hat{\sigma}_{ab\to\gamma,N}+\sum_c \hat{\sigma}_{ab\to c,N}\, d_{c/\gamma,2N+3}\right\}. \tag{2}$$

What prevails for large N depends only upon the large N behaviour of the parton densities and of the fragmentation function. We have

$$\frac{d}{dt}\begin{vmatrix} S_N \\ G_N \end{vmatrix}=\begin{vmatrix} A_N^{qq} & A_N^{qg} \\ A_N^{gq} & A_N^{gg} \end{vmatrix}\begin{vmatrix} S_N \\ G_N \end{vmatrix}, \quad t=\frac{1}{2\pi}\int d\log Q^2\alpha(Q^2)=\frac{1}{2\pi b_0}\log\frac{\alpha(Q_0^2)}{\alpha(Q^2)}, \tag{3}$$

S_N and G_N are moments of the singlet and the gluon parton densities, and A_N^{ab} are the moments of the Altarelli–Parisi splitting functions, that, for large N, have the behaviour

$$A_N^{qq}\approx -2C_{\mathrm{f}}\log N,\quad A_N^{gg}\approx -2C_{\mathrm{A}}\log N, A_N^{gq}\approx\frac{C_{\mathrm{f}}}{N},\quad A_N^{qg}\approx\frac{n_{\mathrm{f}}}{N}. \tag{4}$$

Assuming that at the initial condition G_N is not larger than S_N, it is easy to get large N asymptotic limit

$$\frac{G_N(t)}{S_N(t)}\approx\frac{A_N^{gq}}{A_N^{qq}-A_N^{gg}}\approx\frac{1}{N\log N}\frac{C_{\mathrm{f}}}{2(C_{\mathrm{A}}-C_{\mathrm{f}})}=\frac{4}{10}\frac{1}{N\log N}. \tag{5}$$

An analogous argument gives for the asymtotics of the fragmentation function

$$D_N^{\gamma q}(t)=\frac{e_q^2}{2C_{\mathrm{f}}}\frac{1}{N\log N}\frac{\alpha_{\mathrm{em}}}{\alpha(t)} \tag{6}$$

Let us consider the case of proton-antiproton collisions. The large N behaviour of the $q\bar{q}$, qg and fragmentation subprocesses is given by

$$\sigma_N^{q\bar{q}}\approx C_N\alpha\alpha_{\mathrm{em}}S_N^2,\quad \sigma_N^{qg}\approx C_N\alpha\alpha_{\mathrm{em}}S_N G_N,\quad \sigma_N^{q\to\gamma}\approx C_N\alpha^2 S_N^2 D_N^{\gamma q} \tag{7}$$

where C_N is the large N behaviour of the partonic cross sections. So: the qg and fragmentation components have the same asymptotic behaviour. In the case of proton-antiproton collisions, the $q\bar{q}$ mechanism dominates. However, in the interesting case of proton-proton collisions it does not, and one is left with the qg and the fragmentation process that compete at the same level. One may ask if Sudakov effects may decide which mechanism dominates. However this is not enough. Double logarithms do not appear in the diagonal splitting functions, but they may appear in the off diagonal ones

$$A_N^{gq} \to A_N^{qg}\left(1+\frac{\alpha}{2\pi}(C_{\rm A}-C_{\rm f})\log^2 N\right)\,,\;\; A_N^{\gamma q} \to A_N^{\gamma g}\left(1+\frac{\alpha}{2\pi}C_{\rm f}\log^2 N\right) \tag{8}$$

which propagate into the answer, and compete with Sudakov factors. Other logarithmic enhancements arise in the NL cross section for the process $qq' \to qq'\gamma$. At the end one finds

$$\frac{qg}{\text{fragmentation}} = \frac{\Delta_g}{\Delta_q^2}\left(1-\frac{\alpha(t)}{2\pi}(C_{\rm A}-2C_{\rm f})\log^2 N\right)\,,\;\; \text{with}\; \Delta_{g/q} \approx \exp\left(\frac{\alpha_{\rm S}}{2\pi}2C_{A/F}\log^2 N\right)\,.$$

Thus the Sudakov effects seem to be reduced by other kinds of double logarithms. In general, the Sudakov formalism we use now does not seem to be capable of answering questions that depend upon the large N behaviour of the parton densities.

REFERENCES

1. Gorishnii, S. G., Kataev, A. L., and Larin, S. A., *Phys. Lett.*, **B212**, 238–244 (1988).
2. Gorishnii, S. G., Kataev, A. L., and Larin, S. A., *Phys. Lett.*, **B259**, 144–150 (1991).
3. van Ritbergen, T., Vermaseren, J. A. M., and Larin, S. A., *Phys. Lett.*, **B400**, 379–384 (1997).
4. Larin, S. A., Nogueira, P., van Ritbergen, T., and Vermaseren, J. A. M., *Nucl. Phys.*, **B492**, 338–378 (1997).
5. Retey, A., and Vermaseren, J. A. M., *Nucl. Phys.*, **B604**, 281–311 (2001).
6. van Neerven, W. L., and Zijlstra, E. B., *Phys. Lett.*, **B272**, 127–133 (1991).
7. Gorishnii, S. G., Larin, S. A., Surguladze, L. R., and Tkachov, F. V., *Comput. Phys. Commun.*, **55**, 381 (1989).
8. Larin, S. A., Tkachov, F. V., and Vermaseren, J. A. M. (1991), nIKHEF-H-91-18.
9. Bethke, S., *J. Phys.*, **G26**, R27 (2000).
10. Barate, R., et al., *Eur. Phys. J.*, **C4**, 409–431 (1998).
11. Ackerstaff, K., et al., *Eur. Phys. J.*, **C7**, 571–593 (1999).
12. Kataev, A. L., Kotikov, A. V., Parente, G., and Sidorov, A. V., *Phys. Lett.*, **B417**, 374–384 (1998).
13. Kataev, A. L., Parente, G., and Sidorov, A. V., *Nucl. Phys.*, **B573**, 405–433 (2000).
14. Kataev, A. L., Parente, G., and Sidorov, A. V. (2001).
15. Santiago, J., and Yndurain, F. J. (2001).
16. Hamberg, R., van Neerven, W. L., and Matsuura, T., *Nucl. Phys.*, **B359**, 343–405 (1991).
17. Ellis, R. K., Ross, D. A., and Terrano, A. E., *Nucl. Phys.*, **B178**, 421 (1981).
18. Altarelli, G., Ellis, R. K., and Martinelli, G., *Nucl. Phys.*, **B143**, 521 (1978).
19. Altarelli, G., Ellis, R. K., and Martinelli, G., *Nucl. Phys.*, **B157**, 461 (1979).
20. Aurenche, P., Douiri, A., Baier, R., Fontannaz, M., and Schiff, D., *Phys. Lett.*, **B140**, 87 (1984).
21. Nason, P., Dawson, S., and Ellis, R. K., *Nucl. Phys.*, **B303**, 607 (1988).
22. Nason, P., Dawson, S., and Ellis, R. K., *Nucl. Phys.*, **B327**, 49–92 (1989).
23. Mangano, M. L., Nason, P., and Ridolfi, G., *Nucl. Phys.*, **B373**, 295–345 (1992).
24. Ellis, R. K., and Sexton, J. C., *Nucl. Phys.*, **B269**, 445 (1986).
25. Aversa, F., Chiappetta, P., Greco, M., and Guillet, J. P., *Nucl. Phys.*, **B327**, 105 (1989).

26. Ellis, S. D., Kunszt, Z., and Soper, D. E., *Phys. Rev. Lett.*, **64**, 2121 (1990).
27. Graudenz, D., Spira, M., and Zerwas, P. M., *Phys. Rev. Lett.*, **70**, 1372–1375 (1993).
28. Kunszt, Z., Nason, P., Marchesini, G., and Webber, B. R. (1989), to appear in the Proceedings of the 1989 LEP Physics Workshop, Geneva, Swizterland, Feb 20, 1989.
29. Mele, B., Nason, P., and Ridolfi, G., *Nucl. Phys.*, **B357**, 409–438 (1991).
30. Blunda, A., and Nason, P. (1992), unpublished.
31. Bertram, I. (2001), talk given at the International Europhysics Conference on High Energy Physics July 12-18, 2001, Budapest, HUNGARY.
32. van Neerven, W. L., and Vogt, A., *Nucl. Phys.*, **B603**, 42–68 (2001).
33. Moch, S., Vermaseren, J. A. M., and Zhou, M. (2001).
34. Berends, F. A., and Giele, W. T., *Nucl. Phys.*, **B313**, 595 (1989).
35. Campbell, J. M., and Glover, E. W. N., *Nucl. Phys.*, **B527**, 264–288 (1998).
36. Catani, S., and Grazzini, M., *Nucl. Phys.*, **B570**, 287–325 (2000).
37. Del Duca, V., Frizzo, A., and Maltoni, F., *Nucl. Phys.*, **B568**, 211–262 (2000).
38. Bern, Z., Dixon, L. J., and Kosower, D. A., *Phys. Rev. Lett.*, **70**, 2677–2680 (1993).
39. Kunszt, Z., Signer, A., and Trocsanyi, Z., *Phys. Lett.*, **B336**, 529–536 (1994).
40. Bern, Z., Dixon, L. J., Dunbar, D. C., and Kosower, D. A., *Nucl. Phys.*, **B425**, 217–260 (1994).
41. Kosower, D. A., *Nucl. Phys.*, **B552**, 319–336 (1999).
42. Kosower, D. A., and Uwer, P., *Nucl. Phys.*, **B563**, 477–505 (1999).
43. Bern, Z., Del Duca, V., and Schmidt, C. R., *Phys. Lett.*, **B445**, 168–177 (1998).
44. Bern, Z., Del Duca, V., Kilgore, W. B., and Schmidt, C. R., *Phys. Rev.*, **D60**, 116001 (1999).
45. Catani, S., and Grazzini, M., *Nucl. Phys.*, **B591**, 435–454 (2000).
46. Anastasiou, C., Glover, E. W. N., Oleari, C., and Tejeda-Yeomans, M. E., *Nucl. Phys.*, **B601**, 318–340 (2001).
47. Anastasiou, C., Glover, E. W. N., Oleari, C., and Tejeda-Yeomans, M. E., *Nucl. Phys.*, **B601**, 341–360 (2001).
48. Glover, E. W. N., Oleari, C., and Tejeda-Yeomans, M. E., *Nucl. Phys.*, **B605**, 467–485 (2001).
49. Smirnov, V. A., *Phys. Lett.*, **B460**, 397–404 (1999).
50. Tausk, J. B., *Phys. Lett.*, **B469**, 225–234 (1999).
51. Catani, S., *Phys. Lett.*, **B427**, 161–171 (1998).
52. Dobbs, M., and Lefebvre, M., *Phys. Rev.*, **D63**, 053011 (2001).
53. Webber, B. R. (2000).
54. Corcella, G., and Seymour, M. H., *Phys. Lett.*, **B442**, 417–426 (1998).
55. Collins, J. C., *JHEP*, **05**, 004 (2000).
56. Sterman, G., *Nucl. Phys.*, **B281**, 310 (1987).
57. Catani, S., and Trentadue, L., *Nucl. Phys.*, **B327**, 323 (1989).
58. Catani, S., Mangano, M. L., Nason, P., and Trentadue, L., *Nucl. Phys.*, **B478**, 273–310 (1996).
59. Bonciani, R., Catani, S., Mangano, M. L., and Nason, P., *Nucl. Phys.*, **B529**, 424–450 (1998).
60. Kidonakis, N., and Sterman, G., *Nucl. Phys.*, **B505**, 321–348 (1997).
61. Catani, S., Mangano, M. L., Nason, P., Oleari, C., and Vogelsang, W., *JHEP*, **03**, 025 (1999).
62. Catani, S., Mangano, M. L., and Nason, P., *JHEP*, **07**, 024 (1998).
63. Laenen, E., Oderda, G., and Sterman, G., *Phys. Lett.*, **B438**, 173–183 (1998).
64. Sterman, G., and Vogelsang, W., *JHEP*, **02**, 016 (2001).
65. Catani, S., Mangano, M. L., Nason, P., and Vogelsang, W. (1999), unpublished.

Theoretical Aspects of HERA Physics

Stefano Forte[1]

INFN, Sezione di Roma III, via della Vasca Navale 84, I–00146 Roma, Italy

Abstract. I discuss the theoretical underpinnings for the extraordinary success of perturbative QCD in the description of HERA data. In particular, I examine recent progress in the understanding of perturbative QCD at small x. I explain the relation between evolution equations in Q^2 and x, and how they can be used for simultaneous resummation of the relevant large logs at HERA. I show that while the HERA data can be understood within our current knowledge of the perturbative expansion of QCD, they pose stringent constraints on the perturbatively unaccessible behaviour of QCD in the Regge limit.

PERTURBATIVE QCD AT HERA

QCD has been tested at HERA [1, 2] over the last several years to an accuracy which is now comparable to that of tests of the electroweak sector at LEP: perturbative QCD turns out to provide an embarrassingly successful description of the HERA data, even in kinematic regions where simple fixed–order perturbative predictions should fail. This success is most strikingly demonstrated by the comparison with the data of the scaling violations of structure functions predicted by the QCD evolution equations [3, 4]: the data agree with the theory over five orders of magnitude in both x and Q^2.

The significance of this sort of result is somewhat obscured by the need to fit the shape of parton distributions at a reference scale, which might suggest that deviations from the predicted behaviour could be accommodated by changing the shape of the parton distribution. However, this is not true because of the predictive nature of the QCD result: given the shape of partons at one scale, there is no freedom left to fit the data at other scales. This predictivity is strikingly seen in the small x region, where the fixed–order QCD result actually becomes asymptotically independent of the parton distribution, apart from an overall normalization. Indeed, the data for $\ln F_2$ plotted versus the variable $\sigma \equiv \ln\frac{x_0}{x}\ln\frac{\alpha_s(Q_0^2)}{\alpha_s(Q^2)}$ are predicted to lie on a straight line, with universal slope $2\gamma = 12/\sqrt{33-2n_f}$ (double asymptotic scaling [5, 6]). The predicted scaling is spectacularly borne out by the data, as shown in fig. 1: in fact, the data are now so accurate that one can see the change in slope when passing the b threshold, and indeed double scaling is only manifest if one separates data in the regions where α_s runs with $N_f = 4$ from those with $N_f = 5$.[2] Equally good agreement with fixed–order perturbation

---

[1] On leave from INFN, Sezione di Torino, Italy

[2] The fact that the observed slope is somewhat smaller than the predicted one, especially at low Q^2, is due to NLO corrections [7] as well as corrections due to the "small" eigenvalue of perturbative evolution [8].

CP602, *QCD@Work: International Workshop on Quantum Chromodynamics*
edited by P. Colangelo and G. Nardulli

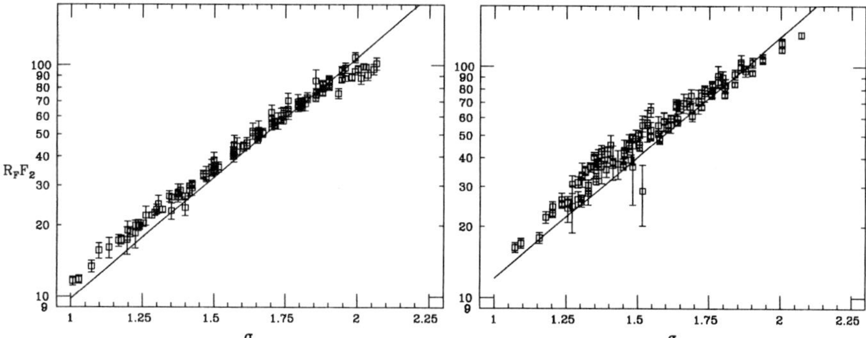

FIGURE 1. Double asymptotic scaling of the H1 data [4]. The scaling variable $\sigma \equiv \ln(x_0/x)\ln(\alpha_s(Q_0^2)/\alpha_s(Q^2))$ is defined with $x_0 = 0.1, Q_0 = 1$ GeV; the rescaling factor R_F is as in Ref. [5]. Only data with $\rho \geq 1$, $\sigma \geq 1$, $Q^2 \geq 4$ GeV^2; $x \leq 0.03$ are plotted. Left: $Q^2 \leq m_b^2$; right: $Q^2 > m_b^2$. The straight line is the asymptotic prediction.

theory is seen when considering less inclusive observables.

This agreement of the data with fixed–order perturbative QCD computations is very surprising, in that the perturbative expansion receives contributions of order $\alpha_s \ln \frac{1}{x}$ so one would expect higher–order corrections to be non–negligible whenever $\alpha_s \ln \frac{1}{x} \gtrsim 1$, i.e. in most of the HERA region. As is well known, the resummation of leading $\ln \frac{1}{x}$ (LLx) contributions to gluon–gluon scattering, and thus to a wide class of hard processes, including small x scaling violations of structure functions, is accomplished by means of the BFKL evolution equation [9, 10, 11]. Matching the BFKL approach to standard perturbative computation, however, is nontrivial [12, 13], while the BFKL equation itself seems to be unstable towards the inclusion of higher order corrections [14]. Hence, the main problem in understanding HERA physics, i.e. perturbative QCD at small x is that of establishing "consistency of the BFKL approach with the more standard DGLAP [15, 16] evolution equations" [17], which embody the leading $\ln Q^2$ (LLQ^2) resummation on which perturbative QCD is based. This problem is now solved [18, 19, 20], and on the basis of this solution it is possible to combine the available information on perturbation theory at small x, and use it to explain the unexpected success of fixed–order calculations.

DUALITY

Let us for definiteness consider the prototype problem of the description of small x scaling violations of parton distributions. For simplicity, consider the case of a single parton distribution $G(x,Q^2)$, which can be thought of as the dominant eigenvector of perturbative evolution. Scaling violations are then described by the Altarelli-Parisi equation satisfied by $G(x,Q^2)$, and thus summarized by the Altarelli–Parisi splitting function $P(x,\alpha_s)$ [15].

The basic result which allows the study of scaling violations at small x is *duality* of perturbative evolution [21, 19, 22], namely, the fact that, because the Altarelli-Parisi equation is an integro–differential equation in the two variables $t \equiv \ln Q^2/\Lambda^2$ and $\xi \equiv 1/x$, it can be equivalently cast in the form of a differential equation in t satisfied by the x–Mellin transform

$$G(N,t) = \int_0^\infty d\xi e^{-N\xi}\, G(\xi,t), \tag{1}$$

or a differential equation in ξ satisfied by the Q^2–Mellin transform

$$G(\xi,M) = \int_{-\infty}^\infty dt\, e^{-Mt}\, G(\xi,t) \tag{2}$$

of the parton distribution. The pair of dual evolution equations are

$$\frac{d}{dt}G(N,t) = \gamma(N,\alpha_s)\, G(N,t) \tag{3}$$

$$\frac{d}{d\xi}G(\xi,M) = \chi(M,\alpha_s)\, G(\xi,M), \tag{4}$$

where eq. (3) is the standard renormalization–group equation, with anomalous dimension $\gamma(N,t)$, and eq. (4) is essentially the BFKL equation. Duality is the statement that the solutions of these two equations coincide to all perturbative orders, up to power suppressed corrections, provided their kernels are related by

$$\chi(\gamma(N,\alpha_s),\alpha_s) = N. \tag{5}$$

This means that the BFKL and Altarelli-Parisi equations describe the same physics: it is the choice of the kernel to be used in the evolution equation which determines which is the large scale which is resummed. We can then discuss the construction and resummation of the kernel irrespective of the specific evolution equation where it is used, with the understanding that the kernel can be equivalently viewed as a $\gamma(N,\alpha_s)$ or a $\chi(M,\alpha_s)$, the two being related by eq. (5). Before doing this, we sketch how duality can be proven order by order in perturbation theory.

Fixed coupling

Perturbative duality is most easy to prove when the coupling does not run, since in this case the two scales t and ξ appear in the Altarelli–Parisi equation in a completely symmetric way. It is convenient to introduce the double–Mellin transform $G(N,M)$ of the parton distribution. The solution to the Altarelli–Parisi equation in M,N space has the form (which can be e.g. obtained by performing an M–mellin transform eq. (2) of the solution to the renormalization–group eq. (3))

$$G(N,M) = \frac{G_0(N)}{M - \gamma(N,\alpha_s)}, \tag{6}$$

where $G_0(N)$ is a boundary condition at a reference scale μ^2.

The inverse Mellin transform of eq. (6) coincides with the residue of the simple pole in the M plane of $e^{tM}G(N,M)$, and thus its scale dependence is entirely determined by the location of the simple pole of $G(N,M)$ (6) , namely, the solution to the equation

$$M = \gamma(N, \alpha_s). \tag{7}$$

The pole condition Eq. (7) can be equivalently viewed as an implicit equation for N: $N = \chi(M, \alpha_s)$, where χ is related to γ by eq. (5). Hence, the function

$$G(N,M) = \frac{F_0(N)}{N - \chi(M, \alpha_s)}, \tag{8}$$

corresponds to the same $G(t,x)$ as eq.(6), because the location of the respective poles in the M plane are the same, while the residues are also the same, provided the boundary conditions are matched by

$$G_0(N) = -\frac{F_0(\gamma(\alpha_s, N))}{\chi'(\gamma(\alpha_s, N))}. \tag{9}$$

Eq. (8) is immediately recognized as the N-Mellin of the solution to the evolution equation (8) with boundary condition $F_0(M)$ (at some reference $x = x_0$), which is what we set out to prove. In general, the analytic continuation of the function χ defined by eq. (5) will be such that eq. (7) has more than one solution (i.e. γ is multivalued). In this case, poles further to the left in the M plane correspond to power–suppressed contributions, while poles to the right correspond to contributions beyond perturbation theory (they do not contribute when the inverse M–Mellin integral is computed along the integration path which corresponds to the perturbative region).

It is easy to see that upon duality the leading–order $\chi = \alpha_s \chi_0$ is mapped onto the leading singular $\gamma = \gamma_s(\alpha_s/N)$, and conversely the leading–order $\gamma = \alpha_s \gamma_0$ is mapped onto the leading singular $\chi = \chi_s(\alpha_s/M)$. In general, the expansion of χ in powers of α_s at fixed M is mapped onto the expansion of γ in powers of α_s at fixed α_s/N, and conversely. So in particular at LLQ^2 it is enough to consider γ_0 or χ_s, and at LLx it is enough to consider γ_s or χ_0. The running of the coupling is a LLQ^2 but NLLx effect, so beyond LLx the discussion given so far is insufficient.

Running coupling

The generalization of duality to the running coupling case is nontrivial because the running of the coupling breaks the symmetry of the two scales ξ and t in the Altarelli–Parisi equation. Indeed, upon M–Mellin transform (2) the usual one–loop running coupling becomes the differential operator

$$\widehat{\alpha}_s = \frac{\alpha_s}{1 - \beta_0 \alpha_s \frac{d}{dM}} + \cdots, \tag{10}$$

where $d\alpha_s/dt = -\beta_0 \alpha_s^2$.

Consider for simplicity the LLx x–evolution equation, i.e. eq. (4) with $\chi = \alpha_s \chi_0(M)$, and include running coupling effects by replacing α_s with the differential operator eq. (10). We can solve the equation perturbatively by expanding the solution in powers of α_s at fixed α_s/N: the leading–order solution is given by eq. (8), the next–to–leading order is obtained by substituting this back into the equation and retaining terms up to order $\beta_0\alpha_s$, and so on [18]. We can then determine the associate $G(N,t)$ by inverting the M–Mellin, and try to see whether this $G(N,t)$ could be obtained as the solution of a renormalization group (RG) equation (3).

The inverse Mellin is again given by the residue of the pole of $e^{tM}G(N,M)$ in the M–plane, where $G(N,M)$ is now the perturbative solution. When trying to identify this with a solution to eq. (3) there are two potential sources of trouble. The first is that now the perturbative solution at order $(\alpha_s\beta_0)^n$ has a $(2n+1)$–st order pole. Therefore, the scale–dependence of the inverse Mellin is now a function of both α_s and t, whereas the solution of a RG equation depends on t only through the running of α_s. Hence it is not obvious that a dual anomalous dimension will exist at all. The second is that even if a dual γ does exist, it is not obvious that it will depend only on χ and not also on the boundary condition $F_0(M)$ eq. (8): in such case, the running of the coupling in the ξ–evolution equation would entail a breaking of factorization.

However, explicit calculation shows that it is possible to match the anomalous dimension and the boundary condition order by order in perturbation theory in such a way that both duality and factorization are respected. Namely, the solution to the leading–twist running coupling x–evolution eq. (4) with kernel $\widehat{\alpha}_s\chi_0$ and boundary condition $G_0(M)$ is the same as that of the renormalization group eq. (3) with boundary conditions and anomalous dimension given by

$$\begin{aligned}\gamma(\alpha_s(t),\alpha_s(t)/N) = {} & \gamma_s(\alpha_s(t)/N) + \alpha_s(t)\beta_0\Delta\gamma_{ss}(\alpha_s(t)/N) + \\ & +(\alpha_s(t)\beta_0)^2\Delta\gamma_{sss}(\alpha_s(t)/N) + O(\alpha_s(t)\beta_0)^3 \end{aligned} \tag{11}$$

$$G_0(\alpha_s,N) = G_0(N) + \alpha_s\beta_0\Delta^{(1)}G_0(N) + (\alpha_s\beta_0)^2\Delta^{(2)}G_0(N) + O(\alpha_s\beta_0)^3, \tag{12}$$

where the leading terms γ_s and $G_0(N)$ are given by eqs. (5) and (9) respectively. The subleading corrections are

$$\Delta\gamma_{ss} = -\beta_0 \frac{\chi_0''\chi_0}{2\chi_0'^2} \tag{13}$$

$$\Delta^{(1)}G_0(N) = \frac{2\chi_0'^2 F_0 - \chi_0\left(F_0'\chi_0'' - \chi_0' F_0''\right)}{2\chi_0'^3}, \tag{14}$$

where all derivatives are with respect to the arguments of $\chi_0(M)$ and $F_0(M)$, which are then evaluated as functions of $\gamma_s(\alpha_s/N)$. The sub–subleading correction to the anomalous dimension is

$$\Delta\gamma_{sss} = -\chi_0^2 \frac{15\chi_0''^3 - 16\chi_0'\chi_0''\chi_0''' + 3\chi_0'^2\chi_0''''}{24\chi_0'^5}, \tag{15}$$

and we omit the very lengthy expression for $\Delta^{(2)}G_0(N)$. The fact that duality and factorization hold up to NNLLx is nontrivial, and suggests that they should hold to all orders. An all–order proof can be in fact constructed [23].

Once the corrections to duality eq. (12) are determined, they can be formally reinterpreted as additional contributions to χ: namely, one can impose that the duality eq. (5) be respected, in which case the kernel to be used in it is an "effective" χ, obtained from the kernel of the x–evolution eq. (4) by adding to it running coupling corrections order by order in perturbation theory: χ_0 will be free of such correction, χ_1 will receive a correction

$$\Delta\chi_1 = \beta_0 \frac{1}{2} \frac{\chi_0(M)\chi_0''(M)}{\chi_0'^2(M)}, \tag{16}$$

and so forth. Applying duality to the known one–loop anomalous dimensions γ_0 thus gives us the resummation of the all–order singular contributions $\chi(\alpha_s/M)$ to this effective χ, which include the running coupling correction eq. (12) and its higher–order generalizations.

RESUMMATION

Because the first two orders of the expansion of χ in powers of α_s at fixed M and of the expansion of γ in powers of α_s at fixed N are known, it is possible to exploit duality of perturbative evolution to combine this information into anomalous dimension which accomplish the simultaneous resummation of leading and next–to–leading logs of x and Q^2. In fact, it turns out that both a small M and a small N resummation of anomalous dimensions are necessary in order to obtain a stable perturbative expansion, while unresummed anomalous dimensions leads to instabilities. Both sources of instability are generic consequences of the structure of the perturbative expansion, and could have been predicted before the actual explicit computation [14] of subleading small-x corrections.

Small M

The perturbative expansion of χ at fixed M is very badly behaved in the vicinity of $M \sim 0$: at $M = 0$, χ_0 has a simple pole, χ_1 has a double pole and so on. In practice, this spoils the behaviour of χ in most of the physical region $0 \leq M \leq 1$. Because $1/M^k$ is the Mellin transform of $\frac{\Lambda^2}{Q^2}\frac{1}{k!}\ln^{k-1}(Q^2/\Lambda^2)$, these singularities correspond to logs of Q^2 which are left unresummed in a LLx or NLLx approach [24].

The origin of these contributions is a straightforward consequence of momentum conservation, which implies that $\gamma(1,\alpha_s) = 0$ (note our definition of the N–Mellin transform (1), and also that γ is to be identified with the large eigenvector of the anomalous dimension matrix). The duality eq. (5) then implies that a momentum–conserving χ must satisfy $\chi(0,\alpha_s) = 1$. This, together with the requirement that χ admits a perturbative expansion in powers of α_s, implies that in the vicinity of $M = 0$, the kernel must behave as

$$\chi_s \underset{M\to 0}{\sim} \frac{\alpha_s}{\alpha_s+M} = \frac{\alpha_s}{M} - \frac{\alpha_s^2}{M^2} + \frac{\alpha_s^3}{M^3} + \dots \quad . \tag{17}$$

Hence we understand the origin of this series of poles at $M = 0$, and also that the series sums up to a regular behaviour. In fact, we can systematically resum singular contributions to χ to all orders in α_s by including in χ the terms $\chi_s(\alpha_s/M)$ derived from the leading order $\gamma_0(N)$, and similarly at next–to–leading order, and so on. Because the usual anomalous dimension automatically respects momentum conservation order by order in α_s, in order to remove the small M instability of the expansion of χ at fixed M, it is sufficient to improve the expansion by promoting it to a "double leading" expansion which combines the expansions in powers of α_s at fixed M and at fixed α_s/M [19]. For example, at leading order $\chi = \alpha_s\chi_0(M) + \chi_s(\alpha_s/M) -$ d.c., where the subtraction refers to the double–counting of the α_s/M term which is present both in $\alpha_s\chi_0$ and in $\chi_s(\alpha_s/M)$. This expansion of χ is dual eq. (5) to an analogous expansion of γ, where at leading order $\gamma = \alpha_s\gamma_0(M) + \gamma_s(\alpha_s/M) -$ d.c., and so forth. Both expansions are well behaved at small M, i.e. large N. At this level, it is already clear that the impact of the inclusion of small-x corrections is moderate: indeed, it turns out that the double–leading kernel is quite close to the usual two–loop kernel, except at the smallest values of N, i.e. in the neighbourhood of the minimum of $\chi(M)$ [19].

Small N

The improved double–leading expansion of the anomalous dimension still requires resummation at small N. This is because, even though the next–to–leading correction to the double–leading evolution kernel is small for all fixed M, it is actually large if N is fixed and small. This in turn follow from the fact that the leading χ kernel has a minimum, so the small $N = \chi$ region corresponds by duality eq. (5) to the vicinity of the minimum where the kernel is almost parallel to the $\gamma = M$ axis.

At small N, unlike at small M, there is no principle like momentum conservation which may provide a fixed point of the expansion and thus fix the all–order behaviour. The only way out is thus to treat this all–order behaviour as a free parameter. Namely, we introduce a parameter λ which is equal to the value of the all–order kernel χ at its minimum, and then we expand about this all–order minimum. In practice, this means that we reorganize the expansion of χ according to [18]

$$\begin{aligned}\chi(M,\alpha_s) &= \alpha_s\chi_0(M) + \alpha_s^2\chi_1(M) + \dots \\ &= \alpha_s\tilde{\chi}_0(M) + \alpha_s^2\tilde{\chi}_1(M) + \dots,\end{aligned} \tag{18}$$

where

$$\alpha_s\tilde{\chi}_0(M,\alpha_s) \equiv \alpha_s\chi_0(M) + \sum_{n=1}^{\infty} \alpha_s^{n+1} c_n, \qquad \tilde{\chi}_i(M) \equiv \chi_i(M) - c_i, \tag{19}$$

and the constants c_i are chosen in such a way that

$$\lambda \equiv \alpha_s\tilde{\chi}_0(\tfrac{1}{2}) = \alpha_s\chi_0(\tfrac{1}{2}) + \Delta\lambda. \tag{20}$$

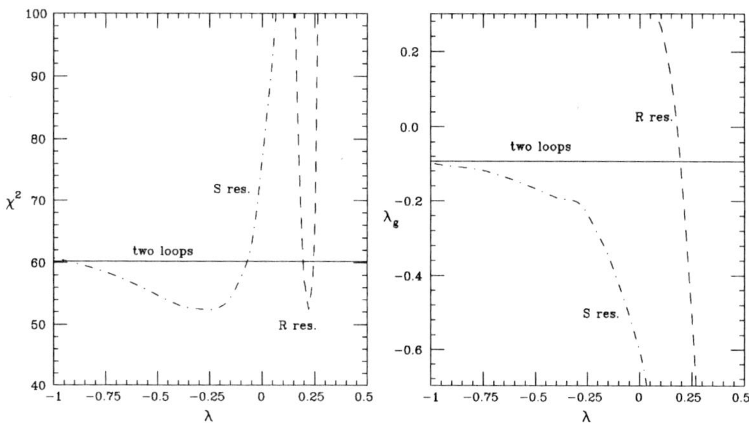

FIGURE 2. χ^2 (left) and starting gluon slope $G(x, 4\text{ GeV}^2) \sim x^{-\lambda_g}$ (right) for the fit [22] to the 95 H1 data [4] as a function of the resummation parameter λ eq. (20), for the two resummation prescriptions discussed in text. The fits are performed with $\alpha_s(M_z) = 0.119$.

is the all–order minimum of χ. Of course, in practice phenomenological predictions will only be sensitive to the value of λ in the region where very small values of N are probed, i.e. at very small x.

PHENOMENOLOGY

Using duality and the resummation discussed above, one can construct resummed expressions for anomalous dimensions and coefficient functions, and wind up with resummed expressions for physical observables which may be directly compared to the data. The need to resum the small N behaviour entails that phenomenological predictions will necessarily depend on the parameter λ eq. (20). When the resummed double–leading expansion is constructed, a further ambiguity arises in the treatment of double–counting terms. This ambiguity is related to the nature of the small N singularities of the anomalous dimension, which control the asymptotic small x behaviour. Specifically, according to the way the double–counting is treated, the $N = 0$ poles of the one– and two–loop result may survive in the resummed result ('S–resummation') or not ('R–resummation'). Both alternatives are compatible with the known low–order information on the evolution kernel, and can be taken as two extreme resummation schemes which parametrize our ignorance of higher order perturbative terms. Since the resummed terms also have a cut starting at $N = \lambda$, whether or not these low–N poles are present only makes a difference if λ turns out to be small, $\lambda \lesssim 0.3$.

The χ^2 and starting gluon slope for a fit [22] to the recent H1 data [4] for the deep–inelastic cross section are shown in figure 2, as a function of λ and for the two different resummation prescriptions. It is clear that if the perturbative $N = 0$ poles do not survive the resummation (R resummation) then only a fine–tuned value of $\lambda \approx 0.2$ is acceptable, whereas if they do survive (S resummation) essentially any $\lambda \lesssim 0$ gives a good fit.

Figure 2 demonstrates that it is possible to accommodate the success of simple fixed–order approach within a fully resummed scheme, and in fact the resummed calculation is in somewhat better agreement with the data than the fixed order one. Even though the effects of the resummation are necessarily small (otherwise the success of the fixed order prediction could not be explained) they do have a significant impact in the extraction of the parton distribution: the gluon comes out to be significantly more valence–like than in an unresummed fit. Hence, the use of resummed perturbation theory is crucial for the extraction of reliable parton distributions at small x.

From a theoretical point of view, we see that current data already pose very stringent constraints on the unknown high–orders of the perturbative expansion: only a rather soft high–energy behaviour of the deep-inelastic cross–section is compatible with the data. Further progress in the understanding of the Regge limit is likely to require either genuinely nonperturbative input, or an extension of the standard perturbative domain [18].

ACKNOWLEDGMENTS

This paper is mostly based on work done in collaboration with G. Altarelli and R. Ball. I thank G. Nardulli for organizing a very stimulating workshop, and P. Nason for interesting discussions during the workshop. This work was supported in part by EU TMR contract FMRX-CT98-0194 (DG 12 - MIHT).

REFERENCES

1. Forte, S., hep-ph/9910397 (1999).
2. Chekelian, V., hep-ph/0107053 (2001).
3. Chekanov, S., et al., ZEUS Coll., hep-ex/0105090 (2001).
4. Adloff, C., et al., H1 Coll., hep-ex/0012053 (2000).
5. Ball, R. D., and Forte, S., *Phys. Lett.*, **B335**, 77–86 (1994).
6. Rujula, A. D., et al., *Phys. Rev.*, **D10**, 1649 (1974).
7. Forte, S., and Ball, R. D., *Acta Phys. Polon.*, **B26**, 2097–2134 (1995).
8. Mankiewicz, L., Saalfeld, A., and Weigl, T., *Phys. Lett.*, **B393**, 175–180 (1997).
9. Lipatov, L. N., *Sov. J. Nucl. Phys.*, **23**, 338–345 (1976).
10. Fadin, V. S., Kuraev, E. A., and Lipatov, L. N., *Phys. Lett.*, **B60**, 50–52 (1975).
11. Kuraev, E. A., Lipatov, L. N., and Fadin, V. S., *Sov. Phys. JETP*, **44**, 443–450 (1976).
12. Ball, R. D., and Forte, S., *Phys. Lett.*, **B351**, 313–324 (1995).
13. Ellis, R., Hautmann, F., and Webber, B., *Phys. Lett.*, **B348**, 582–588 (1995).
14. Fadin, V. S., and Lipatov, L. N., *Phys. Lett.*, **B429**, 127–134 (1998).
15. Altarelli, G., and Parisi, G., *Nucl. Phys.*, **B126**, 298 (1977).
16. Gribov, V. N., and Lipatov, L. N., *Yad. Fiz.*, **15**, 781–807 (1972).
17. McLerran, L., hep-ph/0104285 (2001).
18. Ball, R. D., and Forte, S., *Phys. Lett.*, **B465**, 271–281 (1999).
19. Altarelli, G., Ball, R. D., and Forte, S., *Nucl. Phys.*, **B575**, 313–329 (2000).
20. Ciafaloni, M., Colferai, D., and Salam, G. P., *Phys. Rev.*, **D60**, 114036 (1999).
21. Ball, R. D., and Forte, S., *Phys. Lett.*, **B405**, 317–326 (1997).
22. Altarelli, G., Ball, R. D., and Forte, S., *Nucl. Phys.*, **B599**, 383–423 (2001).
23. Altarelli, G., Ball, R. D., and Forte, S., *in preparation* (2001).
24. Salam, G. P., *JHEP*, **07**, 019 (1998).

Deep Inelastic Scattering Experiments with Unpolarised Beams

A. De Roeck

CERN, 1211 Geneva 23, Switzerland

Abstract. Recent Deep Inelastic Scattering data are reviewed, in particular the latest HERA data. Emphasis is put on structure function measurements, but also aspects of hadronic final states and diffraction are discussed. The prospects for measurements in the high Q^2 region are recalled.

INTRODUCTION

Deep Inelastic Scattering (DIS) data have been very instrumental in unveiling the structure of matter in the last 30 years. After the explorative experiments of the late 60's and 70's, the precision fixed target experiments had there days of glory in the 80's and beginning of the 90's, using electron, muon and neutrino beams, and reaching maximum Center of Mass System (CMS) energies in the range of 3 to 20 GeV. The maximum resolution, or Q^2, reached corresponds to 400 GeV^2. The fixed target legacy is now continued by CCFR, HERMES, NuTeV and possibly CHORUS, which will all still provide useful data in the near future.

In 1992 the first electron-proton collider HERA came into operation in Hamburg, Germany. HERA collides a 27.5 GeV electron or positron beam on a 820 GeV (and since 1998 a 920 GeV) proton beam. Hence the CMS energy is now 320 GeV, and the largest reachable Q^2 is 100000 GeV^2. This corresponds to a 50 TeV beam on a fixed target. The experiments H1 and ZEUS record and analyse these ep collisions.

The available data sets collected by H1 and ZEUS are shown in Tab. 1. In the period of 1992-2000, which covers the so called HERA phase I period, the experiments collected about 120 pb^{-1} each.

TABLE 1. Luminosity collected at different CMS energies at HERA by H1 and ZEUS.

Beams/CMS	H1 (pb^{-1})	ZEUS (pb^{-1})
$e^+p(\sqrt{s}=300\text{GeV})$:	36.6	47.7
$e^+p(\sqrt{s}=320\text{GeV})$:	65	70
$e^-p(\sqrt{s}=320\text{GeV})$:	16.4	17

CP602, *QCD@Work: International Workshop on Quantum Chromodynamics*
edited by P. Colangelo and G. Nardulli

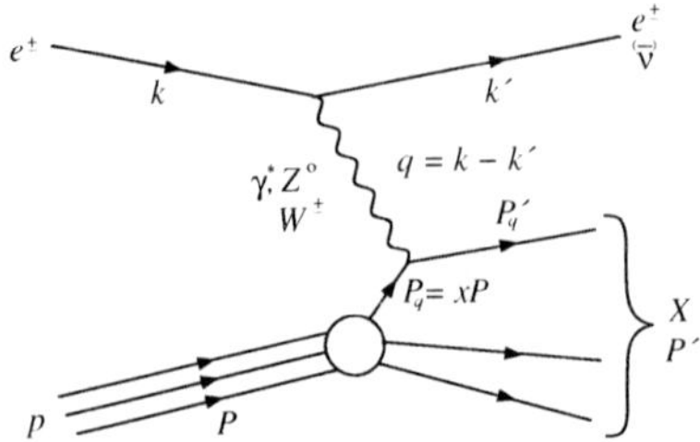

FIGURE 1. Diagram for the Deep Inelastic Scattering process.

The diagram for Deep Inelastic Scattering (DIS) is shown in Fig. 1. The following kinematical variables are defined to analyse the data: $Q^2 = -q^2$ (virtuality of the exchanged γ^*, Z^0 and W); $x = Q^2/2pq$ (Bjorken-x variable); $y = Pq/pk$ (inelasticity); and $W^2 = (p+q)^2$ (hadronic energy). Note that $Q^2 = xys$, with $\sqrt{s} =$ CMS energy.

STRUCTURE FUNCTION MEASUREMENTS

For Neutral Current (NC) – i.e. γ^*/Z^0 exchange– the differential cross-section can be written as a sum of three structure functions:

$$\frac{d^2\sigma_{NC}^{e\pm p}}{dxdQ^2} = \frac{2\pi\alpha^2}{xQ^4}(Y_+\tilde{F}_2(x,Q^2) \mp Y_- x\tilde{F}_3(x,Q^2) - y^2\tilde{F}_L(x,Q^2)), \tag{1}$$

with $Y_\pm = 1 \pm (1-y)^2$. In lowest order we have $\tilde{F}_2 = x\sum A_i(q_i + \overline{q}_i)$, $\tilde{F}_3 = x\sum B_i(q_i - \overline{q}_i)$, $\tilde{F}_L = 0$. In the region where γ exchange is the only significant contribution (i.e. roughly when $Q^2 < 1000\,\text{GeV}^2$), the coefficients A_i are the quark charges squared, and $\tilde{F}_2$ reduces to the more familiar known F_2 structure function.

Similarly for Charged Current (NC) – i.e. W exchange– we have :

$$\frac{d^2\sigma_{CC}^{\pm}}{dxdQ^2} = \frac{G_F^2 M_W^4}{2\pi x}\frac{1}{(Q^2 + M_W^2)^2}\tilde{\sigma}_{CC}^{\pm}(x,Q^2) \tag{2}$$

In lowest order $\tilde{\sigma}_{CC}^{+} = x((\overline{u} + \overline{c}) + (1-y)^2(d+s))$; $\tilde{\sigma}_{CC}^{-} = x((u+c) + (1-y)^2(\overline{d} + \overline{s}))$.

The covered region in x and Q^2 of the DIS measurements is shown in Fig. 2 for HERA and the fixed target experiments.

Figure 3 gives an overview of the quality of the structure function measurements as reached today. It shows the precise fixed target data and data from H1 (based on 100 pb^{-1} (preliminary) [1]) and ZEUS (based on 40 pb^{-1} (published) [3]). The celebrated

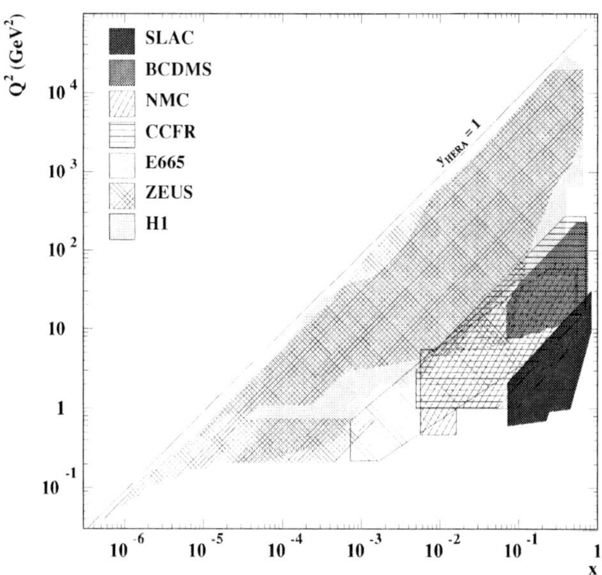

FIGURE 2. Kinematic plane for measurements at HERA and fixed target experiments.

scaling violations as expected from QCD, are clearly visible. The full and dashed lines are QCD fits (see below) to the data, and they are found to describe the data well.

Plotted as function of x one can clearly notice the rise of F_2 at small x, as was already discovered by the first measurements of the structure function back in 1992. Fig. 4 shows as an example one Q^2 bin [4]. Note that the precision in that part of the kinematic region has reached 1% (statistics) and 3% (systematics), and it will be hard to improve!

This rise is visible for all Q^2 bins, but is reduced strongly for low Q^2 as is shown in Fig.5. These low Q^2 measurements were made with purposely built beam-pipe calorimeters to explore this kinematic region [5]. The very low Q^2 region cannot be described by a QCD fit any longer, but can be well described by a Regge/GVDM inspired fit. In fact the data in this region show a similar energy dependence on the γp CMS energy (=$W^2 \sim 1/x$) as total hadronic cross-sections.

A lot of speculation has occured the last 1-2 years on possible signatures of parton saturation effects in the low Q^2 data, however it has remained inconclusive so far. What seems to happen is that the data show a smooth transition to soft physics in the low Q^2 region. The detailed understanding of this region and the transition does remain a challenge for phenomenologists, and model builders. Various kinds of models are being explored; for an overview see [6].

The fixed target experiments CCFR and HERMES have recently produced new results. HERMES is a spectrometer designed to measure polarization asymmetries, and

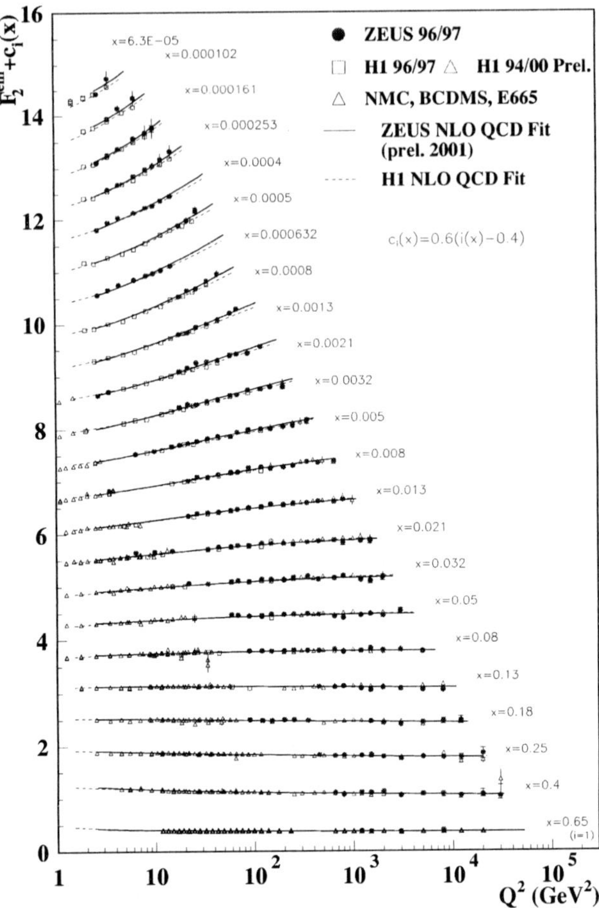

FIGURE 3. State of the art plot with structure function measurements from HERA and fixed target experiments.

thus polarized structure functions and other related quantities. They operate at the HERA ring, making use of the 27.5 GeV polarised electron beam, and a polarised fixed target. However part of the time a high statistics unpolarised data run is taken.

A particularly interesting result from HERMES is the measurement of the nuclear effects on the ratio $R = \sigma_L/\sigma_T$ [7]. This ratio is studied relative to Deuterium, for a nucleus $A : R_A/R_D$. The data show values of R_A/R_D much larger than 1 for x values between 0.01 and 0.03. The deviations are larger for increasing A. These data seem to tell us, intriguingly, that there are enhanced quark-gluon correlations in nuclei.

CCFR recorded data at the neutrino beam in FNAL. First they have reported during the last year a 'physics model' independent reanalysis of the F_2^{ν} data which now show good agreement with the NMC μp data, thus removing a 5 year old unexplained discrepancy

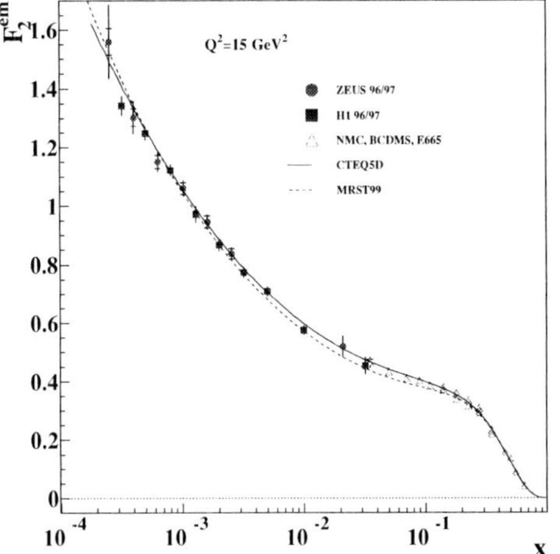

FIGURE 4. The structure function F_2 versus x for $Q^2 = 15$ GeV^2.

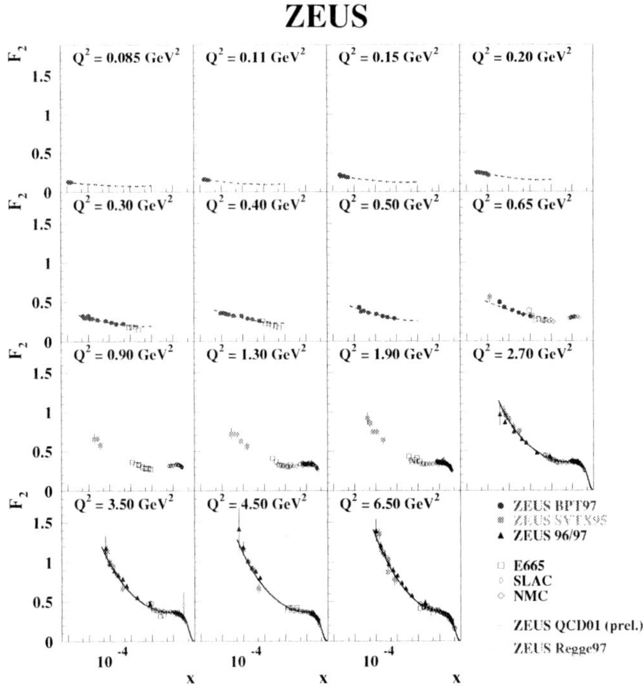

FIGURE 5. F_2 data for the lowest x and Q^2 values reached.

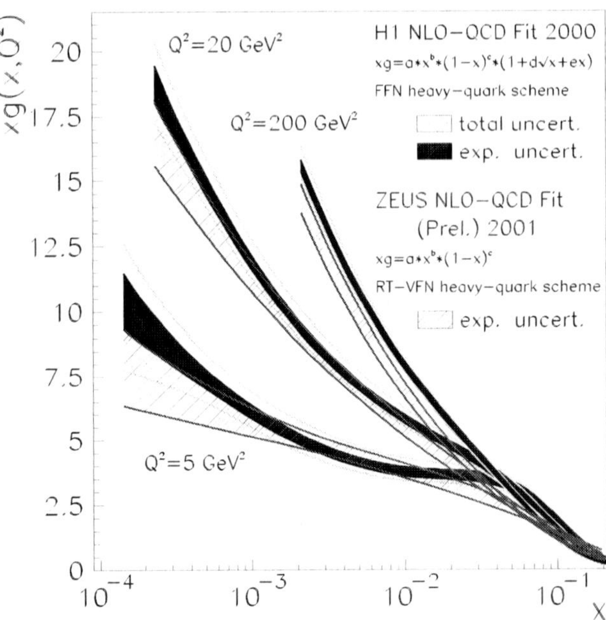

FIGURE 6. Gluon distribution extracted from NLO DGLAP fits from F_2 data by H1 and ZEUS.

between these experiments. Next they have measured $\Delta xF_3 = xF_3^{\nu} - xF_3^{\overline{\nu}} = 4(s-c)$ [8]. Interestingly, but with limited significance, these data are found to be higher than all the predictions!

QCD FITS TO THE F_2 DATA

Fits were made of the DGLAP equations to the structure function data. In [9] Stefano Forte explains why we do not have to worry (too much) on BFKL effects in these fits, so we can go ahead.

The two experiments have a different approach to the fits and use different data sets and selections: H1 selects $3.5 < Q^2 < 3000$ GeV^2, $W^2 > 10$ GeV^2; both H1 + BCDMS (p) ($y > 0.3$) are used. ZEUS selects $Q^2 > 2.5$ GeV^2, $W^2 > 20$ GeV^2; ZEUS and CCFR and NMC(d+p) data are used.

In the NLO fits α_s has been mostly fixed to the world average of 0.118, but in some dedicated fits it was left free. Details on the fits are given in [10, 3]. It remains to be said that lots (!) of systematic checks have been performed before releasing the results.

Fig. 6 shows the final gluon for H1 [4] and the preliminary released gluon of ZEUS [3]. The gluons rise in this Q^2 region with decreasing x. The results shown suggest some differences between the two experiments, but it has been checked [10] that these differences are essentially due to the different assumptions made, such as the different heavy flavour treatment scheme, different functional form at the starting scale, different starting scale, and the use of different data. Hence, within a given set of assumptions one can reach a precision on the gluon distribution of the order of 5-10%, but the differences between results with different assumptions can be of similar magnitude. A HERA task force is now at work to try to find the best way to extract the gluon from such fits.

ZEUS pushes their fits further down to $Q^2 = 1$ GeV^2 (but note that they do not use data in that range, so it is an extrapolation). The result shows the astonishing effect that the gluon turns negative around $Q^2 = 1$ GeV^2. What does that mean? Breakdown of pQCD at this scale, which would be no big surprise? need for higher twists? for NNLO? While the fact that the gluon is negative is worrisome, but not dramatic, the problem sets in when realizing that the true measurable quantity, namely F_L also turns negative around this Q^2 value. That is clearly not acceptable.

The quantity F_L has not been measured directly yet at HERA. It would at least need two beam energy settings to measure the cross-section for a given x and Q^2 value at different y values. H1 uses two different methods to get a determination of F_L having only data at one energy[4], which are based on extrapolating fits made in the low y region (where the effect of F_L is negligible) to the high y region of the data. The results are shown in Fig. 7, compared to fixed target data results. The data show a rise of F_L with respect to the lower energy data, and are consistent with the QCD prediction based on the fit from scaling violations. The lowest Q^2 value that could be reached is however 2.2 GeV^2, so the 'negative F_L prediction' from ZEUS is not yet challenged.

An important quantity to measure in pQCD is the coupling constant α_s. Both experiments have quoted a value based on fits of F_2 data where α_s is left as a free parameter in the fit: (H1) $\alpha_s(M_Z) = 0.1150 \pm 0.0017(exp)^{+0.0009}_{-0.0005}(model) \pm 0.005(scale)$; (ZEUS/prelim.) $\alpha_s(M_Z) = 0.1172 \pm 0.0008(stat.) \pm 0.0054(syst.)$. The latter value does not include the effect of the scale uncertainty, which is a large contribution to the error. It has been shown that NNLO corrections could reduce this error with a factor two or so. If that happens then measurements of α_s from F_2 fits will be among the most precise ones in the world!

HADRONIC FINAL STATES

In the region of Q^2 larger than a few GeV^2, the gluon distribution discussed above is found to be 'universal': several other processes, making use of specific hadronic final states in DIS interactions, yield the same gluon. A first process is measuring the gluon distribution in DIS via charm production, which proceeds via the boson-gluon fusion process. Another is di-jet production which at low-x is also mostly driven by the gluon. The results are shown to agree with the gluon from the QCD fit in Fig. 8.

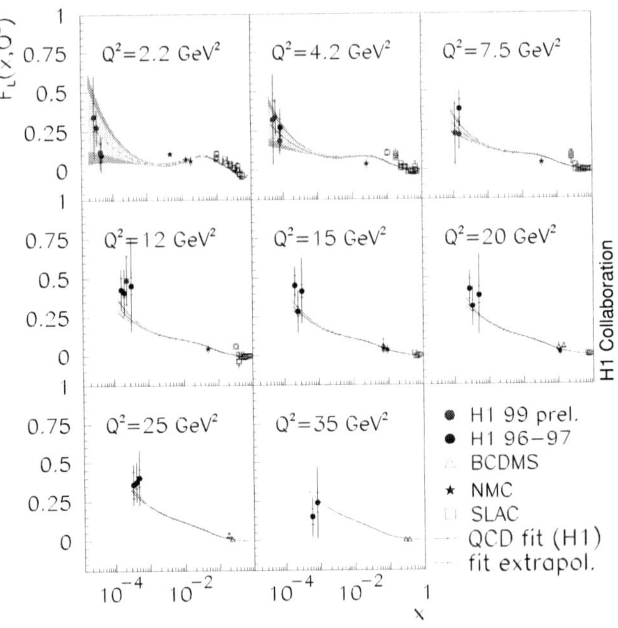

FIGURE 7. Determination of F_L by H1 and measurements at fixed target experiments.

Charm data has been used to extract the charm contribution to F_2, the so called charm structure function $F_2^c(x,Q^2)$. Typically, such measurements use the decays $D^* \rightarrow D^0\pi_{slow} \rightarrow K\pi\pi_{slow}$ to tag and reconstruct the charm. One measures: $\frac{d\sigma^c}{dxdQ^2} = \frac{2\pi\alpha^2}{Q^4x}(1+(1-y)^2)F_2^c(x,Q^2)$. New measurements from both experiments [11] show with better precision that the charm contribution to F_2 is up to 25-30% and that it is in agreement with the prediction based on the gluon obtained from scaling violations.

Di-jet events have been also used to make simultaneous fits of both α_s and the gluon distribution. The followed strategies so far were [12, 13]:

- Fit α_s, $g(x)$ and $q(x)$ from jet and inclusive DIS measurements simultaneously. This however leads to a strong correlations between α_s and $g(x)$.
- Fit $g(x)$ and $q(x)$ from jet and inclusive DIS measurements, and fix α_s
- Fit α_s from jet measurements, taking the gluon and quarks from the PDFs.

Presently the experiments quote the following values; H1 from inclusive jets: 0.1186 ± 0.0059; ZEUS from di-jet rates: $0.1166^{+0.0065}_{-0.0058}$. These results are in agreement with the values obtained from DGLAP fits to the F_2 data and with the world average.

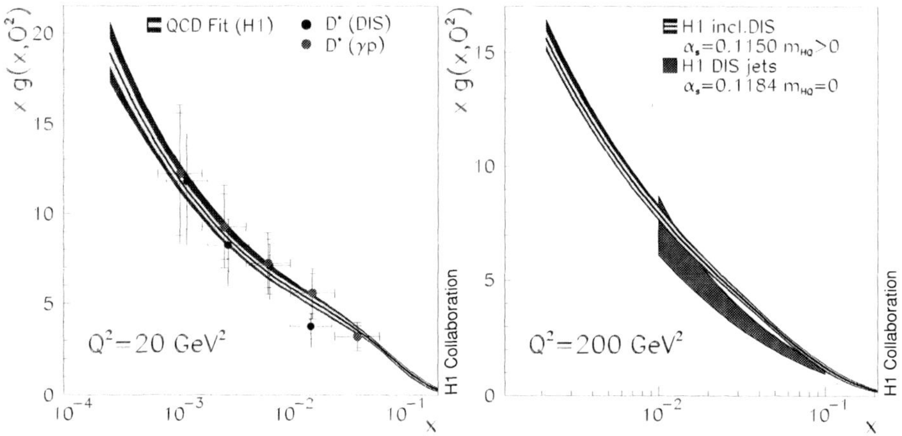

FIGURE 8. Gluon distribution extracted from a QCD fit to F_2 data compared to the one extracted from (left) charm production in DIS and photoproduction and (right) di-jets in DIS.

EVENTS WITH LARGE RAPIDITY GAPS

About 10% of the DIS events at HERA are found to have no activity in the proton direction, except for what may disappear in the beam-pipe. These events have been studied in detail over the last 8 years and are found to agree with what one would expect from diffraction, hence they are now generally called diffractive events. The model often used to describe these events is via the exchange of a colourless object called the pomeron. Fig. 9 shows a diagram for a diffractive event in comparison with a standard DIS event.

In this model DIS scattering of an electron off a proton probes the structure of the "pomeron", emitted from the proton with momentum fraction x_p. Hence one can define a structure function related to the differential cross-section: $\frac{d^3\sigma^D}{d\beta dQ^2 dx_p} = \frac{4\pi\alpha^2}{\beta Q^2}(1-y+\frac{y^2}{2})F_2^{D(3)}(\beta,Q^2,x_p)$ where $F_2^D \sim f(x_p,(t)) \cdot F_2^{pom}(\beta,Q^2)$ and β is x/x_p, which is a measure of the momentum fraction of the parton in the pomeron. The most detailed measurement of $F_2^{pom}(\beta,Q^2)$ stems still from '94 data [14], an example is shown in Fig. 10. One observes that $F_2^{pom}(\beta,Q^2)$ rises with Q^2 for all β values, i.e. the scaling violations of F_2^D are positive for all x, contrary to the proton (see Fig. 3). This suggest that the pomeron is dominantly gluonic. Indeed, performing DGLAP fits

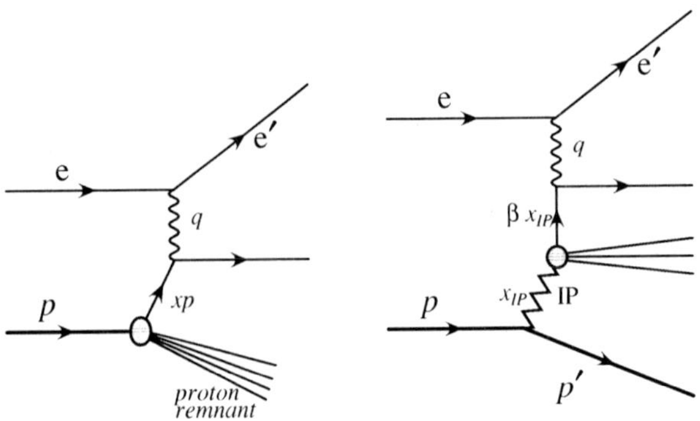

FIGURE 9. Example of a standard DIS event (left) and a diffractive DIS event (right), with definition of the additional variables x_p and β.

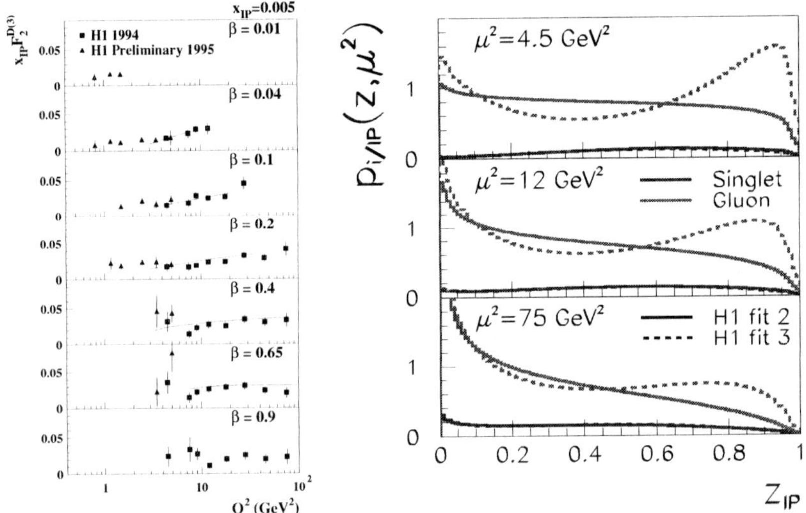

FIGURE 10. (left) Example of $F_2^{pom}(\beta,Q^2)$ at $x_p = 0.005$; (right) extracted parton distributions from DGLAP fits to the $F_2^{pom}(\beta,Q^2)$ data.

(without momentum sum rule constraint) to the $F_2^{pom}(\beta,Q^2)$ data gives parton distributions as shown in Fig. 10: the gluon dominates in both fit examples. It is now crucial to check if that picture holds from studies from hadronic final states.

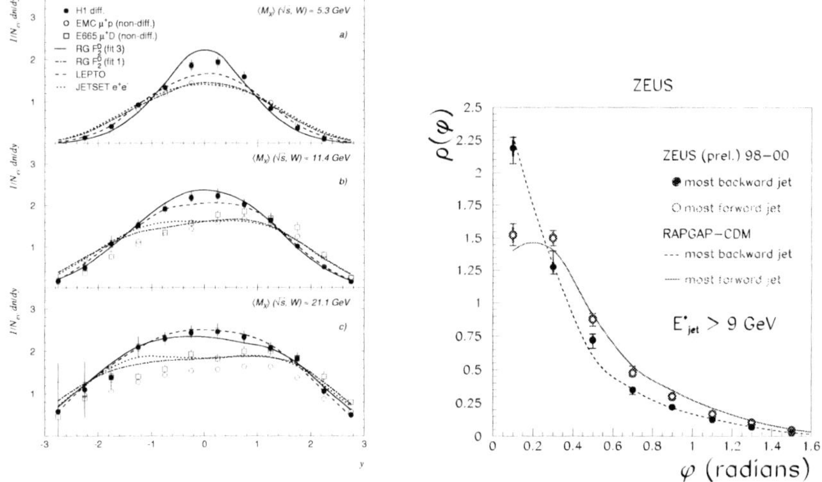

FIGURE 11. (left) Energy flow measurement in diffractive DIS and fixed target DIS events; (right) jet broadness for jets in the photon (backward) and pomeron (forward) hemisphere of the γ-pomeron system.

Over the last two years there have been numerous studies on the final states: energy flows, sphericity behaviour, jets, charm, and so far there is compelling evidence that these phenomena can be described well by this picture. Fig. 11 shows two examples of final state measurements: energy flows [15] and jet broadness [16]. The diffractive data have a higher density in the central rapidity region, than μp data, and are thus different in nature. Their behaviour agrees with predictions of a 'gluonic object'. Jets in the pomeron (forward) hemisphere in the γ pomeron system, are broader, hence more gluonic than the ones in the photon hemisphere. This adds support for $q\bar{q}g$ states where the gluon is aligned with the pomeron remnant.

In all, rapidity gap events in DIS are dominantly 'gluonic' states

Finally a new interesting process with a rapidity gap is the so called Deeply Virtual Compton Scattering process $ep \rightarrow ep\gamma$. This is recognized to be a clean QCD process and allows to access so called skewed parton distributions. Both experiments have shown that this process is clearly observable at HERA, but the first physics results still need to be extracted [17].

HIGH Q^2 REGION

When the mass of the exchanged boson, Q^2, approaches the mass of the W and Z bosons, their exchange becomes of comparable strength to the one of the photon, and the cross-sections will be in part driven by electroweak effects. Fig. 12 shows the cross-sections versus Q^2 for neutral and charged current events. The CC cross-section becomes as large as the NC cross-section for Q^2 values around 10^4 GeV^2. The differential cross-section $d\sigma/dQ^2$ falls by 7 orders of magnitude for the NC case, but follows the EW theory

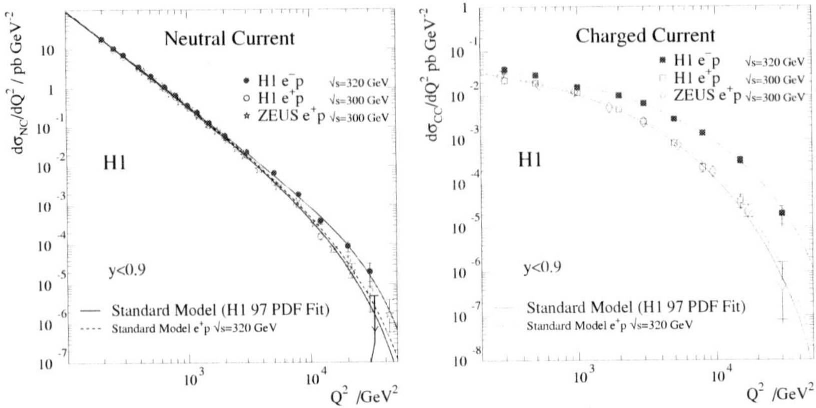

FIGURE 12. Differential cross-sections $d\sigma/dQ^2$ for NC (left) and CC (right) events.

predictions. At HERA the discovery potential for new physics is at highest Q^2 values.

Charged Current cross-sections in LO are given by $\tilde{\sigma}^+ = x((\bar{u}+\bar{c})+(1-y)^2(d+s))$ and $\tilde{\sigma}^- = x((u+c)+(1-y)^2(\bar{d}+\bar{s}))$ with $\tilde{\sigma}_{NC}^{e\pm p}(x,Q^2) = \tilde{F}_2 \mp \frac{Y_-}{Y_+} x\tilde{F}_3$. Hence one expects at low x (low Q^2) $\rightarrow$ sea dominance; at high x (high Q^2) $\rightarrow$ valence dominance. This is clearly born out from the data as shown in Fig. 13.

The data have allowed for the first measurements of the parity violating structure function F_3, and the extraction of xu_v and xd_v distributions. Both results are still statistics limited.

The CC data have also been used to extract the W propagator mass via $\frac{d^2\sigma_{cc}}{dxdQ^2} \sim G_{CC}^2\left(\frac{M_{prop}^2}{Q^2+M_{prop}^2}\right)^2$. The normalization is given by the couplings $G_{CC}(G_F)$ and the shape given by the propagator mass $M_{prop}(M_W)$. From constrained fits to the CC cross-section taking $G_{CC} = G_F$ one finds [1]:

e^+p ZEUS $\quad M_W = 81.4^{+2.7}_{-2.6}(stat) \pm 2.0(syst)^{+3.3}_{-3.0}(pdf)$ GeV
e^+p H1 $\quad M_W = 80.9 \pm 3.3(stat) \pm 1.7(syst) \pm 3.7(pdf)$ GeV
e^-p H1 $\quad M_W = 79.9 \pm 2.2(stat) \pm 0.9(syst) \pm 2.1(pdf)$ GeV

Note that this is a measurement in the space-like regime. The results are in agreement with the time-like (LEP/TEVATRON) measurements.

Generally the data are found be in good agreement with the expectation from theory in the high Q^2 region. Former reported deviations seem to have gotten ironed out. There is still one intriguing class of events namely events with a high p_T lepton and missing E_T [18]. H1 found 10 events with either a muon or electron isolated with more than 10 GeV p_T, a total missing p_T of the event above 12 GeV, with additionally a p_T of the

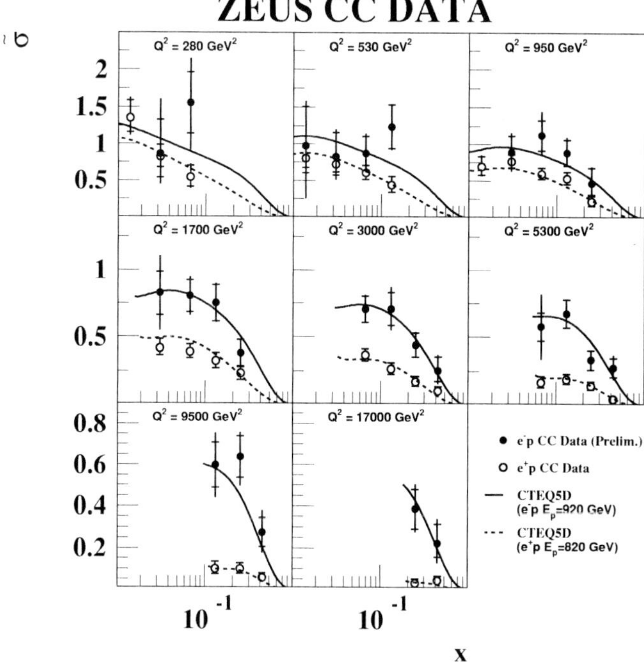

FIGURE 13. Scaled cross-sections for charged current events.

hadronic system of 25 GeV, while expecting 2.8 events mostly from W decays. ZEUS however sees for a similar analysis only 2 events, expecting 2.4. New physics channels which could give these signals include stop and bottom production and also FCNC single top production. The increased luminosity with the future HERA upgrade will reveal if this excess persists.

SUMMARY

The HERA phase I period is now finished: HERA(1992-2000) has accumulated $\simeq 120$ pb^{-1} for analysis per experiment. These data have been shown to be a true laboratory for QCD at high energy! An anthology of the results contains:

- Inclusive cross-section measurements from $Q^2 = 0.045$ GeV^2 to $Q^2 = 30000$ GeV^2 and from $x \simeq 10^{-6}$ to $x \simeq 1$ have reached a high precision (1% stat., 2-3% syst.) in a large region. This has resulted in a good picture of the proton structure
- For the region of QCD ($Q^2 > 1$ GeV^2): All gluons extractions and $\alpha_s(M_Z)$ extractions (inclusive DIS, charm, jets) give consistent results; H1: $\alpha_s = 0.1150 \pm 0.0017(exp.)^{+0.0009}_{-0.0005}(model) \pm 0.005(scale)$; ZEUS: $\alpha_s = 0.1172 \pm 0.0008(stat.) \pm$

0.0054(*sys.*)

- DIS with rapidity gaps (hard diffraction) can be described as DIS on a dominantly gluonic object.
- Fixed target experiments continue to add intriguing results: e.g. from CCFR (ΔxF_3) and HERMES (σ_L/σ_T).

Next will follow HERA phase II which has been prepared over the last year while the machine was shut-down. The expectation is for HERA II (2001-2006): 150 pb^{-1}/year/experiment and a longitudinal polarised electron beam.

Already now one has to think of a possible HERA phase III (2006-?), since time is needed to prepare it. The hottest candidates are a fully polarised HERA, i.e. having polarised protons, or having beams of light and heavy nuclei. These are certainly two very good options, but the next years will tell if the community interested in doing these experiments will be strong enough to have phase III become a reality. I personally hope it is!

ACKNOWLEDGMENTS

I thank the organizers for inviting me to this charming place in the south of Italy, and congratulate them for their excellent organization of the conference.

REFERENCES

1. V. Shekelyan, talk at 15th Rencontres de Physique de la Vallee d'Aoste (2001).
2. ZEUS Collaboration; J.Breitweg et al., Eur. Phys. J. **C11** (1999) 427.
3. K. Nagano, talk at DIS2001, Bologna (2001).
4. H1 Collaboration, C. Adloff et al., Eur. Phys. J. **C21** (2001) 1.
5. ZEUS Collaboration, J. Breitweg et al., Phys. Lett. **B487** (2000) 53.
6. A.M. Cooper-Sarkar, R.C.E. Devenish, A. De Roeck, Int. J. Mod. Phys. **A13** (1998) 3385.
7. HERMES Collaboration, K. Ackerstaff et al., Phys. Lett. **B475** (2000) 386.
8. CCFR/NuTeV Collaboration, U.K. Yang et al., Phys. Rev. Lett. **86** (2001) 2742.
9. S. Forte, these proceedings.
10. R. Walny, talk at DIS2001, Bologna (2001).
11. S. Mohrdieck and S. Schagen, talk at DIS2001, Bologna (2001).
12. G. Grindhammer, talk at DIS2001, Bologna (2001).
13. ZEUS Collaboration, J. Breitweg et al., Phys. Lett. **B507** (2001) 70.
14. H1 Collaboration, C. Adloff et al., Z. Phys. **C76** (1997) 613.
15. H1 Collaboration, C. Adloff et al., Phys. Lett. **B428** (1998) 206.
16. ZEUS Collaboration, S. Chekanov et al., DESY-01-092 (June 2001).
17. L. Favart, talk at DIS2001, Bologna (2001).
18. C. Diaconu and T. Matsushita, talks at DIS2001, Bologna (2001).

HERMES Past and Future: 1995 - 2000 and 2001 - 2005

E.C. Aschenauer [a] on behalf of the HERMES Collaboration

[a] *DESY Zeuthen, Platanenallee 6, 15738 Zeuthen, Germany*

Abstract. This report highlights results obtained with the HERMES experiment during the last 5 years on the spin structure of the nucleon. The inclusive spin-structure function $g_1(x, Q^2)$ is described, followed by measurements of the flavor-separated quark polarizations based on semi-inclusive data. Next, first measurements are presented that indicate that both the gluon polarization and the transversity distribution $h_1(x)$ are non-zero. Finally, the subject of parton-parton correlations is introduced, along with new data related to higher twist effects and generalized parton distributions.

I INTRODUCTION

Deep-inelastic scattering of polarized leptons from polarized targets has been used successfully for more than a decade to yield information about the spin structure of the nucleon. The central question at issue is the manner in which the partonic components of the nucleon manage to produce its overall spin of 1/2 $\hbar$. The expression

$$\frac{1}{2} = \frac{1}{2}\Delta\Sigma + \Delta g + L_q + L_g \tag{1}$$

illustrates how the total spin of the nucleon must arise from a combination of three sources: the helicity distribution of the quarks ($\Delta\Sigma$), the helicity distribution of the gluons (Δg), and the orbital angular momentum of quarks and gluons (L_q, L_g). The notation $\Delta q(x)$ denotes the *polarized* counterpart of the familiar parton distribution functions (PDF) $\Delta q(x) \equiv q^{\uparrow}(x) - q^{\downarrow}(x)$. Here, x is the Bjorken scaling variable, and $q^{\uparrow}(q^{\downarrow})$ represents quarks whose spins are parallel (anti-parallel) to the spin of the nucleon.

To obtain some feeling for the expected magnitudes of these various sources of the nucleon spin, one may turn to phenomenological models. If one considers the current quarks seen by dynamical QCD processes, relativistic effects must be taken into account because of the light quark masses. Calculations performed in, e.g., the relativistic MIT bag model suggest that $\Delta\Sigma \simeq 0.60$ - 0.75 [1]; the remainder of the nucleon spin is accounted for by the orbital angular momentum of the moving

CP602, *QCD@Work: International Workshop on Quantum Chromodynamics*
edited by P. Colangelo and G. Nardulli

quarks. To include sea quarks in the picture, one may turn to meson cloud models, for example. In such models, the nucleon is represented as a bare, valence object which may emit a pseudo-scalar meson. It is through the emission of such mesons that the quark sea is generated. One calculation performed in such a framework indicates that the sea quarks are negatively polarized, while the anti-quarks in the sea are unpolarized; all anti-quarks in this picture are part of a spin-0 meson, and therefore carry no polarization [2]. An alternative picture of the nucleon comes from the chiral-quark soliton model, where the nucleon appears as a chiral soliton in a pion field. A recent calculation of this type [3], performed in the large N_c limit, suggests that the $\overline{u}$ and $\overline{d}$ quarks do carry a significant polarization, but of opposite sign: $\Delta\overline{u} \simeq -\Delta\overline{d}$.

This brief sample of phenomenological models demonstrates that progress in our understanding of the spin structure of the nucleon requires precise experimental input.

II INCLUSIVE MEASUREMENTS

The longitudinal spin structure function $g_1(x, Q^2)$ has been measured by a number of experiments, using longitudinally polarized lepton beams and longitudinally polarized nuclear targets [4]. In the quark-parton model and in leading order QCD, this structure function is given by the charge weighted sum of polarized quark spin distributions Δq of flavor q: $g_1(x) = \frac{1}{2}\sum_q e_q^2 \Delta q(x)$. In the framework of next-to-leading order (NLO) QCD, the relationship between $g_1(x, Q^2)$ and the polarized PDF's is more complex, and includes a contribution from the gluon spin distribution $\Delta g(x)$. The singlet ($\Delta\Sigma$), non-singlet ($\Delta q_3, \Delta q_8$), and gluon (Δg) spin distributions are each prefixed with a coefficient function of different Q^2 dependence. Fits to the data in NLO seek to obtain information about the polarizations of these different flavor combinations by exploiting their distinct Q^2 dependences.

Numerous spin-dependent NLO fits exist in the literature. One example is the global fit performed by the SMC collaboration [5], an analysis distinguished by a careful treatment of uncertainties. The results indicate that the quark spins account for a relatively small fraction of the total nucleon spin, of order 20 - 40% (the value depends on the NLO factorization scheme used in the analysis). Furthermore, the strange sea would seem to have a negative polarization, of order -10%. The gluon polarization Δg, by comparison, is only poorly constrained by the inclusive data but there is some indication that it is positive.

Although the body of data on $g_1(x)$ is still being expanded (see, e.g. the new low-x HERMES data [6]) the precision and kinematic coverage of these measurements is still rather limited. Present NLO analyzes also use information from baryon β-decay to constrain the first moments of the non-singlet distributions (Δq_3 and Δq_8). However, the use of the constraint on Δq_8 is based on SU(3)-symmetry among baryons, which is known to be inexact. A recent study [7] explored the dependence of extracted quark polarizations on SU(3)-symmetry breaking effects. The influence of such effects on the quark polarization $\Delta\Sigma$ was found to be small. However, the gluon and strange quark polarizations were found to vary significantly: e.g. values of Δs from -0.02 to -0.15 were obtained.

III QUARK POLARIZATIONS FROM SEMI-INCLUSIVE MEASUREMENTS

To proceed further in our knowledge of the polarized parton distributions, experimenters have turned to *semi-inclusive* asymmetry measurements. In these measurements, a hadron is detected in coincidence with the scattered beam lepton. Through the agency of the fragmentation functions, a probabilistic relation exists between the flavor of the struck quark and the flavor content of the hadrons generated in the final state. By measuring the spin asymmetry A_1^h for hadrons of various types, one may separate the polarized parton distributions by flavor.

While the inclusive measurements only yielded the first moment of the flavor-separated spin distributions, the full x-dependence of the spin distributions can be extracted from semi-inclusive data. Furthermore, inclusive measurements are not able to distinguish between quark and antiquark polarizations, since the inclusive cross-section depends on the square of the charge of the struck quark.

Semi-inclusive asymmetries for positive and negative hadron production have been measured and analyzed by both the HERMES [8,9] and SMC experiments [10]. The SMC measurements were performed on a proton target, while HERMES has used deuterium (1998-2000), hydrogen (1996-1997) and ^{3}He (1995) targets. As compared to the inclusive data, the precision of the semi-inclusive asymmetries published to date is limitied. Nevertheless, current semi- inclusive analyses can be used to impose constraints on the relations between the various flavor components of the sea polarization. Two alternatives have been considered: either the polarization $\Delta q_s(x)/q_s(x)$ or the spin distribution $\Delta q_s(x)$ of the sea quarks is assumed to be independent of flavor. The choice between these two hypotheses is found to have a negligible influence on the results.

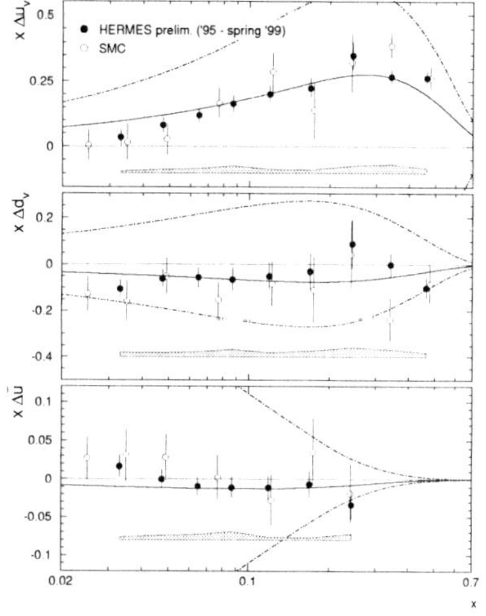

FIGURE 1. The polarized valence quark and $\bar{u}$-quark distributions as extracted from HERMES (solid circles) and SMC (open circles) measurements.

Figure 1 shows the HERMES and SMC results for the extracted valence $(\Delta u_v, \Delta d_v)$ and sea (Δq_s) quark polarizations. All data points are evolved to a common Q^2 of 2.5 GeV/c^2. The dash-dotted lines represent the positivity limit $|\Delta q(x)| \leq q(x)$, while the solid lines show the 'Gluon A' leading order parameterization of Gehrmann and Stirling [11]. As this latter parameterization is based on fits to inclusive measurements, one sees that the results derived from inclusive and semi-inclusive measurements are in good agreement.

IV GLUON POLARIZATION FROM PAIRS OF HIGH-P_T HADRONS

As described above, the gluon polarization is only poorly constrained by existing data. One way to measure $\Delta g(x)$ directly is via the photon gluon fusion process, where a gluon from the nucleon participates in the hard scattering process. A useful experimental signature of this process is the production of jets with high transverse momentum p_T. While the transverse momentum produced in the fragmentation process is small, two back-to-back jets with high p_T are characteristic for the high p_T of the quark and anti-quark produced in the photon gluon fusion process. At the energies of fixed target experiments, high-p_T hadrons can be used in place of jets [13].

HERMES has recently measured the spin asymmetry in the photoproduction ($Q^2 \simeq 0$) of pairs of high-p_T hadrons, using a polarized hydrogen target [12]. Events were selected that contained at least one positively charged hadron h^+ and at least one negatively charged hadron h^-. In the kinematic region $p_T^{h_1} > 1.5$ GeV/c and $p_T^{h_2} > 1.0$ GeV/c, the spin asymmetry is found to have a negative sign (fig. 2a). The symbols h_1 and h_2 denote respectively the hadrons with the highest and next-to-highest p_T in each event.

The measured asymmetry was interpreted using the PYTHIA Monte Carlo generator at leading order. In this Monte Carlo simulation, several different processes contribute to the two-hadron cross section. However, the only presently known process that might contribute to a *negative* asymmetry is the photon-gluon fusion (PGF) process: PGF has an analyzing power of -1 (for massless quarks), and so will produce a negative asymmetry given a positive gluon polarization. In the same region of phase space where a negative asymmetry is observed, the PYTHIA model indicates that the cross section is indeed dominated by photon gluon fusion. The best fit value obtained for the gluon polarization is $\langle \Delta g/g \rangle = 0.41 \pm 0.18$ (stat.) $\pm$ 0.03 (syst.), at an average $\langle x_g \rangle = 0.17$. The systematic uncertainty represents the experimental contribution only (fig. 2b).

It must be pointed out that this measurement is subject to theoretical uncertainties of an as-yet undetermined magnitude. At the kinematics of this measurement, no spin-dependent analyzes of higher order QCD processes are available, for example. However, to alter the principal conclusion of this analysis, i.e., that $\langle \Delta g/g \rangle$ at $\langle x_g \rangle = 0.17$ is positive, a significant contribution from a neglected process with a large negative spin asymmetry would be needed.

The excellent performance of HERA during the 2000 data taking period allowed to quadruplicate the statistics from 2.4 Million DIS events to 9 Million DIS events using a polarized deuterium target. Performing the same analysis as in the hydrogen case the asymmetry $A_{||}$ for pairs of high-p_T hadrons was obtained to be -0.083 ± 0.0665. To enhance the sensitivity to the PGF process, pairs of high-p_T kaons have been selected. The production of strange hadrons in fragmentation is suppressed as compared to non-strange hadrons, unless there is already an s quark in the initial state. This effect is parameterized with a strangeness suppression factor γ_s, which ranges between 0.2 and 0.3 [15]. The production of a high-p_T correlated kaon pair will therefore be strongly suppressed, unless there is already a fragmenting $s\bar{s}$ pair. This selection, however, reduces the event yields to compensate for the reduced event yields, the p_T-cut on the kaons was lowered from 1.5 GeV to 1.0 GeV. A new extraction of $\langle \Delta g/g \rangle$ is currently underway using the improved model included in PYTHIA 6.1X [16].

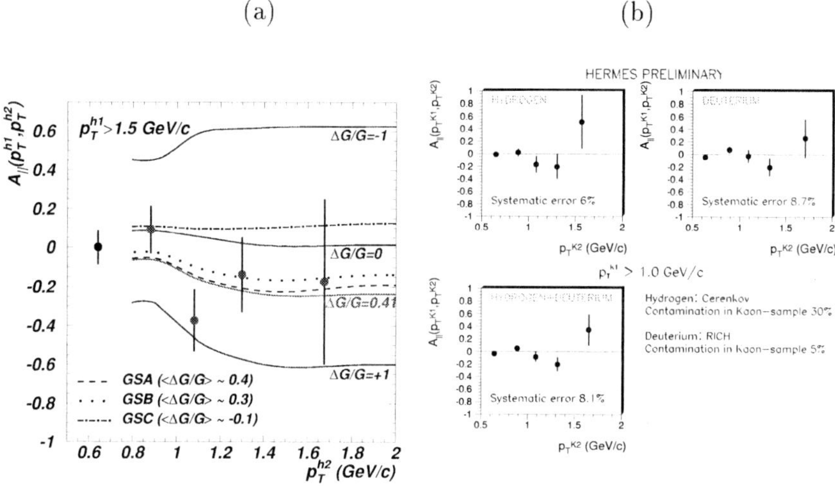

FIGURE 2. (a) $A_{||}$ for high-p_T hadron pair production measured at HERMES compared with Monte Carlo predictions for $\Delta G/G = \pm 1$ (lower/upper solid curves), $\Delta G/G = 0$ (middle solid curve), and the phenomenological LO QCD fits of Ref. [11] (dashed, dotted and dot-dashed curves). (b) $A_{||}$ is shown for the production of pairs of high-p_T kaons on a polarized hydrogen and deuterium target.

V HARD EXCLUSIVE PROCESSES

Much recent interest has surrounded the analysis of hard exclusive processes, where a virtual photon at high Q^2 scatters from a parton in a nucleon target. The final state contains only the recoiling nucleon along with a single additional particle (such as a meson, or a real photon). Theoretical analysis of such exclusive processes has led to the introduction of the Generalized Parton Distribution (GPD) formalism (see, for example, Ref. [28]). These GPD's represent a generalization of the familiar structure functions: in the forward limit of the GPD's, one recovers the parton distributions of deep-inelastic scattering; taking the first moments of their x-dependence, one obtains the elastic form factors of the nucleon. The description of hard exclusive processes involves the so-called handbag diagram, where two quarks of different momentum are exchanged with the nucleon in the same event. Such processes thus provide information on parton-parton correlations in the nucleon wave function. Perhaps most intriguing of all, the second moments of the generalized parton distributions are expected to be sensitive to the unknown orbital angular momentum L_q of the quarks.

The cleanest process to study these new parton distributions is deeply virtual compton scattering (DVCS), where a real photon is produced in the exclusive limit. Following Ref. [21] the ep cross section with unpolarized protons and leptons can be written as

$$\frac{d\sigma}{d\Phi dt dQ^2 dx_B} = \frac{1}{32\cdot(2\pi)^2}\frac{x_B y^2}{Q^4}\frac{1}{\sqrt{1+4x_B^2 m^2/Q^2}}\cdot|\tau_{BH}+\tau_{DVCS}|^2. \tag{2}$$

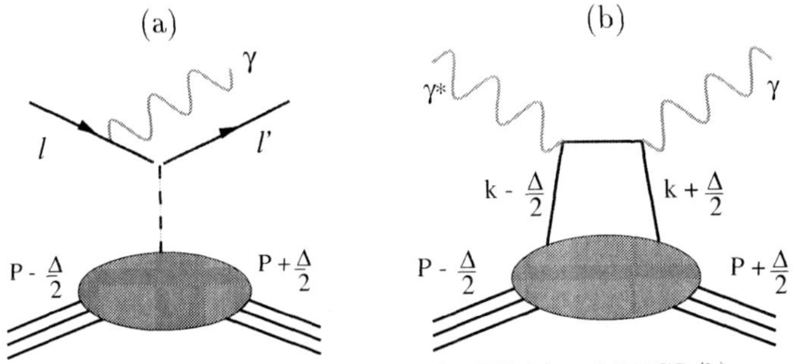

FIGURE 3. Feynman diagram for BH (a) and DVCS (b)

As can be seen from Eq. 2 and figure 3 there are 2 processes contributing to the cross section: the Bethe-Heitler (BH) process and the DVCS process. It is very difficult to separate the two processes on the basis of cross section measurements alone, as the final states are identical. It is easier to study the BH-DVCS interference term (equation 3), which can be expressed in terms of helicity amplitudes $M_{h,h'}^{\lambda\lambda'}(Q^2, x_B, \Delta_T)$ where $\lambda(\lambda')$ is the helicity of the initial (final) state photon, $h(h')$ that of the initial (final) state proton and Δ_T is the transverse momentum transfer from the initial proton to the final one. The interference term consists of two main parts. In Eq. 3 the helicity independent-part is shown, which can be accessed experimentally by measuring the asymmetry of the cross section with respect to the charge of the lepton beam.

$$\tau_{BH}^*\tau_{DVCS} + \tau_{DVCS}^*\tau_{BH} = \frac{e^6}{t}\frac{m}{Q}\cdot\frac{4\sqrt{2}}{x_{bj}}\cdot\frac{1}{\sqrt{1-x_{bj}}}[cos(\Phi)\frac{1}{\sqrt{\epsilon(1-\epsilon)}}Re\tilde{M}^{1,1}$$

$$-cos(2\Phi)\sqrt{\frac{1+\epsilon}{1-\epsilon}}Re\tilde{M}^{0,1} - cos(3\Phi)\sqrt{\frac{\epsilon}{1-\epsilon}}Re\tilde{M}^{-1,1}] + O(\frac{1}{Q^2}) \quad (3)$$

The absorptive part of the DVCS amplitudes, can be probed by using longitudinally polarized lepton beams. By measuring the azimuthal asymmetry with respect to the beam helicity $h_e = \pm 1/2$, one obtains information on the imaginary part of the matrix elements, i.e. $\mathrm{Im}M^{\lambda\lambda'}_{h,h'}$.

$$\tau^*_{BH}\tau_{DVCS} + \tau^*_{DVCS}\tau_{BH} \sim \frac{e^6}{t}\frac{m}{Q} \cdot \frac{4\sqrt{2}}{x_{bj}} \cdot \frac{1}{\sqrt{1-x_{bj}}} \cdot 2 \cdot h_e[-sin(\Phi)\sqrt{\frac{1+\epsilon}{\epsilon}}Im\tilde{M}^{1,1} + sin(2\Phi)Im\tilde{M}^{0,1}] + O(\frac{1}{Q^2}) \quad (4)$$

HERMES [22] has found a significant negative beam spin-azimuthal asymmetry analyzing $\gamma^* p \longrightarrow \gamma p$ data with different beam helicities in the exclusive limit $M_x \to 0$.

$$A^{sin\Phi}_{LU} = \frac{2\int_0^{2\pi} sin(\Phi)d\Phi(d\sigma^+/d\Phi - d\sigma^-/d\Phi)}{<|P_b|> \int_0^{2\pi} d\Phi(d\sigma^+/d\Phi + d\sigma^-/d\Phi)} \quad (5)$$

Here, Φ represents the azimuthal angle of the real photon around the virtual photon direction, with respect to the lepton scattering plane. The subscript LU for the asymmetry refers to the use of a longitudinal polarized beam and an unpolarized target. The single-spin asymmetry (compare Eqn. 5) as a function of missing mass is shown in Figure 4. As expected for hard exclusive processes, the signal peaks at $M_x \sim 1$ GeV.

HERMES has also studied hard exclusive pion production from a polarized proton target. Specifically, the measurement of the $sin(\Phi)$ moment of the target-spin asymmetry $A^{\sin\Phi}_{UL}$ was extended into the exclusive limit $z \to 1$ [23]. A striking behavior is observed: in the semi-inclusive region ($0.2 < z < 0.7$) a small positive value of $A^{\sin\Phi}_{UL}$ is seen for π^+ production, while a large negative value is observed at $z \simeq 1$. By comparison, the asymmetry for π^0 production remains positive but becomes larger in magnitude in the exclusive limit. These measurements have yet to be compared with theoretical calculations.

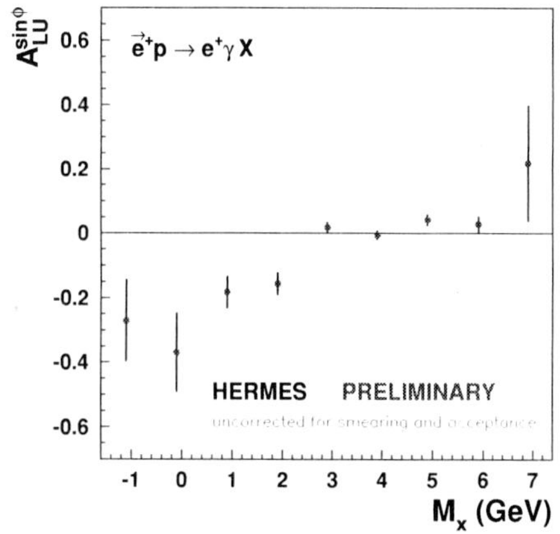

FIGURE 4. The $sin(\Phi)$ moment of the target-spin asymmetry $A^{\sin\Phi}_{LU}$ plotted versus missing mass for the process $\gamma^* p \longrightarrow \gamma p$

VI FUTURE TRANSVERSITY MEASUREMENTS

A complete description of the structure of the nucleon at leading twist requires a third structure function, beyond the already-measured $F_1(x)$ and $g_1(x)$. This structure function is termed *transversity*, and has been assigned the symbol $h_1(x)$. In the same manner as F_1 and g_1 are related respectively to the parton distribution functions $q(x)$ and $\Delta q(x)$, h_1 is related to $\delta q(x)$ – the distribution of *transverse* quark spin in a nucleon polarized transverse to its (infinite) momentum [17]. The function $h_1(x)$ has several interesting properties. In the simplest non-relativistic picture of the nucleon, $\delta q(x)$ is simply equal to the longitudinal helicity distribution $\Delta q(x)$. However, unlike in the case of g_1, the gluon polarization does not mix with quark polarization in h_1, leading to a rather different Q^2 evolution. Furthermore, quarks and anti-quarks contribute to h_1 with opposite sign, making transversity a pure *valence quark* object. The first moment of h_1 represents the tensor polarization of valence quarks, an object which can be compared to lattice QCD calculations.

Transversity is as yet unmeasured because it has the unusual property of being chirally-odd; hence, h_1 is not directly observable in inclusive lepton-nucleon scattering experiments. However, it has been suggested that experimental information on $h_1(x)$ can be obtained by the semi-inclusive production of pions with modest $p_\perp$ [18]. In such reactions, a chiral-odd fragmentation function known as the Collins function $H_1^\perp(z)$ also appears, thus providing a net chirally-even cross-section. The prediction of the Collins function is that leading pions are distributed preferentially above or below a plane that is formed by the virtual photon and the transverse spin direction of the struck quark.

HERMES has made a first measurement of single-spin asymmetries for semi-inclusive pion production in DIS, using an unpolarized beam and a *longitudinally* polarized proton target [19]. A clear $\sin\Phi$ dependence is observed for π^+ mesons, indicating that a different out-of-plane direction is indeed preferred in one target orientation compared with the other. No such behavior is seen for π^- production. The $\sin\Phi$ moment of the asymmetry (denoted $A_{UL}^{\sin\Phi}$) can be interpreted in terms of chiral-odd quark spin-distribution functions closely related to transversity [20].

The obtained non-zero single-spin asymmetries for semi-inclusive pion production with a longitudinal polarized target implies that the foreseen measurement of $h_1(x)$ [24] using a transversely polarized target in the 2001-2002 running period at HERMES is very promising.

To evaluate the level of accuracy that could be obtained when scattering 'unpolarized' leptons off transversely polarized protons at HERMES, simulations were done assuming a target polarization of $P_T = 75\%$ and 7.0 Million reconstructed DIS events. Assuming up-quark dominance for π^+ production, the $\sin\Phi$ asymmetry moment is given by

$$A_T^{\pi^+}(x,y,z) = P_T \cdot D_{nn} \cdot \frac{\delta u(x)}{u(x)} \cdot \frac{\mathrm{H}_1^{\perp(1)u}(z)}{\mathrm{D}_1^u(z)}, \tag{6}$$

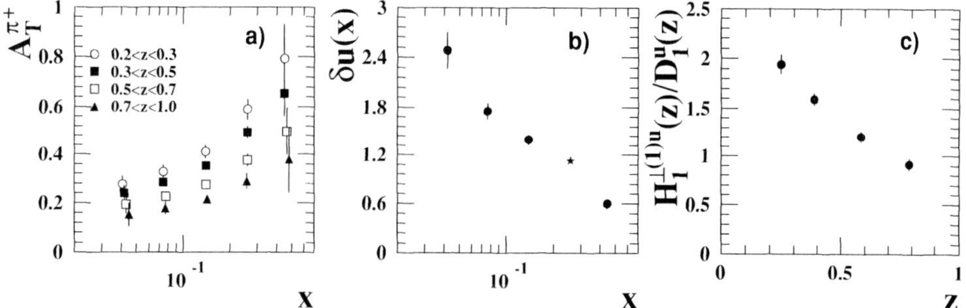

FIGURE 5. *a)* The weighted asymmetry $A_T^{\pi^+}(x)$ as a function of x; *b)* the transversity distribution $\delta u(x)$ as a function of x, and *c)* ratio of the fragmentation functions $\mathrm{H}_1^{\perp(1)u}(z)$ to $\mathrm{D}_1^u(z)$ as a function of z, as they would be measured by HERMES.

where P_T is the target polarization, D_{nn} is the transverse spin transfer (kinematic) coefficient, and $\mathrm{D}_1^u(z)$ is the familiar unpolarized fragmentation function. The magnitude of the observed asymmetry depends on two unknown functions – $\delta u(x)$ and the Collins fragmentation function $\mathrm{H}_1^{\perp(1)u}(z)$.

For the simulations [25], $\delta u(x) = \Delta u(x)$ was assumed, which should be suitable for the relatively small values of Q^2 relevant for the HERMES kinematical region. Values for the function $\mathrm{H}_1^{\perp(1)u}(z)$ were estimated using the model of Kotzinyan and Mulders [26] that was found to be consistent with the HERMES data. The expected asymmetry $A_T^{\pi^+}(x)$ as it would be measured by HERMES is presented as a function of z in Fig. 5*a)*.

The factorised form of the expression in Eq. (6) with respect to the variables x and z allows the reconstruction of a shape for both of the unknown functions $\delta u(x)$ and $\mathrm{H}_1^{\perp(1)u}(z)/\mathrm{D}_1^u(z)$, while the individual normalizations can not be fixed without further assumptions. The normalization ambiguity has been resolved by assuming $\delta u(x = 0.25) = \Delta u(x = 0.25)$, as indicated by the asterisk in Fig. 5*b)*. The expected accuracies for the reconstruction of $\delta u(x)$ and $\mathrm{H}_1^{\perp(1)u}(z)/\mathrm{D}_1^u(z)$ are presented in Figs. 5*b)* and *c)* respectively.

A very good statistical precision is expected for a first measurement of the x- and z-dependence of the transversity distributions $\delta u(x)$ and $\mathrm{H}_1^{\perp(1)u}(z)/\mathrm{D}_1^u(z)$. Given the surprising results found for $\mathrm{g}_1(x)$ a decade ago, we cannot claim to understand the spin structure of the nucleon unless the single remaining leading-twist structure function h_1 has been measured as well.

VII SUMMARY

The past years have seen significant advances in the experimental exploration of spin-dependent processes, and the coming years promise further exciting developments. Continuing studies of semi-inclusive spin asymmetries are providing new information on the flavor-separated quark spin distributions in the nucleon. HERMES had a record data taking year in 2000. In the years 1998-2000 HERMES collected 9 Million DIS events on a polarized deuterium target with a RICH detector, which enables the measurement of charged kaon asymmetries, and will hopefully yield a first direct measurement of the strange quark polarization.

Measurements of the gluon polarization are just beginning: HERMES has taken a first look at the asymmetry for production of pairs of high-p_T hadrons, and measurements of this type will be pursued with greater precision by the upcoming COMPASS and RHIC-SPIN experiments. HERMES has investigated the transversity structure function $h_1(x)$ for the first time, and is expected to collect precise data in coming years with a transversely polarized target program. Finally, (spin-dependent) deep-inelastic scattering experiments are beginning to explore parton-parton correlations in nucleons by extracting signals on hard exclusive processes. Examples of such studies have been presented, including the HERMES data on DVCS and exclusive pion production. Extensions of these studies at higher luminosity will make it possible to study the GPDs in more detail.

REFERENCES

1. R.L. Jaffe and A. Manohar, *Nucl. Phys.* B **337**, 509 (1990).
2. T.P. Cheng and L.-F. Li, *Phys. Lett.* B **336**, 365 (1996).
3. B. Dressler *et al.*, *Eur. Phys. J.* C **14**, 147 (2000).
4. E142 Coll., P.L. Anthony *et al.*, *Phys. Rev.* D **54**, 6620 (1996);
   E154 Coll., K. Abe *et al.*, *Phys. Rev. Lett.* **79**, 26 (1997);
   HERMES Coll., K. Ackerstaff *et al.*, *Phys. Lett.* B **404**, 383 (1997);
   E143 Coll., K. Abe *et al.*, *Phys. Rev.* D **58**, 112003 (1998);
   HERMES Coll., A. Airapetian *et al.*, *Phys. Lett.* B **442**, 484 (1998);
   SM Coll., B. Adeva *et al.*, *Phys. Rev.* D **58**, 112001 (1998);
   E155 Coll., P.L. Anthony *et al.*, *Phys. Lett.* B **463**, 339 (1999).
5. SMC, B. Adeva *et al.*, *Phys. Rev.* D **58**, 112002 (1998).
6. U. Stösslein, Spin 2000 - Proceedings, edited by K. Hatanaka et al. p. 387.
7. E. Leader, A. Sidorov, D. Stamenov, hep-ph/0004106.
8. HERMES Coll., K. Ackerstaff *et al.*, *Phys. Lett.* B **464**, 123 (1999).
9. Th. Lindemann, Spin 2000 - Proceedings, edited by K. Hatanaka et al. p. 392.
10. SMC, B. Adeva *et al.*, *Phys. Lett.* B **420**, 180 (1998).
11. T. Gehrmann and W.J. Stirling, *Phys. Rev.* D **53**, 6100 (1996).
12. HERMES Coll., A. Airapetian *et al.*, *Phys. Rev. Lett.* **84**, 2584 (2000).
13. A. Bravar, D. von Harrach, A. Kotzinian, *Phys. Lett.* B **421**, 349 (1998).
14. M. Glück, E. Reya, M. Stratmann, W. Vogelsang, *Phys. Rev.* D **53**, 4775 (1996)
15. S. Aid et al., Nucl. Phys. B 480 (1996) 3.
    P. Abreu et al., Z.Phys. C 68 (1995) 29.
16. T. Sjöstrand et al., hep-ph 0010017, (2000)
17. J.P. Ralston and P.E. Soper, *Nucl. Phys.* B **152**, 109 (1979);
    R. Jaffe and X. Ji, *Nucl. Phys.* B **375**, 527 (1992).

18. J.C. Collins, *Nucl. Phys.* B **396**, 161 (1993).
19. HERMES Coll., A. Airapetian *et al.*, *Phys. Rev. Lett.* **84**, 4047 (2000).
20. E. De Sanctis, W.-D. Nowak, K.A Oganessyan, *Phys. Lett.* B **483**, 69 (2000); A.V. Efremov *et al.*, *Phys. Lett.* B **478**, 94 (2000).
21. M. Diehl *et al.*, *Phys. Lett.* B **411**, 193 (1997).
22. M. Amarian, CP570, Spin 2000 - Proceedings, edited by K. Hatanaka et al. p. 428. A. Airapetian et al, submitted to PRL and hep-ex/0106068
23. D. Ryckbosch, CP570, Spin 2000 - Proceedings, edited by K. Hatanaka et al. p. 556
24. HERMES Coll.,The HERMES Physics Program and Plans for 2001-2006
25. V.A. Korotkov, W.-D. Nowak and K.A. Oganessyan, DESY-176.
26. A.M. Kotzinian, P.J. Mulders, Phys. Lett. B **406**, 373 (1997).
27. M. Stratmann, *Z. Phys.* C **60**, 763 (1993).
28. M. Vanderhaeghen, P.A.M. Guichon, M. Guidal, *Phys. Rev.* D **60**, 094017 (1999).

QCD Analysis of Polarized Deep Inelastic Scattering Data

J. Blümlein* and H. Böttcher*

*DESY Zeuthen, Platanenallee 6, 15738 Zeuthen, Germany

Abstract. A QCD analysis of the world data on inclusive polarized deep inelastic scattering of leptons on nucleons is presented in leading and next–to–leading order. New parameterizations are derived for the quark and gluon distributions and the value of $\alpha_s(M_Z)$ is determined. Emphasis is put on the derivation of fully correlated error bands for these distributions which are directly applicable to determine experimental errors of other polarized observables. The impact of the variation of both the renormalization and factorization scales on the value of α_s is studied. Finally a factorization–scheme invariant QCD analysis based on the observables $g_1(x,Q^2)$ and $dg_1(x,Q^2)/d\log(Q^2)$ is performed in next–to–leading order, which is compared to the standard analysis.

INTRODUCTION

The remarkable growth of experimental data on inclusive polarized deep inelastic scattering of leptons on nucleons over the last years [1–9] allows to perform refined QCD analyses of polarized structure functions in order to reveal the spin–dependent partonic structure of the nucleon. A number of such analyses has already been worked out. The most recent ones are [10–13] [1]. In this talk results from a new QCD analysis in leading (LO) and next–to–leading (NLO) order [14] [2] are presented. New parameterizations of the polarized quark and gluon distributions are derived including the parameterizations of fully correlated 1σ error bands for these distributions, which are directly applicable to calculate errors of other polarized observables. Furthermore the value of $\alpha_s(M_Z)$ is determined. Finally and for the first time a factorization–scheme independent QCD evolution based on the observables $g_1(x,Q^2)$ and $dg_1(x,Q^2)/d\log(Q^2)$ in next–to–leading order is performed.

FORMALISM

In LO the polarized structure function $g_1(x,Q^2)$ is expressed as the sum of the polarized quark distributions $\Delta q_i(x,Q^2)$ weighted by the square of the quark charges. In NLO the expression for $g_1(x,Q^2)$ involves the polarized singlet $\Delta\Sigma(x,Q^2)$, the gluon $\Delta G(x,Q^2)$, and the non–singlet $\Delta q^{NS}(x,Q^2)$ distributions and reads

---

[1] For a more complete list of references see the references therein and Ref. [14]

[2] All details of the analysis are given in [14].

CP602, *QCD@Work: International Workshop on Quantum Chromodynamics*
edited by P. Colangelo and G. Nardulli

$$g_1(x,Q^2) = \frac{1}{2}\left[\left(\frac{1}{n_f}\sum_{i=1}^{n_f} e_i^2\right)[\delta C_S \otimes \Delta\Sigma + \delta C_G \otimes \Delta G] + \delta C_{NS} \otimes \Delta q^{NS}\right], \quad (1)$$

where n_f is the number of active quark flavors and e_i is the quark charge. The symbol $\otimes$ denotes the Mellin convolution w.r.t. x of the polarized parton densities $\Delta q_i(x,Q^2)$ with the corresponding polarized Wilson coefficient functions $\delta C_i(x,\alpha_s(Q^2))$. The polarized singlet and non–singlet distributions are certain combinations of the polarized quark distributions $\Delta q_i(x,Q^2)$.

The evolution equations used to evolve the parton densities to different Q^2 values contain the polarized splitting functions $\Delta P_{ij}(x,\alpha_s(Q^2))$. Both the polarized Wilson coefficient [15] and the polarized splitting functions [16] are known in the $\overline{MS}$ scheme up to order $O(\alpha_s^2)$.

METHOD

The shape chosen for the parameterization of the polarized parton distributions at the input scale of $Q^2 = 4.0\ GeV^2$ is :

$$x\Delta q_i(x,Q_0^2) = \eta_i A_i x^{a_i}(1-x)^{b_i}(1+\gamma_i x+\rho_i x^{\frac{1}{2}}). \quad (2)$$

The normalization constant A_i is chosen such that η_i is the first moment of $\Delta q_i(x,Q_0^2)$. The densities to be fitted are Δu_v [3], Δd_v, $\Delta\bar{q}$, and ΔG.

Assuming $SU(3)$ flavor symmetry the first moments of Δu_v and Δd_v are determined by the $SU(3)$ parameters F and D measured in neutron and hyperon β–decays and can be fixed to $\eta_{u_v} = 0.926$ and $\eta_{d_v} = -0.341$. In addition we assume a flavor symmetric sea, i.e. only one general sea distribution $\Delta\bar{q}(x,Q^2)$ is required. No assumptions are made concerning positivity and helicity retention. Given the present accuracy of the data we set a number of parameters to zero, namely $\rho_{u_v} = \rho_{d_v} = 0$, $\gamma_{\bar{q}} = \rho_{\bar{q}} = 0$, and $\gamma_G = \rho_G = 0$. This choice reduces the number of parameters to be fitted for each parton distribution to three. In addition the parameter Λ_{QCD} was determined. The relative normalizations of the different data sets were fitted and then fixed. Doing so part of the experimental systematics was taken into account.

RESULTS

The results reported here are based on 433 data points of asymmetry data, i.e. g_1/F_1 or A_1, above $Q^2 = 1.0\ GeV^2$, the world statistics published so far. The QCD fits are performed on g_1 which is evaluated from the asymmetry data using parameterizations for the unpolarized structure functions F_2 [17] and R [18]. We realized that the 4 parameters γ_{u_v}, γ_{d_v}, $b_{\bar{q}}$, and b_G had to be fixed in addition at their values at $\chi^2_{\min}$ since the data do

[3] Note that: $\Delta q + \Delta\bar{q} = \Delta q_v + 2\Delta\bar{q}$.

not constrain these parameters well enough. Only fits with a positive definite covariance matrix were accepted in order to be able to calculate the fully correlated 1σ error bands. The NLO polarized parton densities at the input scale are presented in Fig. 1.

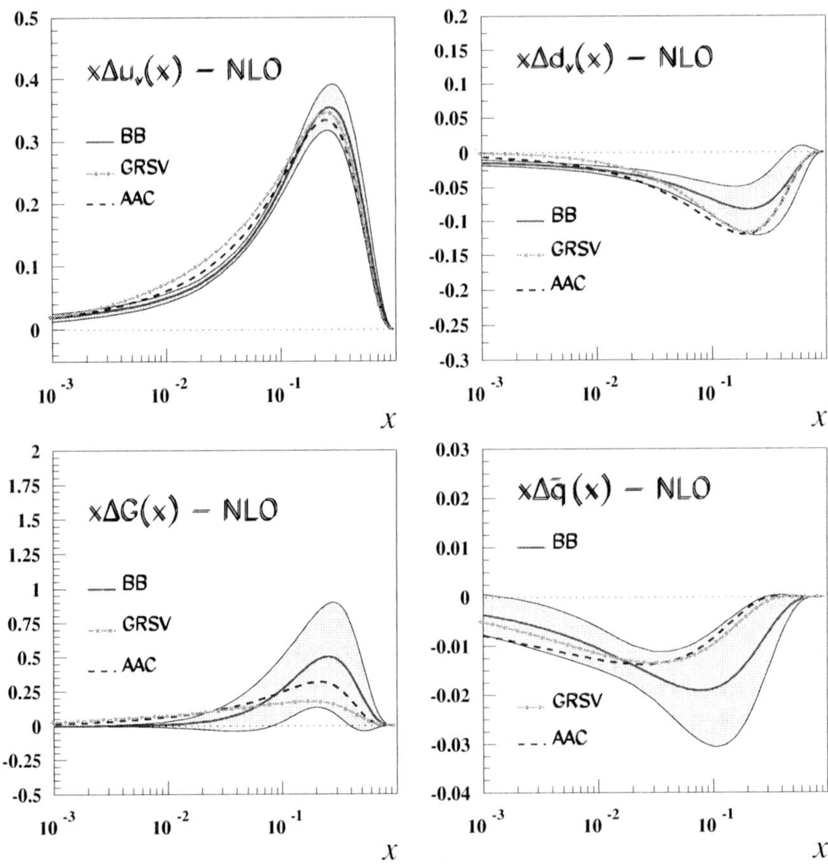

FIGURE 1. Polarized parton distribution at the input scale $Q_0^2 = 4.0\ GeV^2$ (solid line) compared to results obtained by GRSV (dashed–dotted line) [12] and AAC (dashed line) [10]. The shaded areas represent the fully correlated 1σ error bands calculated by Gaussian error propagation, Ref. [14].

While the quality of the data is sufficient to determine Δu_v and Δd_v with good accuracy, ΔG and $\Delta\bar{q}$ have much broader error bands. This is essentially due to the lack of data at low x. The agreement with the results of the analyses of Refs. [10] and [12] is satisfactory within the error bands. The measured structure function g_1^p is well described both as function of x and of Q^2. The derived parton distributions and its error bands have been evolved to Q^2 values up to $10,000\ GeV^2$. As an example the evolution of ΔG is shown in Fig. 2. One observes that even within the error band ΔG stays *positive* up to the highest Q^2 value. It should be mentioned that $\Delta\bar{q}$ develops a trend to change sign and becomes slightly positive towards higher Q^2 values and for $x \gtrsim 0.1$ within the errors.

In determining $\alpha_s(M_z^2)$ the parameter Λ_{QCD} was fitted. The impact of the variation of both the renormalization and factorization scales on the value of α_s was studied. The

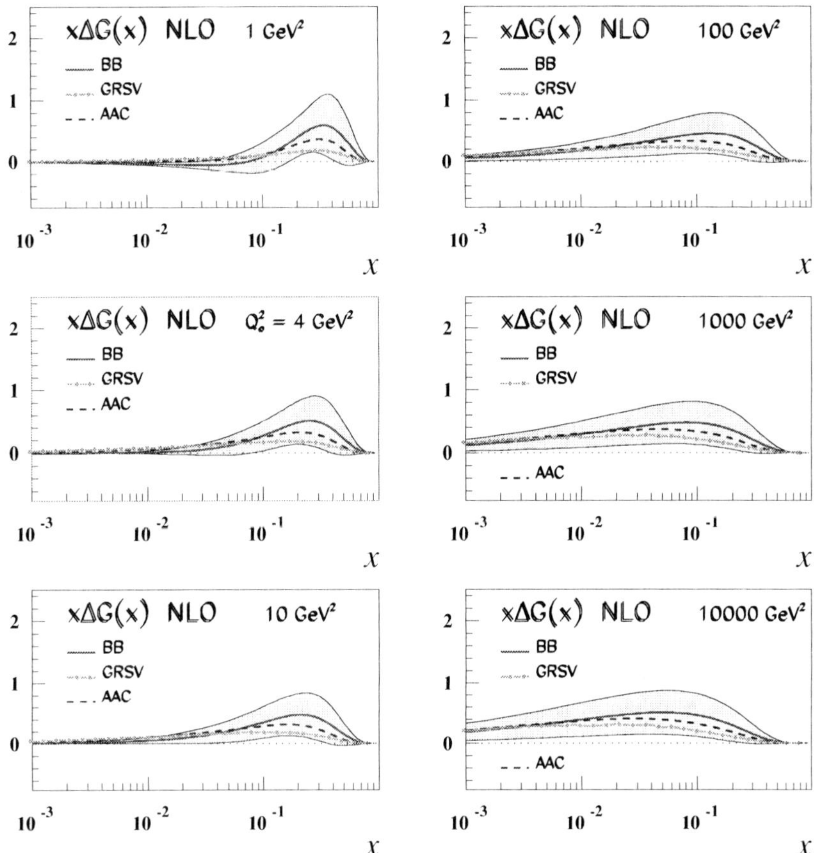

FIGURE 2. The polarized parton distribution ΔG evolved up to Q^2 values up to $Q^2 - 10,000\ Gev^2$ (solid line) compared to results obtained by GRSV (dashed–dotted line) [12] and AAC (dashed line) [10]. The shaded areas represent the fully correlated 1σ error bands calculated by Gaussian error propagation, Ref. [14].

following value for $\Lambda_{\rm QCD}$ was obtained

$$\Lambda^{(4)}_{\rm QCD} = 241 \pm 58\ ({\rm fit})\ {}^{+65}_{-44}\ ({\rm fac})\ {}^{+117}_{-58}\ ({\rm ren})\quad MeV,$$

which results into a value of

$$\alpha_s(M_Z^2) = 0.114\ {}^{+0.004}_{-0.005}\ ({\rm fit})\ {}^{+0.005}_{-0.004}\ ({\rm fac})\ {}^{+0.008}_{-0.005}\ ({\rm ren}).$$

This value of $\alpha_s(M_Z^2)$ is compatible within the errors with the world average of 0.118 ± 0.002 [19] and with values from other QCD analyses [20], although the central value tends to be lower, as also in Ref. [20b].

Finally a factorization–scheme invariant QCD analysis based on the observables $g_1(x,Q^2)$ and $dg_1(x,Q^2)/d\log(Q^2)$ in next-to-leading order was performed. The corresponding evolution equations have been worked out in Ref. [21] [4]. Such an analysis has the advantage of direct control over the input since it comes from measured quantities. The only parameter to be determined is Λ_{QCD}. Unfortunately, the present data do not yet allow to determine the slope $\partial g_1(x,Q^2)/\partial\log(Q^2)$ as an input density from measurements of g_1, but it is derived here from the fit result for g_1 described above. The evolution of the so determined slope is shown in Fig. 3.

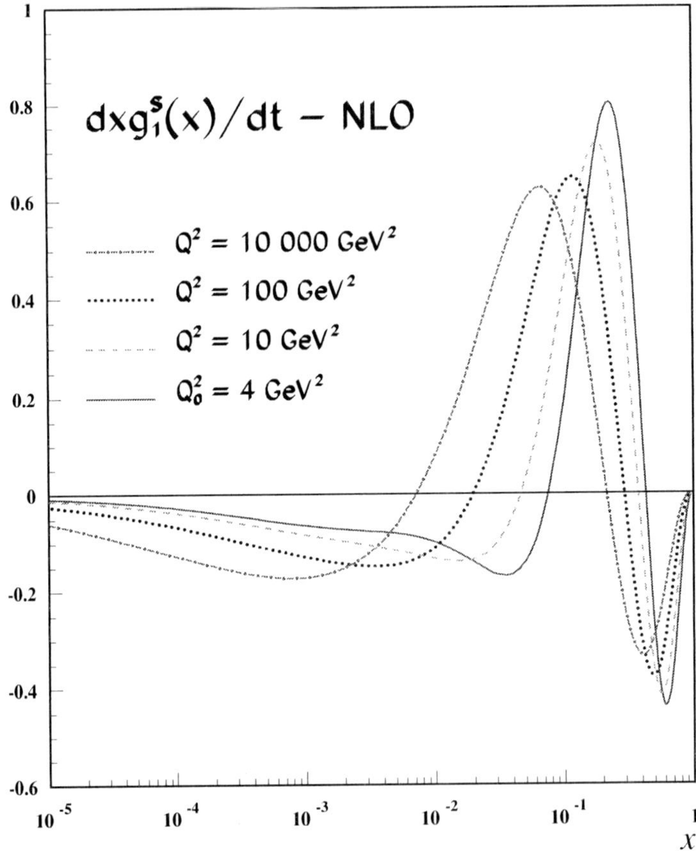

FIGURE 3. The evolution of $dg_1^s(x,Q^2)/dt$ (singlet contribution) with $t = -2/\beta_0 \ln(\alpha_s(Q^2)/\alpha_s(Q_0^2))$. The slope of g_1 was determined from a fitted g_1, see text.

A downward shift of 12 MeV in Λ_{QCD} was found yielding a similar result for $\alpha_s(M_Z^2)$ as obtained in the standard analysis.

---

[4] The same case has already been considered in Ref. [22].

CONCLUSIONS

An LO and NLO QCD Analysis of the current World–Data on Polarized Structure Functions was performed. New parameterizations of the polarized parton densities including their errors were derived. They are available via a fast FORTRAN code for the range: $1 < Q^2 < 10^6$ GeV^2 and $10^{-4} < x < 1$. The value determined for $\alpha_s(M_Z^2)$ is compatible with the world average, although the central value obtained is lower. First steps in a factorization–scheme invariant QCD evolution based on the structure function $g_1(x,Q^2)$ and $\partial g_1(x,Q^2)/\partial \log Q^2$ were performed yielding similar results for $\alpha_s(M_Z^2)$. This latter analysis is a very promising way to proceed in the future, since it allows to extract $\Lambda_{\rm QCD}$ fixing all the input distributions by direct measurements.

ACKNOWLEDGMENTS

This work was supported in part by EU contract FMRX-CT98-0194 (DG 12 - MIHT). For discussions in an early phase of this work we would like to thank A. Vogt.

REFERENCES

1. J. Ashman et al. (EMC), Phys. Lett. **B206** (1988) 364; Nucl. Phys. **B328** (1989) 1.
2. P.L. Anthony et al. (E142), Phys. Rev. **D54** (1996) 6620.
3. K. Ackerstaff et al. (HERMES), Phys. Lett. **B404** (1997) 383.
4. K. Abe et al. (E154), Phys. Rev. Lett. **79** (1997) 26.
5. B. Adeva et al. (SMC), Phys. Rev. **D58** (1998) 112001.
6. K. Abe et al. (E143), Phys. Rev. **D58** (1998) 120003.
7. A. Airapetian et al. (HERMES), Phys. Lett. **B442** (1998) 484,
   D. Hasch, PhD thesis, Humboldt-Univ. Berlin, (1999) and DESY-THESIS-2000-032.
8. P.L. Anthony et al. (E155), Phys. Lett. **B463** (1999) 339.
9. P.L. Anthony et al. (E155), Phys. Lett. **B493** (2000) 19.
10. Y. Goto et al., Phys. Rev. **D62** (2000) 034017.
11. Dilip Kumar Gosh et al., Phys. Rev. **D62** (2000) 094112.
12. M. Glück et al., Phys. Rev. **D63** (2001) 094005.
13. E. Leader et al., hep-ph/0106214, to be published in Proceed. of DIS2001.
14. J. Blümlein and H. Böttcher, DESY 01-087 (2001), in preparation.
15. W. L. van Neerven and E. B. Zijlstra, Nucl. Phys. **B417** (1994) 61, Nucl. Phys. **B426** (1994) 245.
16. R. Mertig and W. L. van Neerven, Z. Phys. **C70** (1996) 637,
   W. Vogelsang, Phys. Rev. **D54** (1996) 2023.
17. M. Arneodo et al., Phys. Lett. **B364** (1995) 107.
18. L. Withlow et al., Phys. Lett. **B250** (1990) 193.
19. Eur. Phys. J. **C15** (2000) .
20. a) G. Altarelli et al., Nucl. Phys. **B496** (1997) 337,
   b) K. Abe et al. (E154), Phys. Lett. **B405** (1997) 180,
   c) B. Adeva et al. (SMC), Phys. Rev. **D58** (1998) 112002.
21. J. Blümlein, V. Ravindran, and W.L. van Neerven, Nucl. Phys. **B586** (2000) 349.
22. W. Furmanski and R. Petronzio, Z. Phys. **C11** (1982) 293.

Experimental Status of α_s Measurements at LEP

R.W.L. Jones

Department of Physics, University of Lancaster,
Lancaster LA1 4YB, UK

Abstract. A summary is given of the current experimental status of α_s measurements at LEP. Particular emphasis is given to the determinations using event-shape observables, and to the outstanding theoretical uncertainties. Recent new approaches to the theoretical uncertainties are discussed.

INTRODUCTION

In all-orders QCD, the only free parameter (other than quark masses) is the strength of the coupling. It is therefore important to measure this parameter in as many processes and at as many scales as possible to test the consistency of the theory, and this program has been followed by many experiments and in many forms of interaction. In recent year, a good consistency between the theoretical predictions for the running of α_s and experiment has been achieved [1,2]; we note that this does not constitute a proof of the running, as that running is assumed in the extraction of the individual α_s values. This paper summarizes the current results of extractions of α_s at LEP, over a broad range of scales, which comprise an important input to the world average value of α_s (M_Z^2)=0.118±0.002[1].

Current techniques cannot deliver all-order predictions of QCD quantities. Thus, all experimental extractions of α_s rely on finite-order predictions. Although resummation may allow some additional higher-order corrections to be included, all extracted α_s values are uncertain in terms of the appropriate renormalization scale, and also in terms of the matching scheme when matched fixed-order and resummed predictions are used.

Inclusive Observables

The value of α_s can be determined without a detailed examination of the final state particle content or topology in e^+e^- annihilations in several ways. For example, the ratio of partial widths $R_l=\Gamma$ (Z→ hadrons)/Γ (Z→ muons)= R_l^{EW} $(1+\delta_{QCD}+\delta_{mass}+\delta_{np})$, where R_l^{EW} is the pure electroweak value, δ_{np} and δ_{mass} are small non-perturbative and quark-mass effects (being suppressed by powers of $1/M_Z^2$), and the QCD perturbative correction $\delta_{QCD}=\Sigma_{n=1}^{3} c_n\alpha_s^n$ is known to NNLO. The current LEP combined value of R_l=20.768±0.024[3] gives α_s (M_Z^2)=0.124±0.004[2]. This extracted value is dominated by the experimental uncertainties, especially in the muon partial width, as

CP602, *QCD@Work: International Workshop on Quantum Chromodynamics*
edited by P. Colangelo and G. Nardulli

well as the renormalization scheme and scale effects, and the limited knowledge of the top quark and Higgs masses in the electroweak sector.

An analogous ratio, R_τ may also be constructed for hadronic and muonic decays of the τ. Events of the type $e^+e^-\rightarrow\tau^+\tau^-$ where one or more decay is hadronic have been exploited by ALEPH and OPAL. They have measured moments of the mass spectrum, the first of which corresponds to R_τ. The use of higher moments also allows effective non-perturbative parameters to be determined along with α_S. The resulting combined α_S value is dominated by theoretical uncertainties, chiefly those arising from the effects of the missing higher order. The combined value is α_s=0.1181±0.0007(experimental)±0.0030(theory)[2].

Event Shapes

Hard gluon radiation gives rise to jet-structures in the event, and the degree of hard gluon radiation depends on α_S. Thus, event shape observables that characterize the topology are good for determining α_S. For several such observables, the cross-section has been calculated at NLO, the observables being infrared and collinear safe. In addition for some, soft gluon radiation has been accounted for in the `two-jet' region by the resummation of large logarithms. A better prediction may be obtained for observables where both sorts of prediction exist by a combination of the two, avoiding double counting by `matching' equivalent terms in both predictions. This matching is not unambiguous, and results are often presented for the so-called R and log(R) matching schemes. The observables used thus far include thrust, C and the differential two-jet rate in the Durham scheme. Other event-shape observables involve a division of the event into hemispheres, including the wide- and total-jet broadenings and the heavy-jet mass. All of these observables have LO predictions proportional to α_S. These observables have been used both at LEP-I and LEP-II, and the combined results are dominated by the theoretical uncertainties. A typical result is that of ALEPH at 206GeV, where α_s (206 GeV)=0.1053±0.0028(exp.)±0.0041(theory). The theory error decomposes into 0.0038 from the scale uncertainty, 0.0016 from the matching scheme ambiguity, and 0.0007 from the hadronization uncertainties.

Scale and Matching Uncertainties

The scale and matching scheme uncertainties currently dominate the extracted α_s from event shapes (and several other determinations). Recent work in the LEP QCD working group has lead to a common, although still arbitrary, treatment of the scale uncertainties. However, the fact that the observables that give by far the poorest overall $\div^2$ per degree of freedom in the fit (the hemisphere-based observables) also give the smallest uncertainties from a simple variation of scales leads to unease as to whether the uncertainties due to missing higher orders are being accounted for properly.

The group has also investigated the matching scheme uncertainties, and revealed differences in the treatment of the application of kinematic and normalization constraints in the matching between the experiment. A common treatment has now been agreed and revised results are expected to be available for averaging soon. However, a true understanding of the effects of the truncation in the predictions awaits NNLO predictions.

It should be noted in passing that various `experimental' determinations of the scale have been attempted [4,5] The scale this obtained is typically orders of magnitude smaller than that expected, and some reduction in the scatter of the α_s values derived has been observed (although not in call cases [6]). However, the fitted scales have a scatter of several orders of magnitude. The interpretation of the results is contentious and awaits further work from the experiments.

Overall, it should be noted that a satisfactory resolution of these uncertainties awaits the availability of NNLO predictions. If these are long delayed, they may be of some use to workers at the Next Linear Collider, but a re-analysis of the LEP data will be difficult, despite best efforts to archive the required tools.

Hadronization Uncertainties

Another source of theoretical uncertainty is hadronization. The theoretical predictions are at the parton level. The experimental results are (after minor corrections for detector efficiencies and resolutions) in terms of the particles in the event. Typically, the parton predictions are corrected to the hadron level before the fitting. This translation is usually done using Monte Carlo models. Several models are in common use (JETSET [7], HERWIG [8] and ARIADNE [9]), and give different correction factors that indicate the degree of systematic uncertainty. It is also possible to obtain different corrections within the same model by using an alternate tuning of that model; this may explain the different factors determined by the various experiments with a given model, but this is under investigation. One can also use different correction techniques (such as a matrix correction or a simple bin-by-bin correction), but studies indicate the differences introduced are small.

Power Law Corrections

A promising alternative approach to the problem of hadronization is that of power-law corrections. These predict a single additive term in the mean value of an event shape observable, and a shift in the distribution. In this formalism, α_s is determined simultaneously with a universal parameter, α_0. Applying this to the available means of the event shapes gives a good description of the mean values of event shapes, with the universality of α_0 good to the 20% level [10]; however, there is poorer consistency when the distributions are studied. These points are evident in Figure 1. Despite this, the approach is worthy of more study, when the effects of resonance and hadron masses may be properly accounted for [11].

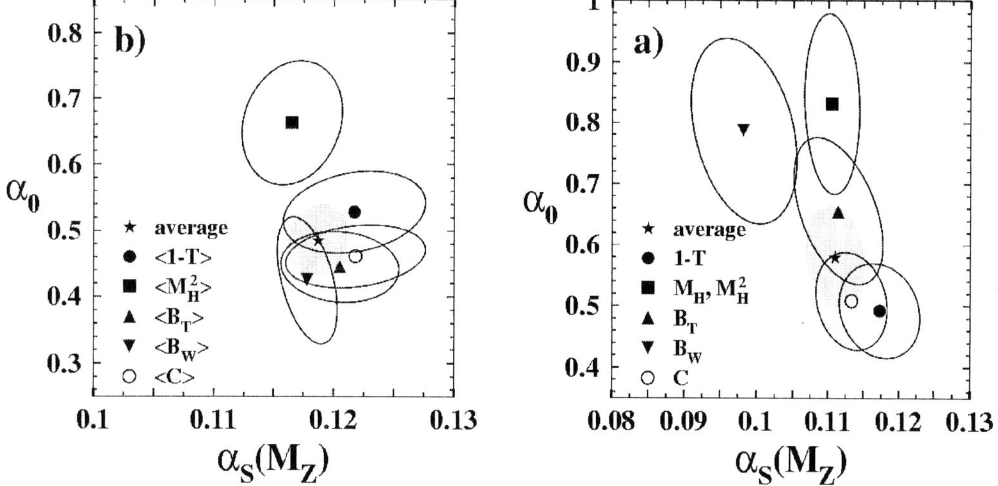

FIGURE 1. The result of a recent analysis [10] of the means (left) and distributions (right) of event shape observables using the power law formalism.

Four-jet observables

Recently, NLO and resummed predictions have also become available for observables whose leading order prediction is proportional to α_s, so-called four-jet observables. Experimental extractions of α_s have been done with matched predictions by OPAL [12] and ALEPH [13], and with NLO predictions by DELPHI [14]. The most widely used observable is the four-jet rate using the Durham jet-finding scheme. These are expected to have smaller higher-order uncertainties of all kinds, being suppressed by powers of 1/Q. This is indeed found to be the case; for instance, in the case of the ALEPH study, the experimentally preferred scale is around x_μ=0.7, within about 2σ (statistical) of the theoretically favored value of unity.

Scaling Violations

Looking further into the detailed structure of the LEP events, α_s may be determined from the scaling violations in the fragmentation function of charged particles. The shape of these functions is not determined in perturbative QCD, but once determined at a given scale, its evolution is known. Thus, if lower-energy e^+e^- data is used to determine the shape, α_s can be determined from the running of that shape up to LEP energies. This has been done by ALEPH and DELPHI using data at the Z pole, see Figure 2; it would be worthy of an update using the LEP-II data and improved NLO calculations. The combined value obtained is α_s=0.126±0.0007(experiment)±0.0009(theory)[2].

CONCLUSIONS

LEP has provided excellent opportunities to determine α_s in a variety of ways. However, in most cases the predictions are limited to NLO, and so the true extent of

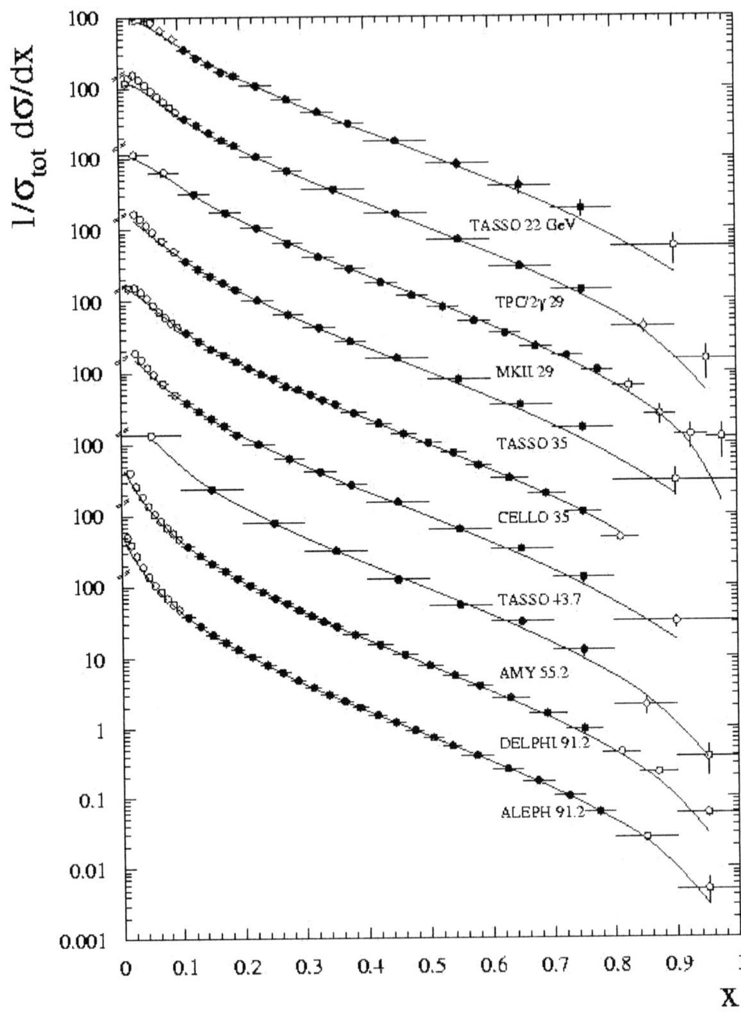

FIGURE 2. The inclusive hadron spectra as used in the LEP scaling violation analyses.

the effect of missing higher orders is not yet clear; NNLO predictions will be extremely welcome. New `four-jet' analyses show a reduced scale dependence, as anticipated, which helps circumvent the problem. Hadronization effects also continue to pose a problem; power-law treatments show promise in dealing with these effects, but there appear to be problems with their description of the distributions of event shape variables (especially those involving the division of the event into hemispheres). Accurate combination of results is still the subjects of much study. However, a recent compilation of results is shown in Figure 3. The running of α_s is illustrated in Figure 4, which shows a recent combination of extracted couplings from event shapes (done by the LEP QCD Working Group), of the α_s from the Z line shape (from the LEP Electroweak Working Group), and the combination of the α_s values determined from τ-decays and from scaling violations [2].

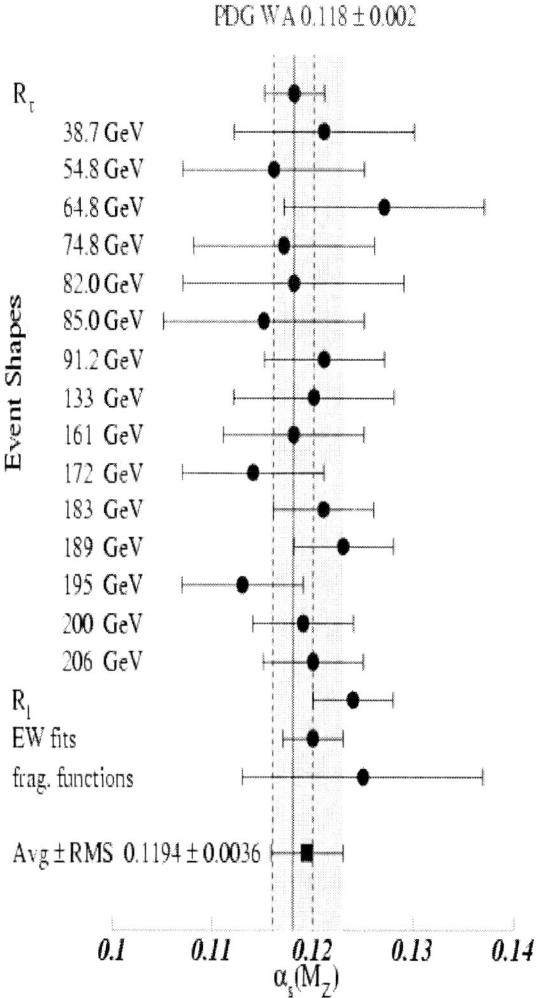

FIGURE 3. A representative selection of α_s values from LEP. A consistent treatment of the theoretical errors is only available for subsets at present, for example the event shape results. However, the RMS scatter may be taken as a broad indication of the overall uncertainties.

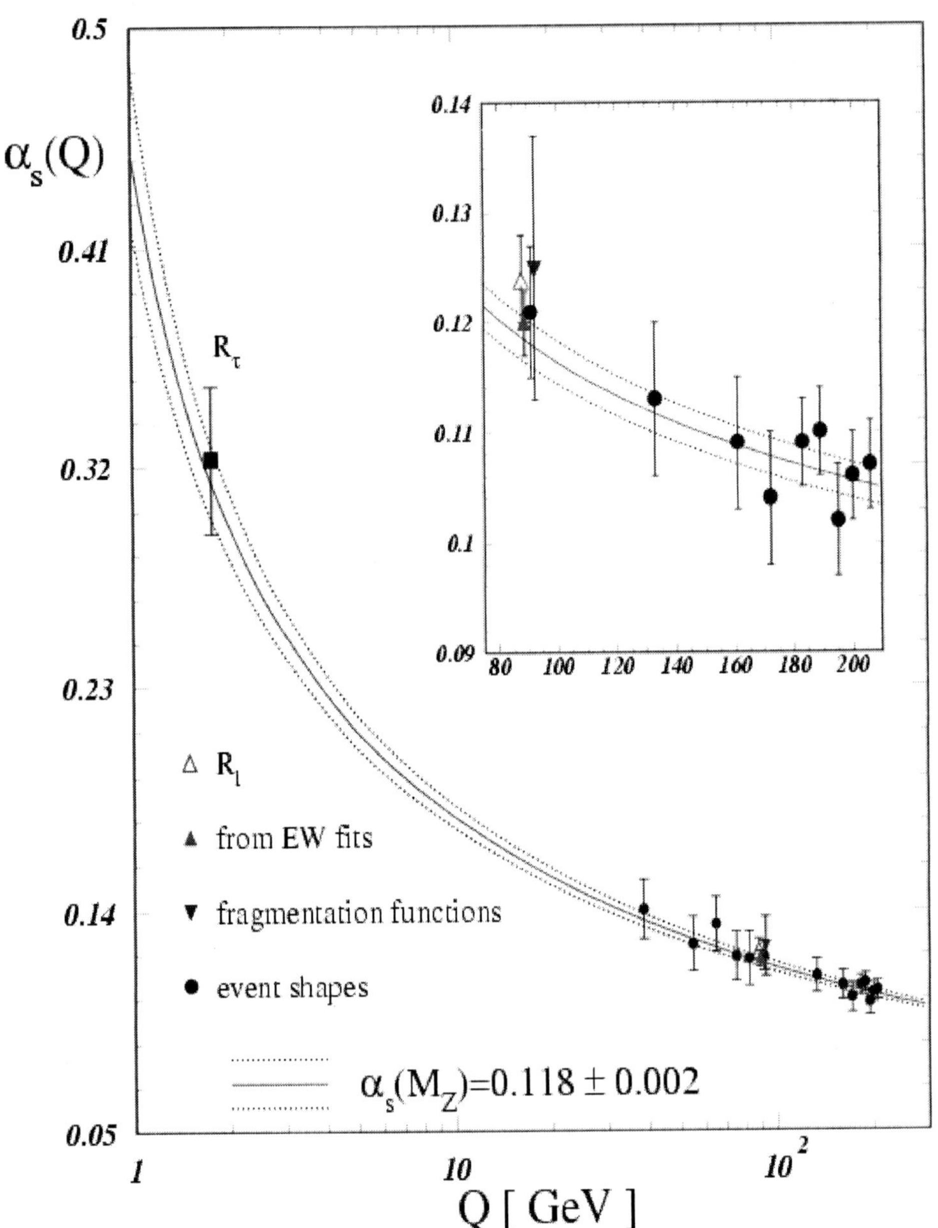

FIGURE 4. The inset figure shows the LEP combined α_s from event shapes. The values from the Z line shape and from scaling violations are also shown. The larger figure shows the same results along with the value determined from τ-decays. Curve and error are the predicted result from the quoted input, not the result of a fit to the data.

ACKNOWLEDGMENTS

I would like to thank the members of the LEP QCD Working Group for their help with this talk, and in particular Guenther Dissertori. As always, I wish to thank the LEP operations group for their unstinting efforts to produce superlative accelerator conditions over the decade of LEP running. Finally, I thank the organizers of the workshop for the invitation to a stimulating and productive meeting.

REFERENCES

1. Groom, D. E. et al., *Eur. Phys. J* **C15**, 1, 2000.
2. Bethke, S., *J. Phys.* **G26**, R27 (2000).
3. The LEP Electroweak Working Group, *CERN* **EP/2000-016**, 2000.
4. OPAL Collaboration, Acton, P. *et al., Z. Phys.* **C55**, 1, 1992.
5. DELPHI Collaboration, *DELPHI 2001-059, conf.* 487.
6. Burrows, P, *et al.*, *Phys. Lett*. **B382**, 157, 1996.
7. Sjöstrand, T., *Comp. Phys. Comm.* **67**, 74, 1994.
8. Marchesini, I. *et al., Comp. Phys. Comm.* **67**, 163, 1992.
9. Lönnblad, L., *Comp. Phys. Comm.* **71**, 15, 1992.
10. Kluth, S., *arXiv:hep-ex/0104016,* 2001.
11. Salam, G.P., and Wicke, D., *arXiv:hep-ph/0102313,* 2001.
12. OPAL Collaboration, Abbendi, G. *et al., CERN* **EP/2001-001**, 2001.
13. ALEPH Collaboration, *conf. note ALEPH* **2001-007,** *CONF* **2001-004**, 2001.
14. *DELPHI Collaboration, DELPHI 2001-060, conf 489.*

Multi-jet event shapes in QCD hard processes [1]

A. Banfi*, G.Marchesini *, G. Smye* and G. Zanderighi†

*Università degli Studi di Milano-Bicocca and INFN, Sezione di Milano, Italy
†Università degli Studi di Pavia and INFN, Sezione di Pavia, Italy

Abstract. We present recent results on the perturbative and non-perturbative QCD analysis of three-jet event shapes in the near-to-planar region in various QCD hard processes.

INTRODUCTION

Event shape variables constitute one of the most attractive fields in QCD. As long as they are collinear and infrared safe (CIS), they can be computed using perturbative (PT) techniques and used to measure α_s. For a number of two-jet observables, it is also possible to estimate the $1/Q$ suppressed non perturbative (NP) corrections, which affect both the mean values and the distributions. Experimental analyses have shown that they are consistent with the 'universality' hypothesis, i.e. they can be expressed in terms of only one NP parameter α_0, and relative coefficients from one observable to the next can be perturbatively computed. The aim of our work is to extend the notion of universality of confinement effects to multi-jet event shapes. This requires a PT treatment of these observables at the same accuracy achieved in the study of two-jet shape variables and the computation of leading power corrections. After a brief review of the known results for event shapes in e^+e^- annihilation, we discuss how these results can be extended to near-to-planar three-jet shapes in hadron-hadron and lepton-hadron collisions. This will open the way to the study of further multi-jet observables.

EVENT SHAPES IN E^+E^- ANNIHILATION

Two-jet event shapes (such as thrust T, broadening B, heavy-jet mass M_H and C-parameter) have been intensively studied both theoretically and experimentally.

The state-of-the-art level of PT analysis of two-jet observables consists of [1]

- resummation of all double- (DL) and single-logarithmic (SL) enhanced contributions due to multiple radiation of soft and collinear partons,
- matching of the resummed result with the exact fixed order result.

---

[1] Talk given by Giuseppe Marchesini.

CP602, *QCD@Work: International Workshop on Quantum Chromodynamics*
edited by P. Colangelo and G. Nardulli

To reach such an accuracy one needs

- soft matrix elements at two-loop order and reconstruction of the running coupling at the proper scale,
- factorization and exponentiation of observable and kinematical constraints,
- proper treatment of hard intra-jet and soft inter-jet radiation.

Besides the pure PT contributions, experimental data (see for instance [2]) have revealed the presence of power suppressed $1/Q$ corrections (Q is the e^+e^- center-of-mass energy), which emerge as a shift both to mean values and differential distributions. It is widely believed [3] that such contributions are of NP origin and arise from the running of the coupling into the infrared domain.

A systematic way to deal with power corrections is provided by the dispersive approach [4], in which a prescription is given to extend the QCD coupling into the confinement region. In terms of this dispersive coupling, the typical large distance contribution to an event shape V can be factorized in an observable dependent coefficient c_V and a universal NP parameter α_0. More precisely, in any 'linear' shape variable V the dependence on the rapidity η_i and transverse momentum k_{ti} of emitted hadron i factorizes,

$$V = \sum_i k_{ti} f_V(\eta_i), \tag{1}$$

so that the NP contribution to V becomes

$$\delta V = C_F\, c_V\, \lambda^{\mathrm{NP}}, \qquad c_V = \int d\eta\, f_V(\eta),$$
$$\lambda^{\mathrm{NP}} = \frac{\mu_I}{Q}\frac{4}{\pi^2}\left(\alpha_0(\mu_I) + O(\alpha_s(Q))\right), \qquad \alpha_0(\mu_I) = \frac{1}{\mu_I}\int_0^{\mu_I} dk\, \alpha_s(k). \tag{2}$$

The NP parameter α_0 is the average of the dispersive coupling in the region $k \leq \mu_I$ and measures the interaction strength in the confinement region. The $O(\alpha_s(Q))$ piece in (2) is needed in order to merge PT and NP results in a renormalon free manner and ensures that the final answer is independent of the infra-red matching scale μ_I. The NP parameter $\alpha_0(2\mathrm{GeV})$ has been measured and appears to be consistent with the universality hypothesis within a 10-20% accuracy, as shown in Fig.1.

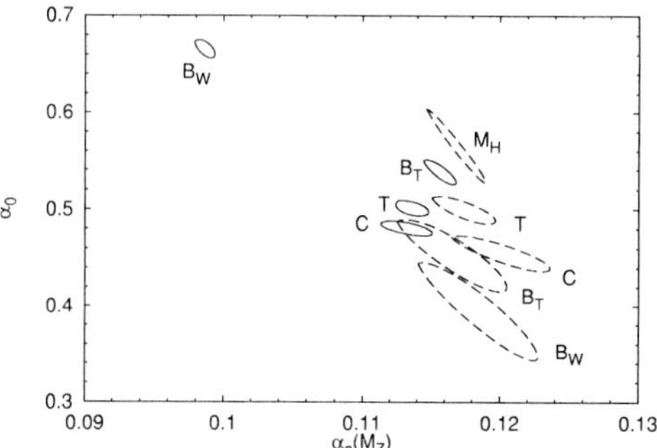

FIGURE 1. 2-σ contours for fits to various two-jet observables [5]. Solid curves indicate fits to distributions, while dashed lines indicate fits to mean values.

Only very recently these techniques have been extended to the case of three-jet observables in the near-to-planar region. In particular the mean values and distributions of the thrust minor T_m [6] and the D-parameter [7] have been computed. Both of them are a measure of QCD radiation out of the event plane. This study has revealed an unexpectedly rich geometry and colour dependence both of the PT result and of the NP corrections. The crucial point is the rapidity dependence of the observables. Actually, T_m is independent of rapidity, while D is exponentially damped.

On the PT side, this implies that hard parton recoil affects the observable in the T_m case, while it does not contribute to D. Therefore the T_m distribution is sensitive not only to the underlying hard event geometry (the angles between the jets), but also on its colour configuration through the kinematical constraints which define the event plane.

As far as NP corrections are concerned, one finds that the shift to the D distribution is simply geometry dependent, while the T_m distribution is also squeezed, since the shift depends logarithmically on T_m. This behaviour (already encountered for the B in two-jet events) results from a complicated interplay between PT and NP effects.

The same physical features appearing in the NP shift to the T_m distribution are also present in hadron-hadron and lepton-hadron collisions, as we are going to discuss in the following.

K_{OUT} IN HADRON-HADRON COLLISIONS

We consider hadron-hadron collisions and study an observable similar to T_m [8]. We select events in which a Z_0 is produced with large transverse momentum. We can then define an 'event plane' as the one containing the Z_0 momentum and the beam axis and introduce a measure of out-of-plane radiation

$$K_{out} = \sum_h{}' |p_h^{out}|\,. \qquad (3)$$

Here p_h^{out} is the out-of-plane momentum of the hadron h and the sum extends to all hadrons with rapidity in the range $|\eta_h| > \eta_0$, in order to avoid measurements in the beam region. At Born level one has two incoming partons p_1 and p_2 and an additional hard parton p_3 recoiling against the vector boson q (see fig. 2a).

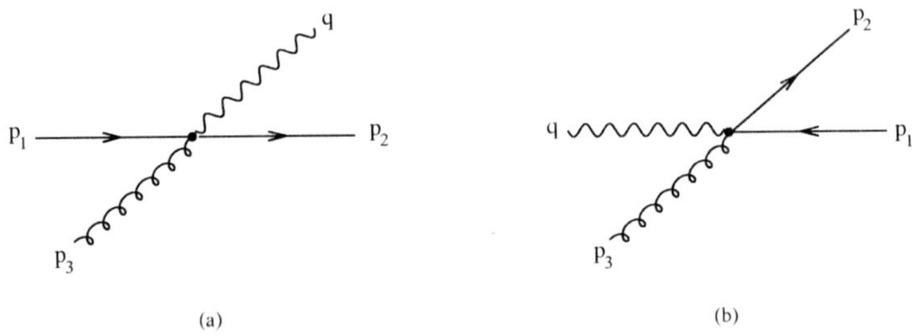

FIGURE 2. A particular Born configuration (a) for the process $p\bar{p} \to Z_0 + jet$ (b) for lepton-hadron scattering.

The event-plane definition gives rise to the conservation law

$$p_3^{out} + \sum_i k_i^{out} = 0\,, \tag{4}$$

so that only p_3 can take an out-of-plane recoil, while the remaining hard momenta are fixed in the event plane.

The main difference between hadronic and e^+e^- collisions is the presence of initial state radiation. Its contribution has to be factorized in order to make our observable CIS and reconstructs the proper hard scale for the incoming parton distributions, which turns out to be essentially K_{out}.

The PT distribution turns out to be (roughly) the product of the initial state parton distributions $P_{inc}(K_{out})$ with a radiation factor

$$\Sigma(K_{out}) \sim P_{inc}(K_{out}) \cdot e^{-R(K_{out})} \cdot S(R')\,. \tag{5}$$

The radiator R is the same as that of T_m, it resums all DL contributions and accounts for SL effects due to hard intra-jet and soft inter-jet radiation. It is given in terms of three hard parton 'antennae', each one proportional to the colour charge C_a of emitting parton p_a (C_a equals C_F for a quark and C_A for a gluon)

$$R(K_{out}) = \sum_{a=1}^{3} C_a \int_{K_{out}}^{Q_a} \frac{dk}{k} \frac{2\alpha_s(2k)}{\pi} \ln\frac{Q_a}{2k} \simeq \sum_a C_a \frac{\alpha_s}{\pi} \ln^2 \frac{Q_a}{2K_{out}}\,. \tag{6}$$

The hard scales Q_a are determined essentially by soft radiation at large angles, which is a characteristic of multi-jet observables. In particular, apart from a factor due to hard collinear splitting, the scale for the quarks is the invariant mass of the $q\bar{q}$ system, for the gluon is its invariant transverse momentum with respect to the $q\bar{q}$ pair.

The SL function S depends on the logarithmic derivative of the radiator R' and accounts for all effects coming from multiple secondary radiation.

The NP correction is proportional to the same parameter $\lambda^{\rm NP}$ which enters the power corrections to e^+e^- event shapes in (2) and is given by

$$\delta K_{out} = \frac{2}{\pi}\lambda^{\rm NP}\left(C_1(\eta_0 - \eta_3) + C_2(\eta_0 + \eta_3) + C_3 \ln\frac{Q_t}{|p_3^{out}|}\right)\,, \tag{7}$$

where Q_t is the transverse momentum of the emitted boson. Such a shift arises from integration over the rapidity of a emitted gluons (see eq.(2)). Since K^{out} is independent of rapidity, one has to carefully consider the effective rapidity cutoff. For radiation from the incoming partons p_1 and p_2 it is given by the distance between η_3 (the rapidity of p_3) and the experimental resolution η_0. On the contrary, for a NP gluon emitted from p_3, it is the hard parton recoil momentum p_3^{out} which provides the needed cutoff. Moreover, since p_3 always takes recoil, one has $p_3^{out} \sim K_{out}$. This interplay between PT and NP emissions makes the shift logarithmically dependent on K_{out}.

Unfortunately, this is not the end of the story. In hadronic collisions one has to add a soft contribution due to the beam remnant interactions. This is the so-called 'soft underlying event', which was introduced for the first time in the study of the 'pedestal height' in hadronic jet production [9]. Therefore the analysis of K_{out} is also important to understand the physics of soft collisions.

K_{OUT} IN LEPTON-HADRON COLLISIONS

Even in DIS one can define an event plane and measure the out-of-plane radiation. In the Breit frame, we define the thrust major T_M in analogy with e^+e^- annihilation

$$T_M Q = \max_{\vec{n}_M \cdot \vec{n}} \sum_h{}' |\vec{p}_h \cdot \vec{n}_M| \,, \tag{8}$$

where $\vec{n}$ and $Q^2 = -q^2$ are the Breit axis and the virtuality of the exchanged boson q respectively. As in the previous case, the sum extends to all hadrons not in the beam direction (with $\eta > \eta_0$). At Born level, we have one incoming parton p_1 which is struck by a vector boson q and produces two hard large angle partons p_2 and p_3 recoiling one against each other (see fig. 2b). We define K_{out} again as the cumulative out-of-event-plane momentum, where the event plane is the one containing $\vec{n}$ and the T_M axis.

The event-plane definition implies the following kinematical constraint

$$p_2^{out} + \sum_{i \in U} k_i^{out} = p_3^{out} + \sum_{i \in D} k_i^{out} \,, \tag{9}$$

with U (D) the region containing parton p_2 (p_3).

The PT K_{out} distribution has the same form as (5), with $P_{inc}(K_{out})$ the structure function of the incoming parton, again at the scale K_{out}. The radiator turns out to be the same as in e^+e^- and in hadron-hadron collisions, due to universality of soft and collinear radiation. However the exact functional form of the SL function S differs between these, due to different event-plane kinematics.

The NP shift is similar to the one in (7)

$$\delta K_{out} = \frac{2}{\pi} \lambda^{\mathrm{NP}} \left(C_1 \ln \frac{Q_1^{\mathrm{NP}} e^{\eta_0}}{Q} + C_2 \ln \frac{Q_2^{\mathrm{NP}}}{|p_2^{out}|} + C_3 \ln \frac{Q_3^{\mathrm{NP}}}{|p_3^{out}|} \right) , \tag{10}$$

with Q_a^{NP} the proper NP hard scales, which are proportional to the ones entering the PT radiator. The crucial difference is that here both p_2 and p_3 are not fixed in the event plane. This implies that the way the NP correction in (10) affects the PT distribution is very different according to which phase space region is considered. Namely one has the two regimes

- $\alpha_s \ln^2 K_{out}/Q \gg 1$: one has well developed QCD radiation, so that all p_a take recoil and $p_a^{out} \sim K_{out}$. This gives rise to a logarithmically enhanced shift

$$\delta K_{out} \sim C_1 \ln \frac{Q_1^{\mathrm{NP}} e^{\eta_0}}{Q} + C_2 \ln \frac{Q_2^{\mathrm{NP}}}{K_{out}} + C_3 \ln \frac{Q_3^{\mathrm{NP}}}{K_{out}} \,, \tag{11}$$

- $\alpha_s \ln^2 K_{out}/Q \ll 1$: radiation from one hard parton dominates. According to the event-plane kinematics in (9), when PT radiation comes from the U region (which happens with probability $(\frac{1}{2}C_1 + C_2)/(2C_F + C_A)$), only p_2 takes recoil, so that that $p_2^{out} \sim K_{out}$, while $p_3^{out} \ll p_2^{out}$. The part of the shift proportional to C_2 gets logarithmically enhanced as in the previous case, while the one proportional to C_3 gives rise to a very singular contribution proportional to $1/\sqrt{\alpha_s}$, coming from the average of $\ln Q/p_3^{out}$ over its DL Sudakov form factor $\exp\left\{-(\frac{1}{2}C_1+C_3)\frac{\alpha_s}{\pi}\ln^2 Q/p_3^{out}\right\}$. This gives

$$\delta K_{out} \sim C_1 \ln\frac{Q_1^{\mathrm{NP}} e^{\eta_0}}{Q} + C_2 \ln\frac{Q_2^{\mathrm{NP}}}{K_{out}} + C_3 \frac{\pi}{2\sqrt{(\frac{1}{2}C_1+C_3)\alpha_s}}. \qquad (12)$$

The converse argument holds for an emission in the D region.

CONCLUSIONS

We have now a promising theoretical method to deal with multi-jet observables. Not only are we able to identify all sources of SL contributions, but we can also relate the NP $1/Q$ power-suppressed corrections with the ones which affect e^+e^- two-jet shapes. A next step will be the extension of the above results to event shapes involving any number of jets. We hope that this analysis will greatly improve the understanding of confinement physics, especially at hadron colliders.

ACKNOWLEDGMENTS

We are grateful to Yuri Dokshitzer, Gavin Salam and Bryan Webber for discussions and suggestions.

REFERENCES

1. S. Catani, L. Trentadue, G. Turnock and B.R. Webber, *Nucl. Phys.* B**407** (1993) 3;
   S. Catani and B.R. Webber, *Phys. Lett.* B**427** (1998) 377 [hep-ph/9801350];
   S. Catani, G. Turnock and B.R. Webber, *Phys. Lett.* B**295** (1992) 269;
   Yu.L. Dokshitzer, A. Lucenti, G. Marchesini and G.P. Salam, *JHEP* **01** (1998) 011 [hep-ph/9801324].
2. P.A. Movilla Fernández, O. Biebel and S. Bethke, hep-ex/9807007; *Phys. Lett.* B**459** (1999) 326 [hep-ex/9903009].
3. G.P. Korchemsky and G. Sterman, *Nucl. Phys.* B**437** (1995) 415 [hep-ph/9411211];
   M. Beneke and V.M. Braun, *Nucl. Phys.* B**454** (1995) 253 [hep-ph/9506452];
   Yu.L. Dokshitzer and B.R. Webber, *Phys. Lett.* B**352** (1995) 451 [hep-ph/9504219];
   R. Akhoury and V.I. Zakharov, *Phys. Lett.* B**357** (1995) 646 [hep-ph/9504248]; *Nucl. Phys.* B**465** (1996) 295 [hep-ph/9507253];
   Yu.L. Dokshitzer, V.A. Khoze and S.I. Troyan, *Phys. Rev.* D**53** (1996) 89 [hep-ph/9506425].
4. Yu.L. Dokshitzer, G. Marchesini and B.R. Webber, *Nucl. Phys.* B**469** (1996) 93 [hep-ph/9512336].
5. G.P. Salam, G. Zanderighi, *Nucl. Phys. Proc. Suppl.* **86** (2000) 430 [hep-ph/9909324].
6. A. Banfi, Yu.L. Dokshitzer, G. Marchesini and G. Zanderighi, *JHEP* **07** (2000) 002 [hep-ph/0004027]; *JHEP* **03** (2001) 007 [hep-ph/0101205]; *Phys. Lett.* B**508** (2001) 269 [hep-ph/0010267].
7. A. Banfi, Yu.L. Dokshitzer, G. Marchesini and G. Zanderighi, *JHEP* **05** (2001) 040 [hep-ph/0104162].
8. A. Banfi, G. Marchesini, G. Smye and G. Zanderighi, *JHEP* **08** (2001) 047 [hep-ph/0106278].
9. G. Marchesini and B.R. Webber, *Phys. Rev.* D**38** (1988) 3419.

Measurements using Multijet Final States from Hadronic Decays of the Z Boson.

H.Jeremie* for the OPAL collaboration

**Université de Montréal, département de physique, CP 6128, MONTREAL H3C 3J7, Canada*

Abstract. We report on two measurements using multijet final states from hadronic decays of the Z^0 obtained with the OPAL detector at LEP.

In the first analysis we extracted the QCD colour factors C_A and C_F together with $\alpha_s(M_z)$ from the comparison of the data to the next-to-leading order predictions of four-jet angular correlations and matched next-to-leading order and next-to-leading logarithmic approximations for multi-jet rates. The measurements are in agreement with SU(3) expectations for C_A and C_F and the world average of $\alpha_s(M_z)$.

In the second analysis the four-jet angular correlations alone are used to investigate the colour factor ratio T_R/C_F, obtained by comparing the data to leading (=2nd) order matrix models. In particular we investigated the distribution of this quantity as a function of the invariant mass of the two least energetic jets. Some features of this distribution like the appearance of a maximum around 20 GeV are qualitatively reproduced by hybrid second order matrix element models coupled with a parton shower mechanism.

INTRODUCTION

The QCD colour factors C_F, C_A and T_F have been subject of various measurements, since they influence the shape of the angular correlations between the jets of a four-jet event (refs. [1, 2]). In particular the value of $T_R = n_f \cdot T_F$, with n_f the number of flavours, has been investigated since its value might be indicative of an increase of the number of effective flavours and hence of the existence of a supersymmetric gluino [3]. Beyond leading order in the theoretical predictions, the strong coupling α_s and the colour factors are interdependent and one has to do a simultaneous measurement of the colour factors and the strong coupling by fitting simultaneously angular correlations and multi-jet rates.

In part 1 of this report we will describe a measurement where for the first time NLO predictions [4, 5] for the four-jet correlations are used in conjunction with the 3-jet and the four-jet rate.

Another theoretical advance is the development of hybrid second-order perturbative (matrix element) calculations, where each of the four produced partons evolves into a parton shower similar to the two- and three-parton Monte Carlo programs already widely in use (refs.[6, 7]). In part 2 of this report we will describe comparisons between such a model and angular correlations by investigating its effect on measurements of T_R. In particular we will investigate the dependency of T_R on the invariant mass of the system of the two least energetic jets.

CP602, *QCD@Work: International Workshop on Quantum Chromodynamics*
edited by P. Colangelo and G. Nardulli

PART 1: A SIMULTANEOUS MEASUREMENT OF THE QCD COLOUR FACTORS AND THE STRONG COUPLING

Observables

We group the emitted particles of a hadronic event into jets using the Durham scheme [8, 9], with jet resolution parameters $y^{n,n+1}$ which separate for each event the region where it is considered either to be a n-jet or a n+1-jet event. It is customary to specify a parameter y_{cut} such that $y^{n,n+1} < y_{\text{cut}} < y^{n-1,n}$ if one is interested in n-jet events.

To perform a simultaneous measurement of the strong coupling and the colour factors we use

a.) the differential two-jet rate, $D_2(y^{23}) = \frac{d\sigma}{dy^{23}}/\sigma_{\text{tot}}$ where y^{23} is the y_{cut} value for which the two- and three-jet configurations are separated in a given event, b.) the four-jet rate, $R_4(y_{\text{cut}}) = \sigma_{4-\text{jet}}(y_{\text{cut}})/\sigma_{\text{tot}}$, c.) the following angular correlations between the jets of a four-jet event: the Bengtsson-Zerwas angle, the modified Nachtmann-Reiter angle, the Körner-Schierholtz-Willrodt angle and the angle between the two lowest energy jets [10]. We imposed on the four jets the conditions: $E_1 > E_2 > E_3 > E_4$, $y_{\text{cut}} = 0.008$, and $E > 3\,\text{GeV}$.

Analysis Procedure and Results

This analysis relies mainly on the reconstruction of charged particle trajectories and on the measurement of energy deposited in the electromagnetic and hadronic calorimeters of the OPAL detector.

It is based on 4.1 million hadronic events recorded within 3 GeV of the Z peak by the OPAL detector between 1991 and 1995. The details of track and event-selection can be found in ref.[2]. The final event sample contained about 250 000 four-jet events.

For the analysis we first combined the distributions of the six variables $\cos\chi_{\text{BZ}}$, $\cos\Theta_{\text{NR}}, \cos\Phi_{\text{KSW}}, \cos\alpha_{34}, D_2$ and R_4 into a one-dimensional distribution, with 128 bins, 20 for each angular correlation and 24 each for the four-jet rates (covering the range of $0.001 < y_{\text{cut}} < 1.0$) and the differential two-jet rates (in the range of $1.0 < -\ln y_{23} < 5.8$). Next, using computed Monte Carlo events, we applied standard bin-by-bin corrections for detector distortions and hadronization, in order to obtain distributions corrected to the parton level to be compared with the theoretical predictions.

Having prepared the corrected distribution of six variables, we then performed a χ^2 minimization to determine the most probable values of the variables $\eta = \alpha_s \text{C}_\text{F}/(2\pi)$, $x = \text{C}_\text{A}/\text{C}_\text{F}$ and $y = \text{T}_\text{F}/\text{C}_\text{F}$ (see ref.[2]) with the program MINUIT [11].

The central values of the fit results for the parameter η and colour factor ratios x and y are listed in the upper section of Table 1. The renormalization scale was fixed at $x_\mu = 1$.

We observe a strong correlation between x and y, it was taken into account when the fit results were converted to the standard QCD parameters $\alpha_s(M_Z)$ and colour factors, using $\text{T}_\text{F} = 1/2$. These latter results are given in the lower section of Table 1.

TABLE 1. Results of the default fit. The errors and correlations are statistical only.

$\eta = \alpha_s C_F/(2\pi)$	0.0256±0.0003
$x = C_A/C_F$	2.25±0.08
$y = T_F/C_F$	0.37±0.04
$\rho_{\eta x}$	-0.33
$\rho_{\eta y}$	-0.11
ρ_{xy}	0.90
χ^2/d.o.f.	98.5/79
α_s	0.120±0.011
C_A	3.02±0.25
C_F	1.34±0.13

Systematic uncertainties

The errors obtained from the different systematic checks are summarized in Table 2. The theoretical uncertainty refers to the larger of the errors from the renormalization scale uncertainty of the jet-related variables, the matching ambiguity (both of those estimate the effect of unknown higher order perturbative contributions), and the scale uncertainty of the four-jet angular variables. The different sources were added in quadrature to define the total systematic error. The errors are dominated by the theoretical uncertainty.

TABLE 2. Contributions to the total systematic error.

	$\Delta\eta$	Δx	Δy	$\rho_{\eta x}$	$\rho_{\eta y}$	ρ_{xy}
Detector effects	0.0000	0.004	0.002	-0.87	-0.89	0.95
Hadronization	0.0003	0.056	0.021	-0.86	-0.71	0.92
Theoretical uncertainty	0.0013	0.096	0.038	-1.0	1.0	-1.0
Fit range	0.0001	0.033	0.013	-1.0	-1.0	1.0
5-parton background	0.0001	0.070	0.039	-1.0	-1.0	1.0
y_{cut}	0.0000	0.005	0.002	1.0	1.0	1.0
Total systematic uncertainties	0.0013	0.136	0.060	-0.83	0.06	0.38
	$\Delta\alpha_s$	ΔC_A	ΔC_F			
Total systematic uncertainties	0.020	0.49	0.22			

Summary of Part 1

A test of perturbative QCD at LEP has been done by comparing next-to-leading order perturbative predictions for four-jet angular variables, and the resummation improved next-to-leading order perturbative predictions for the multi-jet rates, to OPAL data taken at the Z peak, with the following results: $C_A = 3.02 \pm 0.25(\text{stat.}) \pm 0.49(\text{syst.})$, $C_F = 1.34 \pm 0.13(\text{stat.}) \pm 0.22(\text{syst.})$ and $\alpha_s = 0.120 \pm 0.011(\text{stat.}) \pm 0.020(\text{syst.})$, in agreement with the world average. The existence of a light gluino is disfavoured with more than 95% confidence.

PART 2: A STUDY OF FOUR-JET EVENTS

Observables

Here we use similar quality criteria for event selection and the same jetfinding technique as in part 1, but the selection of the four-jet events is different: we used $y^{34} > 0.015$, $y^{45} < 0.006$, $E_2 > E_3 + 0.5\sqrt{E_3}$ and $\theta_{12}, \theta_{34} < 160°$. With these cuts we obtain 4.4×10^4 energy ordered four-jet events.

The theoretical distributions we used for the analysis were obtained from the second-order ($O(\alpha_s^2)$) QCD matrix element calculation by Ellis, Ross and Terrano (ERT) [12], as implemented in the JETSET simulation package [13], which includes hadronization of the partons. These events were then passed through the OPAL detector simulation, allowing us to perform our analysis directly at the detector level.

More details about this and other sections can be found in ref. [14]

Analysis Procedure

Here we concentrate on only one of the QCD colour factors, namely T_R, which is approximately proportional to the fraction f_q of four-quark events contained in the event sample. We quote our fit results as R_{4q}, the ratio of the experimentally fitted fraction of four-quark events to the value predicted by the ERT model. [1]

Global Results

The result of a fit at the detector level, which optimizes the agreement between data and second order theory by varying f_q, using all the available events, is $R_{4q} = 1.28 \pm 0.12(\text{stat})^{+0.97}_{-0.95}(\text{syst})$, χ^2 per d.o.f = 0.61. Assuming five effective flavours, this corresponds to $T_F/C_F = 0.49 \pm .43$.

In table 3 we show the differences between the default value of R_{4q} and the values extracted from the fit, for the various systematic uncertainties studied. Only variations larger than 0.1 are shown in this table [2].

[1] It can happen that the data differ so much from the model that the fitted fraction can actually become negative, which has no meaning other than a discrepancy between shapes of distributions.

[2] The cut $\theta_{12}, \theta_{34} < 160°$ was not applied in the case of the N-R correlation

TABLE 3. Systematic errors of R_{4q}.

cut	ΔR_{4q}	χ^2/d.o.f	% change in number of events
a.) $y^{34} > 0.018$	+0.20	0.79	−28
b.) $E_2 > E_3 + \sqrt{E_3}$	−0.14	1.13	−27
c.) without first bin of χ_{BZ}	−0.11	0.28	−6
d.) Nachtmann-Reiter correlation	−0.75	0.60	+32
e.) $y^{45} < 0.012$	+0.42	0.66	+24
f.) only tracks and EM-clusters	−0.16	0.72	
g.) correction to parton level	+0.12	0.71	
h.) fragmentation	−0.28	0.65	
i.) statistical uncertainty	±0.12		
total systematic uncertainty	$^{+0.97}_{-0.95}$		
total error	$^{+0.98}_{-0.96}$		

Results as a Function of the Invariant Mass of Jets Three and Four

The measured value of $1.28^{+0.98}_{-0.96}$ for R_{4q} is compatible with unity within the errors. We now proceed to calculate the Bengtsson-Zerwas distribution separately for different bins of m_{34}, in steps of 3 GeV, for data, ERT Monte Carlo $q\bar{q}q\bar{q}$ and $q\bar{q}gg$ events. For each bin we optimize the agreement between data and Monte Carlo by varying the corresponding four-quark fraction, extracting thus a value of f_q^{exp} for each bin. The corresponding ratios R_{4q} of experimental to theoretical f_q are shown in fig.1.

Variations of the apparent R_{4q} as a function of m_{34}, with values both larger and smaller than one are observed, with a maximum at around 20 GeV. The two dash-dotted histograms in correspond to distributions with the minimum and maximum observed value of the global R_{4q}, namely systematic errors "d" and "e" of table 3, and are therefore indicative of the possible variations of these distributions. The full line represents a fit where events calculated with the JETSET hybrid model combining matrix elements with parton showers, also incorporated into the package of ref.[13], were analyzed in the same way as the data. Although generally higher than the data, this curve also features a maximum in the same range of invariant masses as the data. The dotted line in fig.1 corresponds to a mixture of 4- and 5-parton events (69% vs 31%), where the five-parton events were calculated with the program of ref.[15], and the trend is again to higher values of R_{4q} than the data.

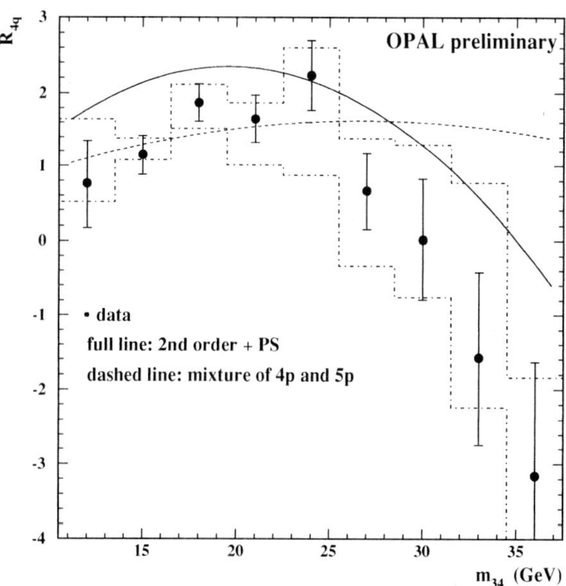

FIGURE 1. R_{4q} for data (full circles) and some theoretical predictions (errors statistical). The two dash-dotted histograms correspond to $y^{45} < 0.012$ (upper histogram) and the N-R correlation (lower histogram). The full line represents the JETSET hybrid matrix element model, while the dashed line represents a mixture of four- and five-parton events (69% vs 31%). The ERT model was used as theoretical reference everywhere, and all distributions were obtained directly from the detector level.

Summary of Part 2

The ERT [12] second-order matrix element calculation coupled with the LUND string fragmentation model has been compared with data using four-jet correlations from hadronic decays of the Z^0. From a global comparison we derive a value of $1.28^{+0.98}_{-0.96}$ for the ratio of the apparent four quark content of the data over the model, with larger systematic uncertainties than have been previously quoted. The main effects contributing to these larger errors are five- and more-parton events reconstructed as four jets, and the difference between results obtained using the Bengtsson-Zerwas and the modified Nachtmann-Reiter [10] correlations.

Differences in shape between the Bengtsson-Zerwas correlations of the data and the second-order theory have been observed as a function of m_{34}, the invariant mass formed by the system of the two least energetic jets. These differences, expressed by the value of $R_{4q}(m_{34})$, exhibit a maximum in the region of 20 GeV and become negative at higher values of m_{34}. Some of these features can also be qualitatively observed in models using higher order effects such as a combination of matrix element with a parton shower.

REFERENCES

1. ALEPH collaboration, D.Decamp et al., Phys. Lett. **B 284** (1992) 151; DELPHI collaboration, P.Abreu et al.,Phys. Lett. **B 255** (1991) 466;
   DELPHI collaboration, P.Abreu et al., Zeit. f. Phys. **C 59** (1993) 357;
   DELPHI collaboration, P.Abreu et al., Phys. Lett. **B 414** (1997) 401;
   L3 collaboration, L3 note 1805, contributed paper eps0119 to the HEP95 EPS conference in Brussels, 27 July-2 August;
   OPAL collaboration, contributed paper eps0320 to the HEP95 EPS conference in Brussels, 27 July-2 August, OPAL physics note PN 176;
   OPAL collaboration, R. Akers et al., Zeit. f. Phys. **C 68** (1995) 519;
   OPAL collaboration, R. Akers et al., Zeit. f. Phys. **C 65** (1995) 367;
   ALEPH collaboration, R.Barate et al., Zeit. f. Phys. **C 76** (1997) 1.
2. OPAL collaboration, G.Abbiendi et al., CERN-EP-2001-001, Eur. Phys. J. C20 (2001) 601;
3. G. Farrar, Phys. Rev. D **51**, 3904 (1995); hep-ph/9707467 .
4. Adrian Signer, Comp. Phys. Comm. **106** (1977) 125;
   A.Signer and L.Dixon, Phys. Rev. Lett. **78** (1997) 811;
   L.Dixon and A.Signer, SLAC-PUB-7528, hep-ph/9706285.
5. Z.Trocsányi and Z.Nagy, Phys. Rev.**D 57** (1998) 5793;
   Z.Trocsányi and Z.Nagy, Phys. Rev. **D 59** (1999) 014020.
6. J.André and T.Sjöstrand, Phys. Rev. **D 57** (1998) 5767.
7. G.Corcella et al., CAVENDISH-HEP-99-17;
   G.Corcella and M.H.Seymour, Phys. Lett. **B 442** (1998) 417.
8. S. Catani, Yu.L. Dokshitzer, M. Olsson, G. Turnock and B.R. Webber, Phys. Lett. B **269**, 432 (1991).
9. S. Catani, in QCD at 200 TeV, Proc. 17th INFN Eloisatron Project Workshop, ed.: L. Cifarelli and Yu.L. Dokshitzer, Plenum Press, New York (1992).
10. M. Bengtsson and P.M. Zerwas, Phys. Lett. B **208**, 306 (1988);
    O. Nachtmann and A. Reiter, Zeit. Phys. C **16**, 45 (1982);
    J.G. Körner, G. Schierholz and J. Willrodt, Nucl. Phys. B **185**, 365 (1981);
    DELPHI Collaboration, P. Abreu *et al.*, Phys. Lett. B **255**, 466 (1991).
11. F. James and M. Roos, Computer program MINUIT, CERN program library, writeup CERN D506 (1989).
12. R.K.Ellis, D.A. Ross and A.E. Terrano, Nucl. Phys. **B 178** (1981) 421.
13. T.Sjöstrand, Comp. Phys. Comm. **39** (1986) 347;
    T.Sjöstrand and M.Bengtsson, Computer Phys. Comm. **43** (1987) 367.
14. Opal collaboration, Physics Note 473, June 2001.
15. K.Hagiwara and D.Zeppenfeld, Nucl. Phys. **B 313** (1989) 560 ;
    F.Wäckerle, Diplomarbeit Karlsruhe 1994, IEKP-KA/93-19.

Measurements of Color Transparency

Steven Heppelmann [1]

Penn State University, University Park, Pa. 16801,USA

Abstract. We present new results on the transparency of the carbon nucleus for (p,2p) quasi-elastic scattering. The new data are obtained with beam momenta ranging from 5.9 to 14.5 GeV/c, corresponding to momentum transfer Q^2 ranging from about 5 to 15 GeV^2. The new result confirms the surprising trend observed previously, a large transparency near the beam energy of 9 GeV and a return to Glauber levels at lower and higher energy.

HISTORY OF COLOR TRANSPARENCY

Color transparency refers to a QCD phenomena, independently predicted in 1982 by Brodsky [1] and Mueller [2], involving reduction of secondary absorption in proton-nucleus quasi-elastic scattering. These theorists deduced from QCD that when a proton traversing the nucleus experiences a hard collision, a special quantum state is selected. The special state represents the part of the proton wave function that is very shock-resistant and that will tend to survive the hard collision without breaking up or radiating a gluon. However, the state is also expected to have a reduced interaction with the spectators in the target nucleus. The state is predicted to involve a rare component of the proton wave function that should be dominated by 3 valence quarks at small transverse spatial separation. The color transparency prediction of QCD is that the fraction of nuclear protons contributing to (p,2p) quasi-elastic scattering from nuclei should increase from a nominal level consistent with Glauber absorption [3],[4] at low Q^2 to a larger fraction at higher Q^2. This fraction (or transparency T) is predicted to increase toward unity at very high Q^2.

The first color transparency experiment involved (p,2p) quasi-elastic scattering and was published in 1988. The trend observed in that measurement suggested that transparency was increasing above the Glauber levels as the incident beam momentum ramped up from 6 to 10 GeV/c but beyond that point the trend seemed to reverse[5].

This result inspired corresponding measurements with electron beams. The results from the NE18, an (e,e'p) experiment at SLAC, were published in 1995 [6]. NE18 covered the low end of the momentum transfer range reported in the (p,2p) measurement, the region where an increase in (p,2p) transparency had been seen. They reported no observable uptrend in transparency. There has been much discussion about how these experiments should be compared. Some have claimed that there is no clear disagreement because, for reasonable variations of the models, effects show up at somewhat larger

[1] For the E850 (EVA) collaboration.

CP602, *QCD@Work: International Workshop on Quantum Chromodynamics*
edited by P. Colangelo and G. Nardulli

energy and momentum transfer with (e,e'p) kinematics [4, 7].

A newer result, reported from our E850 group in 1998 at beam energies of 6 and 7.5 GeV, confirmed the uptrend in transparency for scattering at 90° c.m. (center of mass). However, the result was surprisingly sensitive to the c.m. scattering angle with the uptrend tending to disappear for scattering angles nearer to 80° c.m.[8].

NEW TRANSPARENCY MEASUREMENT

We now present a new measurement of transparency for (p,2p) scattering in carbon over a wider range of beam energies. The experiment was done at the Brookhaven AGS with the E850 spectrometer, which has been described elsewhere [9]. The measurement covered a wider range of beam energies and Q^2 (four momentum transfer squared) than the 1988 measurement. With an improved detector, these results differed from the 1988 experiment in that all four components of missing momentum and energy associated with the (p,2p) hypothesis were measured. This allowed a more constrained measurement of the quasi-elastic event.

Elastic and Quasi-elastic Event Selection

The traditional method for extraction of the quasi-elastic cross section involves a measurement of two final state particles with cuts on missing momentum, perhaps implicit cuts reflecting the acceptance. In the impulse approximation, this corresponds to a selection of some range of nuclear momentum for the struck proton. The missing energy distribution is then analyzed to obtain the quasi-elastic contribution.

In the E850 detector, the missing energy distribution is measured only with resolution of 200-300 MeV/c and that measurement resolution varies with incident beam energy. Because of the resolution, a significant background subtraction is required to extract the signal. For this reason we introduce a new method for extraction of quasi-elastic signal from a distribution in a variable P^4 defined to be

$$P^4 = p_T^4 + MM^4 \tag{1}$$

where p_T is the missing transverse momentum and MM is the missing mass for the quasi-elastic event.

The Missing Mass and Missing Transverse Momentum Distributions

Applying the (p,2p) hypothesis, event candidates with two measured final state protons will be characterized in terms of their missing energy and missing vector momentum. Rather than considering primarily the missing energy distribution for identification of quasi-elastics, in this analysis event candidates will be studied through the distributions of missing mass (MM) and missing transverse momentum ($\vec{P}_T$). We use these variables for event selection because the fundamental quasi-elastic cross section depends

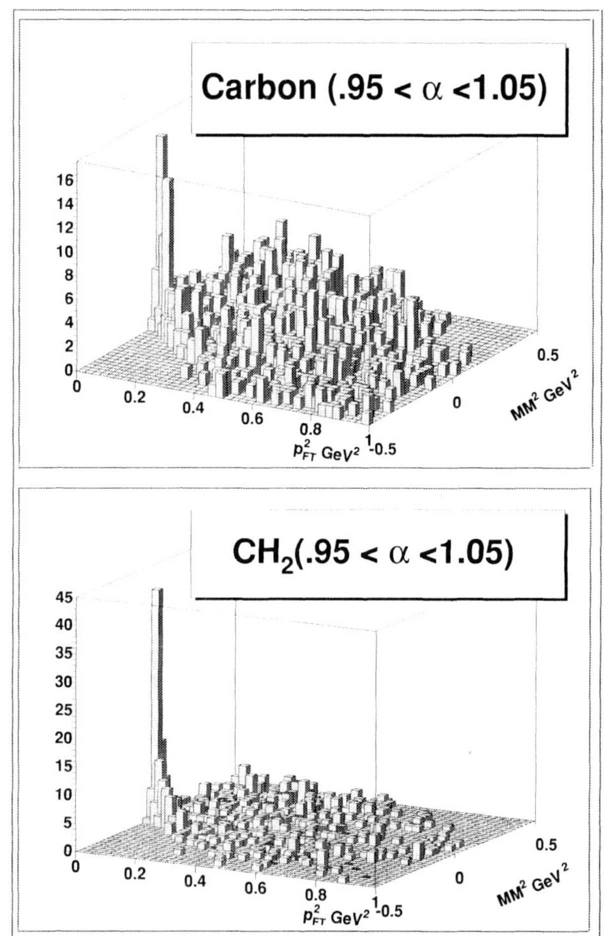

FIGURE 1. The missing transverse momentum squared P^2_{FT} vs missing mass squared MM^2 distribution for elastic or quasi-elastic event candidates at a beam momentum of 5.9 GeV/c with light cone momentum fraction near 1 ($.95 < \alpha < 1.05$). **Top:** The distribution of missing mass squared vs missing transverse momentum squared for scattering from carbon. **Bottom:** The same distribution for scattering from CH_2.

most critically on c.m. energy. In turn, the c.m. energy depends primarily on longitudinal missing momentum and only weakly upon P_T and MM.

In Fig. 1 we show a lego plot of the number of events versus P_T^2 and MM^2 for both carbon and CH_2 data at 6 GeV beam energy. The missing longitudinal momentum will be represented by a light cone variable α defined in terms of the missing energy E_F, the proton mass m_p and missing longitudinal momentum P_{Fz},

$$\alpha \equiv A\frac{(E_F - P_{Fz})}{M_A} \simeq 1 - \frac{P_{Fz}}{m_p}. \qquad (2)$$

where A and M_A are the atomic number and mass of the nucleus.

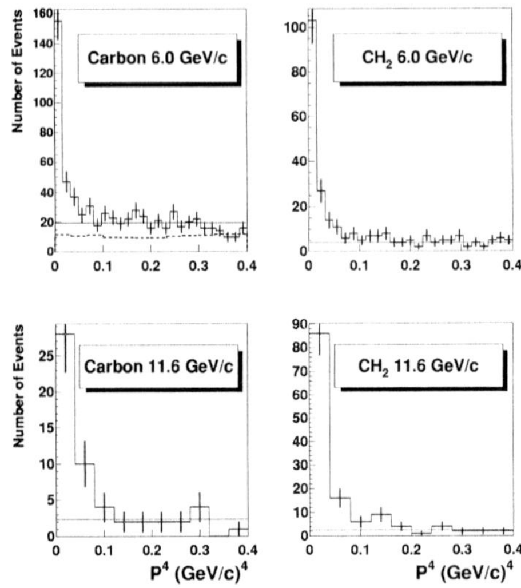

FIGURE 2. The P4 Distribution for candidate events passing the cuts of Eq. 3 at beam momenta of 5.9 and 11.6 GeV/c. The solid line indicates the fit for extraction of signal from background. The dashed line on the 5.9 Gev/c carbon distribution represents the shape of the distribution for tagged background events. The background sample shows that there is no kinematic enhancement under the quasi-elastic peak at $P^4 = 0$ and that the shape of the background distribution is also likely to be quite flat.

The detector was set up to have acceptance for quasi-elastic events associated with center of mass scattering angles in the range from approximately $86° - 90°$. Within this acceptance, the cuts used to define event candidates are summarized as follows:

$$|P_{Fx}| < 0.5\frac{GeV}{c}; \ |P_{Fy}| < 0.3\frac{GeV}{c}; \ |1-\alpha| < 0.05. \tag{3}$$

Quasi-elastic data obtained at values of α away from unity can also be analyzed but the analysis involves more controversial assumptions. An analysis of large and small α data has also been done and the results of that analysis will also be shown here.

For both carbon and CH_2 targets shown in Fig. 1, we see a strong peak at the origin. This corresponds to the quasi-elastic events from carbon nuclei elastically scattering from the free protons of the CH_2 target. The width of the peak for CH_2 targets is primarily due to resolution while the width of the peak in carbon reflects also the nuclear momentum. We reduce this distribution in two variables to a distribution in one variable by introducing P^4 (defined in Eq. 1) as the square of the radial distance from the origin in the $P_T^2 \times MM^2$ plane.

One advantage of this analysis is that for suitable longitudinal cuts, each bin in Fig. 1 corresponds to an equal volume of the missing four momentum space. A uniform population of the background in Fig. 1 also leads to a uniform background distribution in the P^4 distribution.

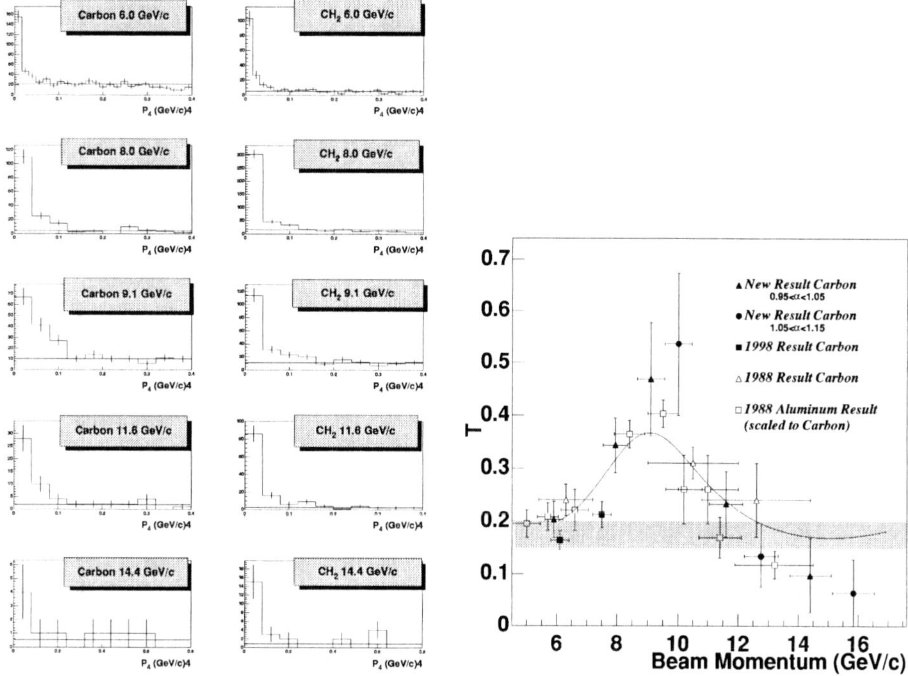

FIGURE 3. **Left:** The raw P4 distribution (similar to Fig. 2) for 5 incident beam momenta and for both carbon and CH_2 targets. The vertical axis represents the number of observed events in each P^4 bin that pass the cuts of Eq. 3. **Right:** The Transparency determined in this new analysis is displayed along with previously published measurements. Note that for data points corresponding to an α slice away from $\alpha = 1$, the beam momentum refers to the effective momentum, calculated in the rest frame of the moving target nucleon. The errors shown for new data are statistical and the statistical errors dominate over systematic errors. The indicated band represents the prediction of Glauber models with allowance for variations in the assumptions used to calculate the constant level. The curve through the data points is the curve R(s) defined in Eq. 4.

In Fig. 2 we show the P^4 distribution for events passing the cuts of Eq. 3 for four data sets, 5.9 and 11.6 GeV/c data on the two target choices, carbon and CH_2. To satisfy any possible concerns that this analysis generates a false peak at the origin of the P^4 distribution, we include for the case of the 5.9 GeV/c carbon data set, a comparison to the distribution obtained from a set of tagged background events. Our tagged background data set is obtained by selecting events that satisfy all required cuts for quasi-elastic events but for which additional soft charged tracks are detected in coincidence with the two track measurement.

We see that the P^4 distribution of tagged background appears very flat. To extract quasi-elastic events, we assume a constant background level, which is determined by the event candidate distribution away from the peak.

Having extracted the quasi-elastic signal in a particular α range, along with the hydrogen signal from CH_2 targets, the transparency can only be calculated with knowledge of the nuclear momentum distribution. In this analysis, a recent parameterization of the carbon distribution was used[10].

The results shown in Fig. 3 indicate a clear trend. We see that the transparency does depend upon beam momentum, in contrast to the Glauber prediction of a constant transparency. The peak in transparency is at about 9 GeV/c and we see a return to Glauber levels near 6 and 12 GeV/c. There is no indication of transparency above Glauber levels for beam momentum in the 12 to 16 GeV/c range.

One interesting observation is that the shape of this T distribution is closely connected to the deviation of the elastic cross section from the dimensional scaling prediction. The measured pp cross section near 90° depends on c.m. energy ($E_{cm} = \sqrt{s}$) according to

$$\frac{d\sigma}{dt} = \frac{s^{-10}}{R(s)}. \tag{4}$$

The dimensional scaling rule requires $R(s)$ to be a constant. The measurement cross section, however, shows that $R(s)$ depends somewhat upon s. In Fig. 3, the curve $R(s)$ as measured in many experiments is superimposed over the transparency. While the normalization is arbitrary, the shapes are similar. Several theoretical models have been proposed to explain the relationship between the energy dependence of the pp large angle cross section and the energy dependence of T[11, 12]. A consensus as to the origin of this trend does not yet exist.

ACKNOWLEDGMENTS

This research was supported by the U.S. - Israel Binational Science Foundation, the Israel Science Foundation founded by the Israel Academy of Sciences and Humanities, NSF grants PHY-9804015, PHY-0072240 and the U.S. Department of Energy grant DEFG0290ER40553.

REFERENCES

1. S.J. Brodsky, *Proceedings of the XIII International Symposium on Multi-particle Dynamics-1982*, eds. W. Kittel, W. Metzger and A. Stergiou, World Scientific, Singapore (1983), p. 963.
2. A. Mueller, *Proceedings of the XVII Rencontre de Moriond*, ed. J. Tran Thanh Van, Editions Frontieres, Gif-sur-Yvette, France (1992), p. 13.
3. R. L. Glauber, *Lectures in Theoretical Physics*, ed. W.E. Britin *et al.*, Interscience, New York (1959).
4. L. L. Frankfurt, M. I. Strikman and M. B. Zhalov, Phys. Rev. C **50**, 2189 (1994).
5. A. S. Carroll *et al.*, Phys. Rev. Lett. **61**, 1698 (1988).
6. T. G. O'Neill *et al.*, Phys. Lett. **B351**, 87 (1995).
7. L. L. Frankfurt, G. A. Miller and M. Strikman, Ann. Rev. Nucl. Part. Sci. **44**, 501 (1994).
8. I. Mardor *et al.*, Phys. Rev. Lett. **81**, 5085 (1998).
9. J. Wu, E. D. Minor, J. E. Passaneau, S. F. Heppelmann, C. Ng, G. Bunce and I. Mardor, Nucl. Instrum. Meth. A **349**, 183 (1994).
10. C. Ciofi degli Atti and S. Simula, Phys. Rev. C **53**, 1689 (1996).
11. J. P. Ralston and B. Pire, Phys. Rev. Lett. **61**, 1823 (1988).
12. S. J. Brodsky and G. F. de Teramond, Phys. Rev. Lett. **60**, 1924 (1988).

Soft Gluons and the Energy Dependence of Total Cross-Sections

R. M. Godbole*, A. Grau†, G. Pancheri** and Y. N. Srivastava‡

*Centre for Theoretical Studies, Indian Institute of Science, Bangalore, 560 012, India
†Centro Andaluz de Física de Partículas Elementales and Departamento de Física Teórica y del Cosmos, Universidad de Granada, Spain
**INFN Frascati National Laboratories, Via E. Fermi 40, I00044 Frascati, Italy
‡INFN, Physics Department, University of Perugia, Perugia, Italy

Abstract. We discuss the high energy behaviour of total cross-sections for protons and photons, in a QCD based framework with particular emphasis on the role played by soft gluons.

INTRODUCTION

Energy dependence of hadronic total cross-sections has fascinated particle physicists for decades now. In this talk we address a number of questions which arise when studying total hadronic cross-sections, namely

- Is it possible to study the energy dependence of the cross-sections for pp, $p\bar{p}$, γp and $\gamma\gamma \to hadrons$ in the same phenomenological/theoretical framework?
- what governs the energy dependence of these total cross-sections?
- what is the role played by the electromagnetic form factors in the description of the total cross-section?

The first question about treating *together* the $pp, p\bar{p}$ case on the one hand and the $\gamma p, \gamma\gamma$ case on the other, arises naturally as the 'hadronic' structure [1] of the photon has now been established in both e^+e^- and ep experiments conclusively [2]. Further, the photonic partons seem to have nontrivial effects on the photon-induced processes at high energies [3]. Equally importantly, along with the data already available for the $pp, p\bar{p}$ case [4], data have become available on total cross-sections for photon-induced processes reaching up to high γ energies, γp and $\gamma\gamma$ processes being studied in ep[5, 6, 7, 8], and e^+e^- [9, 10] collisions respectively. In Fig.1 we show a compilation of these proton and photon total cross sections, including cosmic ray data as well [11]. In order to put all the data on the same scale[12, 13], we have used a multiplication factor suggested by quark counting and Vector Meson Dominance[14], namely a factor $2/3\sum_{V=\rho,\omega,\pi}\left(4\pi\alpha_{QED}/f_V^2\right)$. Using a running α_{QED}, the VMD factor ranges from $1/250$ at low energy to $1/240$ at HERA energy. Square of this factor enters the photon-photon cross-sections.

At first glance, these data raise two questions: (i) whether the $\gamma\gamma$ total cross-section rises faster than the others and (ii) whether these various sets of data are mutually con-

CP602, *QCD@Work: International Workshop on Quantum Chromodynamics*
edited by P. Colangelo and G. Nardulli

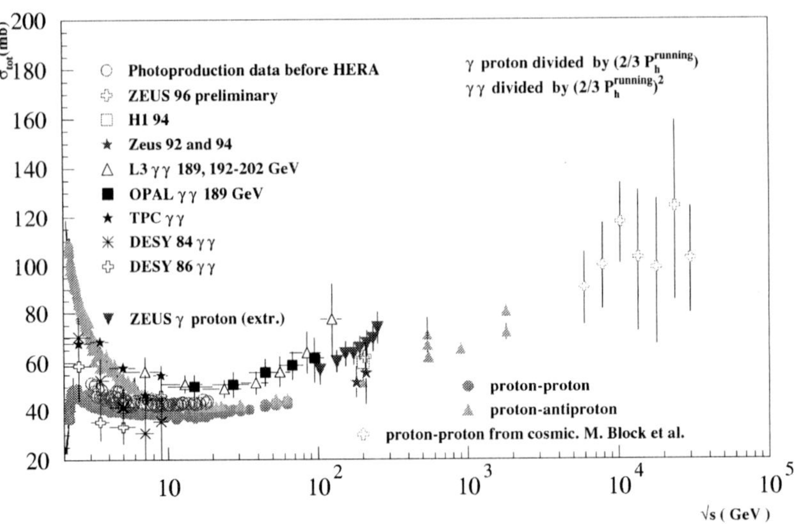

FIGURE 1. A compilation of pp,$p\bar{p}$,γp and $\gamma\gamma$ total cross sections with scaling factors described in the text.

sistent (at least at low energies) with the factorization hypothesis [15]. The uncertainty in the normalization of photon processes does not yet allow for a definite answer, but the photon-photon cross-sections do seem to be rather different, both from the point of view of the normalization [15] as well as the rise [13, 16, 17].

The next question is whether and how can we understand these data with our present means to deal with QCD. It appears that not all but many of the observed features are quantitatively obtainable from QCD. Our present goal is to obtain a QCD description of the initial decrease and the final increase of total cross-sections through soft gluon summation (via Bloch-Nordsieck Model) and mini-jets. Thus, our physical picture includes multiple parton collisions and soft gluons dressing each collision. We shall describe in the following sections details of the theoretical model proposed.

A QCD APPROACH

The task of describing the energy behaviour of total cross-sections can be broken down into three parts:

- the rise
- the initial decrease
- the normalization

The rise [18] can be obtained using the QCD calculable contribution from the parton-parton cross-section, whose total yield increases with energy, as shown in Fig.(2), where the jet cross-sections for proton-proton, γp and $\gamma\gamma$ are scaled by a common factor α.

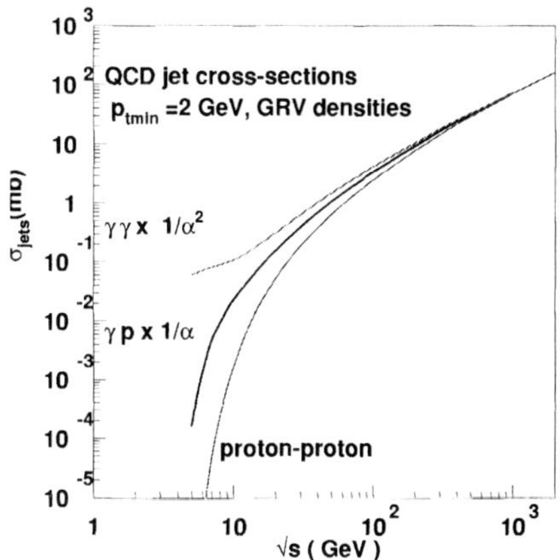

FIGURE 2. Minijets: Integrated jet cross-sections

In all cases, in particular for the proton case (where there are no direct scattering terms), one observes that σ_{jet} rises too fast for the observed values of σ_{tot} (less than 100 mb at the Tevatron) and that other terms, due to soft interactions, are missing. For a unitary description, the jet cross-sections are embedded into the eikonal formalism [19], namely one writes

$$\sigma^{\rm tot}_{pp(\bar{p})} = 2\int d^2\vec{b}[1-e^{-\chi_I(b,s)}cos(\chi_R)] \tag{1}$$

where the eikonal function $\chi = \chi_R + i\chi_I$ contains both the energy and the transverse momentum dependence of matter distribution in the colliding particles, through the impact parameter distribution in b-space[20]. The simplest formulation with minijets to drive the rise, in conjunction with eikonalization to ensure unitarity, is:

$$2\chi_I(b,s) \equiv n(b,s) = A(b)[\sigma_{soft} + \sigma_{jet}] \tag{2}$$

The normalization depends both upon σ_{soft} and the b-distribution. A very first working hypothesis is that the impact parameter distribution follows the matter distribution inside hadrons, namely that it is given by the Fourier transform of the electromagnetic form factors of the colliding particles, i.e.

$$A_{ab}(b) \equiv A(b;k_a,k_b) = \frac{1}{(2\pi)^2}\int d^2\vec{q}e^{iq\cdot b}\mathcal{F}_a(q,k_a)\mathcal{F}_b(q,k_b) \tag{3}$$

With such hypothesis, it is possible to describe the early rise, which takes place around $10\div 50\ GeV$ for proton-proton and proton-antiproton scattering, using GRV [21] densities for the protons and a transverse momentum cut-off in the jet cross-sections, $p_{tmin} \simeq 1\ GeV$, but then the cross-sections begin to rise too rapidly. One needs a

$p_{tmin} \approx 2\ GeV$ in order to reproduce the Tevatron data, with the drawback, however, that one misses the early rise. In Fig.(3) we show a straightforward application of the Eikonal Minijet Model (EMM), with different values of p_{tmin}, to illustrate this feature.

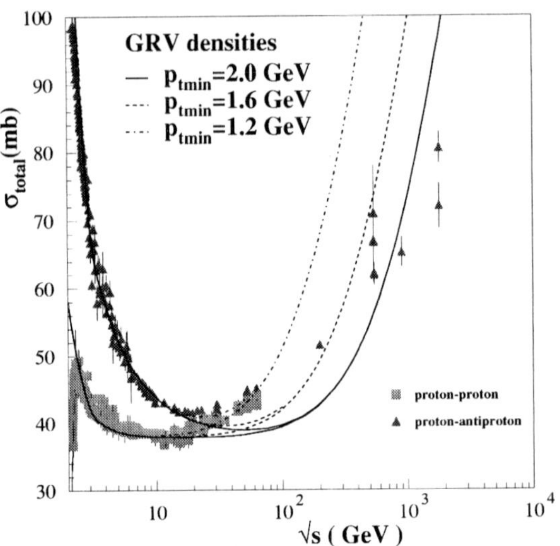

FIGURE 3. Total cross sections for pp and $p\bar{p}$ from EMM for various p_{tmin}.

A possible way to circumvent this problem lies in the use of soft gluons instead of form factors, but before turning to the issue of how to reproduce the early rise in proton-proton as well as the further Tevatron data points, we discuss the question of the photon cross-sections.

PHOTON PROCESSES AND MINIJETS

Photo-production and extrapolated data from Deep Inelastic Scattering (DIS) can be described through the same simple eikonal minijet model, with the relevant parton densities for the jet cross-sections, scaling [22] the non perturbative part given by σ_{soft} with the VMD and quark counting factor discussed above. The minijet cross-sections are then embedded into the eikonal formalism, with proper choice of impact parameter distribution. One needs a b-distribution of partons in the photon, which can be chosen to be a meson-like form factor.

The result is shown in Fig.(4), where the band corresponds to different sets of model parameters, with both GRV [23] and GRS [24] densities for the photon, and the dotted line corresponds to the predictions of the so-called Aspen Model[15]. The low energy region is obtained using quark counting and VMD from the proton data, while the high energy part is obtained from the QCD minijet cross-section and the impact parameter distribution from proton and pion-like form factors. As discussed in [12], the scale parameter k_0 in the photon form factor is allowed to vary in the range $0.4 \div 0.66\ GeV$.

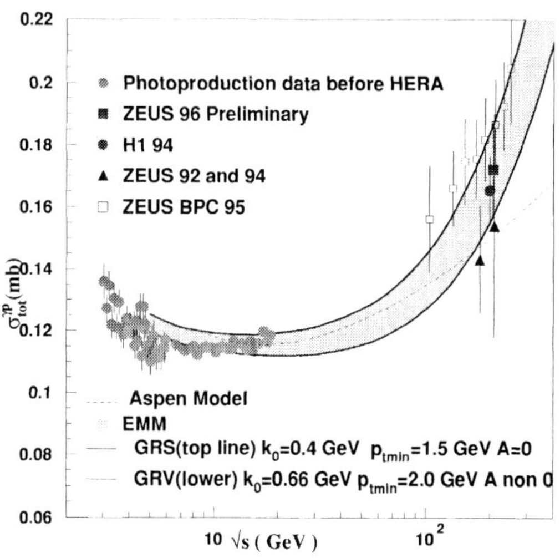

FIGURE 4. Description of photoproduction data with the EMM (band) and Aspen model (dotted line)

One encounters the same problem as in proton-proton case, albeit in a less severe form. When the parameters of the EMM are chosen so as to reproduce the low as well as the high energy data, the early rise is not well described. Modelling of $\gamma\, p$ data is further complicated, however, by the existence of data extrapolated from DIS which lie above, but within 1σ, from recent photoproduction measurements. Using a set of parameters consistent with those used to obtained the band of Fig.4, one can now attempt a description of photon-photon collisions and make predictions for future linear and photon colliders.

As before, one starts with the mini-jet cross-sections, for various parton densities and different values of p_{tmin}, as shown in Fig.(5). These minijets are them embedded into the eikonal, with parameters consistent [25] with the $\gamma\, p$ band shown in Fig.(4). Present LEP data are shown in Fig.(6) where EMM predictions [12, 13, 25] are compared with those from various models [15, 26, 27, 28, 29, 30] which have been proposed to describe $\gamma\gamma$ total cross-sections. The uncertainty in the predictions of photon-photon collisions is reflected in the uncertainty in $e^+e^- \to hadrons$, albeit, in such case, the difference between the predictions of different models for $\gamma\gamma$ total cross-section is, at the end, at most a factor 2, even at TESLA energies[25, 31].

THE TAMING OF THE RISE THROUGH SOFT GLUON SUMMATION

The fast rise due to mini-jets and the increasing number of gluon-gluon collisions as the energy increases, can be reduced if one takes into account that soft gluons, emit-

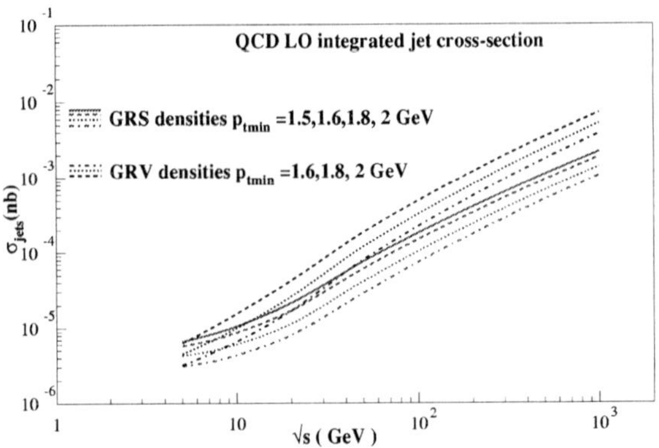

FIGURE 5. Minijets in photon-photon collisions

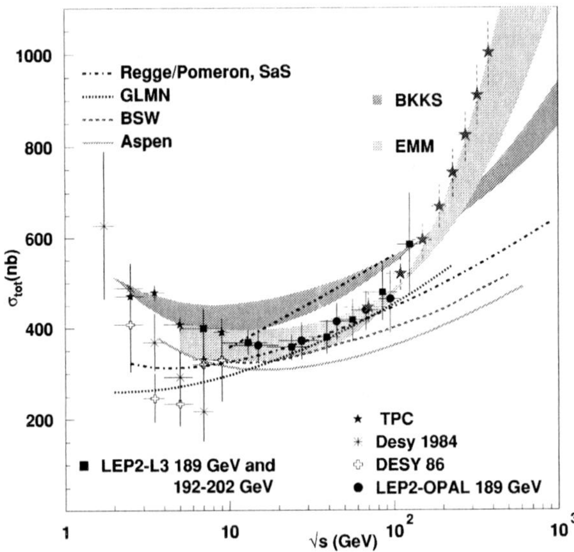

FIGURE 6. Photon photon total cross section data compared with various models.The stars at high photon-photon energies correspond to pseudo-data points extrapolated [31] from EMM predictions

ted mostly by the initial state valence quarks, determine an acollinearity between the partons which reduces the overall parton-parton luminosity. That is, as the energy increases, the larger phase space available for soft gluon emission implies more and more acollinearity and thus a reduced collision probability. This is the physical picture underlying the eikonal minijet model with Bloch-Nordsieck resummation[20]. In this model, the impact parameter distribution of partons is the (normalized) Fourier transform of the total transverse momentum distribution of valence quaks, obtained through soft gluon resummation, i.e.

$$A(b,s) = \frac{e^{-h(b,s)}}{\int d^2\vec{b}\; e^{-h(b,s)}} \tag{4}$$

with

$$h(b,s) = \int_{k_{min}}^{k_{max}} d^3\bar{n}(k)[1 - e^{-i\vec{k}_\perp \cdot \vec{b}}] \tag{5}$$

where $d^3\bar{n}(k)$ is the single soft gluon differential distribution and the integral runs, in principle, from zero to the maximum kinematic limit. Phenomenological applications of this expression encounter two main problems, one of theoretical origin, the other more of a phenomenological nature, namely, on the one side, a lack of our knowledge of the infrared behaviour of α_s , and, on the other, the unavailabily of reliable unintegrated parton distributions, i.e. parton distributions before the integration of their initial transverse momentum. The second difficulty can be phenomenologically overcome by averaging the function $A(b,s)$ over the parton densities to obtain the total number of collisions as

$$n(b,s) = A_{soft}(b)\sigma_{soft} + A_{PQCD}(b,s)\sigma_{jet}^{LO} \tag{6}$$

with $A_{soft}(b)$ as in the simpler EMM (form factors), and $A_{PQCD}(b,s)$ given by eqs.(4,5). The maximum energy for single soft gluon emission is obtained by averaging over the valence parton densities, i.e,

$$M \equiv < k_{max}(s) > = \frac{\sqrt{s}}{2} \frac{\sum_{i,j} \int \frac{dx_1}{x_1} f_{i/a}(x_1) \int \frac{dx_2}{x_2} f_{j/b}(x_2) \sqrt{x_1 x_2} \int dz(1-z)}{\sum_{i,j} \int \frac{dx_1}{x_1} f_{i/a}(x_1) \int \frac{dx_2}{x_2} f_{j/b}(x_2) \int (dz)}$$

with $z_{min} = 4p_{tmin}^2/(sx_1x_2)$. The quantity M can be calculated as a function of s for different values of p_{tmin}. For a p_{tmin} between 1 and 2 GeV, it ranges between 700 MeV and 3 GeV at $\sqrt{s} = 20\, GeV$ and $10\, TeV$, respectively.

To proceed further, one also needs to specify the lower limit of integration, or, if the value zero is assumed, the behaviour of $\alpha_s(k_t)$ as $k_t \to 0$. Our model assumes $k_{min} = 0$ and two different trial behaviours are utilized for the above limit, a frozen α_s model i.e. $\alpha_s(0) = constant$ and a model in which α_s is singular, but integrable[32, 33]. Since a single soft gluon is never observed, one only needs integrated quantities and, at least phenomenologically, this model seems adequate. As discussed elsewhere [20], the effect of soft gluon summation is mostly to introduce an energy dependence in the large b-behavior. In the frozen α_s case, the large b-behaviour is not depressed enough, compared to the form factor case, thus indicating the need to introduce an intrinsic transverse momentum cut off, namely a gaussian decrease in the b-variable. Different is the singular α_s case, where the expression [32] $\alpha_s(k_\perp) = \frac{12\pi}{(33-2N_f)} \frac{p}{\ln[1+p(\frac{k_\perp}{\Lambda})^{2p}]}$, produces an increasingly faster falloff in the b-distribution as the energy increases. The s-behavior

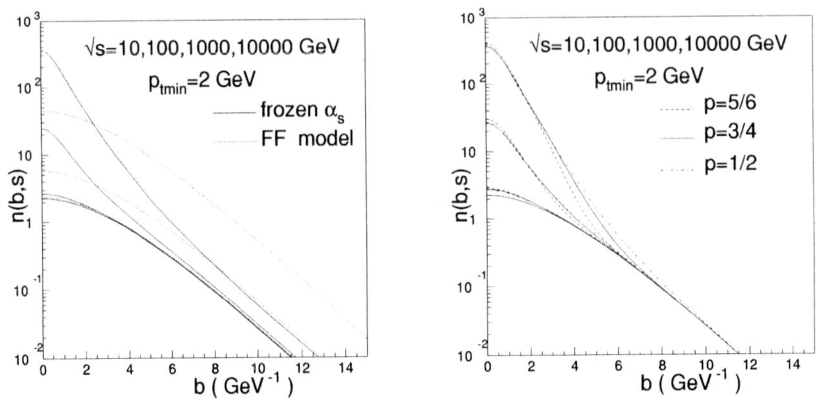

FIGURE 7. The average number of collisions for the frozen α_s case in comparison with the FF model (left) and the singular α_s case (right) for different values of the singularity parameter p.

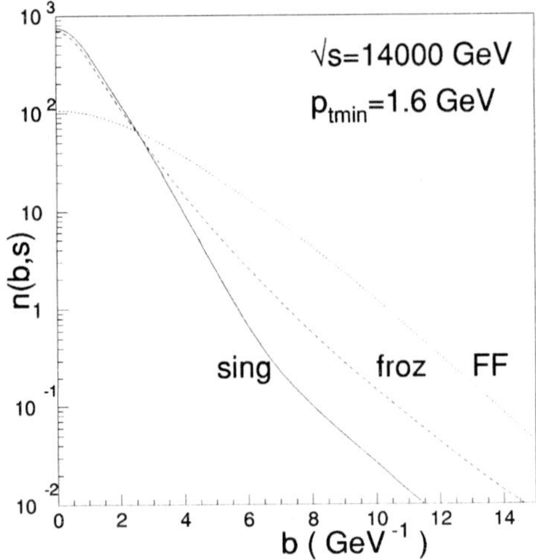

FIGURE 8. The average number of collisions in the form factor model and the Bloch Nordsieck model, at LHC energy

of the b-distribution modifies strongly the energy behaviour of the average number of collisions, as one can see from Figs.(7,8).

As the energy increases, the average number of collisions, relative to the form factor model, is strongly depressed at large b, thus smaller b-values contribute to the total cross-section, and the cross-section remains in general smaller than in the form factor case. In Fig.(9), we show how the integrand of eq.(1) behaves as a function of b, for different energy values, in the three models examined here. The integrand is peaked at different b-values as the energy increases, but also as the model for $A(b)$ changes. The rise with energy of the area under the curve, i.e. the cross-section, at the same energy shrinks for

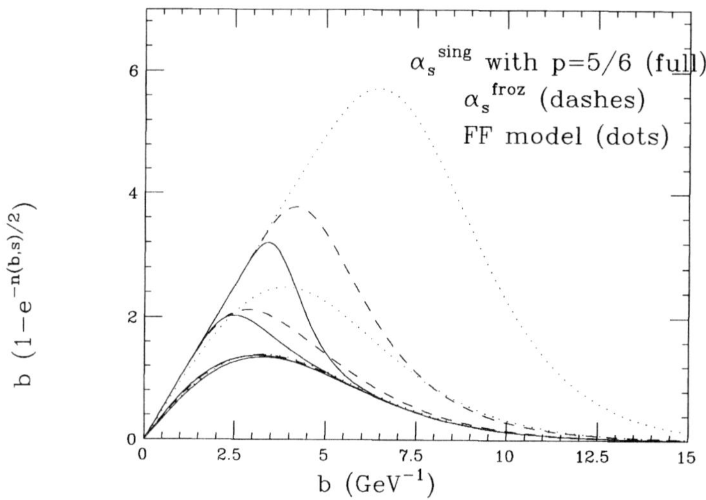

FIGURE 9. Integrand of the eikonal function for σ_{tot} in the three different models

the more singular α_s case. All the above considerations are exhibited in the plots given in the left panel of Fig. (10), where the effect of soft gluon summation in the singular α_s model is shown to reproduce quite well the early rise and the asymptotic softening. In comparison, the frozen α_s model appears almost as bad as the form factor model. The Bloch-Nordsieck model is practically indistinguishable from more conventional curves obtained through the Regge-Pomeron exchange [26] or the QCD inspired Aspen model[15], labelled BGHP in Fig.(10).

The analysis of proton collisions implies that straightforward applications of the minijet model through form factors are unable to describe correctly the large energy rise of total cross-sections. On the other hand, we have seen that the EMM can reproduce well the rise observed in $\gamma\gamma$ collisions in the present energy range, $\sqrt{s_{\gamma\gamma}} \approx 50 \div 100\ GeV$. But is the trend predicted by the EMM for photon-photon scattering correct at larger c.m. energies? It is quite possible that photon-photon data are only showing the early rapid rise, and that the rise at higher energies needs further corrections of the type we have described. Soft gluons are probably necessary in order to extrapolate to the higher energies of future electron-positron colliders such as TESLA, CLIC, NLC or Photon Colliders. An application of the Bloch-Nordsieck method to the case of photon-photon collisions is shown in the right panel of Fig.(10).

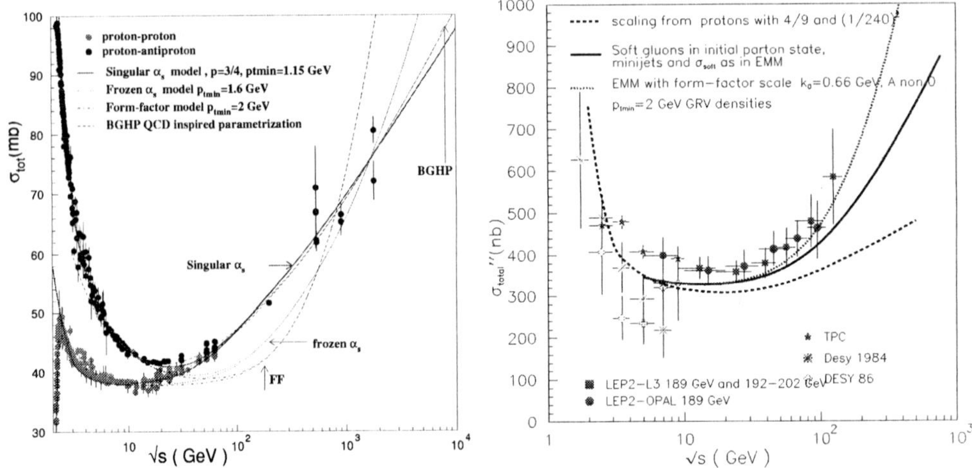

FIGURE 10. Total cross sections for pp and $p\bar{p}$ with frozen and singular α_s and FF models (left panel). LEP data are compared with EMM with and without soft gluons and with a curve scaled from protons (right panel).

CONCLUSIONS

We have described a unified approach to the calculation of total cross-sections for protons and photons. In all cases, the driving cause for the rise of total cross-sections is the energy dependent perturbative QCD parton-parton cross-section. For photon induced processes the model seems to describe the rise adequately. However for all proton processes it gives a rise which appears way too strong. Taming of the rise can be accomplished by an energy dependent impact parameter distribution, and different models for the infrared behaviour of α_s in the soft gluon summation have been explored. Our phenomenological analysis indicates a distinct preference for a singular but integrable α_s which automatically produces the desired effect of an initial intrinsic tranverse momentum of partons in the hadrons. The resulting physical picture is that of multiple scattering between partons, implemented by initial state soft gluon bremmstrahlung.

ACKNOWLEDGEMENTS

We ackowledge partial support through EEC Contract EURODAPHNE, TMR0169, and from Ministerio de Ciencia y Tecnología project FPA2000-1558.

REFERENCES

1. R.M. Godbole, *Pramana* **51**, 217 (1998), **hep-ph/9807402**; M. Drees and R.M. Godbole , *J.Phys. G* **21**, 1559 (1995), **hep-ph/9508221**.
2. M. Krawczyk, A. Zembrzuski and M. Staszel,Phys. Rept. **345**, 265 (2001) **hep-ph/0011083**.
3. M. Drees and R.M. Godbole, *Phys. Rev. Lett.* **67**, 1189 (1991); R.M. Godbole, Proceedings of the Workshop on*Quantum Aspects of Beam Physics, Jan. 5 1998 - Jan. 9 1998, Monterey, U.S.A.*, 404-416, Edited by P. Chen, World Scientific, 1999; **hep-ph/9807379**.
4. CDF Collaboration, Abe, F., et al *Phys. Rev. D* **50**, 5550 (1994).
5. H1 Collaboration, Aid, S., et al., *Zeit. Phys. C* **69**, 27 (1995), **hep-ex/9405006**.
6. ZEUS Collaboration, Derrick, M., et al., *Phys. Lett. B* **293**, 465 (1992); Derrick, M., et al., *Zeit. Phys. C* **63**, 391 (1994).
7. Breitweg, J., et al., ZEUS collaboration, **DESY-00-071**, **hep-ex/0005018**.
8. ZEUS Collaboration (C. Ginsburg et al.), Proc. 8th International Workshop on Deep Inelastic Scattering, April 2000, Liverpool,
9. L3 Collaboration, Paper 519 submitted to *ICHEP'98*, Vancouver, July 1998; Acciarri, M., et al., *Phys. Lett. B* **408**, 450 (1997); L3 Collaboration, Csilling, A., *Nucl.Phys.Proc.Suppl. B* **82**, 239 (2000); Acciari, M., et al,, **CERN-EP/2001-012**, *To appear in Phys. Lett. B*, **hep-ex/0102025**.
10. OPAL Collaboration. Waeckerle, F., *Multiparticle Dynamics 1997, Nucl. Phys. Proc. Suppl.B* **71**, 381 (1999) edited by G. Capon, V. Khoze, G. Pancheri and A. Sansoni; Söldner-Rembold, S., **hep-ex/9810011**, To appear in the proceedings of the *ICHEP'98*, Vancouver, July 1998; Abbiendi, G., et al., *Eur.Phys.J.C* **14**, 199 (2000), **hep-ex/9906039**.
11. M.M. Block, F. Halzen and T. Stanev, *Phys.Rev.* **D62**, 077501 (2000). M.M. Block, F. Halzen, G. Pancheri and T. Stanev, hep-ph/0003226, *25th Pamir-Chacaltaya Collaboration Workshop*, Lodz, Poland, November 1999.
12. R. M. Godbole and G. Pancheri, *Phys. Lett. B* **435** 441 (1998), **hep-ph/9807236**
13. R.M. Godbole and G. Pancheri, *Eur.Phys.J.C*, **19**, 129 (2001), **hep-ph/0010104**
14. R.S. Fletcher , T.K. Gaisser and F.Halzen, Phys. Rev. **D 45** (1992) 377;
15. M. Block, E. Gregores, F. Halzen and G. Pancheri, Phys.Rev.**D60** (1999) 054024.
16. R. M. Godbole and G. Pancheri, hep-ph/9903331. Proceedings of the *LUND workshop on photon interactions and photon structure*, Aug. 1998, 217-227, Eds. T. Sjostrand and J. Jarsklog,
17. A. De Roeck, Nucl. Phys. Proc. Suppl. **99A**, 144 (2001) In the proceedings of *DIFFRACTION 2000: International Workshop on Diffraction in High-energy and Nuclear Physics, Cetraro, Cosenza, Italy, 2-7 Sep 2000.* [hep-ph/0101076].
18. D. Cline, F. Halzen and J. Luthe, *Phys. Rev. Lett.* **31**, 491 (1973). G. Pancheri and C. Rubbia, *Nucl. Phys.* **A 418**, 117c (1984). T.Gaisser and F.Halzen, *Phys. Rev. Lett.* **54**, 1754 (1985). G.Pancheri and Y.N.Srivastava, *Phys. Lett.* **B 158**, 402 (1986).
19. L. Durand and H. Pi, *Phys. Rev. Lett.* **58**, 58 (1987). A. Capella, J. Kwiecinsky, J. Tran Thanh, *Phys. Rev. Lett.* **58**, 2015 (1987). M.M. Block, F. Halzen, B. Margolis, *Phys. Rev.* **D 45**, 839 (1992). A. Capella and J. Tran Thanh Van, *Z. Phys.* **C 23**, 168 (1984). P. l'Heureux, B. Margolis and P. Valin, *Phys. Rev.* **D 32**, 1681 (1985).
20. A. Grau, G. Pancheri and Y. N. Srivastava, PR **D60** (1999) 114020.
21. Glück, M., Reya, E., and Vogt, A., *Zeit. Physik C* **67**, 433 (1994).
22. J.C. Collins and G.A. Ladinsky, Phys. Rev. **D 43** (1991) 2847.
23. Glück, M., Reya, E., Vogt, A., *Phys. Rev. D* **46**, 1973 (1992).
24. Glück, M., Reya, E., and Schienbein, I., *Phys.Rev. D* **60**, 054019 (1999); Erratum, *ibid.* **62**, 019902 (2000).
25. R.M. Godbole and G. Pancheri, **hep-ph/0102188**, To appear in the Proceedings of *5th International Linear Collider Workshop (LCWS 2000), Fermilab, Batavia, Illinois, 24-28 Oct 2000.*
26. Donnachie, A., and Landshoff, P.V., *Phys. Lett. B* **296**, 227 (1992), **hep-ph/9209205**.
27. Schuler, G.A., and Sjöstrand, T., *Z.Phys.C* **73**, 677 (1997), **hep-ph/9605240**.
28. Gotsman, E., Levin, E., Maor, U., and Naftali, E., *Eur. Phys. J. C* **14**, 511 (2000), **hep-ph/0001080**.
29. Bourrelly, C., Soffer, J., and Wu, T.T., *Mod.Phys.Lett. A* **15**, 9 (2000).
30. B. Badelek, B., M. Krawczyk, J. Kwiecinski and A.M. Stasto, **hep-ph/0001161**.
31. R.M. Godbole, G. Pancheri and A. de Roeck, **LC-TH-2001-030**.
32. A. Nakamura, G. Pancheri and Y. Srivastava, *Z. Phys* **C21** (1984) 243.

33. Y. Srivastava, S. Pacetti, G. Pancheri and A. Widom, **hep-ph/0106005.** Proceedings of Workshop on e^+e^- interactions at low and medium energies, SLAC, 2001

The Diagonal Ghost Equation Ward Identity for Yang-Mills Theories in the Maximal Abelian Gauge

A.R. Fazio

Dipartimento di Fisica, Università Statale di Milano and INFN
via Celoria 16, I-20133, Milano, Italy

Abstract. A BRST perturbative analysis of $SU(N)$ Yang-Mills theory in a class of maximal Abelian gauges is presented. We point out the existence of a new nonintegrated renormalizable Ward identity which allows to control the dependence of the theory from the diagonal ghosts. This identity, called the diagonal ghost equation, plays a crucial role for the stability of the model under radiative corrections. Moreover, the Ward identity corresponding to the Abelian Cartan subgroup is easily derived from the diagonal ghost equation. A possible mechanism for the decoupling of the diagonal ghosts, valid in the fondamental modular Gribov region, is suggested.

INTRODUCTION

The understanding of the color confinement has been a challenging issue in theoretical physics since long time, being at present subject of intensive research. Certainly, the idea that confinement could be interpreted as a dual Meissner effect for type II superconductors [1, 2] is both attractive and very promising.

An important ingredient in order to implement this program is the mechanism of the Abelian projection introduced by 't Hooft [2], which consists of reducing the gauge group to an Abelian subgroup, commonly identified with the Cartan subgroup, by means of a partial gauge fixing. This procedure starts by decomposing the gauge field into its diagonal and off-diagonal parts. The diagonal components correspond to the generators of the Cartan subgroup and behave as photons. The off-diagonal components are charged with respect to the Abelian residual subgroup and can become massive, being not protected by gauge invariance anymore. The appearance of this mass scale allows for the decoupling of the off-diagonal fields at low energy. Moreover, the Abelian projected theory turns out to contain magnetic monopoles, whose condensation should account for the confinement of all chromoelectric charges.

Lattice calculations [3, 4] have provided evidences for the Abelian dominance hypothesis, according to which QCD in the low energy regime is described by an effective Abelian theory. This supports the realization of confinement through a dual Meissner effect, although the infrared Abelian dominance in lattice simulations seems not to be a general feature of any Abelian gauge [5].

Furthermore, many conceptual points remain to be clarified in order to achieve an analytic derivation of the Abelian dominance directly from the QCD Lagrangian. One

CP602, *QCD@Work: International Workshop on Quantum Chromodynamics*
edited by P. Colangelo and G. Nardulli

crucial question is to understand how the off-diagonal fields acquire mass, so that they can decouple at low energy.

Recently, the authors [6, 7] have proposed an original mechanism for generating a mass term for the off-diagonal fields. They make use of the maximal Abelian gauge (MAG) which amounts to choose a nonlinear gauge condition for the off-diagonal components of the gauge field. Therefore, the introduction of a four-ghost self-interaction term [8] is required for renormalizability. As discussed in [6, 7], the four-ghost term is responsible for the existence of a nonperturbative ghost-antighost condensation which provides a dynamical mass generation for all off-diagonal gauge and ghost fields. It is worth underlining that this mechanism shares great analogy with the BCS gap equation for superconductivity [9]. Another peculiar aspect of the gauge fixing adopted in [6, 7] is the presence of a unique gauge parameter instead of the usual pair of independent parameters associated respectively to the MAG gauge fixing condition and to the four-ghost self-interaction. This particular choice of the gauge parameters allows for both BRST and anti-BRST invariances, which may provide a possible consistent interpretation of the ghost-antighost condensation in terms of a spontaneous breakdown of a global $SL(2,R)$ symmetry present in the theory [7].

The aim of this contribution,based on [10], is to point out some new properties of the gauge proposed in [6, 7]. We shall be able to prove indeed that, as a consequence of having a unique gauge parameter, a new Ward identity arises, which allows to control the dependence of the theory from the diagonal ghosts. A remarkable feature of this Ward identity, which we shall call the *diagonal ghost equation*, relies on the fact that it holds at the nonintegrated level, a property which will have far reaching consequences. For instance, the Abelian Ward identity corresponding to the Cartan subgroup follows by commuting the Slavnov-Taylor identity with the diagonal ghost equation. Furthermore, this equation is easily extended at the quantum level implying the stability under radiative corrections of the gauge fixing condition. Finally we show that in absence of Gribov ambiguity the diagonal ghosts are decoupled as in QED. It is another argument [10] in favour of the Abelian dominance in this full scaled analysis in Yang-Mills theory in the maximal Abelian gauge with only one gauge parameter.

The paper is organized as follows. In section 2 we discuss the $SU(N)$ Yang-Mills in the MAG, we derive the diagonal ghost equation Ward identity and its consequences. Section 3 is devoted to the decoupling of the diagonal ghosts.

SU(N) YANG-MILLS GAUGE THEORY IN THE MAG

Let $\mathcal{A}_\mu$ be the Lie algebra valued connection for the gauge group $SU(N)$, whose generators T^A $(A = 1,..,N^2-1)$ are chosen to be antihermitean and orthonormal. Following [2], we decompose the gauge field into its off-diagonal and diagonal parts

$$\mathcal{A}_\mu = \mathcal{A}_\mu^A T^A = A_\mu^a T^a + A_\mu^i T^i, \tag{1}$$

where the index i labels the generators T^i of the Cartan subgroup $U(1)^{N-1}$ and runs from 1 to $N-1$. The remaining off-diagonal generators T^a will be labelled by the index a running from 1 to $N(N-1)$.

Analogously, decomposing the field strength we obtain

$$\mathcal{F}_{\mu\nu} = \mathcal{F}^A_{\mu\nu} T^A = F^a_{\mu\nu} T^a + F^i_{\mu\nu} T^i \,, \tag{2}$$

with the off-diagonal and diagonal parts given respectively by

$$\begin{aligned} F^a_{\mu\nu} &= D^{ab}_\mu A^b_\nu - D^{ab}_\nu A^b_\mu + f^{abc} A^b_\mu A^c_\nu \,, \\ F^i_{\mu\nu} &= \partial_\mu A^i_\nu - \partial_\nu A^i_\mu + f^{abi} A^a_\mu A^b_\nu \,, \end{aligned} \tag{3}$$

where the covariant derivative D^{ab}_μ is defined with respect to the diagonal components A^i_μ

$$D^{ab}_\mu \equiv \partial_\mu \delta^{ab} - f^{abi} A^i_\mu \;. \tag{4}$$

Rewriting the Yang-Mills action in terms of the off-diagonal and diagonal fields we get

$$S_{\mathrm{YM}} = -\frac{1}{4g^2} \int d^4x \left(F^a_{\mu\nu} F^{a\mu\nu} + F^i_{\mu\nu} F^{i\mu\nu} \right) \,. \tag{5}$$

In order to quantize the theory, we shall adopt the so called MAG condition [3], which amounts to fix the value of the covariant derivative $(D^{ab}_\mu A^{b\mu})$ of the off-diagonal components. However, this condition being nonlinear, a four-ghost self-interaction term is needed to guarantee the perturbative renormalizability [8]. Therefore, the corresponding gauge fixing term is found to be

$$S_{\mathrm{MAG}} = s \int d^4x \left(\bar{c}^a \left(D^{ab}_\mu A^{b\mu} + \frac{\xi}{2} b^a \right) - \frac{\xi}{2} f^{abi} \bar{c}^a \bar{c}^b c^i - \frac{\xi}{4} f^{abc} c^a \bar{c}^b \bar{c}^c \right) \,. \tag{6}$$

Notice that we have restricted the MAG condition in eq.(6) to contain the unique gauge parameter ξ. In the present case, the BRST transformations read

$$\begin{aligned} sA^a_\mu &= -\left(D^{ab}_\mu c^b + f^{abc} A^b_\mu c^c + f^{abi} A^b_\mu c^i \right), \quad & sA^i_\mu &= -\left(\partial_\mu c^i + f^{iab} A^a_\mu c^b \right) \,, \\ sc^a &= f^{abi} c^b c^i + \frac{1}{2} f^{abc} c^b c^c, & sc^i &= \frac{1}{2} f^{iab} c^a c^b, \\ s\bar{c}^a &= b^a \,, & s\bar{c}^i &= b^i \,, \\ sb^a &= 0 \,, & sb^i &= 0 \,. \end{aligned} \tag{7}$$

It should be remarked that the use of the MAG condition allows for the existence of a residual local $U(1)$ invariance with respect to the diagonal subgroup, which has to be fixed by imposing a suitable further gauge condition on the diagonal component A^i_μ of the gauge field. Adopting without loss of generality a linear Landau condition, the remaining gauge fixing term is given by

$$S_{\mathrm{diag}} = s \int d^4x \, \bar{c}^i \partial^\mu A^i_\mu \tag{8}$$

where $\bar{c}, b$ are the antighost and Lagrange multiplier fields.

To write down the Slavnov-Taylor identity we introduce external classical sources

$$S_{\text{ext}} = \int d^4x \left(A^{a*}_{\mu} sA^{a\mu} + A^{i*}_{\mu} sA^{i\mu} + c^{a*} sc^a + c^{i*} sc^i \right) , \tag{9}$$

so that the complete action

$$\Sigma = S_{\text{YM}} + S_{\text{MAG}} + S_{\text{diag}} + S_{\text{ext}} \tag{10}$$

obeys the Slavnov-Taylor identity

$$\mathcal{S}(\Sigma) = 0 , \tag{11}$$

with

$$\mathcal{S}(\Sigma) = \int d^4x \left(\frac{\delta\Sigma}{\delta A^{a*}_{\mu}} \frac{\delta\Sigma}{\delta A^{a\mu}} + \frac{\delta\Sigma}{\delta A^{i*}_{\mu}} \frac{\delta\Sigma}{\delta A^{i\mu}} + \frac{\delta\Sigma}{\delta c^{a*}} \frac{\delta\Sigma}{\delta c^{a}} + \frac{\delta\Sigma}{\delta c^{i*}} \frac{\delta\Sigma}{\delta c^{i}} + b^a \frac{\delta\Sigma}{\delta \bar{c}^a} + b^i \frac{\delta\Sigma}{\delta \bar{c}^i} \right) . \tag{12}$$

Applying the functional operator

$$\mathcal{G}^i = \frac{\delta}{\delta c^i} + f^{abi} \bar{c}^a \frac{\delta}{\delta b^b} \tag{13}$$

to the complete action Σ and using then the Jacobi identities we obtain

$$\mathcal{G}^i \Sigma = \Delta^i_{\text{cl}} , \tag{14}$$

where

$$\Delta^i_{\text{cl}} = -\partial^2 \bar{c}^i + f^{abi} A^{a*}_{\mu} A^{b\mu} - \partial^{\mu} A^{i*}_{\mu} - f^{abi} c^{a*} c^b . \tag{15}$$

Equation (14) is called the diagonal ghost equation. In the MAG with only one parameter the breaking term (15) is purely linear in the quantum fields and will not be affected by radiative corrections [12].

It is worth underlining that the diagonal Ward identity corresponding to the $U(1)^{N-1}$ Cartan subgroup follows from anticommuting the ghost equation (14) with the Slavnov-Taylor identity (11), *i.e.*

$$\mathcal{G}^i \mathcal{S}(\Sigma) = 0 \quad \Rightarrow \quad \mathcal{W}^i \Sigma = -\partial^2 b^i , \tag{16}$$

where $\mathcal{W}^i$ is the Ward operator

$$\mathcal{W}^i = \partial_{\mu} \frac{\delta}{\delta A^i_{\mu}} + f^{abi} \left(A^a_{\mu} \frac{\delta}{\delta A^b_{\mu}} + c^a \frac{\delta}{\delta c^b} + b^a \frac{\delta}{\delta b^b} + \bar{c}^a \frac{\delta}{\delta \bar{c}^b} + A^{a*}_{\mu} \frac{\delta}{\delta A^{b*}_{\mu}} + c^{a*} \frac{\delta}{\delta c^{b*}} \right) \tag{17}$$

of the residual $U(1)^{N-1}$ subgroup. The diagonal components A^i_{μ} behave as photons, while all off-diagonal fields play the role of charged matter.

Let us end up this section by remarking that the whole set of Ward identities is easily extended to the quantum level [12]. They imply the stability of the model under radiative corrections implying, in particular, the vanishing of the anomalous dimensions of the diagonal ghost c^i and of A^i_μ [10].

DECOUPLING OF THE DIAGONAL GHOSTS

The Diagonal Ghost Equation Ward identity provides a strong indication of the decoupling of the diagonal ghosts. Let $W(A)$ be any gauge invariant observable depending only on the gauge connection, for instance a Wilson loop. The decoupling of the diagonal ghosts means that the vacuum expectation value $\langle W(A)\rangle$ does not receive contributions from the diagonal ghosts. More precisely, there are no contributions coming from Feynman diagrams with diagonal ghosts propagating in the internal lines of the diagram. At path integral level that amounts to integrate the diagonal ghosts and antighosts.

In *QED* the diagonal ghost and antighost are decoupled. In fact the lagrangian ghost term

$$\int d^4x\bar{c}\partial_\mu\partial^\mu c$$

can be written as the determinant of the operator $\partial_\mu\partial^\mu$ in the functional space defined by the boundary conditions on c and $\bar{c}$,

$$\int [D\bar{c}][Dc]exp(-i\int d^4x\bar{c}\partial_\mu\partial^\mu c) = \det\partial_\mu\partial^\mu .$$

In the present case, this determinant does not depend on the gauge field A_μ and may be absorbed in the normalization .

A very similar analysis can be performed in the MAG with only one parameter. In fact the diagonal ghost equation implies that the vertex functional Γ depends on c^i and the off-diagonal Lagrange multipliers b^a through the combination $\widehat{b^a} = (b^a - f^{abi}\bar{c}^b c^i)$. Then, performing in the path integral the local change of variables $b^a \to \widehat{b^a}$, $\bar{c}^a \to \bar{c}^a$, $c^i \to c^i$, whose Jacobian is equal to one, it turns out that the diagonal ghosts c^i contribute to the transformed action through the bilinear term $\bar{c}^i\partial^2 c^i$, up to terms in the external classical sources. It is only the diagonal antighosts $\bar{c}^i$ which appears in the interaction terms. Due to the quartic ghost interaction, linearized by the introduction of a set of auxiliary scalar fields ϕ^i, as

$$-\frac{\xi}{4}f^{abi}f^{cdi}\bar{c}^a\bar{c}^b c^c c^d \to \frac{1}{2\xi}\phi^i\phi^i + if^{abi}\phi^i\bar{c}^a c^b \tag{18}$$

we are able to integrate out the diagonal ghosts fields as in QED. We obtain an effective action which possess fewer ghosts and which is well defined non-perturbatively also in presence of Gribov copies[14].

REFERENCES

1. Y. Nambu, *Phys. Rev.* **D10** (1974) 4262;
   G. 't Hooft, *High Energy Physics EPS Int. Conference,* Palermo 1975, ed. A. Zichichi;
   S. Mandelstam, *Phys. Rept.* **23** (1976) 245.
2. G. 't Hooft, *Nucl. Phys.* **B190** [FS3] (1981) 455.
3. A. Kronfeld, G. Schierholz and U.-J. Wiese, *Nucl. Phys.* **B293** (1987) 461;
   A. Kronfeld, M. Laursen, G. Schierholz and U.-J. Wiese, *Phys. Lett.* **B198** (1987) 516;
4. T. Susuki and I. Yotsuyanagi, *Phys. Rev.* **D42** (1990) 4257;
   A. Di Giacomo, *Nucl. Phys. Proc. Suppl.* **47** (1996) 136;
   M. Polikarpov, *Nucl. Phys. Proc. Suppl.* **B53** (1997) 134.
5. A. Di Giacomo *Prog.Theor.Phys.Suppl.* **131** (1998) 161, hep-th/9603029;
   M.I. Polikarpov *Nucl.Phys.Proc.Suppl.* **53** (1997) 134;
   G.S. Bali, hep-ph/9809351.
6. K.I. Kondo, *Phys. Rev.* **D58** (1998) 105019;
   K.I. Kondo and T. Shinohara, *Phys. Lett.* **B491** (2000) 263.
7. M. Schaden, hep-th/9909011, hep-th/0003030.
8. H. Min, T. Lee and P.Y. Pac, *Phys. Rev.* **D32** (1985) 440.
9. See for instance:
   H. Kleinert, *Gauge Fields in Condensed Matter, Vol. I: Superflow and Vortex Lines* , World Scientific Publishing Co., Singapore, 1989;
   B. Sakita, *Quantum Theory of Many-Variable Systems and Fields,* World Scientific Publishing Co., Singapore, 1985.
10. A.R. Fazio, V.E.R. Lemes, M.S. Sarandy, S.P. Sorella, hep-th/0105060 accepted for publication in Phys. Rev. D
11. M. Quandt and H. Reinhardt, *Phys. Lett.* **B424** (1998) 115, *Int. J. Mod. Phys.* **A13** (1998) 4049.
12. O. Piguet and S.P. Sorella, *Algebraic Renormalization*, Monograph series **m28**, Springer Verlag, 1995.
13. G. Barnich, F. Brandt and M. Henneaux, *Phys. Rept.* **338** (2000) 439
14. A.R. Fazio, V.E.R. Lemes, M.S. Sarandy, S.P. Sorella, in preparation.

SUSY Scaling Violations and UHECR

Claudio Coriano'* and Alon E. Faraggi†

*Dipartimento di Fisica Universita' di Lecce, I.N.F.N. Sezione di Lecce Via Arnesano, 73100 Lecce, Italy
†Theoretical Physics Department, University of Oxford, Oxford, OX1 3NP, United Kingdom, and Theory Division, CERN, CH–1211 Geneva, Switzerland

Abstract. Advancing QCD toward astroparticle applications generates new challenges for perturbation theory, such as the presence of large evolution scales with sizeable scaling violations involving both the initial and the final state of a collision. Possible applications in the context of Ultra High Energy Cosmic Rays (UHECR) of these effects are discussed.

INTRODUCTION

Nowadays intriguing theoretical extensions of the Standard Model are being explored, while the experimental results continue to confirm the validity of the model and constrain its extensions. Given the limited energy range available at colliders, it is therefore vital to develop experimental probes that can climb up the energy ladder. A complementary way to analize these extensions while waiting for colliders of the next generations is provided by cosmic rays, which exceed the energy scales currently attainable.

We have undertaken the preliminary steps in a program that aims to utilize cosmic rays as an experimental probe of theoretical generalizations of the Standard Model. We have in mind possible applications of cosmic rays for the study of supersymmetry. The first is in the context of top-down models of Ultra–High–Energy Cosmic Rays (UHECR), around and above the Greisen–Zatsepin–Kuzmin cutoff. In these models the primary cosmic rays originate from the decay of a metastable superheavy particle which decay at rest, fragmenting into ordinary hadrons and photons. The dynamics of these decays can be modelled using standard QCD tools on which we elaborate below. We propose to analize supersymmetric effects in the decay of these metastable states using 2 scales:

A High Energy Λ_F fragmentation Scale $\approx 10^{11}$ (decay of a metastable state $\rightarrow$ primary protons)

A collision scale Λ_{coll} due to the interaction of surviving primaries with air-nuclei ($E_{\rm CoM} \approx 10-400$ TeV)

At both scales supersymmetric scaling violations should be included and the multiplicities of the spectrum analized.

CP602, *QCD@Work: International Workshop on Quantum Chromodynamics*
edited by P. Colangelo and G. Nardulli

THE TIMELIKE EVOLUTION EQUATIONS

Scaling violations induced by the supersymmetric evolution couple to the standard QCD scaling violations. Even from vanishing boundary conditions supersymmetric distributions are *radiatively* generated. The first analysis in the literature of the fragmentation functions using evolution methods can be found in [1] and we briefly describe it here.

Introducing functions $D_f^h(x,Q^2)$ to denote the fragmentation of a parton f - quarks (q), gluons (G), left- and right- squarks ($\tilde{q}_{L,R}$) and gluinos (λ) - into a hadron h, the equations for the timelike evolution are given by

$$\frac{d}{d\log(Q^2)} D_g^h(x,Q^2) = \frac{\alpha_s}{2\pi}\left(P_{gg}\otimes D_g^h + P_{\lambda g}\otimes D_\lambda^h + P_{qg}\otimes\sum_i\left(D_{q_i}^h + D_{\bar{q}_i}^h\right) + P_{\tilde{q}g}\otimes\sum_{i=1}^{n_f}(\tilde{q}_{iL}+\tilde{q}_{iR}+\bar{\tilde{q}}_{iL}+\bar{\tilde{q}}_{iR})\right) \quad (1)$$

$$\frac{d}{d\log(Q^2)} D_\lambda^h(x,Q^2) = \frac{\alpha_s}{2\pi}\left(P_{g\lambda}\otimes D_g^h + P_{\lambda\lambda}\otimes D_\lambda^h + P_{q\lambda}\otimes\sum_i\left(D_{q_i}^h + D_{\bar{q}_i}^h\right) + P_{\tilde{q}\lambda}\otimes\sum_{i=1}^{n_f}(\tilde{q}_{iL}+\tilde{q}_{iR}+\bar{\tilde{q}}_{iL}+\bar{\tilde{q}}_{iR})\right) \quad (2)$$

$$\frac{d}{d\log(Q^2)} D_{q_i}^h(x,Q^2) = \frac{\alpha_s}{2\pi}\left(\frac{1}{2n_f}P_{gq}\otimes D_g^h + \frac{1}{2n_f}P_{\lambda q}\otimes D_\lambda^h + P_{\tilde{q}q}\otimes\left(D_{\tilde{q}_{iL}}^h + D_{\tilde{q}_{iR}}^h\right) + P_{qq}\otimes D_{q_i}^h\right) \quad (3)$$

$$\frac{d}{d\log(Q^2)} D_{\tilde{q}_{iL}}^h(x,Q^2) = \frac{\alpha_s}{2\pi}\left(\frac{1}{4n_f}P_{g\tilde{q}}\otimes D_g^h + \frac{1}{4n_f}P_{\lambda\tilde{q}}\otimes D_\lambda^h + \frac{1}{2}P_{q\tilde{q}}\otimes D_{q_i}^h + P_{\tilde{q}\tilde{q}}\otimes D_{\tilde{q}_{iL}}^h\right) \quad (4)$$

$$\frac{d}{d\log(Q^2)} D_{\tilde{q}_{iR}}^h(x,Q^2) = \frac{\alpha_s}{2\pi}\left(\frac{1}{4n_f}P_{g\tilde{q}}\otimes D_g^h + \frac{1}{4n_f}P_{\lambda\tilde{q}}\otimes D_\lambda^h + \frac{1}{2}P_{q\tilde{q}}\otimes D_{q_i}^h + P_{\tilde{q}\tilde{q}}\otimes D_{\tilde{q}_{iR}}^h\right). \quad (5)$$

We separate these equations into various non-singlet and into one singlet (matrix) equations and solve them numerically using a recursive method [1]. We perform a backward evolution of these equations using as initial conditions known QCD fragmentation functions at lower energies and move from a regular QCD evolution to a supersymmetric evolution as we step into higher Q^2 values.

NUMERICAL RESULTS

As an illustration of the procedure we adopt in our studies, let's consider the decay of a hypothetical massive state of mass 1 TeV into supersymmetric partons. The decay can proceed, for instance, through a regular $q\bar{q}$ channel and a shower is developed starting from the quark pair. The $N = 1$ DGLAP equation describes in the leading logarithmic approximation the evolution of the shower which accompanies the pair, and we are interested in studying the impact of the supersymmetry breaking scale (m_λ) on the fragmentation. In our runs we have chosen the initial set of Ref. [2] .

We parameterize the fragmentation functions as

$$D(x,\mu^2) = Nx^\alpha(1-x)^\beta\left(1+\frac{\gamma}{x}\right) \tag{6}$$

Typical fragmentation functions in QCD involve final states with p, $\bar{p}$, $\pi^\pm,\pi^0$ and kaons $k^\pm$. We have chosen an initial evolution scale of 10 GeV and varied both the mass of the SUSY partners (we assume for simplicity that these are all degenerate) and the final evolution scale. In general the effects of supersymmetric evolution are small within the range described by the factorization scales Q_f and Q_i ($Q_f = 10^3$ GeV, $Q_i = 200$ GeV). We mention that Q_f is the starting scale (the highest scale) at which the decay of the supersymmetric partons starts. Q_i is fixed by the gluino/squark masses and coincides with them.

The situation appears to be completely different for the gluon fragmentation functions (f.f's) (Fig. 1). The regular and the SQCD evolved f.f.'s differ largely in the diffractive region, and this clearly will show up in the spectrum of the primary protons if the decaying state has a supersymmetric content. As we raise the final evolution scale we start seeing more pronounced differences between regular and supersymmetric distributions. We have shown in Fig. 2 the squark f.f.'s for all the flavours and the one of the gluino for comparison. The scalar charm distribution appear to grow slightly faster then the remaining scalar ones. The gluino f.f. is still the fastest growing at small-x values. As we have mentioned in the introduction, the second important scale appearing in the analysis of UHECR concerns the interaction of the primary protons with the nuclei in the athmosphere. These interactions, estimated to be in the several TeV's, may require a supersymmetric analysis. Fig. 3 shows the impact of the supersymmetric evolution on the parton distributions of gluons, squarks and gluinos. The small-x rapid grwth of the gluino distribution is visible and points toward an interesting effect in the hadronic cross section of the primaries.

SUMMARY

In a few years several experiments, including the Pierre Auger experiment [3], will start collecting data from cosmic rays. The issue of the origin of UHECR will be -hopefully- clarified. While the link of UHECR to Active Galacti Nuclei (AGN) has been disfavored on the basis of a quite homogeneous distributions, the local origin of these events remains an open possibility. Potential meta–stable superheavy string relics have been suggested as dark matter candidates, as well as potential sources for the UHECR [4]. A QCD/SQCD analysis of these events is in progress. With the forthcoming experimental data [3], and improved theoretical analysis, along the lines discussed here, cosmic ray physics enters an exciting new era, with potentially ground–breaking discoveries.

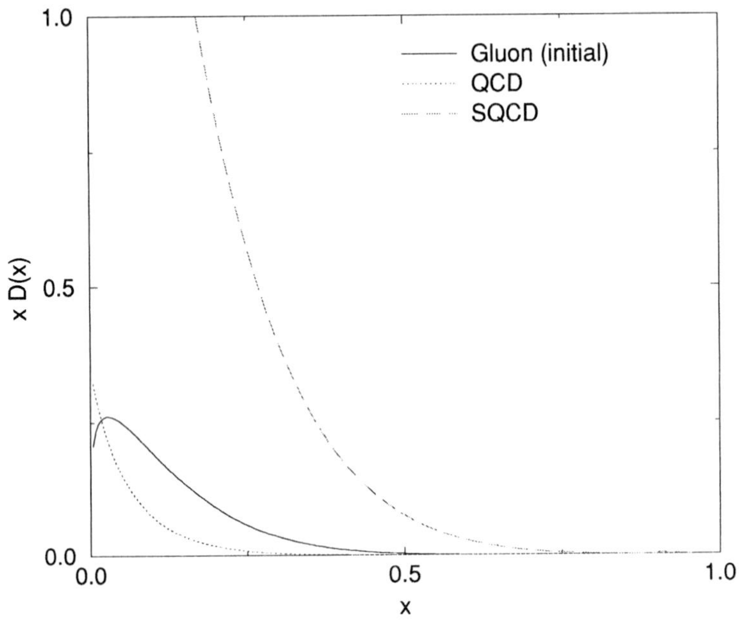

FIGURE 1. The gluon fragmentation function $xD_g^{p,\bar{p}}(x,Q^2)$ at the lowest scale (input) $Q_0 = 10$ GeV, and its evolved QCD (regular) and SQCD/QCD evolutions with $Q_f = 10^3$ GeV.The SUSY fragmentation scale is chosen to be 200 GeV.

ACKNOWLEDGMENTS

This work is dedicated to the memory of Prof. Nathan Isgur.

A.F. thanks the CERN theory division for hospitality. The work of C.C. is supported in part by INFN (iniziativa specifica BARI-21) and by MURST. The work of A.F. is supported by PPARC.

REFERENCES

1. C. Corianò and A.E. Faraggi, **hep-ph/0106326**.
2. B.A. Kniehl, G. Kramer, B. Pötter, *Nucl. Phys.* **B582** (2000) 514
3. The Pierre Auger observatory, The Auger Collaboration, *Nucl. Phys. Proc. Suppl.* **B85** (2000) 324.

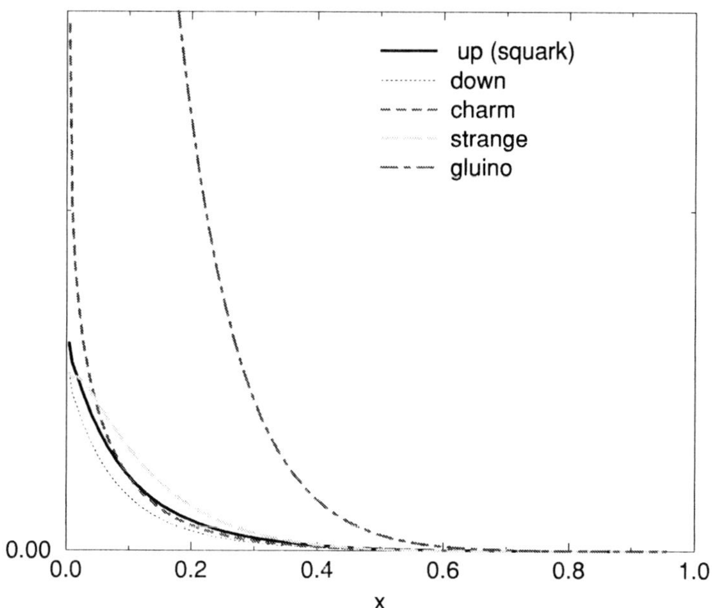

FIGURE 2. The fragmentation functions of squarks and gluino at the lowest scale (input) $Q_0 = 10$ GeV, with $Q_f = 10^3$ GeV and SUSY scale 200 GeV.

4. S. Chang, C. Corianò, and A.E. Faraggi, *Phys. Lett.* **B397** (1997) 76; *Nucl. Phys.* **B477** (1996) 65; C. Corianò, A.E. Faraggi and M. Plumacher **hep-ph/0107053**, to appear on *Nucl. Phys.***B**.

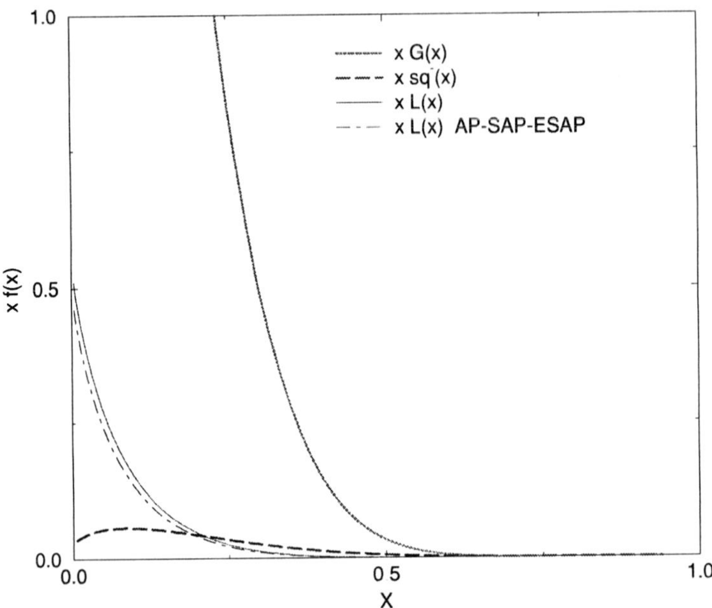

FIGURE 3. $xf(x)$ in the AP-ESAP evolution with a very large final scale $Q_f = 10^3$ GeV and with a squark mass $m_{\tilde{q}} = 100$ GeV. Shown are the non-singlet squark, the gluon and the gluino distributions for the AP-ESAP evolution. The gluino distribution for the AP-SAP-ESAP evolution is also shown (with $m_{2\lambda} = 40$ GeV).

HEAVY QUARKS

Charmless 2-body B decays: a way to α and γ

Fernando Ferroni
INFN Roma and Università di Roma La Sapienza

Abstract. The first year of running of the two B-factories, KEKB and PEPII has yielded a copious mess of results on 2-body charmless B-decays. The outstanding operations of the two machines combined with Belle and BaBar detectors excellent performances allow the claim that we are on the right track to be able soon to measure the two angles (α and γ) of the Unitarity Triangle.

I INTRODUCTION

The study of B meson decays into charmless hadronic final states plays an important role in the understanding of CP violation. In the Standard Model, all CP-violating phenomena are a consequence of a single complex phase in the Cabibbo-Kobayashi-Maskawa (CKM) quark-mixing matrix [1]. Measurements of the rates and charge asymmetries for B decays into the charmless final states $\pi\pi$ and $K\pi$ can be used to constrain the angles α and γ of the Unitarity Triangle.

The successful exploitation of the recently constructed B-factories has already given clear indication that this goal is achievable though an intense effort both from the experimental and theoretical side will be needed. I will briefly review the status of the accelerators, mention the crucial feature of the detectors with respect to this physics, describe the analysis technique, quote the results and throw a glance into the future perspectives.

II MACHINE PERFORMANCES

PEPII and KEKB have been conceived as B-factories. The concept requires the achievement at the same time of a very high peak luminosity, a very fast re-filling of the machine with the technique of the top-off and a detector able to take data with an efficiency as close to one as possible. Minimal time given to machine development and still the capability of constantly improving the accelerator performances and a detector not requiring major intervention for the entire run period, typically 10 months. It looks a very stringent set of requirements but amazingly indeed both the accelerators and their detectors (BaBar [2] and Belle [3]) have got into

CP602, *QCD@Work: International Workshop on Quantum Chromodynamics*
edited by P. Colangelo and G. Nardulli

this operation mode in an incredibly short time, matter of few months. Their performances at the time of this conference are illustrated in Fig. 1 and Fig. 2.

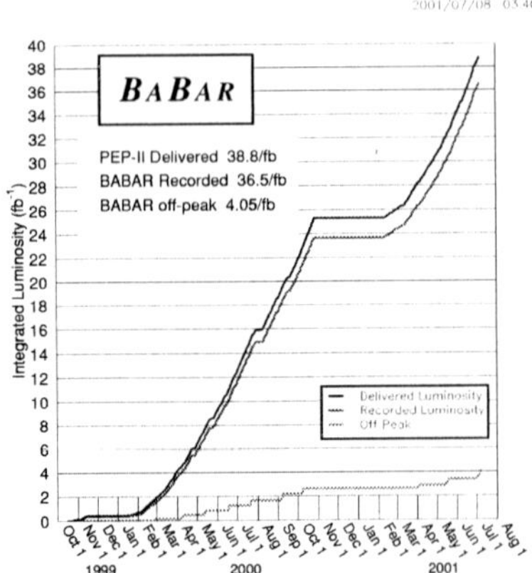

FIGURE 1. BaBar performances as a function of time.

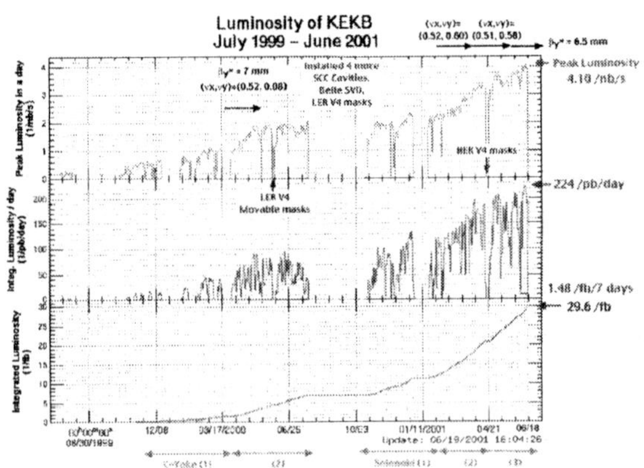

FIGURE 2. Belle performances as a function of time.

The peak luminosity in KEKB has exceeded 4×10^{33} cm^{-2} s^{-1} and both BaBar and Belle have routinely integrated 150-200 pb^{-1}/day. Every previous performance of any electron-positron collider has been surpassed and there is a clear indication for future improvements.

III DETECTOR PERFORMANCES

The crucial detector issue in this type of analysis is the capability of separating kaons from pions at relatively high energies (2-4 GeV). The two competing experiments both achieve this goal with comparable performances though with completely different detector technologies. Although the physics principle is the same, based on Cherenkov effect, Belle makes use of a set of threshold counters made by Aerogel (ACC [3]) while BaBar has developed an innovative differential counter (DIRC [4]) which makes use of the internal reflection of the Cherenkov light into synthetic quartz bars. The performances are shown in Fig. 3 for Belle and in Fig. 4 for BaBar. The knowledge of efficiencies and mis-identification rates is greatly facilitated by the use of the tagged decay of $D^{*+} \to D^0\pi^+$ ($D^0 \to K^-\pi^+$) which is an almost pure sample where the sign of the kaon is unambiguosly determined (up to DCSD) by the one of the soft pion accompanying the D^0.

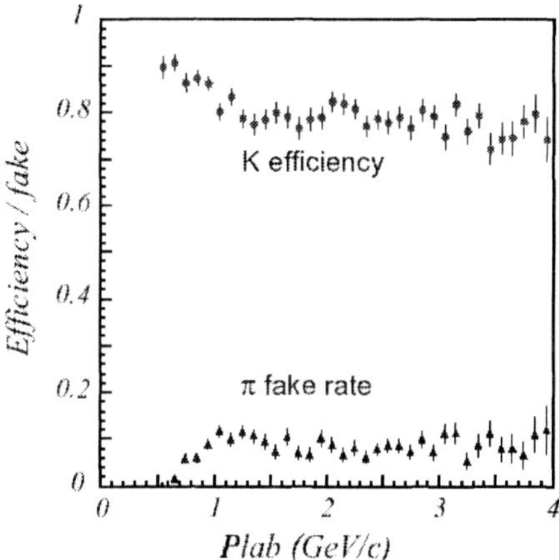

FIGURE 3. Belle K/π separation as a function of momentum.

IV ANALYSIS TECHNIQUE

The dominant background to these modes is from continuum $q\bar{q}$ process. The suppression therefore is mainly performed by exploiting the event topology which is spherical for $b\bar{b}$ and jet-like for $q\bar{q}$ in $\Upsilon(4S)$ rest frame. Typical variables include the event sphericity, the angle between the B candidate thrust axis and the thrust axis of the rest of the event, and the Fox-Wolfram moments. These variables are used to perform an event preselection. After this step the strategies for extracting the still tiny signal slightly differ between the two experiments. The B meson candidates

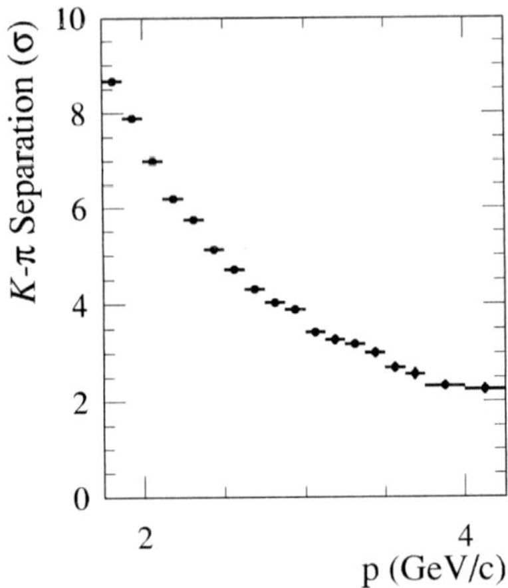

FIGURE 4. BaBar K/π separation as a function of momentum in the range of interest for this measurement.

are reconstructed using the beam constrained mass $m_{ES} = \sqrt{E_{\rm beam}^2 - p_B^2}$ and the energy difference , $\Delta E = E_B - E_{\rm beam}$ where $E_{\rm beam} = \sqrt{s}/2$.

A BaBar

An unbinned maximum likelihood fit, following CLEO [5] pioneeering work, is employed to determine signal yields. A total of eight parameters in the most difficult of the cases, namely the h'^+h^-, is used:

- $N_{\pi\pi}$, the number of $B \to \pi^+\pi^-$ decays;
- $N_{K\pi}$, the number of $B \to K^+\pi^-$ decays;
- $A_{K\pi}$, the observed asymmetry between $B \to K^+\pi^-$ and $B \to K^-\pi^+$ decays, $(N_{K^+\pi^-} - N_{K^-\pi^+})/(N_{K^+\pi^-} + N_{K^-\pi^+})$;
- N_{KK}, the number of $B \to K^+K^-$ decays;
- $N_{b\pi\pi}$, the number of background $\pi^+\pi^-$ candidates;
- $N_{bK\pi}$, the number of background $K^\pm\pi^\mp$ candidates;
- $A_{bK\pi}$, the observed asymmetry between the number of background $K^+\pi^-$ and $K^-\pi^+$ candidates;
- N_{bKK}, the number of background K^+K^- candidates.

Five quantities from each h'^+h^- candidate are used as inputs to the fit: m_{ES}, ΔE, $\mathcal{F}$ and θ_+ and θ_-, the θ_c values measured by the DIRC for the h'^+ and h^- track, respectively. $\mathcal{F}$ is constructed from the scalar sum of the CM momenta of all tracks and photons, excluding the B candidate, entering nine concentric cones centered on the thrust axis of the B_{hh} candidate. In all other cases a simpler form for the likelihood is used. The functional forms of the PDFs are derived from data samples that are independent of the sample used in the fit. These include: off-resonance data, on-resonance data from ΔE sidebands, control samples of fully reconstructed $B^- \to D\pi^-$ decays, control samples of $c\bar{c}\, toD^{*+}X$ decays and Monte Carlo simulated events. The PDF for each component is the product of the PDFs for each of the fit input variables:

$$\mathcal{P}^k = \mathcal{P}^k_{m_{ES}} \mathcal{P}^k_{\Delta E} \mathcal{P}^k_{\mathcal{F}} \mathcal{P}^k_{\theta_+} \mathcal{P}^k_{\theta_-}$$

.

The likelihood is thus:

$$L = e^{-N'} (N')^N \prod_{i=1}^{N} \mathcal{P}_i \, .$$

B Belle

Also in Belle a likelihood is constructed, by using a slightly different set of variables. Main differences with BaBar are the use of $cos\theta_B$ (the direction of B flight direction) and $cos\theta_{hh}$ (the decay axis direction) and the replacement of $\mathcal{F}$ with a Super Fox-Wolfram algorithm. The other difference is the cut on the π(K) likelihood ratio $\mathcal{R}_{\pi(\mathrm{K})} = \mathcal{L}_{\pi(\mathrm{K})}/(\mathcal{L}_\pi + \mathcal{L}_K)$ mainly obtained by ACC response but also including the dE/dx measured in the Drift Chamber and Time of Flight information.

V RESULTS

Branching fractions are measured, or upper limits are set for most of the charmless two body modes both by BaBar [6] (Table 1, Fig. 5) and Belle (Table 2, Fig. 6, Fig. 7)

The analysis technology is therefore established and the only important missing mode is $B^0 \to \pi^0\pi^0$ which combines two difficulties: the possibly very low yield and the worse energy resolution of all. Much more luminosity is needed for this mode.

VI ANGLE EXTRACTION

In general, CP violation is observable in the neutral B system through interference in mixing or decay, or in the interference *between* mixing and decay. Defining

TABLE 1. Summary of results for detection efficiencies (ε), fitted signal yields (N_S), statistical significances (S), measured branching fractions ($\mathcal{B}$), and charge asymmetries. The efficiencies include the branching fractions for $K^0 \to K_S^0 \to \pi^+\pi^-$ and $\pi^0 \to \gamma\gamma$. The 90% confidence level (C.L.) intervals for the charge asymmetries include the systematic uncertainties, which have been added in quadrature with the statistical errors.

Mode	ε (%)	N_S	S (σ)	$\mathcal{B}(10^{-6})$	A_{CP}	A_{CP}90% C.L.
$\pi^+\pi^-$	45	$41 \pm 10 \pm 7$	4.7	$4.1 \pm 1.0 \pm 0.7$		
$K^+\pi^-$	45	$169 \pm 17 \pm 13$	15.8	$16.7 \pm 1.6 \pm 1.3$	$-0.19 \pm 0.10 \pm 0.03$	$[-0.35, -0.03]$
K^+K^-	43	$8.2^{+7.8}_{-6.4} \pm 3.5$	1.3	< 2.5 (90% C.L.)		
$\pi^+\pi^0$	32	$37 \pm 14 \pm 6$	3.4	< 9.6 (90% C.L.)		
$K^+\pi^0$	31	$75 \pm 14 \pm 7$	8.0	$10.8^{+2.1}_{-1.9} \pm 1.0$	$0.00 \pm 0.18 \pm 0.04$	$[-0.30, +0.30]$
$K^0\pi^+$	14	$59^{+11}_{-10} \pm 6$	9.8	$18.2^{+3.3}_{-3.0} \pm 2.0$	$-0.21 \pm 0.18 \pm 0.03$	$[-0.51, +0.09]$
$\overline{K}^0K^+$	14	$-4.1^{+4.5}_{-3.8} \pm 2.3$	–	< 2.4 (90% C.L.)		
$K^0\pi^0$	10	$17.9^{+6.8}_{-5.8} \pm 1.9$	4.5	$8.2^{+3.1}_{-2.7} \pm 1.2$		

TABLE 2. Summary of the Belle results. The obtained signal yield (N_s), statistical significance (Σ), efficiency (ϵ), charge averaged branching fraction ($\mathcal{B}$) and its 90% confidence level upper limit (U.L.) are shown. In the calculation of $\mathcal{B}$, the production rates of B^+B^- and $B^0\overline{B}^0$ pairs are assumed to be equal. In the modes with K^0 mesons, N_s and ϵ are quoted for K_S^0, while $\mathcal{B}$ and U.L. are for K^0. Submode branching fractions for $K_S^0 \to \pi^+\pi^-$ and $\pi^0 \to \gamma\gamma$ are included in ϵ. The first and second errors in N_s and $\mathcal{B}$ are statistical and systematic errors, respectively.

Mode	N_s	Σ	ϵ [%]	$\mathcal{B}$ [$\times 10^{-5}$]	U.L. [$\times 10^{-5}$]
$B^0 \to \pi^+\pi^-$	$17.7\ ^{+7.1}_{-6.4}\ ^{+0.3}_{-1.1}$	3.1	28.1	$0.56\ ^{+0.23}_{-0.20} \pm 0.04$	–
$B^+ \to \pi^+\pi^0$	$10.4\ ^{+5.1}_{-4.3}\ ^{+1.2}_{-1.6}$	2.7	12.0	$0.78\ ^{+0.38}_{-0.32}\ ^{+0.08}_{-0.12}$	1.34
$B^0 \to K^+\pi^-$	$60.3\ ^{+10.6}_{-9.9}\ ^{+2.7}_{-1.1}$	7.8	28.0	$1.93\ ^{+0.34}_{-0.32}\ ^{+0.15}_{-0.06}$	–
$B^+ \to K^+\pi^0$	$34.9\ ^{+7.6}_{-7.0}\ ^{+0.6}_{-2.0}$	7.2	19.2	$1.63\ ^{+0.35}_{-0.33}\ ^{+0.16}_{-0.18}$	–
$B^+ \to K^0\pi^+$	$10.3\ ^{+4.3}_{-3.6}\ ^{+0.4}_{-0.1}$	3.5	13.5	$1.37\ ^{+0.57}_{-0.48}\ ^{+0.19}_{-0.18}$	–
$B^0 \to K^0\pi^0$	$8.4\ ^{+3.8}_{-3.1}\ ^{+0.4}_{-0.6}$	3.9	9.4	$1.60\ ^{+0.72}_{-0.59}\ ^{+0.25}_{-0.27}$	–
$B^0 \to K^+K^-$	$0.2\ ^{+3.8}_{-0.2}$	–	24.0	–	0.27
$B^+ \to K^+\overline{K}^0$	$0.0\ ^{+0.9}_{-0.0}$	–	12.1	–	0.50

$\Delta t = t_{CP} - t_{tag}$, where t_{CP} and t_{tag} are the proper decay times of the CP and tagged B's, respectively, the normalized decay rate distribution f_+ (f_-) for $B_{CP} \to f$ when B_{tag} is a B^0 ($\overline{B}^0$) is given by

$$f_\pm(\Delta t) = \frac{e^{-|\Delta t|/\tau}}{4\tau} \left[1 \pm S_f \sin(\Delta m_d \Delta t) \mp C_f \cos(\Delta m_d \Delta t)\right],$$

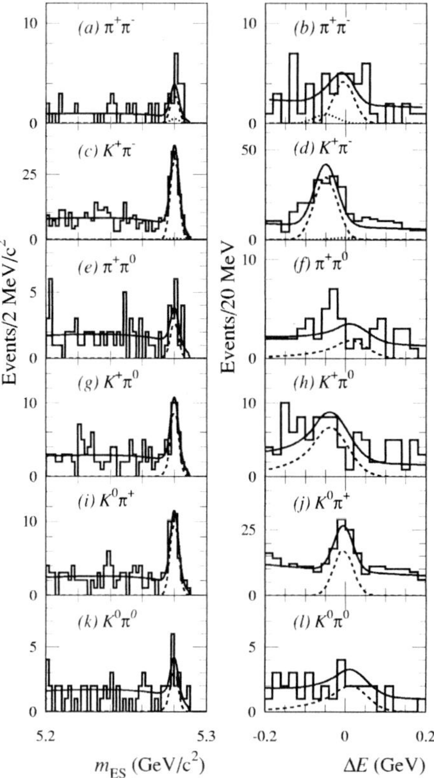

FIGURE 5. BaBar: the m_{ES} and ΔE distributions for the various modes, using likelihood ratio requirements described in the text. The solid curves represent the fit predictions for both signal and background; the dashed curve represents the given signal mode only and the dotted curve represents other modes of the same topology.

where τ is the average B^0 lifetime, Δm_d is the mixing frequency, and

$$S_f = \frac{2Im|\lambda|}{1+|\lambda|^2} \text{ and } C_f = \frac{1-|\lambda|^2}{1+|\lambda|^2}.$$

All interference effects are encapsulated in the physical quantity λ, defined as

$$\lambda \equiv \frac{q}{p}\frac{\bar{A}_f}{A_f} = \eta_f e^{-2i\beta}\frac{\bar{A}_{\bar{f}}}{A_f},$$

where $e^{-2i\beta}$ is the mixing phase, $A_f\,(\bar{A}_{\bar{f}})$ is the amplitude for the decay $B^0 \to f\,(\overline{B}^0 \to \bar{f})$, η_f is the CP eigenvalue of the final state, and the assumption of

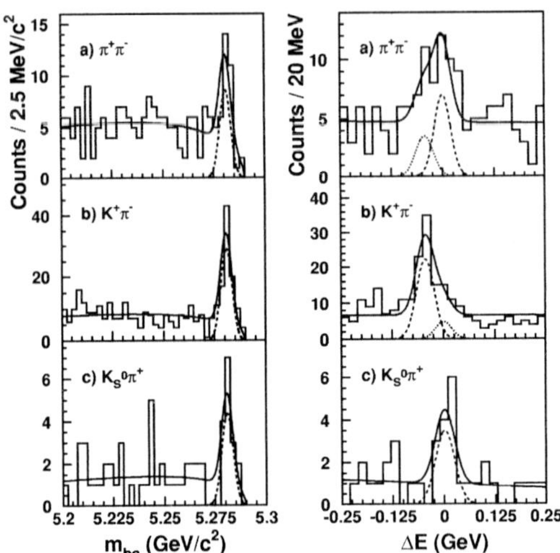

FIGURE 6. Belle: the m_{bc} (left) and ΔE (right) distributions, in the signal region of the other variable, for $B \to$ a) $\pi^+\pi^-$, b) $K^+\pi^-$ and c) $K_S^0\pi^+$. The fit function and its signal component are shown by the solid and dashed curve, respectively. In the $\pi^+\pi^-$ and $K^+\pi^-$ fits, the cross-talk components are shown by dotted curves.

no CP violation in mixing ($|q/p| = 1$) is implicit. Since the decay amplitudes are complex in general, observable CP violation effects can arise from interference between different decay amplitudes ($\left|\bar{A}_{\bar{f}}/A_f\right| \neq 1$) and interference between the mixing and decay weak phases.

For decays involving $b \to c\bar{c}s$ transitions, the dominant tree and penguin amplitudes carry the same weak phase, which is real in the Wolfenstein parameterization [8]. In this case λ simplifies to the familiar form

$$|\lambda| = 1, \; Im|\lambda| = -\eta_f sin2\beta,$$

where $\eta_f = -1\,(+1)$ for $J/\psi\, K_S^0\,(K_L^0)$. For decays involving $b \to u\bar{u}d$ transitions, the tree amplitude carries the weak phase $\gamma = \arg(V_{ub}^*)$ and λ becomes

$$\lambda = \eta_f e^{-2i(\beta+\gamma)} = \eta_f e^{2i\alpha},$$

assuming $\alpha + \beta + \gamma = \pi$ and ignoring the penguin contribution.

For the decay $B^0 \to \pi^+\pi^-$, $\eta_f = +1$ and therefore $Im|\lambda| = sin2\alpha$. However, the penguin amplitude carries the weak phase $-\beta = \arg(V_{td})$ and, in general, modifies both $|\lambda|$ and $Im|\lambda|$ [9]. The total decay amplitude can be written as

$$A_{\pi^+\pi^-} = Te^{i\gamma}e^{i\delta_T} + Pe^{-i\beta}e^{i\delta_P}, \; \bar{A}_{\pi^+\pi^-} = Te^{-i\gamma}e^{i\delta_T} + Pe^{i\beta}e^{i\delta_P},$$

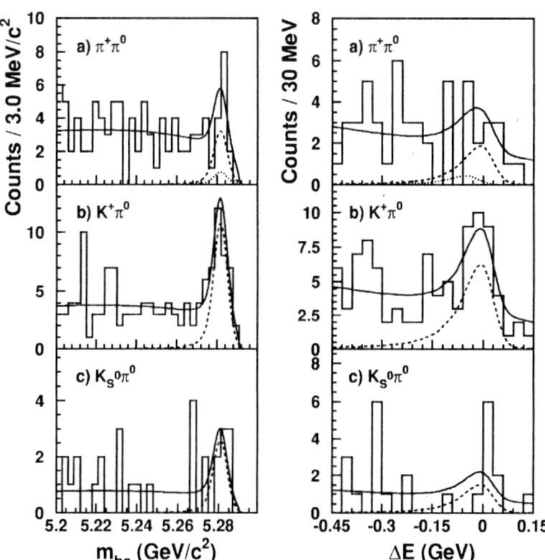

FIGURE 7. Belle: the m_{bc} (left) and ΔE (right) projections for $B \to$ a) $\pi^+\pi^0$, b) $K^+\pi^0$ and c) $K^0_S\pi^0$. For $K^+\pi^0$, a K mass is assumed for the charged particle. The projection of the two-dimensional fit onto each variable and its signal component are shown by the solid and dashed curve, respectively. In the $\pi^+\pi^0$ fit, the cross-talk from $K^+\pi^0$ is indicated by a dotted curve.

where T and P are the (real) tree and penguin amplitudes, respectively, and $\delta_{T(P)}$ are the corresponding strong phases. Defining $r = P/T$ and $\delta = \delta_P - \delta_T$, the expression for λ becomes [10]

$$\lambda_{\pi\pi} = e^{-2i\beta}\frac{e^{-i\gamma} + re^{i\beta}e^{i\delta}}{e^{i\gamma} + re^{-i\beta}e^{i\delta}} = e^{2i\alpha}\frac{1 - re^{i(\delta-\alpha)}}{1 - re^{i(\delta+\alpha)}} \equiv |\lambda_{\pi\pi}|\, e^{2i\alpha_{\text{eff}}}.$$

It is possible to extract α in the presence of penguins with little or no theoretical error using an isospin analysis [11]. In addition to $S_{\pi\pi}$ and $C_{\pi\pi}$, this analysis requires measurements of the separate branching fractions for $B^0 \to \pi^0\pi^0$ and $\overline{B}^0 \to \pi^0\pi^0$, as well as the charge-averaged branching fraction for $B^\pm \to \pi^\pm\pi^0$. However, unless the $\pi^0\pi^0$ rate is surprisingly large, it will be several years before such an analysis is possible. So, the best chance in the near future might just be to calculate the shift in α using factorization in the heavy quark limit [12], with assumptions about the values of β and $|V_{ub}/V_{cb}|$. This approach remains however largely untested.

Concerning γ the hope is to parameterize the dependence of the measured branching fractions and the observed direct CP asymmetries of each individual decay as a function of this angle, following a model based on an improved factorization and be eventually able to perform a global fit to the data yielding a consistent result for this parameter. Needless to say that also this approach is by now untested. An example of the prediction [13] in this field is given in Fig. 8. The field is very active and other predictions [14] are available.

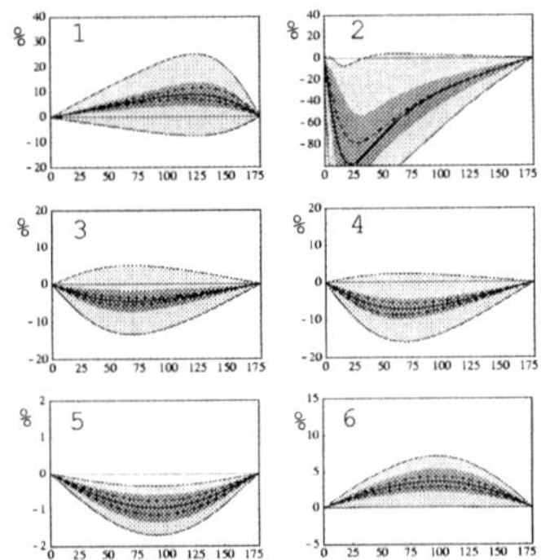

FIGURE 8. Direct CP asymmetries expected [13] in several charmless 2-body B-decays as a function of the angle γ (deg) of the Unitarity Triangle. 1)$A_{CP}(\pi^+\pi^-)$ 2) $A_{CP}(\pi^0\pi^0)$ 3)$A_{CP}(\pi^-K^+)$ 4)$A_{CP}(\pi^0K^+)$ 5) $A_{CP}(\pi^+K^0)$ 6)$A_{CP}(\pi^0K^0)$.

VII EXTRAPOLATION TO NEAR FUTURE

Results will improve with luminosity. Systematics will almost never limit these measurements. Extrapolation to future results are therefore simple up to the extent that the luminosity improvement program of the two competing projects will keep the promised pace. In terms of direct CP asymmetries, taking as a base the existing results, shown in Fig. 9, it is easy to predict (with the above mentioned caveat) that normalizing to BaBar (20 fb^{-1}) the combined Belle/BaBar harvest will give in the following years:

- 2001: divide errors by 2
- 2002: divide errors by 3+/- 0.5
- 2005: divide errors by 6+/-1

So that deviation from symmetry at 1% level will be testable. The same scaling law will apply to Branching Fraction determination. Much more challenging will be the determination of α. The error from the time-dependent fit to the asymmetry of $B^0 \to \pi^+\pi^-$ will yield an error on $sin2\alpha_{eff}$ about 4 times worse that the one found on $sin2\beta$, also scaling with luminosity. The translation $\alpha \to \alpha_{eff}$ as we have seen is still work in progress. The concluding message is therefore: stay tuned, exciting times are coming for this physics.

	CLEO	Belle	BaBar	Average
	Branching fractions			
$\pi^+\pi^-$	4.3+/-1.7	5.6+/-2.3	4.1+/-1.2	**4.4+/-0.9**
$K^+\pi^-$	17.2+/-2.8	19.3+/-3.7	16.7+/-2.1	**17.3+/-1.5**
$K^0\pi^+$	18.2+/-4.9	13.7+/-6.0	18.2+/-3.9	**17.3+/-2.7**
$K^+\pi^0$	11.6+/-3.3	16.3+/-3.8	10.8+/-2.3	**12.1+/-1.7**
$\pi^+\pi^0$	5.6+/-3.1	7.8+/-3.9	5.1+/-2.2	**5.7+/-1.6**
$K^0\pi^0$	14.6+/-6.4	16.0+/-7.6	8.2+/-3.3	**10.4+/-2.7**
	Direct CP Asymmetries			
$K^+\pi^-$	-0.04+/-0.16	0.04+/-0.18	-0.19+/-0.10	-0.11+/-0.08
$K^0\pi^+$	0.18+/-0.24		-0.21+/0.18	-0.07+/-0.14
$K^+\pi^0$	-0.29+/-0.23	0.02+/-0.22	0.00+/0.18	-0.07+/-0.12

FIGURE 9. Results from CLEO [15], Belle [16] and BaBar [6] averaged together.

REFERENCES

1. N. Cabbibo, Phys. Rev. Lett. **10** (1963) 531; M. Kobayashi and T. Maskawa, Prog. Theor. Phys. **49**, (1973) 652.
2. B. Aubert et al, BaBar Collaboration, hep-ex/0105044, submitted to Nucl.Instrum.Meth. , 2001.
3. A.Abashian et al. Belle Collaboration, KEK Progress Report 2000-4, to appear in Nucl.Instrum.Meth., 2000.
4. I. Adam et al., Nucl. Phys. Proc. Suppl. **93** (2001) 340.
5. R. Godang et al., CLEO Collaboration, Phys. Rev. Lett. **80**, (1998) 3456.
6. B. Aubert et al., BaBar Collaboration, hep-ex/0105061, to appear in Phys. Rev. Lett., (2001).
7. K. Abe et al., Belle Collaboration , Phys. Rev. Lett. **87**, (2001) 101801.
8. L. Wolfenstein, Phys. Rev. Lett. **51**(1983) 1945.
9. For a review see Y. Nir and H.R. Quinn, Ann. Rev. Nucl. Part. Sci. **42** (1992) 211, and references therein.
10. Y. Grossman and H.R. Quinn, Phys. Rev. D **56**, (1997) 7259.
11. M. Gronau and D. London, Phys. Rev. Lett. **65**, (1990) 3381 ; M. Gronau, J.L. Rosner and D. London, Phys. Rev. Lett. **73** (1994) 21.
12. M. Beneke, G. Buchalla, M. Neubert, and C.T. Sachrajda, Nucl. Phys. B **606** (2001) 245 .
13. M. Neubert, talk at this conference.
14. M. Ciuchini et al., Phys. Lett. B **515** (2001) 33; C. Isola et al., Phys. Rev. D **64** (2001) 014029; Y.Y. Keum et al., Phys.Rev. D **63** (2001) 074006; A.I. Sanda, Nucl. Instrum. Meth. A **462** (2001) 39.

15. S. Chen et al., CLEO Collaboration, Phys. Rev. Lett. **85** (2000) 525.
16. T. Iijima for the Belle Collaboration, hep-ex/0105005, (2000).

CP Violation at the B Factories

L. Lanceri

Università di Trieste, Dipartimento di Fisica and INFN, Trieste, Italy

Abstract. Recent results and future prospects are reviewed for the measurement of time-dependent *CP*-violating asymmetries in neutral *B* decays at the PEP-II and KEKB asymmetric energy e^+e^- colliders, with emphasis on the determination of the Standard Model *CP* violation parameter $\sin 2\beta$.

After becoming operational in 1999, the asymmetric energy e^+e^- colliders PEP-II and KEKB recently reached peak luminosities in excess of 4×10^{33} $\mathrm{cm}^{-2}\mathrm{s}^{-1}$. The *BABAR* and Belle Collaborations have been very efficient in collecting data at the $\Upsilon(4S)$ resonance; they recorded the largest samples now available of events with decays of $B\bar{B}$ meson pairs.

Both experiments pursue a wide range of heavy quark physics studies, having their main focus at present on rare *B* decays and on the search for *CP* violation, previously observed only in the neutral *K* meson system [1].

Preliminary results from the two Collaborations were reviewed at this workshop, with emphasis on the analysis of neutral *B* decays to *CP* eigenstates containing charmonium. The time-dependent *CP* asymmetries of these final states are the result of the interference between mixing and decay amplitudes, and allow a theoretically and experimentally clean test of the explanation of *CP* violation proposed by Kobayashi and Maskawa, as a phase in the three-generation CKM quark-mixing matrix [2]. From these measurements the Standard Model *CP* violation parameter $\sin 2\beta$ can be determined [3].

Both experiments require the reconstruction of the decay of one *B* meson to a *CP* eigenstate, the inclusive determination of the flavor of the accompanying (tagging) *B* meson, and the measurement of the time difference Δt between the two decays, from the separation of the decay vertices along the Lorentz boost of the $\Upsilon(4S)$. The $\sin 2\beta$ parameter is estimated fitting the flavor-tagged Δt distributions with likelihood methods. Detailed descriptions of experimental resolution functions and of tagging efficiencies are obtained from control samples of fully reconstructed *B* decays to flavor eigenstates.

Shortly after this workshop was held, the *BABAR* Collaboration announced the observation of *CP* violation in the decays of neutral *B* mesons[4], soon followed by Belle [5]. These exciting results are based on data samples of 32 and 31.3 million $B\bar{B}$ pairs respectively, corresponding to an integrated luminosity of almost 30 fb^{-1} per experiment.

In the *BABAR* analysis, the sample of 803 tagged events includes as *CP* eigenstate modes $J/\psi K^0_S$, $\psi(2S)K^0_S$, $\chi_{c1}K^0_S$, $J/\psi K^0_L$, and $J/\psi K^{*0}(K^{*0}\to K^0_S\pi^0)$, reconstructed with an average purity of 80%. The effective tagging efficiency, including the dilution due to wrong tags, is about 26%. The control samples include about 7600 (6800) events with a fully reconstructed neutral (charged) *B* decay. The simultaneous fit to all *CP* modes

CP602, *QCD@Work: International Workshop on Quantum Chromodynamics*
edited by P. Colangelo and G. Nardulli

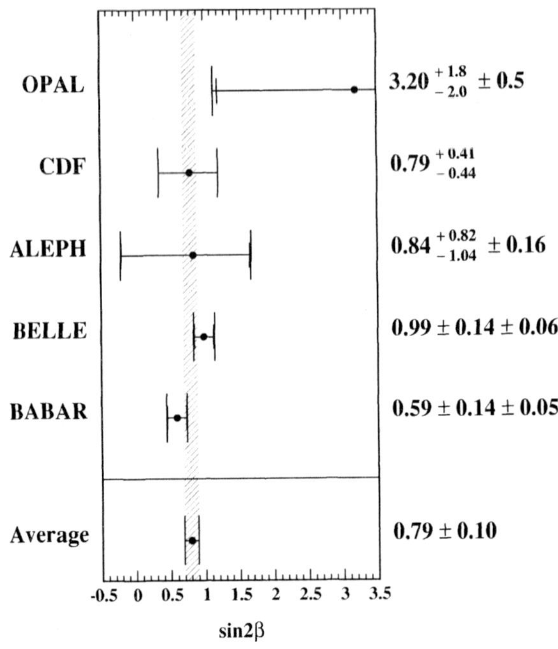

FIGURE 1. Summary of experimental results on $\sin 2\beta$.

yields:

$$\sin 2\beta = 0.59 \pm 0.14(\text{stat}) \pm 0.05(\text{syst}). \tag{1}$$

The similar analysis performed by Belle, including also the $\eta_c K_S^0$ *CP* mode, gives:

$$\sin 2\beta = 0.99 \pm 0.14(\text{stat}) \pm 0.06(\text{syst}). \tag{2}$$

In both analyses, the dominant sources of systematic errors are uncertainties in the parameterization of Δt resolution functions, in the wrong tag fractions, and in the level, composition and possible *CP* asymmetries of the backgrounds; the overall systematic uncertainty can be expected to decrease, as more data will become available for systematics studies based on control samples.

An overview of experimental results [6], and the present world average for $\sin 2\beta$, are shown in Fig.1. Both PEP-II and KEKB are planning substantial improvements in peak luminosity. According to the present projections, each experiment should be able to integrate about 500 fb^{-1} by year 2005: with respect to the results discussed above, this correponds to an increase by more than an order of magnitude. Besides producing a substantial improvement in the present determination of $\sin 2\beta$, *B* Factories can be expected to be at the forefront of *B* physics and of indirect searches for new physics for the next few years.

REFERENCES

1. J.H. Christensen *et al.*, Phys. Rev. Lett. **13**, 138 (1964).
2. N. Cabibbo, Phys. Rev. Lett. **10**, 531 (1963);
   M. Kobayashi and T. Maskawa, Prog. Th. Phys. **49**, 652 (1973).
3. A.B. Carter and A.I. Sanda, Phys. Rev. D **23**, 1567 (1981);
   I.I. Bigi and A. Sanda, Nucl. Phys. B **193**, 85 (1981).
4. The *BABAR* Collaboration, B. Aubert *et al.*, Phys. Rev. Lett. **87**, 091801 (2001).
5. The Belle Collaboration, K. Abe *et al.*, Phys. Rev. Lett. **87**, 091802 (2001).
6. The OPAL Collaboration, K. Ackerstaff *et al.*, Eur. Phys. Jour. C **5**, 379 (1998);
   The ALEPH Collaboration, R. Barate *et al.*, Phys. Lett. B **492**, 259 (2000);
   The CDF Collaboration, T. Affolder *et al.*, Phys. Rev. D **61**, 072005 (2000).

Aspects of QCD Factorization

Matthias Neubert

Newman Laboratory of Nuclear Studies, Cornell University, Ithaca, NY 14853, USA

Abstract. The QCD factorization approach provides the theoretical basis for a systematic analysis of nonleptonic decay amplitudes of B mesons in the heavy-quark limit. After recalling the basic ideas underlying this formalism, several tests of QCD factorization in the decays $B \to D^{(*)}L$, $B \to K^*\gamma$, and $B \to \pi K, \pi\pi$ are discussed. It is then illustrated how factorization can be used to obtain new constraints on the parameters of the unitarity triangle.

INTRODUCTION

In many years of intense experimental and theoretical investigations the flavor sector of the Standard Model has been explored in great detail by studying mixing and weak decays of B mesons and kaons. CP violation has been observed in K–$\bar{K}$ mixing (1964), $K \to \pi\pi$ decays (1999), and most recently in the interference of mixing and decay in $B \to J/\psi K$ (2001). There is now compelling evidence that the Cabibbo–Kobayashi–Maskawa (CKM) mechanism accounts for the dominant source of CP violation in low-energy hadronic weak interactions. Most notably, the discovery of a large CP asymmetry in the B system has established that CP is not an approximate symmetry of Nature. Rather, the smallness of CP-violating effects in kaon (and charm) physics reflects the hierarchy of CKM matrix elements.

Measurements of $|V_{cb}|$ and $|V_{ub}|$ in semileptonic B decays and of the magnitude and phase of V_{td} in K–$\bar{K}$ mixing, $B_{d,s}$–$\bar{B}_{d,s}$ mixing, and $B \to J/\psi K$ decays has helped to determine the parameters of the unitarity triangle $V_{ub}^* V_{ud} + V_{cb}^* V_{cd} + V_{tb}^* V_{td} = 0$ with good accuracy. The current values obtained at 95% confidence level are $\bar{\rho} = 0.21 \pm 0.12$, $\bar{\eta} = 0.38 \pm 0.11$ for the coordinates of the apex of the (rescaled) triangle, and $\sin 2\beta = 0.74 \pm 0.15$, $\sin 2\alpha = -0.14 \pm 0.57$, $\gamma = (61 \pm 16)^\circ$ for its angles [1]. These studies have established the existence of a CP-violating phase in the top sector of the CKM matrix, i.e., $\mathrm{Im}(V_{td}^2) \neq 0$. The next step in testing the CKM paradigm must be to explore the CP-violating phase in the bottom sector, i.e., $\gamma = \arg(V_{ub}^*) \neq 0$. In the Standard Model the two phases are, of course, related to each other. However, there is still plenty of room for New Physics to affect the magnitude of flavor violations in both mixing and weak decays (see, e.g., [2]). In particular, the present upper bound on γ is derived from the experimental limit on B_s–$\bar{B}_s$ mixing, which has not yet been seen experimentally and could well be affected by New Physics.

CP602, *QCD@Work: International Workshop on Quantum Chromodynamics*
edited by P. Colangelo and G. Nardulli

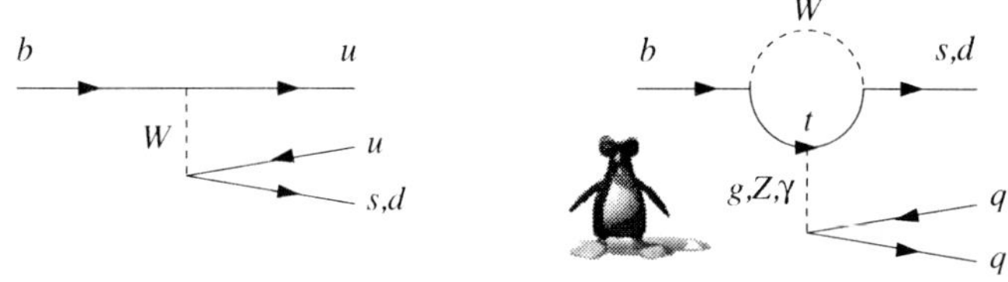

FIGURE 1. Tree and penguin topologies in charmless hadronic B decays.

Common lore says that measurements of γ are difficult. Several "theoretically clean"[1] determinations of this phase have been suggested (see, e.g., [3, 4]), which are extremely challenging experimentally. Likewise, "clean" measurements of $\alpha = \pi - \beta - \gamma$ [5, 6] are very difficult. It is more accessible experimentally to probe γ (and α) via the sizeable tree–penguin interference in charmless hadronic decays such as $B \to \pi K$ and $B \to \pi\pi$. The basic decay topologies contributing to these modes are shown in Figure 1. Experiment shows that the tree-to-penguin ratios in the two cases are roughly $|T/P|_{\pi K} \approx 0.2$ and $|P/T|_{\pi\pi} \approx 0.3$, indicating a sizeable amplitude interference. It is important that the relative weak phase between the two amplitudes can be probed not only via CP asymmetry measurements ($\sim \sin\gamma$), but also via measurements of CP-averaged branching fractions ($\sim \cos\gamma$). Extracting information about CKM parameters from the analysis of nonleptonic B decays is a challenge to theory, since it requires some level of control over hadronic physics, including strong-interaction phases. Such challenges, combined with the importance of the issue, is what triggers theoretical progress.

QCD FACTORIZATION

Hadronic weak decay amplitudes simplify greatly in the heavy-quark limit $m_b \gg \Lambda_{\rm QCD}$. This statement should not surprise those who have followed the dramatic advances in our theoretical understanding of B physics in the past decade. Many areas of B physics, from spectroscopy to exclusive semileptonic decays to inclusive rates and lifetimes, can now be systematically analyzed using heavy-quark expansions. Yet, the more complicated exclusive nonleptonic decays have long resisted any theoretical progress. The technical reason is that, whereas in most other applications of heavy-quark expansions one proceeds by integrating out heavy fields (leading to local operator product expansions), in the case of nonleptonic decays the large scale m_b enters as the energy carried by light fields. Therefore, in addition to hard and soft subprocesses collinear degrees of freedom become important. This complicates the understanding of hadronic decay amplitudes

[1] In this area of flavor physics many practitioners would consider a method to be "theoretically clean" only if it exclusively relies on elementary geometry (amplitude triangles) and, perhaps, isospin symmetry. We adopt the rationale followed in most other branches of high-energy physics and call a method theoretically clean if it relies on systematic expansions in small parameters. The methods discussed later in this talk are theoretically clean in this wider sense.

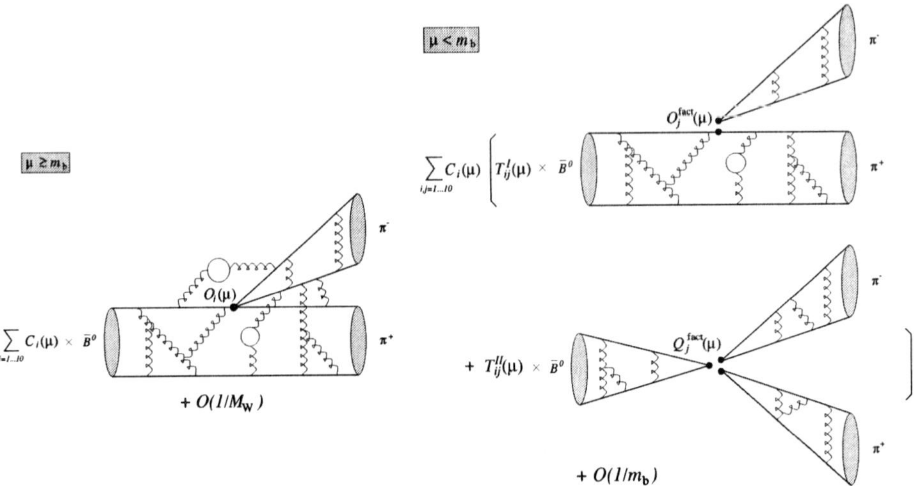

FIGURE 2. Factorization of short- and long-distance contributions in hadronic B decays. Left: Factorization of short-distance effects into Wilson coefficients of the effective weak Hamiltonian. Right: Factorization of hard "nonfactorizable" gluon exchanges into hard-scattering kernels (QCD factorization).

using the language of effective field theory. (Yet, very significant progress towards an effective field-theory description of nonleptonic decays has been made recently with the establishment of a "collinear–soft effective theory" [7]. The reader is referred to these papers for more details on this important development.)

The importance of the heavy-quark limit is based on the physical idea of color transparency [8, 9, 10]. A fast-moving light meson (such as a pion) produced in a point-like source (a local operator in the effective weak Hamiltonian) decouples from soft QCD interactions. More precisely, the couplings of soft gluons to such a system can be analyzed using a multipole expansion, and the first contribution (from the color dipole) is suppressed by a power of $\Lambda_{\rm QCD}/m_b$. The QCD factorization approach provides a systematic, model-independent implementation of this idea [11, 12]. It gives rigorous results in the heavy-quark limit, which are valid to leading power in $\Lambda_{\rm QCD}/m_b$ but to all orders of perturbation theory. Having obtained control over nonleptonic decays in the heavy-quark limit is a tremendous advance. We are now able to talk about power corrections to a well-defined and calculable limiting case, which captures a substantial part of the physics in these complicated processes.

The workings of QCD factorization can best be illustrated with the cartoons shown in Figure 2. The first graph shows the well-known concept of an effective weak Hamiltonian obtained by integrating out the heavy fields of the top quark and weak gauge bosons from the Standard Model Lagrangian. This introduces new effective interactions mediated by local operators $O_i(\mu)$ (typically four-quark operators) multiplied by calculable running coupling constants $C_i(\mu)$ called Wilson coefficients. This reduction in complexity (nonlocal heavy particle exchanges $\to$ local effective interactions) is exact up to corrections suppressed by inverse powers of the heavy mass scales. The resulting picture at scales at or above m_b is, however, still rather complicated, since gluon exchange is possible between any of the quarks in the external meson states. Additional simplifications occur when the renormalization scale μ is lowered below the scale m_b. Then color transparency comes to play and implies systematic cancellations of soft and collinear gluon exchanges. As a result, all "nonfactorizable" exchanges, i.e., gluons connecting the light meson at the "upper" vertex to the remaining mesons, are dominated by virtualities of order m_b and can be calculated. Their effects are absorbed into a new set of running couplings $T_{ij}^{I,II}(\mu)$ called hard-scattering kernels, as shown in the two graphs on the right-hand side. What remains are "factorized" four-quark and six-quark operators $O_j^{\rm fact}(\mu)$ and $Q_j^{\rm fact}(\mu)$, whose matrix elements can be expressed in terms of form factors, decay constants and light-cone distribution amplitudes. As before, the reduction in complexity (local four-quark operators $\to$ "factorized" operators) is exact up to corrections suppressed by inverse powers of the heavy scale, now set by the mass of the b quark.

The factorization formula is valid in all cases where the meson at the "upper" vertex is light, meaning that its mass is much smaller than the b-quark mass. The second term in the factorization formula (the term involving "factorized" six-quark operators) gives a power-suppressed contribution when the final-state meson at the "lower" vertex is a heavy meson (i.e., a charm meson), but its contribution is of leading power if this meson is also light. Aspects of this power counting will be discussed in more detail later.

Factorization is a property of decay amplitudes in the heavy-quark limit. Comparing the magnitude of "nonfactorizable" effects in kaon, charm and beauty decays, there can be little doubt about the relevance of the heavy-quark limit to understanding nonleptonic processes [13]. Yet, for phenomenological applications it is important to explore the structure of at least the leading power-suppressed corrections. While no complete classification of such corrections has been given to date, several classes of power-suppressed terms have been analyzed and their effects estimated. These estimates (with conservative errors) have been implemented in the phenomenological applications to be discussed later in this talk. Specifically, the corrections that have been analyzed are "chirally-enhanced" power corrections [11], weak annihilation contributions [12, 14], and power corrections due to nonfactorizable soft gluon exchange [15, 16, 17]. With the exception of the "chirally-enhanced" terms, no unusually large power corrections (i.e., corrections exceeding the naive expectation of 5–10%) have been identified so far. Nevertheless, it is important to refine and extend the estimates of power corrections. Fortunately, the QCD factorization approach has a wide range of applicability and makes many testable predictions. Ultimately, therefore, the data will give us conclusive evidence on the relevance of power-suppressed effects. Many tests can, in fact, already be done using existing data. Several examples will now be discussed in detail.

Tests of Factorization in $B \to D^{(*)}L$ Decays

In B decays into a heavy–light final state, when the light meson is produced at the "upper" vertex, the factorization formula assumes its simplest form. Then only the form factor term (the first graph on the right-hand side in Figure 2) contributes at leading power. This is also the place where QCD factorization is best established theoretically. In [12], the systematic cancellation of soft and collinear singularities was demonstrated explicitly at two-loop order. The proof of these cancellations has recently been extended to all orders in perturbation theory [18]. In order to complete a rigorous proof of factorization one would still have to show that the hard-scattering kernels are free of endpoint singularities stronger than $1/x$ or $1/(1-x)$ as one of the quarks in the light meson becomes a soft parton. It has been demonstrated that the kernels tend to a constant (modulo logarithms) at the endpoints in the so-called "large-β_0 limit" of QCD, i.e., to order $\beta_0^{n-1}\alpha_s^n$ for arbitrary n in perturbation theory [17]. However, it is an open question whether such a smooth behavior persists in higher orders of full QCD.

Let us first consider the decays $\bar{B}^0 \to D^{(*)+}L^-$, where L denotes a light meson. In this case the flavor content of the final state is such that the light meson can only be produced at the "upper" vertex, so factorization applies. One finds that process-dependent "nonfactorizable" corrections from hard gluon exchange, though present, are numerically very small. All nontrivial QCD effects in the decay amplitudes are then described by a quasi-universal coefficient $|a_1(D^{(*)}L)| = 1.05 \pm 0.02 + O(\Lambda_{\rm QCD}/m_b)$ [12]. For a given decay channel this coefficient can be determined experimentally from the ratio [8]

$$\frac{\Gamma(\bar{B}^0 \to D^{*+}L^-)}{d\Gamma(\bar{B}^0 \to D^{*+}l^-\nu)/dq^2|_{q^2=m_L^2}} = 6\pi^2|V_{ud}|^2 f_L^2\, |a_1(D^{(*)}L)|^2 .$$

Using CLEO data one obtains $|a_1(D^*\pi)| = 1.08 \pm 0.07$, $|a_1(D^*\rho)| = 1.09 \pm 0.10$, and $|a_1(D^*a_1)| = 1.08 \pm 0.11$, in good agreement with theory. This is a first indication that power corrections in these modes are under control, but more precise data are required for a firm conclusion. Other tests of factorization in B decays to heavy–light final states have been discussed in [12, 19, 20].

Recently, the experimental observation of unexpectedly large rates for color-suppressed decays such as $\bar{B}^0 \to D^{0(*)}\pi^0$ [21, 22] has attracted some attention. QCD factorization does not allow us to calculate the amplitudes for these processes in a reliable way. It predicts that these amplitudes are power-suppressed with respect to the corresponding $\bar{B}^0 \to D^{+(*)}\pi^-$ amplitudes, but only by one power of $\Lambda_{\rm QCD}/m_c$. Specifically, the prediction is that a certain ratio of isospin amplitudes approaches unity in the heavy-quark limit: $A_{1/2}/(\sqrt{2}A_{3/2}) = 1 + O(\Lambda_{\rm QCD}/m_c)$ [12]. Considering that charm is not a particularly heavy quark, we find that this scaling law is respected by the experimental data, which give $A_{1/2}/(\sqrt{2}A_{3/2}) = (0.70 \pm 0.11)\, e^{\pm i(27\pm 7)^\circ}$ for $B \to D\pi$ and $(0.72 \pm 0.08)\, e^{\pm i(21\pm 8)^\circ}$ for $B \to D^*\pi$ [13]. Assuming the hierarchy $m_b \gg m_c > \Lambda_{\rm QCD}$, a rough theoretical estimate of the amplitude ratio, $A_{1/2}/(\sqrt{2}A_{3/2}) \sim 0.75\, e^{-15^\circ i}$, had been obtained prior to the observation of the color-suppressed decays [12]. It anticipated the correct order of magnitude of the deviation from the heavy-quark limit.

Tests of Factorization in $B \to K^*\gamma$ Decays

The QCD factorization approach not only applies to nonleptonic decays, but also to other exclusive processes such as $B \to V\gamma$ and $B \to V\, l^+l^-$ (where $V = K^*, \rho, \ldots$ is a vector meson) [23, 24]. The resulting factorization formula is similar (but simpler) to that for B decays into two light mesons. Therefore, the study of exclusive radiative transitions not only extends the range of applicability of the method, but also provides a new testing ground for the factorization idea.

Interestingly, the analysis of isospin-breaking effects in radiative B decays gives a direct probe of power corrections to the factorization formula. Experimentally, it is found that [25, 26, 27]

$$\Delta_{0-} \equiv \frac{\Gamma(\bar{B}^0 \to \bar{K}^{*0}\gamma) - \Gamma(B^- \to \bar{K}^{*-}\gamma)}{\Gamma(\bar{B}^0 \to \bar{K}^{*0}\gamma) + \Gamma(B^- \to \bar{K}^{*-}\gamma)} = 0.11 \pm 0.07\,,$$

indicating (albeit with a large error) that isospin-breaking effects could be as large as 10% at the level of the decay amplitudes. Such effects are absent in the heavy-quark limit. A detailed theoretical analysis of the leading power-suppressed contributions leads to the prediction $\Delta_{0-} = (8.0^{+2.1}_{-3.2})\% \times (0.3/T_1^{B\to K^*})$ [28], where $T_1^{B\to K^*}$ is a tensor form factor, whose value is expected to be close to 0.3. By far the largest contribution to the result comes from an annihilation contribution involving the $(V-A)\otimes(V+A)$ penguin operator O_6 in the effective weak Hamiltonian. Therefore, the quantity Δ_{0-} is a sensitive probe of the magnitude and sign of the ratio $C_6/C_{7\gamma}$ of Wilson coefficients.

The above discussion shows that in the Standard Model one indeed expects a sizeable isospin breaking in the $B \to K^*\gamma$ decay amplitudes, in agreement with the current central experimental value. If the agreement persists as the data become more precise, this would not only test the penguin sector of the effective weak Hamiltonian, but also provide a quantitative test of factorization at the level of power corrections.

Tests of Factorization in $B \to \pi K, \pi\pi$ Decays

The factorization formula for B decays into two light mesons is more complicated because of the presence of the two types of contributions shown in the graphs on the right-hand side in Figure 2. The finding that these two topologies contribute at the same power in $\Lambda_{\mathrm{QCD}}/m_b$ is nontrivial [14] and relies on the heavy-quark scaling law $F^{B\to L}(0) \sim m_b^{-3/2}$ for heavy-to-light form factors. Whereas this scaling law has been obtained from several independent studies (see, e.g., [29, 30, 31]), it is not as rigorously established as the corresponding scaling law for heavy-to-heavy form factors. In the QCD factorization approach the kernels $T_{ij}^I(\mu)$ are of order unity, whereas the kernels $T_{ij}^{II}(\mu)$ contribute first at order α_s. Numerically, the latter ones give corrections of about 10–20% with respect to the leading terms. This is consistent with being of the same power but down by a factor of α_s. Therefore, the scaling laws that form the basis of the QCD factorization formula appear to work well empirically.

The factorization formula for B decays into two light mesons can be tested best by using decays that have negligible amplitude interference. In that way any sensitivity to the value of the weak phase γ is avoided. For a complete theoretical control over charmless hadronic decays one must control the magnitude of the tree topologies, the magnitude of the penguin topologies, and the relative strong-interaction phases between trees and penguins. It is important that these three key features can be tested separately. Once these tests are conclusive (and assuming they are successful), factorization can be used to constrain the parameters of the unitarity triangle. (Of course, alternative schemes such as pQCD [32] and "charming penguins" [33] must face the same tests.)

Magnitude of the Tree Amplitude. The magnitude of the leading $B \to \pi\pi$ tree amplitude can be probed in the decays $B^\pm \to \pi^\pm\pi^0$, which to an excellent approximation do not receive any penguin contributions. The QCD factorization approach makes an absolute prediction for the corresponding branching ratio [14],

$$\text{Br}(B^\pm \to \pi^\pm\pi^0) = \left[5.3^{+0.8}_{-0.4}\,(\text{pars.}) \pm 0.3\,(\text{power})\right] \cdot 10^{-6} \times \left[\frac{|V_{ub}|}{0.0035}\,\frac{F_0^{B\to\pi}(0)}{0.28}\right]^2,$$

which compares well with the experimental result $(5.6 \pm 1.5) \times 10^{-6}$ (see Table 7 in [14] for a compilation of the experimental data on charmless hadronic B decays). The theoretical uncertainties quoted are due to input parameter variations and to the modeling of the leading power corrections. An additional large uncertainty comes from the present error on $|V_{ub}|$ and the semileptonic $B \to \pi$ form factor. The sensitivity to these quantities can be eliminated by taking the ratio

$$\frac{\Gamma(B^\pm \to \pi^\pm\pi^0)}{d\Gamma(\bar{B}^0 \to \pi^+ l^- \bar{\nu})/dq^2|_{q^2=0}} = 3\pi^2 f_\pi^2 \underbrace{|a_1^{(\pi\pi)} + a_2^{(\pi\pi)}|^2}_{1.33^{+0.20}_{-0.11}\,(\text{pars.}) \pm 0.07\,(\text{power})} = (0.68^{+0.11}_{-0.06})\,\text{GeV}^2\,.$$

This prediction includes a sizeable ($\sim 25\%$) contribution of the hard-scattering term in the factorization formula (the lower graph on the right-hand side in Figure 2). Unfortunately, this ratio has not yet been measured experimentally.

Magnitude of the T/P Ratio. The magnitude of the leading $B \to \pi K$ penguin amplitude can be probed in the decays $B^\pm \to \pi^\pm K^0$, which to an excellent approximation do not receive any tree contributions. Combining it with the measurement of the tree amplitude just described, a tree-to-penguin ratio can be determined via the relation

$$\varepsilon_{\text{exp}} = \left|\frac{T}{P}\right| = \tan\theta_C \,\frac{f_K}{f_\pi} \left[\frac{2\text{Br}(B^\pm \to \pi^\pm\pi^0)}{\text{Br}(B^\pm \to \pi^\pm K^0)}\right]^{\frac{1}{2}} = 0.223 \pm 0.034\,.$$

The quoted experimental value of this ratio is in good agreement with the theoretical prediction $\varepsilon_{\text{th}} = 0.24 \pm 0.04\,(\text{pars.}) \pm 0.04\,(\text{power}) \pm 0.05\,(V_{ub})$ [14], which is independent of form factors but proportional to $|V_{ub}/V_{cb}|$. This is a highly nontrivial test of the QCD factorization approach. Recall that when the first measurements of charmless hadronic decays appeared several authors remarked that the penguin amplitudes were much larger than expected based on naive factorization models. We now see that QCD factorization reproduces naturally (i.e., for central values of all input parameters) the correct magnitude of the tree-to-penguin ratio. This observation also shows that there is no need to supplement the QCD factorization predictions in an ad hoc way by adding enhanced phenomenological penguin amplitudes, such as the "nonperturbative charming penguins" introduced in [33]. In their most recent paper [34], the advocates of charming penguins parameterize the effects of these animals in terms of a nonperturbative "bag

TABLE 1. Direct CP asymmetries in $B \to \pi K$ decays

	Experiment [35, 36, 37, 38]	Theory Beneke et al. [14]	Keum et al. [32]	Ciuchini et al. [33]
$A_{\rm CP}(\pi^+K^-)$ (%)	-4.8 ± 6.8	5 ± 9	-18	$\pm(17\pm6)$
$A_{\rm CP}(\pi^0K^-)$ (%)	-9.6 ± 11.9	7 ± 9	-15	$\pm(18\pm6)$
$A_{\rm CP}(\pi^-\bar{K}^0)$ (%)	-4.7 ± 13.9	1 ± 1	-2	$\pm(3\pm3)$

parameter" $\hat{B}_1 = (0.13\pm0.02)\,e^{i(188\pm82)^\circ}$ fitted to the data on charmless decays. By definition, this parameter contains the contribution from the perturbative charm loop, which is calculable in QCD factorization. Using the factorization approach as described in [14] we find that $\hat{B}_1^{\rm fact} = (0.09^{+0.03+0.04}_{-0.02-0.02})\,e^{i(185\pm3\pm21)^\circ}$, where the errors are due to input parameter variations and the estimate of power corrections. The perturbative contribution to the central value is 0.08; the remaining 0.01 is mainly due to weak annihilation. We conclude that, within errors, QCD factorization can account for the "charming penguin bag parameter", which is, in fact, dominated by short-distance physics.

Strong Phase of the T/P Ratio. QCD factorization predicts that (most) strong-interaction phases in charmless hadronic B decays are parametrically suppressed in the heavy-quark limit, i.e., $\sin\phi_{\rm st} = O[\alpha_s(m_b), \Lambda_{\rm QCD}/m_b]$. This implies small direct CP asymmetries since, e.g., $A_{\rm CP}(\pi^+K^-) \simeq -2|\frac{T}{P}|\sin\gamma\sin\phi_{\rm st}$. The suppression results as a consequence of systematic cancellations of soft contributions, which are missed in phenomenological models of final-state interactions. In many other schemes the strong-interaction phases are predicted to be much larger, and therefore larger CP asymmetries are expected. Table 1 shows that first experimental data provide no evidence for large direct CP asymmetries in $B \to \pi K$ decays. However, the errors are still too large to draw a definitive conclusion that would allow us to distinguish between different theoretical predictions.

Remarks on Sudakov Logarithms

In recent years, Li and collaborators have proposed an alternative scheme for calculating nonleptonic B decay amplitudes based on a perturbative hard-scattering approach [32]. From a conceptual point of view, the main difference between QCD factorization and this so-called pQCD approach lies in the latter's assumption that Sudakov form factors effectively suppress soft-gluon exchange in diagrams such as those shown in the graphs on the right-hand side in Figure 2. As a result, the $B \to \pi$ and $B \to K$ form factors are assumed to be perturbatively calculable. This changes the counting of powers of α_s. In particular, the nonfactorizable gluon exchange diagrams included in the QCD factorization approach, which are crucial in order to cancel the scale and scheme-dependence in the predictions for the decay amplitudes, are formally of order α_s^2 in the pQCD scheme and consequently are left out. Thus, to the considered order there are no loop graphs that could give rise to strong-interaction phases in that scheme. (However, in [32] large phases are claimed to arise from on-shell poles of massless propagators in tree diagrams. These phases are entirely dominated by soft physics. Hence, the prediction of large direct CP asymmetries in the pQCD approach rests on assumptions that are strongly model dependent.)

The assumption of Sudakov suppression in hadronic B decays is questionable, because the relevant "large" scale $Q^2 \sim m_b \Lambda_{\mathrm{QCD}} \sim 1\,\mathrm{GeV}^2$ is in fact not large for realistic b-quark masses. Indeed, one finds that the pQCD calculations are very sensitive to details of the $p_\perp$ dependence of the wave functions [39]. This sensitivity to infrared physics invalidates the original assumption of an effective suppression of soft contributions. The argument just presented leaves open the conceptual question whether Sudakov logarithms are relevant in the asymptotic limit $m_b \to \infty$. This question has not yet been answered in a satisfactory way.

NEW CONSTRAINTS ON THE UNITARITY TRIANGLE

The QCD factorization approach, combined with a conservative estimate of power corrections, offers several new strategies to derive constraints on CKM parameters. This has been discussed at length in [14], to which we refer the reader for details. Some of these strategies will be illustrated below. Note that the applications of QCD factorization are not limited to computing branching ratios. The approach is also useful in combination with other ideas based on flavor symmetries and amplitude relations. In this way, strategies can be found for which the residual hadronic uncertainties are simultaneously suppressed by three small parameters, since they vanish in the heavy-quark limit ($\sim \Lambda_{\mathrm{QCD}}/m_b$), the limit of SU(3) flavor symmetry ($\sim (m_s - m_q)/\Lambda_{\mathrm{QCD}}$), and the large-$N_c$ limit ($\sim 1/N_c$).

Determination of γ *with minimal theory input.* Some years ago, Rosner and the present author have derived a bound on γ by combining measurements of the ratios $\varepsilon_{\mathrm{exp}} = |T/P|$ and $R_* = \frac{1}{2}\Gamma(B^\pm \to \pi^\pm K^0)/\Gamma(B^\pm \to \pi^0 K^\pm)$ with the fact that for an arbitrary strong phase $-1 \le \cos\phi_{\mathrm{st}} \le 1$ [40]. The model-independent observation that $\cos\phi_{\mathrm{st}} = 1$ up to second-order corrections to the heavy-quark limit can be used to turn this bound into a determination of γ (once $|V_{ub}|$ is known). The resulting constraints in the $(\bar\rho, \bar\eta)$ plane, obtained under the conservative assumption that $\cos\phi_{\mathrm{st}} > 0.8$ (corresponding to $|\phi_{\mathrm{st}}| < 37°$) are shown in the left-hand plot in Figure 3 for several illustrative values of the ratio R_*. Note that for $0.8 < R_* < 1.1$ (the range preferred by the Standard Model) the theoretical uncertainty reflected by the widths of the bands is smaller than for any other constraint on $(\bar\rho, \bar\eta)$ except for the one derived from the $\sin 2\beta$ measurement. With present data the Standard Model is still in good shape, but it will be interesting to see what happens when the experimental errors are reduced.

Determination of $\sin 2\alpha$. With the help of QCD factorization it is possible to control the "penguin pollution" in the time-dependent CP asymmetry in $B \to \pi^+\pi^-$ decays, defined such that $S_{\pi\pi} = \sin 2\alpha \cdot [1 + O(P/T)]$. This is illustrated in the right-hand plot in Figure 3, which shows the constraints imposed by a measurement of $S_{\pi\pi}$ in the $(\bar\rho, \bar\eta)$ plane. It follows that even a result for $S_{\pi\pi}$ with large experimental errors would imply a useful constraint on the unitarity triangle. A first, preliminary measurement of the

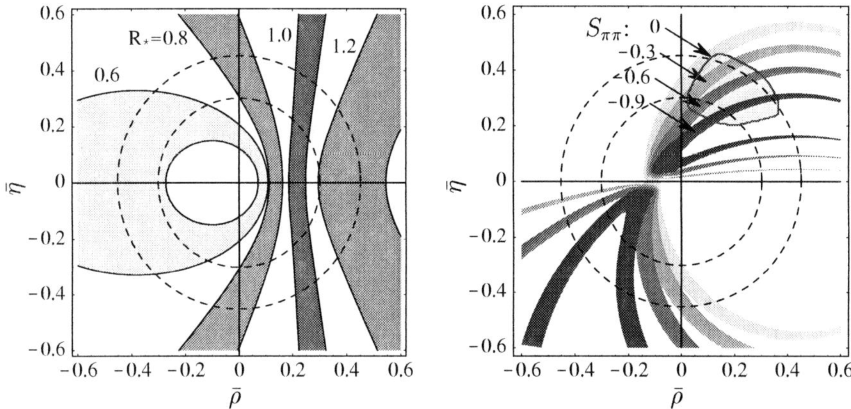

FIGURE 3. Left: Allowed regions in the $(\bar{\rho},\bar{\eta})$ plane corresponding to $\varepsilon_{\rm exp} = 0.22$ and different values of the ratio R_* as indicated. The widths of the bands reflect the total theoretical uncertainty. The current experimental values are $\varepsilon_{\rm exp} = 0.22 \pm 0.03$ and $R_* = 0.71 \pm 0.14$. Right: Allowed regions in the $(\bar{\rho},\bar{\eta})$ plane corresponding to different values of the mixing-induced CP asymmetry $S_{\pi\pi}$. The widths of the bands reflect the total theoretical uncertainty. The corresponding bands for positive values of $S_{\pi\pi}$ are obtained by a reflection about the $\bar{\rho}$ axis. The bounded light area is the allowed region obtained from the standard global fit of the unitarity triangle [1].

asymmetry has been presented by the BaBar Collaboration this summer. Their result is $S_{\pi\pi} = 0.03^{+0.53}_{-0.56} \pm 0.11$ [38].

Global Fit to $B \to \pi K, \pi\pi$ Branching Ratios. Various ratios of CP-averaged $B \to \pi K, \pi\pi$ branching fractions exhibit a strong dependence on γ and $|V_{ub}|$, or equivalently, on the parameters $\bar{\rho}$ and $\bar{\eta}$ of the unitarity triangle. From a global analysis of the experimental data in the context of the QCD factorization approach it is possible to derive constraints in the $(\bar{\rho},\bar{\eta})$ plane in the form of regions allowed at various confidence levels. The results are shown in Figure 4. The best fit of the QCD factorization theory to the data yields an excellent $\chi^2/n_{\rm dof}$ of less than 0.5. (We should add at this point that we disagree with the implementation of our approach presented in [34] and, in particular, with the numerical results labeled "BBNS" in Table II of that paper, which led the authors to the premature conclusion that the "theory of QCD factorization ... is insufficient to fit the data". Even restricting $(\bar{\rho},\bar{\eta})$ to lie within the narrow ranges adopted by these authors, we can find parameter sets for which QCD factorization fits the data with a good $\chi^2/n_{\rm dof}$ of less than 1.5.)

The results of this global fit are compatible with the standard CKM fit using semileptonic decays, K–$\bar{K}$ mixing and B–$\bar{B}$ mixing ($|V_{ub}|$, $|V_{cb}|$, ε_K, Δm_d, Δm_s, $\sin 2\beta$), although the fit prefers a slightly larger value of γ and/or a smaller value of $|V_{ub}|$. The combination of the results from rare hadronic B decays with $|V_{ub}|$ from semileptonic decays excludes $\bar{\eta} = 0$ at 95% CL, thus showing first evidence for the existence of a CP-violating phase in the bottom sector. In the near future, when the data become more precise, this will provide a powerful test of the CKM paradigm.

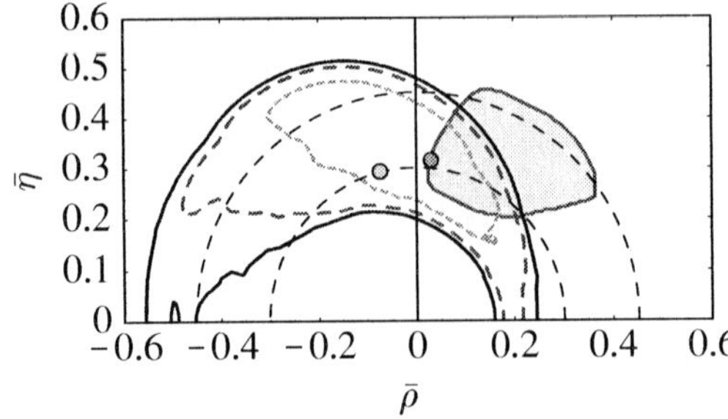

Decay Mode	Best Fit	Exp. Average
$B^0 \to \pi^+\pi^-$	4.6	4.4 ± 0.9
$B^\pm \to \pi^\pm\pi^0$	5.3	5.6 ± 1.5
$B^0 \to \pi^\mp K^\pm$	17.9	17.2 ± 1.5
$B^\pm \to \pi^0 K^\pm$	11.3	12.1 ± 1.7
$B^\pm \to \pi^\pm K^0$	17.7	17.2 ± 2.5
$B^0 \to \pi^0 K^0$	7.1	10.3 ± 2.5

FIGURE 4. 95% (solid), 90% (dashed) and 68% (short-dashed) confidence level contours in the $(\bar{\rho}, \bar{\eta})$ plane obtained from a global fit of QCD factorization results to the CP-averaged $B \to \pi K, \pi\pi$ branching fractions. The dark dot shows the overall best fit, whereas the light dot indicates the best fit for the default choice of all theory input parameters. The table compares the best fit values for the various CP-averaged branching fractions (in units of 10^{-6}) with the world average data.

OUTLOOK

The QCD factorization approach provides the theoretical framework for a systematic analysis of hadronic and radiative exclusive B decay amplitudes based on the heavy-quark expansion. This theory has already passed successfully several nontrivial tests, and will be tested more thoroughly with more precise data. A new effective field-theory language appropriate to QCD factorization is emerging in the form of the collinear–soft effective theory. Ultimately, the developments reviewed in this talk may lead to theoretical control over a vast variety of exclusive B decays, giving us new constraints on the unitarity triangle.

ACKNOWLEDGMENTS

I wish to thank the organizers of *QCD@Work* for arranging a splendid workshop in a fabulous setting, and for their generous support. This work was supported in part by the National Science Foundation.

REFERENCES

1. A. Höcker, H. Lacker, S. Laplace and F. Le Diberder, Eur. Phys. J. C **21**, 225 (2001) [hep-ph/0104062]; for an updated analysis, see `http://www.slac.stanford.edu/~laplace/ckmfitter.html`.
2. A. L. Kagan and M. Neubert, Phys. Lett. B **492**, 115 (2000) [hep-ph/0007360];
   Y. Grossman, M. Neubert and A. L. Kagan, JHEP **9910**, 029 (1999) [hep-ph/9909297].
3. D. Atwood, I. Dunietz and A. Soni, Phys. Rev. Lett. **78**, 3257 (1997) [hep-ph/9612433].
4. I. Dunietz and R. G. Sachs, Phys. Rev. D **37**, 3186 (1988) [Erratum-ibid. D **39**, 3515 (1988)];
   I. Dunietz, Phys. Lett. B **427**, 179 (1998) [hep-ph/9712401].
5. M. Gronau and D. London, Phys. Rev. Lett. **65**, 3381 (1990).
6. A. E. Snyder and H. R. Quinn, Phys. Rev. D **48**, 2139 (1993).

7. C. W. Bauer, S. Fleming, D. Pirjol and I. W. Stewart, Phys. Rev. D **63**, 114020 (2001) [hep-ph/0011336];
C. W. Bauer and I. W. Stewart, Phys. Lett. B **516**, 134 (2001) [hep-ph/0107001];
C. W. Bauer, D. Pirjol and I. W. Stewart, hep-ph/0109045.
8. J. D. Bjorken, Nucl. Phys. Proc. Suppl. **11**, 325 (1989).
9. M. J. Dugan and B. Grinstein, Phys. Lett. B **255**, 583 (1991).
10. H. D. Politzer and M. B. Wise, Phys. Lett. B **257**, 399 (1991).
11. M. Beneke, G. Buchalla, M. Neubert and C. T. Sachrajda, Phys. Rev. Lett. **83**, 1914 (1999) [hep-ph/9905312].
12. M. Beneke, G. Buchalla, M. Neubert and C. T. Sachrajda, Nucl. Phys. B **591**, 313 (2000) [hep-ph/0006124].
13. M. Neubert and A. A. Petrov, hep-ph/0108103.
14. M. Beneke, G. Buchalla, M. Neubert and C. T. Sachrajda, Nucl. Phys. B **606**, 245 (2001) [hep-ph/0104110].
15. A. Khodjamirian, Nucl. Phys. B **605**, 558 (2001) [hep-ph/0012271].
16. C. N. Burrell and A. R. Williamson, Phys. Rev. D **64**, 034009 (2001) [hep-ph/0101190].
17. T. Becher, M. Neubert and B. D. Pecjak, hep-ph/0102219.
18. C. W. Bauer, D. Pirjol and I. W. Stewart, hep-ph/0107002.
19. Z. Ligeti, M. Luke and M. B. Wise, Phys. Lett. B **507**, 142 (2001) [hep-ph/0103020].
20. M. Diehl and G. Hiller, JHEP **0106**, 067 (2001) [hep-ph/0105194].
21. E. von Toerne (CLEO Collaboration), talk at the International Europhysics Conference on High Energy Physics, Budapest, Hungary, July 2001.
22. K. Abe *et al.* [Belle Collaboration], hep-ex/0107048.
23. M. Beneke, T. Feldmann and D. Seidel, hep-ph/0106067.
24. S. W. Bosch and G. Buchalla, hep-ph/0106081.
25. T. E. Coan *et al.* [CLEO Collaboration], Phys. Rev. Lett. **84**, 5283 (2000) [hep-ex/9912057].
26. Y. Ushiroda [Belle Collaboration], hep-ex/0104045.
27. J. Nash [BaBar Collaboration], talk at the 20th International Symposium on Lepton and Photon Interactions, Rome, Italy, July 2001.
28. A. L. Kagan and M. Neubert, hep-ph/0110078.
29. V. L. Chernyak and I. R. Zhitnitsky, Nucl. Phys. B **345**, 137 (1990).
30. A. Ali, V. M. Braun and H. Simma, Z. Phys. C **63**, 437 (1994) [hep-ph/9401277].
31. E. Bagan, P. Ball and V. M. Braun, Phys. Lett. B **417**, 154 (1998) [hep-ph/9709243];
P. Ball and V. M. Braun, Phys. Rev. D **58**, 094016 (1998) [hep-ph/9805422].
32. Y. Keum, H. Li and A. I. Sanda, Phys. Lett. B **504**, 6 (2001) [hep-ph/0004004]; Phys. Rev. D **63**, 054008 (2001) [hep-ph/0004173];
Y. Keum and H. Li, Phys. Rev. D **63**, 074006 (2001) [hep-ph/0006001].
33. M. Ciuchini, E. Franco, G. Martinelli and L. Silvestrini, Nucl. Phys. B **501**, 271 (1997) [hep-ph/9703353];
M. Ciuchini, E. Franco, G. Martinelli, M. Pierini and L. Silvestrini, Phys. Lett. B **515**, 33 (2001) [hep-ph/0104126].
34. M. Ciuchini, E. Franco, G. Martinelli, M. Pierini and L. Silvestrini, hep-ph/0110022.
35. S. Chen *et al.* [CLEO Collaboration], Phys. Rev. Lett. **85**, 525 (2000) [hep-ex/0001009].
36. K. Abe *et al.* [Belle Collaboration], Phys. Rev. D **64**, 071101 (2001) [hep-ex/0106095].
37. B. Aubert *et al.* [BaBar Collaboration], hep-ex/0105061.
38. B. Aubert *et al.* [BaBar Collaboration], hep-ex/0107074.
39. S. Descotes and C. T. Sachrajda, hep-ph/010926.
40. M. Neubert and J. L. Rosner, Phys. Lett. B **441**, 403 (1998) [hep-ph/9808493];
M. Neubert, JHEP **9902**, 014 (1999) [hep-ph/9812396].

Two Body B Decays, Factorization and Λ_{QCD}/m_b Corrections[1]

M. Ciuchini*, E. Franco†, G. Martinelli†, M. Pierini† and L. Silvestrini†

*Dip. di Fisica, Univ. di Roma III and INFN, Sezione di Roma Tre, Via della Vasca Navale 84, I-00146 Roma, Italy
†Dip. di Fisica, Univ. "La Sapienza" and INFN, Sezione di Roma, P.le A. Moro, I-00185 Rome, Italy

Abstract. By using the recent experimental measurements of $B \to \pi\pi$ and $B \to K\pi$ branching ratios, we find that the amplitudes computed at the leading order of the Λ_{QCD}/m_b expansion disagree with the observed BRs, even taking into account the uncertainties of the input parameters. Beyond the leading order, Charming and GIM penguins allow to reconcile the theoretical predictions with the data. Because of these large effects, we conclude, however, that it is not possible, with the present theoretical and experimental accuracy, to determine the CP violation angle γ from these decays. We compare our results with those obtained with the parametrization of the chirally enhanced non-perturbative contributions by BBNS. We also predict large asymmetries for several of the particle–antiparticle BRs, in particular $BR(B^+ \to K^+\pi^0)$, $BR(B_d \to K^+\pi^-)$ and $BR(B_d \to \pi^+\pi^-)$.

INTRODUCTION

The advent of the B factories, Babar [1] and Belle [2], opens new perspectives for very precise measurements of non-leptonic B-decays [3] and calls for a significant improvement of the theoretical predictions. In this respect, important progress has been recently achieved by systematic studies of factorization in B decays [4, 5]. These studies confirmed the physical idea [6] that factorization holds in heavy hadron decays, $m_Q \gg \Lambda_{QCD}$, for the leading terms of the Λ_{QCD}/m_Q expansion. They leave open, however, the central question of whether i) the leading terms predict with sufficient accuracy the relevant B-meson decay rates or ii) the power-suppressed corrections, which cannot be evaluated without some model assumption, are phenomenologically important. This problem has been addressed in a series of papers [7]–[13]. In particular, the main conclusion of ref. [12] is that *non-perturbative* Λ_{QCD}/m_b corrections from the leading operators of the effective weak Hamiltonian, conventionally called Charming (or GIM) penguins, are very important in cases where the factorized amplitudes are either colour or Cabibbo suppressed. One of the consequences of this observation is that factorization at the leading order in Λ_{QCD}/m_b is unable to reproduce the observed $B \to \pi\pi$ and $B \to K\pi$ BRs even taking into account the uncertainties of the input parameters.

The results of ref. [12] seem to be at variance with the conclusions of ref. [13] where it is stated that, in the QCD factorization approach, "an acceptable fit to the branching

[1] Talk given by G. Martinelli

CP602, *QCD@Work: International Workshop on Quantum Chromodynamics*
edited by P. Colangelo and G. Nardulli

fractions is obtained even if we impose $\gamma < 90^o$ as implied by the standard constraints on the unitarity triangle" [2].

Given the contradictory conclusions of the different studies, it is very important to clarify the situation by comparing input parameters, methods of analysis and results. In order to help the debate on this subject, we discuss in the following:

- The theoretical framework of factorization and the calculation of the amplitudes at the leading order of the Λ_{QCD}/m_b expansion;
- A comparison of the leading terms, including their uncertainties, with the measured $B \to \pi\pi$ and $B \to K\pi$ branching ratios;
- Results for the $B \to \pi\pi$ and $B \to K\pi$ BRs including charming (and GIM) penguins;
- A comparison of the results obtained with charming and GIM penguins [12] with the model of the Λ_{QCD}/m_b corrections adopted in [13].

FACTORIZATION

In this section we recall the basic ingredients necessary to compute the relevant amplitudes either whithin factorization or including the corrections which arise at higher order in the Λ_{QCD}/m_b expansion.

The physical amplitudes for $B \to K\pi$ and $B \to \pi\pi$ can be conveniently written in terms of RG-invariant parameters built using the Wick contractions of the effective Hamiltonian [16]. In the heavy quark limit, following the approach of ref. [5], it is possible to compute these RG invariant parameters using factorization. The formalism of ref. [5] has been developed so that it is also possible to include the perturbative corrections to order α_s, i.e. at the next-to-leading order in perturbation theory [3].

For the sake of discussion, it is instructive to start from the explicit expression of the $B_d \to K^+\pi^-$ amplitude. In terms of the parameters defined in [16], this amplitude reads

$$\mathcal{A}(B_d \to K^+\pi^-) = -V_{us}V_{ub}^*\Big(E_1(s,u,u;B_d,K^+,\pi^-) - P_1^{\rm GIM}(s,u;B_d,K^+,\pi^-)\Big) \\ + V_{ts}V_{tb}^* P_1(s,u;B_d,K^+,\pi^-)\,. \qquad (1)$$

We have

$$\begin{aligned} E_1(s,u,u;B_d,K^+,\pi^-) &= a_1^u(K\pi)\langle Q_1^u\rangle_{\rm fact} + a_2^u(K\pi)\langle Q_2^u\rangle_{\rm fact} + \tilde{E}_1 \\ P_1(s,u;B_d,K^+,\pi^-) &= \sum_{i=3}^{10} a_i^c(K\pi)\langle Q_i\rangle_{\rm fact} + \tilde{P}_1 \\ P_1^{\rm GIM}(s,u;B_d,K^+,\pi^-) &= \sum_{i=3}^{10}(a_i^c(K\pi) - a_i^u(K\pi))\langle Q_i\rangle_{\rm fact} + \tilde{P}_1^{\rm GIM}\,, \end{aligned} \qquad (2)$$

---

[2] Indeed the analysis of ref. [14] gives $\gamma = (54.8 \pm 6.2)^o$ and ref. [15] quotes $34^o < \gamma < 82^o$.

[3] An alternative framework, provided by the approach of ref. [4], will not be discussed here. For a recent discussion see also [17].

where $\langle Q_i \rangle_{\text{fact}}$ denotes the factorized matrix elements and a_i^f the parameters introduced in [5]. Eqs. (1) and (2) are exact. Unfortunately, similar equations require the knowledge of several non-perturbative parameters, which at present cannot be extracted from the data. To be predictive, we will then use our physical intuition to reduce their number. In eqs. (1) and (2):

1. The CKM matrix elements can either be taken from other experimental measurements or from a fit to the non-leptonic *BR*s, assuming that factorization is accurate enough;
2. The coefficients $a_i^f(M_1M_2)$ (e.g. $a_1^u(K\pi)$) are renormalization group invariant, as the corresponding factorized matrix elements, and have been computed perturbatively at the NLO in refs. [5, 13];
3. The coefficients a_6^f and a_8^f, which have also been computed to one-loop order, are instead scheme dependent. Their scheme-dependence is cancelled by the hadronic matrix elements of the penguin operators Q_6 and Q_8 respectively. Assuming factorization for the chirally enhanced contributions [5], the latter can be expressed in terms of the ratio
$$r_\chi^K(\mu) = \frac{2m_K^2}{\overline{m}_b(\mu)(\overline{m}_s(\mu)+\overline{m}_q(\mu))}, \quad (3)$$
which is formally of $O(\Lambda_{QCD}/m_b)$ but numerically important (an analogous parameter $r_\chi^\pi(\mu)$ can be defined for $\pi\pi$ decays). We will discuss the rôle of these terms below.
4. The leading amplitudes $\langle Q_i \rangle_{\text{fact}}$ are computed in terms of decay constants and semileptonic form factors. The form factors can either be taken from theoretical calculations [18, 19] or fitted from the experimental *BR*s (the possibility of extracting them from the corresponding $B \to \pi$ semileptonic form factor at small momentum transfer is at present rather remote).
5. The tilded parameters, namely $\tilde{P}_1$, $\tilde{P}_1^{\text{GIM}}$ and $\tilde{E}_1$, are genuine, non-perturbative Λ_{QCD}/m_b corrections which cannot be computed at present. If we neglect Zweig-suppressed contributions, by $SU(2)$ symmetry one can show that all the Cabibbo-enhanced Λ_{QCD}/m_b corrections to $B \to K\pi$ decays can be reabsorbed in $\tilde{P}_1$. Several corrections are contained in $\tilde{P}_1$: this parameter includes not only the charming penguin contributions, but also annihilation and penguin contractions of penguin operators. It does not include leading emission amplitudes of penguin operators (Q_3–Q_{10}) which have been explicitly evaluated using factorization. Had we included these terms, this contribution would exactly correspond to the parameter P_1 of ref. [16]. The parameter $\tilde{P}_1$ (P_1) encodes automatically not only the effect of the annihilation diagrams considered in [20], but all the other contributions of $O(\Lambda_{QCD}/m_b)$ with the same quantum numbers of the charming penguins. In this respect it is the most general parameterization of all the perturbative and non-perturbative contributions of the operators Q_5 and Q_6 (Q_3 and Q_4), including the worrying higher-twist infrared divergent contribution to annihilation discussed in ref. [13, 21]. The parameter $\tilde{P}_1$ has the same quantum numbers and physical effects as the original charming penguins proposed in [7], although it has a more general meaning.

6. If one also includes $B \longrightarrow \pi\pi$ decays we have several other parameters, for example P_1^{GIM} and P_3, in the formalism of ref. [16]. A closer look to P_3 shows that this term is due either to Zweig suppressed annihilation diagrams (called CPA and DPA in ref. [7]) or to annihilation diagrams which are colour suppressed with respect to those entering $\tilde{P}_1$. For this reason in ref. [12] P_3 was taken to be zero. $\tilde{P}_1$ is equal to the corresponding parameter in $K\pi$ decays if $SU(3)$ symmetry is assumed. In our analysis we have used the same value of $\tilde{P}_1$ for all $K\pi$ and $\pi\pi$ channels. P_1^{GIM} will be discussed later on.

We are now ready to discuss *factorization* for the leading terms of the Λ_{QCD}/m_b expansion. Factorization is the theory of non leptonic decays which is obtained in the limit $m_b \longrightarrow \infty$. Thus it consists in neglecting *all* terms of $O(\Lambda_{QCD}/m_b)$ ($\tilde{E}_1 = 0$, $\tilde{P}_1 = 0$, $r_\chi^{K,\pi} = 0$, etc.). At lowest order in perturbation theory, called also naïve factorization, the a_i^f are simple combinations of the Wilson coefficients and do not depend on the hadron wave functions. The inconvenience of naïve factorization is that physical amplitudes still have a marked dependence on the renormalization scale because, contrary to the Wilson coefficients, the factorized matrix elements are scale independent.

The scale dependence is reduced by working at $O(\alpha_s)$, both for the Wilson coefficients and the matrix elements. In refs. [5] it has been shown that, at this order, all the dangerous infrared divergences can be reabsorbed in the definition of the hadronic wave functions. For the leading terms in the Λ_{QCD}/m_b expansion there are strong arguments to support the idea that this will remain true at all orders in α_s, see also ref. [26]. Thus, in the limit $m_b \longrightarrow \infty$, it is likely that factorization is preserved by strong interactions. At $O(\alpha_s)$ or higher, the coefficients a_i^f depend on the specific detail of the hadron wave-fuctions. For this reason, the uncertainties relative to the wave functions, as the residual renormalization scale dependence, must be taken into account in the evaluation of the uncertainties for the theoretical predictions. The approximation in which we neglect all the Λ_{QCD}/m_b corrections, but include the perturbative corrections to the leading contribution, is called QCD factorization or simply factorization. Factorization implies an important consequence: predictions of non-leptonic decay rates are *model independent* to the extent that the few relevant hadronic parameters, namely the kaon and pion decay constants, $f_{K,\pi}$, the semileptonic form factors, $f_{K,\pi}(0)$, and the hadronic wave functions are known.

In cases like $B \longrightarrow K\pi$ decays, where the factorized amplitudes are Cabibbo suppressed, the corrections of $O(\Lambda_{QCD}/m_b)$, which unfortunately are *model dependent*, become important. At lowest order in α_s, the chirally enhanced terms proportional to r_χ^K (r_χ^π) are computable by assuming that factorization can be applied beyond the leading order. A substantial difficulty arises, however, at $O(\alpha_s\Lambda_{QCD}/m_b)$. Although the chirally enhanced corrections from $Q_{6,8}$ are infrared finite, other contributions of the same order from different operators are infrared divergent, signaling that they belong to the class of the non-perturbative contributions which appear beyond factorization. These cannot be predicted using the same hadronic quantities of the factorized amplitudes. For this reason, any phenomenological analysis which aims at including in a coeherent way the terms of $O(\alpha_s\Lambda_{QCD}/m_b)$ is forced to introduce extra model-dependent non-perturbative parameters besides $f_{K,\pi}$, $f_{K,\pi}(0)$ and the hadronic wave functions. This implies that, *at*

$O(\Lambda_{QCD}/m_b)$, *model dependence is unavoidable* (even in the subsector of the chirally enhanced contributions) and it is present in both the analyses of ref. [12] and ref. [13], which we will compare below.

Different model-dependent assumptions were made in the two approaches:

1. In ref. [12], $\tilde{E}_1$ and $\tilde{P}_3$ were neglected and $SU(2)$ symmetry was assumed for $\tilde{P}_1$ ($SU(3)$ when it was used for $B \to \pi\pi$ decays). The same approximations were made for $\tilde{P}_1^{\mathrm{GIM}}$. The complex parameter $\tilde{P}_1$ ($\tilde{P}_1^{\mathrm{GIM}}$) was then fitted to reproduce the experimental *BR*s.
2. In ref. [13] the effects of the chirally enhanced Λ_{QCD}/m_b corrections were either computed perturbatively or encoded in two complex phenomenological parameters called X_H and X_A. An uncertainty of 100 % to the "default" values (e.g. $X_H = 2.4$) was assigned to these parameters in order to determine allowed bands for the predicted *BR*s. The bands include all other sources of uncertainties.

We now compare the leading amplitudes to the experimental results in order to *test factorization.* Before using predictions based on factorization to test the Standard Model and look for signals of new physics, it is crucial to check how large are the errors induced by our ignorance of the $O(\Lambda_{QCD}/m_b)$ corrections which we are unable to compute. Our position, indeed, is that we have more confidence in the SM rather than in factorization. In order to test factorization, we ought to use all the information that we have from other measurements. Thus, for example, whereas the size of the error on $|V_{cb}|$ and $|V_{ub}|$ can be debated, there is no question that these experimental inputs must be included in any analysis that aims at testing (or using) factorization. We also stress that the value of $|V_{ub}|$ is not expected to be affected by the presence of new physics beyond the SM.

The CP parameter γ does in general change if there is physics beyond the SM. It remains an interesting exercise, however, to verify whether, by taking the value of γ from the Unitarity Triangle Analysis (UTA) in the SM, the predicted *BR*s are in agreement with the data. If, by using γ from the UTA, one is unable to reproduce the experimental $B \to \pi\pi$ and $B \to K\pi$ *BR*s, this implies that "either there is new physics or Λ_{QCD}/m_b corrections are important" [12].

In our analysis we have used the likelihood method which has been described in all details in ref. [14]. Without entering in the "ideological" controversy about frequentistic and bayesian methods, we only note here that in [14] it has been shown that, at 95 % C.L., the Bayesian analysis give the same results as the frequentistic Babar Scanning method (and its variations) when the same inputs are used. Thus we will present our contour plots, corresponding to fig. 17 of ref. [13], both with factorization plus chirally enhanced contributions and with the non-factorizable charming (and GIM) penguin corrections. Besides this, we will also give tables with the relevant *BR*s, both in the factorization approximation with chirally enhanced terms and with the charming (and GIM) penguin corrections included.

We end this section with some remarks. In our approach, we have first checked that, within factorization and the SM, it is impossible to fit the experimental *BR*s. The Λ_{QCD}/m_b terms, that we are then forced to include in order to reproduce the experimental results, are non-perturbative quantities, infrared divergent in perturbation theory, on which we do not have any knowledge *a priori*. For this reason we decided to fit them on the data. The experimental numbers are nicely reproduced and the corrections to factorization are well consistent with the expected size (i.e. $\tilde{P}_1$ is of $O(\Lambda_{QCD}/m_b)$ with respect to the leading contributions).

In ref. [13] the subleading power corrections are varied in predefined intervals and the change in the predicted *BR*s is interpreted as uncertainty on the factorized amplitudes. In our opinion the uncertanties on factorization are only those coming from the CKM matrix elements or the form factors, etc. The Λ_{QCD}/m_b terms instead are really contributions beyond factorization: if they are necessary to reproduce the data then it is not possible to make model independent predictions. This would remain true even if we knew without any uncertainty the hadronic parameters entering at the lowest order of the Λ_{QCD}/m_b expansion [4].

Thus we are in the Bermuda triangle: i) without the $O(r_\chi)$ and $O(\alpha_s r_\chi)$ terms, that is within QCD factorization, we cannot reproduce the data; ii) the inclusion of the computable subset of $O(\alpha_s r_\chi)$ terms only is inconsistent since there is no reason to exclude the other non-perturbative non-computable contributions of the same order. In any case, we will show that, by using all the available experimental information, also this case is very difficult to reconcile with the data; iii) the complete set of $O(\alpha_s r_\chi)$ corrections leads us beyond factorization and the results are model dependent. Indeed there is no proof that the one-loop finite chirally enhanced terms remain infrared finite at higher orders in α_s. Moreover, if the corrections of $O(\alpha_s r_\chi \sim \alpha_s \Lambda_{QCD}/m_b)$ are phenomenologicaly important, it is difficult to understand why, at the same level of numerical accuracy, other non-chirally enhanced non-perturbative Λ_{QCD}/m_b terms should not also be taken into account.

The sad conclusion is that the very nice theory of QCD factorization developed in [5] is insufficient to fit the data because power corrections, which are model dependent, are important in $B \to K\pi$ and $B \to \pi\pi$ decays. Finally, model dependence does not implies that we are unable to make any prediction. If the assumptions made in our approach are reasonable [5], by fitting $\tilde{P}_1$ to the data and with the increasing experimental precision we may hope to extract also the value of γ or to constrain $\sin 2\alpha$.

RESULTS

In this section we present our analysis and a detailed comparison with the results of ref. [13]. In order to obtain our results we used the likelihood method as in [12]. The input parameters, given in table 1, are also the same but for a few differences:

- We added the $O(\alpha_s)$ corrections to the coefficients of the penguin and electropenguin operators computed in [13], which appeared after the completion of [12]. In this respect the criticisms of ref. [27] do not apply to the present analysis. We will see below that, even including these new ingredients, the main physics conclusions of ref. [12] are confirmed.

[4] Note that some *BR*s are dominated by the Λ_{QCD}/m_b corrections.

[5] For example that we may neglect $\tilde{E}_1$.

- As for the matrix elements of $Q_{6,8}$, we include the "computable" factorizable chirally enhanced terms in the definition of $\tilde{P}_1$, which in this way contains all the possible Cabibbo enhanced, model-dependent corrections of $O(\Lambda_{QCD}/m_b)$. In practice this corresponds to fit $\tilde{P}_1$ as in ref. [12] with $r_\chi^{K,\pi}=0$. There is no substantial difference between the old choice and the new one, since the contribution of the chirally enhanced terms and of $\tilde{P}_1$ have exactly the same quantum numbers.
- Although this is a minor source of uncertainty we also allow a variation of the renormalization scale and of the parameters of the hadronic wave functions.

We have also verified that by using the inputs of ref. [13] we obtain essentially the same results, and hence arrive to the same physics conclusions.

In this section we present:

1. A brief discussion of the results obtained in QCD factorization, namely with all the terms of $O(\Lambda_{QCD}/m_b)$ set to zero.
2. The results obtained in our approach by fitting $\tilde{P}_1$ and using γ as determined from the Unitary Triangle Analysis in ref. [14]

$$\gamma = (54.8 \pm 6.2)^o. \tag{4}$$

On the basis of this study we predict the B–$\bar{B}$ asymmetries of the BRs, such as

$$\mathcal{A}(B_d \to \pi^+\pi^-) = \frac{BR(\bar{B}_d^0 \to \pi^+\pi^-) - BR(B_d^0 \to \pi^+\pi^-)}{BR(\bar{B}_d^0 \to \pi^+\pi^-) + BR(B_d^0 \to \pi^+\pi^-)}; \tag{5}$$

3. The results obtained by letting γ free and a comparison of our results with those of ref. [13].

Results in QCD factorization

In this subsection we compare the model independent results obtained with QCD factorization, namely including only the terms which survive when $m_b \to \infty$, with the experimental data. In this case there is still a residual model dependence due to our ignorance of the semileptonic form factors $f_{K,\pi}(0)$ and, at order α_s, to the ignorance of the hadron wave functions. We vary the semileptonic form factors with flat p.d.f. in the intervals given in table 1 and the parameters of the hadron wave functions in the intervals given in table 2 of ref. [13]. We also vary the renormalization scale between $m_b/2$ and $2m_b$, see table 1 of [13]. The data show a generalized disagreement with the QCD factorization predictions. In particular the allowed region in the ρ–η plane and the value of γ do not have any overlap with the corresponding ones from the unitarity triangle analysis [14]. This remains true even if we double the uncertainty on $|V_{ub}|$. We conclude that QCD factorization cannot be reconciled with data.

TABLE 1. Input values used in the numerical analysis. The form factors are taken from refs. [18, 19], the CKM parameters from ref. [14] and the BRs correspond to our average of CLEO, BaBar and Belle results [23, 24, 25]. All the *BR*s are given in units of 10^{-6}.

$f_\pi(0)$	0.27 ± 0.08	$f_K(0)/f_\pi(0)$	1.2 ± 0.1
ρ	0.224 ± 0.038	η	0.317 ± 0.040
$BR(B_d \to K^0\pi^0)$	10.3 ± 2.6	$BR(B^+ \to K^+\pi^0)$	12.0 ± 1.7
$BR(B^+ \to K^0\pi^+)$	17.4 ± 2.6	$BR(B_d \to K^+\pi^-)$	17.3 ± 1.6
$BR(B_d \to \pi^+\pi^-)$	4.4 ± 0.9	$BR(B^+ \to \pi^+\pi^0)$	5.3 ± 1.7

TABLE 2. *BR*s with Charming ad GIM penguins, with QCD factorization and the infrared-finite chirally enhanced corrections only and with the BBNS model which includes $X_{A,H}$. All the *BR*s are given in units of 10^{-6}.

BR	Charming + GIM	chirally enhanced	BBNS	*BR*	Charming + GIM	chirally enhanced	BBNS
$K^0\pi^0$	8.6 ± 0.9	3.6 ± 1.5	4.1 ± 1.8	$K^+\pi^0$	9.8 ± 1.0	5.7 ± 2.3	6.2 ± 2.6
$K^0\pi^+$	18.7 ± 1.6	10.2 ± 4.2	10.9 ± 4.8	$K^+\pi^-$	17.9 ± 1.4	8.2 ± 3.4	9.2 ± 4.0
$\pi^+\pi^-$	4.9 ± 0.8	9.2 ± 3.8	9.2 ± 3.8	$\pi^+\pi^0$	3.5 ± 0.8	5.7 ± 2.2	6.5 ± 2.5
$\pi^0\pi^0$	0.6 ± 0.2	0.2 ± 0.1	0.4 ± 0.3				

Factorization with Charming and GIM penguins

We now discuss the effects of charming and GIM penguins, parameterized by $\tilde{P}_1$ and $\tilde{P}_1^{\rm GIM}$. $\tilde{P}_1$ is a complex amplitude that we fit on the $B \to K\pi$ *BR*s. In order to have a reference scale for its size, we introduce a suitable "Bag" parameter, $\hat{B}_1$, by writing

$$\tilde{P}_1 = \frac{G_F}{\sqrt{2}} f_\pi f_\pi(0)\, g_1 \hat{B}_1 \,, \tag{6}$$

where G_F is the Fermi constant, g_1 is a Clebsh-Gordan parameter depending on the final $K\pi$ ($\pi\pi$) channel and $B_1 = |B_1|\exp(i\phi)$. Note that $\hat{B}_1$ differs from the parameter defined in ref. [12] because it now includes all the chirally enhanced Λ_{QCD}/m_b corrections, part of which were previously explicitly calculated using factorization. In a similar way we introduce $\hat{B}_1^{\rm GIM}$.

We fit all the *BR*s given in table 1 with GIM and charming penguins included and taking the value of γ determined from the UTA, see eq. (4). We find

$$\begin{aligned} |\hat{B}_1| &= 0.13 \pm 0.02 \,, \qquad \phi = (188 \pm 82)^o \,, \\ |\hat{B}_1^{\rm GIM}| &= 0.17 \pm 0.08 \,, \qquad \phi^{\rm GIM} = (181 \pm 59)^o \,, \end{aligned} \tag{7}$$

where the notation is self-explaining. Note that the size of the charming and GIM penguin effects is of the expected magnitude. Note that $|\hat{B}_1^{\rm GIM}|$ is very poorly determined.

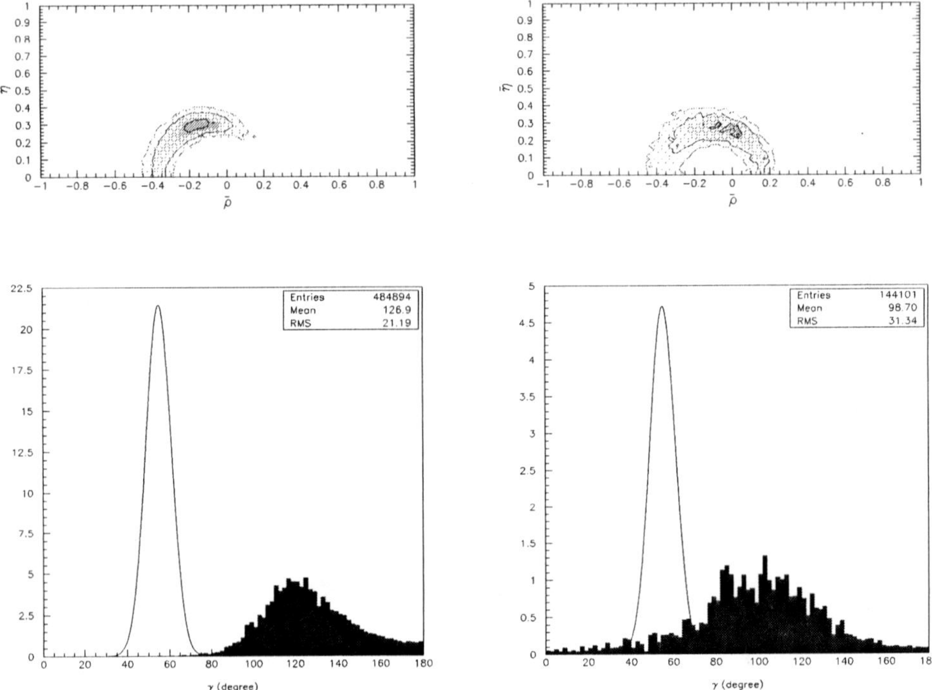

FIGURE 1. ρ–η contour plots obtained with factorization and infrared-finite chirally enhanced terms using $|V_{ub}|$ (up-left) or letting ρ and η free (up-right). We also show a comparison of the p.d.f. of γ with the one from the UTA analysis of ref. [14] in the two cases.

The reason is that only $BR(B_d^0 \to \pi^+\pi^-)$ in the fit is sensitive to GIM penguins. Thus, in practice, we are trying to fit two parameters, namely $|\hat{B}_1^{\rm GIM}|$ and $\phi^{\rm GIM}$, to a single BR.

The results for the BRs can be found in table 2 with the label "Charming+GIM". They show that the extra charming and GIM parameters radically improve the agreement for the measured $B \to K\pi$ and $B \to \pi\pi$ BRs. We do not claim, however, to be able to predict $BR(B_d \to \pi^+\pi^-)$ since many effects of the same order besides charming and GIM contributions, which in this case are not Cabibbo enhanced, were ignored: our results instead show that accurate predictions for $B_d \to \pi\pi$ decays can only be obtained by controlling quantitatively the $O(\Lambda_{QCD}/m_b)$ corrections, which is presently far beyond the theoretical reach. To give a complete information, and for comparison with ref. [13] we also fit the data by letting γ free. In this case we obtain $\gamma = (89 \pm 42)^o$. At present, the precision of the data and the number of free parameters does not allow a useful determination of γ.

The large absolute values of φ, and the sizable effects that penguins have on the BRs, stimulated us to consider whether we could find observable particle-antiparticle asymmetries as the one defined in eq. (5). We find large effects in $BR(B^+ \to K^+\pi^0)$,

TABLE 3. Absolute values of the rate CP asymmetries for $B \to K\pi$ and $B \to \pi\pi$ decays.

$\|\mathcal{A}\|$	Charming + GIM	$\|\mathcal{A}\|$	Charming + GIM
$K^0\pi^0$	0.05 ± 0.03	$K^+\pi^0$	0.16 ± 0.08
$K^0\pi^+$	0.02 ± 0.02	$K^+\pi^-$	0.15 ± 0.07
$\pi^+\pi^-$	0.44 ± 0.21	$\pi^0\pi^0$	0.61 ± 0.29

$BR(B_d \to K^+\pi^-)$ and $BR(B_d \to \pi^+\pi^-)$. As discussed before, for $BR(B_d \to \pi^+\pi^-)$ our predictions suffer from very large uncertainties due to contributions which cannot be fixed theoretically. For this reason, the values of the asymmetry reported in table 3 are only an indication that a large asymmetry could be observed also in this channel. There is a sign ambiguity in $\mathcal{A} \sim \sin\gamma \sin\phi$. This ambiguity can be solved only by an experimental measurement or, but this is extremely remote, by a theoretical calculation of the relevant amplitudes. For each channel, we give the absolute value of the asymmetry in table 3. Note that within factorization all asymmetries would be unobservably small, since the strong phase is a perturbative effect of $O(\alpha_s)$ [5]. The possibility of observing large asymmetries in these decays opens new perspectives. These points will be the subject of a future study.

Comparison with BBNS

In this section we make a critical comparison with the latest analysis of BBNS [13]. We recall that we added all the perturbative corrections and allowed the variations of the non-perturbative parameters which were implemented by BBNS.

We start by discussing the rôle of chirally enhanced terms that do not suffer from infrared divergences at $O(\alpha_s)$. This is an istructive case since, if not for other infrared-divergent contributions of the same order, the inclusion of only these terms, although not justified, would still allow *model independent* predictions, in the sense discussed before. Thus the question is whether these chirally enhanced terms alone, part of which have the same effect as $\tilde{P}_1$, can describe the data. Thus we have repeated the analysis in the UTA case with only factorization and chirally enhanced infrared-finite terms with the results for the *BR*s given in table 2 with the label "chirally enhanced". The combined probability that all the predicted values for the $K\pi$ channels are within two σs from the experimental numbers is 6%. If we relax the constraint on γ we obtain $\gamma = (127 \pm 20)^o$ with a probability of 0.3%that $\gamma < 80^o$ and of 2% that $\gamma < 90^o$. The corresponding ρ–η contour plot is given in fig. 1 (upper–left). We conclude that this model (model in the sense of including only a chosen subset of the chirally enhanced terms) is strongly disfavoured by the data. The reader may be surprised of the difference between fig. 1 and fig. 17 of ref. [13] (where the non-perturbative parameters $X_{A,H}$ were however included). The difference is explained by the fact that we used the experimental measurements of $|V_{ub}|$. If, following ref. [13, 27], we let both ρ and η free, the ρ–η contour plot changes and becomes that shown in fig. 1 (upper-right), which is very similar to fig. 17 of ref. [13]. For completeness, we also give in fig. 1 the p.d.f. of γ in the two cases, together with the

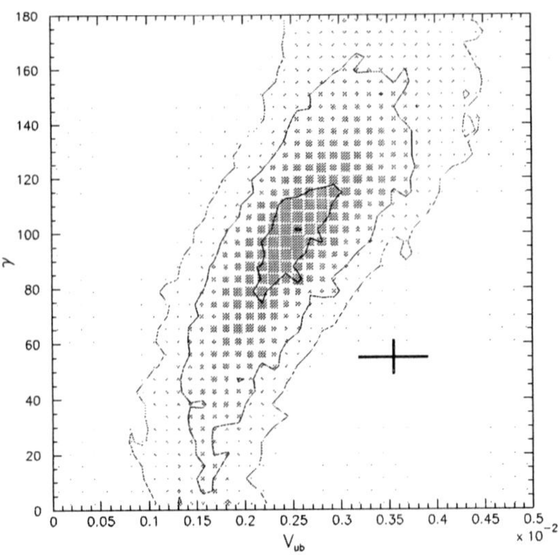

FIGURE 2. Contour plot in the $\gamma - |V_{ub}|$ plane, see text.

p.d.f. obtained with the UTA analysis of ref. [14]. When the experimental information on $|V_{ub}|$ is used the two p.d.f. have no overlap (lower-left). In the other case, it is still possible to find values of γ compatible with the UTA analysis (lower-right), at the price of a rather low value of $|V_{ub}|$. The situation is well illustrated by fig. 2, where the contour plot in the γ-$|V_{ub}|$ plane of the joint p.d.f. from non-leptonic decays is compared with the UTA range for γ and the allowed interval for the measured $|V_{ub}|$, at 1σ. This figure demonstrates that, given the strong correlation between γ and $|V_{ub}|$, it is crucial to take into account the experimental knowledge of $|V_{ub}|$. An important theoretical remark is in order at this point. The coefficient a_6^f is enhanced by the $O(\alpha_s)$ corrections and may play the same rôle of charming penguins. It is very scaring that its actual value is strongly affected by the contribution of the chromomagnetic operator computed at tree level [13]. It is very hard to believe that this contribution which, besides all general considerations, is also non-local, will remain infrared-finite in higher orders and can be really evaluated in this way.

We now discuss the effect of $X_{A,H}$ on the final results. The addition of these parameters, in the range proposed in ref. [13], leaves the situation substantially unaltered as shown in fig. 3. This holds true also for the *BR*s, which are given in table 2 with the label BBNS. The reason is that on the one hand X_A and X_H do not have the same quantum numbers, and hence effects, of charming penguins, on the other the range chosen *a priori*, on the basis on one-loop perturbation theory, is not large enough to improve the agreement with the measured *BR*s (it essentially increases the uncertainty on the predictions). Of course by choosing a low value of the strange quark mass and of $|V_{ub}|$ (without using the experimental information coming from its measurements), a large

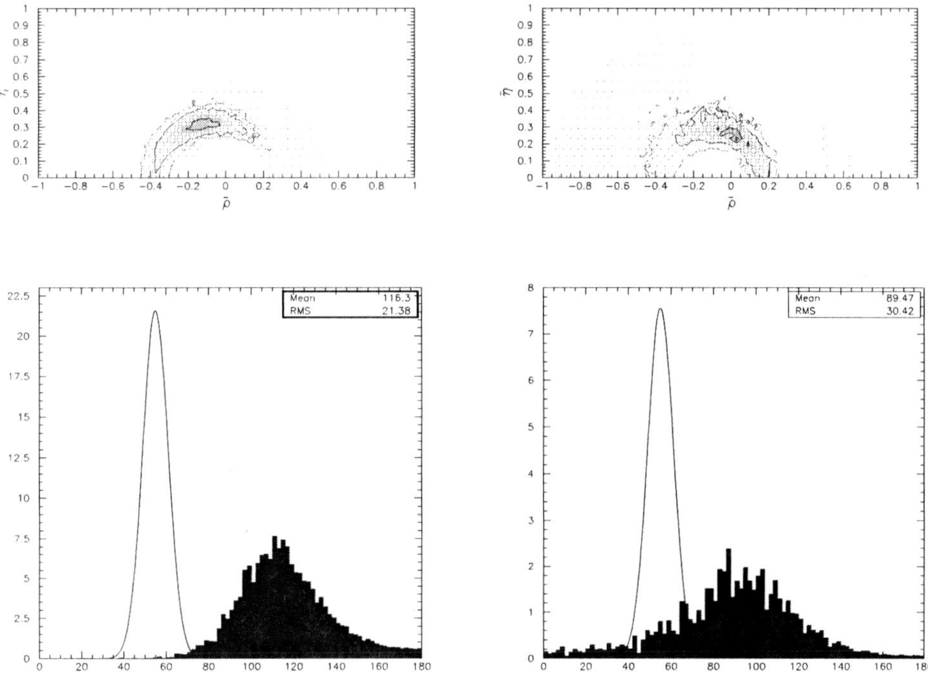

FIGURE 3. ρ–η contour plots obtained with the BBNS model using $|V_{ub}|$ (up-left) or letting ρ and η free (up-right). We also show a comparison of the p.d.f. of γ with the one from the UTA analysis of ref. [14] in the two cases.

value of $f_K(0)$ and a small value for $f_\pi(0)$ etc. it is still possible to find some point in parameter space where the χ^2 is good. That this can be used to fit γ and test the SM is hard to believe though.

The situation would be different if $X_{A,H}$ are let free to vary and fitted to the data. We have done this exercise and found that the preferred value of ρ_A is much larger than the values allowed in the interval chosen in [13], whereas ρ_H is not determined by the fit. We conclude that since X_A and X_H are infrared divergent quantities, the value of which cannot be predicted, and since without the inclusion of non-perturbative contributions of $O(\Lambda_{QCD}/m_b)$ is not possible to reproduce the experimental data, we are bound to use model dependent assumptions in the analysis of non-leptonic $B \to K\pi$ and $B \to \pi\pi$ decays.

CONCLUSION

We have analyzed the predictions of QCD factorization for $B \to \pi\pi$ and $B \to K\pi$ decays. Even taking into account the uncertainties of the input parameters, we find that QCD factorization is unable to reproduce the observed *BR*s. The introduction of charming and GIM penguins [7] allows to reconcile the theoretical predictions with the data. Instead of varying the non-perturbative phenomenological parameters in preassigned ranges, we prefer to try to fit them on the data. With the present theoretical and experimental accuracy, we find that it is still not possible to determine the CP violation angle γ. The situation is expected to improve in the near future with more accurate experimental measurements of the relevant *BR*s.

Contrary to factorization, we predict large asymmetries for several of the particle–antiparticle *BR*s, in particular $BR(B^+ \to K^+\pi^0)$, $BR(B_d \to K^+\pi^-)$ and, possibly, $BR(B_d \to \pi^+\pi^-)$. This opens new perspectives for the study of CP violation in B systems.

ACKNOWLEDGMENTS

We thank C. Sachrajda for useful discussions on this work. G.M. and L.S. thanks the TH division at CERN where part of this work has been done.

REFERENCES

1. *The Babar Physics Book*, edited by P.F. Harrison and H.R. Quinn, SLAC report SLAC-R-504, October 1998.
2. F. Takasaki, in *Proc. of the 19th Intl. Symp. on Photon and Lepton Interactions at High Energy LP99* ed. J.A. Jaros and M.E. Peskin, Int. J. Mod. Phys. A **15S1** (2000) 12, hep-ex/9912004.
3. F. Ferroni, talk given at this Workshop.
4. H-n Li and H.L. Yu, Phys. Rev. Lett. **74** (1995) 4388 and Phys. Rev. D **53** (1996) 2480; C.H. Chang and H-n Li, Phys. Rev. D **55** (1997) 5577; T.W. Yeh and H-n Li, Phys. Rev. D **56** (1997) 1615; H.Y. Cheng, H-n Li and K.C. Yang, Phys. Rev. D **60** (1999) 094005; for a recent review see H-n Li, hep-ph/0103305.
5. M. Beneke, G. Buchalla, M. Neubert and C.T. Sachrajda, Phys. Rev. Lett. **83** (1999) 1914; Nucl. Phys. B **591** (2000) 313; hep-ph/0007256.
6. J.D. Bjorken, Nucl. Phys. B (Proc. Suppl.) **11** (1989) 325;
   M. J. Dugan and B. Grinstein, Phys. Lett. B **255** (1991) 583;
   H. D. Politzer and M. B. Wise, Phys. Lett. B **257** (1991) 399.
7. M. Ciuchini, E. Franco, G. Martinelli and L. Silvestrini, Nucl. Phys. B **501** (1997) 271, hep-ph/9703353. M. Ciuchini, R. Contino, E. Franco, G. Martinelli and L. Silvestrini, Nucl. Phys. B **512** (1998) 3, hep-ph/9708222.
8. P. Colangelo, G. Nardulli, N. Paver and Riazuddin, Z. Phys. C **45** (1990) 575. C. Isola, M. Ladisa, G. Nardulli, T. N. Pham and P. Santorelli, hep-ph/0101118.
9. A. J. Buras and R. Fleischer, Phys. Lett. B **341** (1995) 379, hep-ph/9409244.
10. A. Ali and C. Greub, Phys. Rev. D **57** (1998) 2996, hep-ph/9707251;
    A. Ali, G. Kramer and C. Lu, Phys. Rev. D **58** (1998) 094009, hep-ph/9804363; Phys. Rev. D **59** (1999) 014005, hep-ph/9805403.
11. W. N. Cottingham, H. Mehrban and I. B. Whittingham, hep-ph/0102012.
12. M. Ciuchini, E. Franco, G. Martinelli, M. Pierini and L. Silvestrini, hep-ph/0104126.

13. M. Beneke, G. Buchalla, M. Neubert and C.T. Sachrajda, hep-ph/0104110.
14. M. Ciuchini *et al.*, hep-ph/0012308.
15. A. Hocker, H. Lacker, S. Laplace and F. Le Diberder, hep-ph/0104062.
16. A. J. Buras and L. Silvestrini, Nucl. Phys. B **569** (2000) 3, hep-ph/9812392.
17. S. Descotes and C.T. Sachrajda, SHEP/01-27, hep-ph/0109260.
18. A. Abada, D. Becirevic, P. Boucaud, J. P. Leroy, V. Lubicz and F. Mescia, hep-lat/0011065. See also L. Del Debbio, J. M. Flynn, L. Lellouch and J. Nieves [UKQCD Collaboration], Phys. Lett. B **416** (1998) 392, hep-lat/9708008.
19. A. Khodjamirian, R. Ruckl, S. Weinzierl, C. W. Winhart and O. Yakovlev, Phys. Rev. D **62** (2000) 114002, hep-ph/0001297. For an earlier determination of the form factors, see also P. Ball, JHEP**9809** (1998) 005, hep-ph/9802394.
20. Y. Keum, H. Li and A. I. Sanda, hep-ph/0004004; Y. Y. Keum, H. Li and A. I. Sanda, Phys. Rev. D **63** (2001) 054008, hep-ph/0004173.
21. M. Beneke, hep-ph/0009328.
22. R. Godang *et al.* [CLEO Collaboration], Phys. Rev. Lett. **80** (1998) 3456, hep-ex/9711010.
23. D. Cronin-Hennessy *et al.* [CLEO Collaboration], hep-ex/0001010. D. M. Asner *et al.* [CLEO Collaboration], hep-ex/0103040.
24. G. Cavoto, BaBar Collaboration, talk given at the XXXVI Rencontres de Moriond QCD.
25. B. Casey, Belle Collaboration, talk given at the XXXVI Rencontres de Moriond QCD.
26. C. W. Bauer, D. Pirjol and I. W. Stewart, hep-ph/0107002.
27. M. Beneke, hep-ph/0109243.

QCD Sum Rules for Heavy Flavour Physics

Alexander Khodjamirian [1]

Sektion Physik der Universität München, Thereseinstr. 37, D-80333 München, Germany

Abstract. Uses of QCD sum rules for heavy flavoured hadrons are discussed. "Standard" applications such as the determination of the b, c quark masses, the calculation of f_B, f_D and of the heavy-to-light form factors are overviewed. Furthermore, a new approach to calculate the $B \to \pi\pi$ hadronic matrix elements from QCD light-cone sum rules is described.

INTRODUCTION

Starting from the first work [1] the method of QCD sum rules was frequently applied to various problems of heavy flavour physics. Nowadays, different versions of the sum rule approach are used, all of them based on the general idea of calculating a quark-current correlation function and relating it to the hadronic parameters via dispersion relations.

The original version [1] (often called *SVZ sum rules*) employs the operator-product expansion (OPE) of correlation functions in terms of quark and gluon vacuum condensates. A typical and important application of this technique is the calculation [2, 3, 4] of the B-meson decay constant f_B defined as $\langle 0 \mid m_b \bar{q} i\gamma_5 b \mid B\rangle = f_B m_B^2$, $q = u,d$. One starts from the correlation function of two heavy-light currents

$$\Pi(q^2) = i\int d^4x e^{iqx} \langle 0 \mid T\{m_b \bar{q} i\gamma_5 b(x), m_b \bar{b} i\gamma_5 q(0)\} \mid 0\rangle . \tag{1}$$

Depending on the region of the momentum transfer q, the amplitude $\Pi(q^2)$ represents either a short-distance fluctuation (at q^2 far below m_b^2) or a complicated sum over hadronic states (at $q^2 \geq m_B^2$) starting from the ground-state B meson. At $\mid q^2 - m_b^2 \mid \gg \Lambda_{QCD}^2$, the correlation function (1) is approximately calculated in terms of the condensate expansion including the perturbative part (the loop and $O(\alpha_s)$ correction) and the quark-, gluon- and quark-gluon condensate contributions. A detailed derivation and the resulting expression can be found, e.g. in Ref. [5]. On the other hand, the correlation function (1) obeys the dispersion relation, schematically

$$\Pi(q^2) = \frac{f_B^2 m_B^4}{m_B^2 - q^2} + \sum_{B_h} \frac{f_{B_h}^2 m_{B_h}^4}{m_{B_h}^2 - q^2}, \tag{2}$$

where the contribution of the ground-state B meson is shown explicitly and the sum over B_h represents the excited resonances and the continuum of hadronic states with

[1] *on leave from Yerevan Physics Institute, 375036 Yerevan, Armenia*

CP602, *QCD@Work: International Workshop on Quantum Chromodynamics*
edited by P. Colangelo and G. Nardulli

the B meson quantum numbers. Matching the condensate expansion of $\Pi(q^2)$ with the dispersion relation one obtains the primary sum rule. To achieve practical results one may proceed in different directions. Knowing the values of universal QCD parameters such as α_s, m_b and the condensate densities $\langle G^2\rangle$, $\langle\bar{q}q\rangle$ it is possible to estimate f_B applying the quark-hadron duality approximation to the contribution of higher states in Eq. (2). A reverse way to use QCD sum rules is available in cases when the parameters of few lowest hadronic states in the dispersion relation are measured. Saturating the hadronic part of the sum rule it is then possible to determine the QCD parameters. This kind of analysis is accessible for the two-point correlation functions of $\bar{b}\gamma_\mu b$ or $\bar{c}\gamma_\mu c$ currents, where the hadronic states are Υ- or ψ-resonances, respectively, with measured decay constants and masses.

A different version of QCD sum rules, the so called *light-cone sum rules* (LCSR) [6] are attracting a lot of attention in recent years. LCSR are used to calculate various hadronic transition matrix elements. The outline of this method can be illustrated taking the $B\to\pi$ form factor calculation [7] as an example. The correlation function is in this case a vacuum-to-pion matrix element

$$\begin{aligned}F_\lambda(p,q) &= i\int d^4x e^{ipx}\langle\pi^+(q)\,|\,T\{\bar{u}\gamma_\lambda b(x), m_b\bar{b}i\gamma_5 d(0)\}\,|0\rangle\\ &= F((p+q)^2,p^2)q_\mu+\tilde{F}((p+q)^2,p^2)p_\mu\,, \end{aligned}\tag{3}$$

correlating the $b\to u$ weak current and the heavy-light current interpolating B_d meson. At large virtualities, that is at $|\,(p+q)^2-m_b^2\,|\gg\Lambda^2_{QCD}$ and $p^2\ll m_b^2$ this correlator can be calculated employing OPE near the light-cone $x^2=0$. The result is expressed in terms of pion distribution amplitudes (DA) of growing twists, the most important one being the twist-2 pion DA defined as [8]

$$\langle\pi^+(q)\,|\,\bar{u}(x)\gamma_\mu\gamma_5 d(0)\,|\,0\rangle = -iq_\mu f_\pi\int_0^1 du\, e^{iuqx}\,\varphi_\pi(u). \tag{4}$$

Higher-twist contributions are suppressed by inverse powers of $((p+q)^2-m_b^2)$. Again, using the dispersion relation in the channel of the B-meson current, one obtains

$$F((p+q)^2,p^2)=\frac{2f_Bf^+_{B\pi}(p^2)m_B^2}{m_B^2-(p+q)^2}+\sum_{B_h}\frac{2f_{B_h}f^+_{B_h\pi}(p^2)m^2_{B_h}}{m^2_{B_h}-(p+q)^2}\,, \tag{5}$$

where the ground-state contribution contains a product of f_B and the $B\to\pi$ form factor $f^+_{B\pi}(p^2)$ defined in the standard way: $\langle\pi^+(q)\mid\bar{u}\gamma_\mu b\mid\bar{B}^0(p+q)\rangle = 2f^+_{B\pi}(p^2)q_\mu+\ldots$. Matching the result of the twist expansion for the amplitude F with the dispersion relation one calculates the $B\to\pi$ form factor provided that f_B is obtained from the SVZ sum rule as explained above. Conversely, one may use LCSR to estimate the parameters of the light-cone DA by saturating the sum-rule relations with the experimentally known form factors and decay constants. This is possible, e.g. for various vacuum-to-pion correlators of light-quark currents yielding LCSR for $\gamma\gamma^*\to\pi$ [9] or for the pion e.m. form factor [10]. Again, quark-hadron duality is employed in the same way as in SVZ sum rules. Note that in LCSR the meson DA (such as $\varphi_\pi(u)$ with its normalization factor f_π) play the role of nonperturbative inputs, similar to the role the vacuum condensates play in SVZ sum rules.

For completeness, one has to mention that an independent version of QCD sum rules emerges when one takes the local limit of LCSR. The result is equivalent to the *sum rules in the external field* [11]. In particular, for the pion-to vacuum correlation function (3) the local limit ($q \to 0$) corresponds to the external soft-pion field.

One should always bear in mind that QCD sum rules have a limited accuracy. Both OPE and the duality approximation should be kept under control by performing the Borel transformation or taking the power moments. The working region of the corresponding auxiliary parameters (Borel parameter or the number of moment) has to be restricted, so that the contributions of excited states and higher orders in OPE are simultaneously small.

Why QCD sum rules are in particular advantageous for heavy flavour physics? First of all, the presence of an intrinsic heavy-quark mass scale provides necessary conditions for applying the short-distance or light-cone OPE to the correlation functions. One has to emphasize that no infinite quark mass limit is necessary. The sum rules can be derived in full QCD for finite b and c quark masses. Moreover, since the correlation functions are Lorentz-covariant objects one does not need nonrelativistic approximations. On the other hand, the sum rule method is very flexible and can be applied in the frameworks of the effective QCD theories such as HQET and NRQCD. Finally, since QCD sum rules employ universal inputs (condensates, light-cone DA) there is a possibility to estimate theoretical uncertainties in the determination of heavy-flavour parameters such as f_B or $f^+_{B\pi}$.

In what follows I will overview the status of a few "standard" applications of QCD sum rules: the m_b, m_c determination, the calculation of f_B, f_D and of the $B \to \pi, K, ...$, $D \to \pi, K, ...$ form factors. Furthermore, I describe a new application of the method, the calculation of the $B \to \pi\pi$ matrix elements. I will not discuss many other interesting applications such as B_c meson, $B - \bar{B}$ mixing parameter, properties of heavy baryons, sum rules in HQET. These issues as well as many other details concerning QCD sum rules can be found in the recent review [12].

"STANDARD" APPLICATIONS OF QCD SUM RULES

b and c quark masses

The heavy quark mass m_Q, $Q = b$ or c, can be determined if one considers the two-point correlation function of two $\bar{Q}\gamma_\mu Q$ currents and uses for the hadronic dispersion relation the experimental data on the masses and electronic widths of $J^P = 1^-$ quarkonium levels, Υ or ψ resonances, respectively. In recent years there has been a lot of progress in the determination of the b quark mass employing precise data on six $\Upsilon(nS)$ resonances obtained in e^+e^--annihilation. The emphasize was made on working with the highest possible power moments of the correlation function in which the region of small quark velocities $v_Q = \sqrt{1 - 4m_Q^2/q^2}$ dominates and the NRQCD approximation is valid. In

TABLE 1. b-quark mass determination: QCD sum rules vs other methods; $m_b(\overline{m}_b)$ is the pole ($\overline{MS}$) mass

m_b (GeV)	$\overline{m}_b(\overline{m}_b)$ (GeV)	Ref.	Method
4.72 ± 0.05		[14]	SVZ
4.62 ± 0.02		[15]	"
4.827 ± 0.007		[16]	
4.84 ± 0.08	4.19 ± 0.06	[17]	
	4.20 ± 0.10	[18]	NRQCD
4.88 ± 0.10	4.20 ± 0.06	[19]	SR
4.80 ± 0.06	4.21 ± 0.11	[20]	
4.97 ± 0.17	4.26 ± 0.10	[21]	
5.04 ± 0.09	4.44 ± 0.04	[22]	
–	4.24 ± 0.09	[21]	Υ
–	4.21 ± 0.07	[19]	+NRQCD
	$4.21^{\pm 0.09}_{\mp 0.025}$	[23]	
–	4.26 ± 0.11	[24]	lattice av.

this framework it is possible to sum over Coulomb $O(\alpha_s^n/v_Q^n)$ terms and to include relativistic corrections order by order (for a recent review, see e.g. Ref. [13]). An alternative approach is to stay within full QCD and to employ few first moments of the SVZ sum rules, determining m_b in a purely relativistic way. In this case the Coulomb resummation is not accessible but also not that important. The price to pay is the sensitivity to the nonresonant tail of the hadronic spectral function in the dispersion relation which, in principle, can be reliably estimated from experimental data on the inclusive $e^+e^- \to b\bar{b}$ cross section above resonances. In Table 1 the earlier SVZ and more recent NRQCD sum rule results are compared with the m_b determinations using other approaches. The values of $\overline{m}_b(\overline{m}_b)$ obtained by various methods agree within uncertainties. The potential of SVZ sum rules for b-quarkonium in full QCD is not yet thoroughly exploited, e.g. there is a possibility to include the already available $O(\alpha_s^2)$ corrections to the correlation function [25].

TABLE 2. m_c determination: QCD sum rules vs other methods.

m_c (GeV)	$\overline{m}_c(\overline{m}_c)$ (GeV)	Ref.	method
1.46 ± 0.05		[27]	SVZ
1.42 ± 0.03	$1.23^{+0.02}_{-0.04} \pm 0.03$	[15]	"
1.70 ± 0.13	1.23 ± 0.09	[26]	SVZ+NRQCD
-	1.37 ± 0.09	[28]	FESR
-	$1.21 \pm 0.07^{\pm 0.065}_{\mp 0.045}$	[23]	$m_B - m_D$ + HQET
-	$1.525 \pm 0.040 \pm 0.125$	[29]	lattice QCD
	1.33 ± 0.08	[30]	"
	$1.20 \pm 0.04 \pm 0.11 \pm 0.2$	[31]	latt. NRQCD

The predictions for the c-quark mass obtained from QCD sum rules for the charmonium system are collected in Table 2, in comparison with the results of various other methods. The most recent sum rule analysis [26] combines NRQCD at small v_c with the full QCD spectral function at large v_c. The latter includes $O(\alpha_s)^2$ terms and $d = 6, 8$ gluon condensates. Although the value of $\overline{m}_c(\overline{m}_c)$ is in agreement with earlier sum rule determinations, the pole c quark mass is surprisingly large, being interpreted [26] as a result of large Coulomb corrections. At this point one has to note that for the charmonium levels the applicability of NRQCD is questionable due to the large average values of v_c. An update of the full QCD SVZ sum rule at the $O(\alpha_s^2)$ level remains a task which deserves attention. Summarizing the estimates given in Tables 1,2 I will adopt the following intervals of the pole quark masses obtained from QCD sum rules:

$$m_b = 4.8 \pm 0.1 \text{ GeV}, \quad m_c = 1.3 \pm 0.1 \text{ GeV}. \tag{6}$$

f_B and f_D

Having determined the value of m_b with a certain accuracy it is possible to calculate f_B from the SVZ sum rule based of the dispersion relation (2). For the D-meson decay constant f_D the analogous sum rule is obtained by a simple $b \to c$ ($\bar{B} \to D$) replacement in the f_B sum rule, together with the necessary adjustment of the normalization scale. Moreover, switching to $q = s$ in Eq. (1) it is possible to predict also the ratios f_{B_s}/f_B and f_{D_s}/f_D. Currently, the OPE of the correlation function (1) includes the $O(\alpha_s)$ correction to the perturbative part which is rather large and all $d \leq 6$ condensate contributions.

The values of f_B and f_D determined from SVZ sum rules are quite sensitive to the b and c quark pole masses. Varying the latter in the intervals (6) one typically obtains (see, e.g. the review [12]):

$$f_B = 170 \mp 30 \text{ MeV}, \quad f_D = 180 \mp 30 \text{ MeV},$$
$$f_{Bs}/f_B = 1.16 \pm 0.09, \quad f_{D_s}/f_D = 1.19 \pm 0.08. \tag{7}$$

Here, the normalization scales $\mu_b^2 \sim m_B^2 - m_b^2$ and $\mu_c^2 \sim m_D^2 - m_c^2$, respectively are adopted. These scales are of the order of the corresponding Borel parameters reflecting the average virtualities of the quarks in the correlators. Within uncertainties, the predictions (7) agree with the lattice determinations of the heavy meson decay constants.

Are improvements in $f_{B,D}$ determination still possible? Apart from narrowing the intervals for heavy quark masses one has to cope with the numerically large $O(\alpha_s)$ correction. Studies of $O(\alpha_s^2)$ effects, at least for an accurate fixing of the relevant scale in the $O(\alpha_s)$ term are necessary. An important progress in this direction is the calculation of the three-loop radiative corrections to the heavy-to-light correlator completed just recently [32]. The accuracy of the sum rule could be improved further if the $d = 7$ corrections proportional to the heavy quark mass are calculated including both factorizable $\sim \langle G^2 \rangle \langle \bar{\psi}\psi \rangle$ and nonfactorizable $d = 7$ condensates. On the hadronic side, it is important to get a better control over quark-hadron duality in the B and D channels. For the latter channel a valuable information could be provided by experimental studies of excited D states in the semileptonic $B \to X_c l \nu$ decays. Knowing the positions of

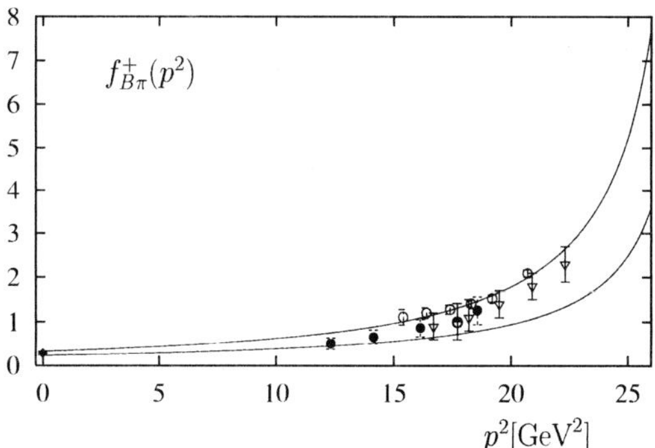

FIGURE 1. *The LCSR prediction for the $B \to \pi$ form factor f^+ [33]. The full curves indicate the theoretical uncertainty, the points represent various lattice QCD calculations.*

excited D resonances one may try various alternative patterns for the hadronic spectral function in the f_D sum rule and thereby test the validity and consistency of the duality approximation.

Heavy-to-light form factors

The procedure of obtaining LCSR for the $B \to \pi$ form factor is briefly outlined in the Introduction. More detailed discussion can be found in Ref. [5]. The most recent LCSR prediction [33] for $f^+_{B\pi}$ is presented in Fig. 1. This calculation includes twist 2 (LO and $O(\alpha_s)$ NLO) and twist 3,4 effects. The inputs used to calculate $f^+_{B\pi}$ from LCSR are: 1) the values of f_B, m_b and the duality threshold, all taken from the SVZ sum rule for f_B and 2) the pion DA of twist 2,3,4. The shapes of the latter are largely fixed by their asymptotic forms whereas the sensitivity of LCSR to the nonasymptotic effects in DA turns out to be very mild.

The LCSR result [33] is parametrized in a form suggested in Ref. [34]:

$$f^+_{B\pi}(p^2) = \frac{f^+_{B\pi}(0)}{(1 - p^2/m^2_{B^*})(1 - \alpha_{B\pi} p^2/m^2_{B^*})} \tag{8}$$

with $f^+_{B\to\pi}(0) = 0.28 \pm 0.05$ and $\alpha_{B\pi} = 0.32^{+0.21}_{-0.07}$. The theoretical uncertainties are estimated by varying all sum rule parameters within allowed regions and adding them up linearly. The accuracy of the form factor calculation can still be improved if the twist 3 $O(\alpha_s)$ correction is calculated and if the pion DA are better constrained, e.g. from LCSR for the pion form factors or from lattice QCD studies.

With the form factor (8) it is possible to calculate the semileptonic $\bar{B}^0 \to \pi^+ l \bar{\nu}_l$ width ($l = e, \mu$) and to extract the CKM parameter $| V_{ub} |$ using the current experimental data: $BR(B \to \pi l \nu) = (1.8 \pm 0.4 \pm 0.4) \cdot 10^{-4}$ (CLEO [35]) and $BR(B \to \pi l \nu) = (1.28 \pm 0.20 \pm 0.26) \cdot 10^{-4}$ (Belle, preliminary [36]). The results are:
$| V_{ub} | = (4.0 \pm 0.6 \pm 0.7) \cdot 10^{-3}$(CLEO), and $| V_{ub} | = (3.4 \pm 0.4 \pm 0.6) \cdot 10^{-3}$(Belle),

where the first error is experimental and the second one is caused by the theoretical uncertainty of the form factor calculation.

Replacing the pion with the kaon in the correlation function (3) leads to LCSR for the $B \to K$ form factor, including the effects of the $SU(3)$-flavour symmetry breaking such as $f_K/f_\pi \neq 1$ and the asymmetry in the kaon DA $\varphi_K(u)$. Interestingly, the predicted ratio [33]

$$f^+_{B\to K}(0)/f^+_{B\to\pi} = 1.28^{+0.18}_{-0.10}, \tag{9}$$

is mainly sensitive to the value of the strange quark mass $m_s(1\text{GeV}) = 150 \mp 50$ MeV. This result indicates that the rate of $SU(3)$ breaking could be quite noticeable, an important message for studies of CP-violation in hadronic B decays where various SU(3) relations are frequently used. The semileptonic form factors in the charmed sector are also predicted from LCSR , e.g. [33] : $f^+_{D\to\pi}(0) = 0.65 \pm 0.11$ and $f^+_{D\to K}(0)/f^+_{D\to\pi} = 1.20$ (at $m_s(1\text{ GeV}) = 150$ MeV), in a good agreement with both experiment and lattice QCD.

To complete our discussion on heavy-to-light form factors one has to mention various $B \to V$ form factors, $V = K^*, \rho, \phi$, relevant for $B \to V l \nu_l$ and $B \to V\gamma$ decays. Their most accurate calculation is in Ref. [37]. Using LCSR it is also possible to estimate the amplitudes of $B \to \rho\gamma$ weak annihilation [38, 39] and the $B \to \mu\nu\gamma$ width [38] employing the photon DA. The list of heavy-to-light semileptonic and radiative processes treated with the help of LCSR can be enlarged to include also $B \to a_{0,1,2}$, $B \to K_0^*, K_1, K_2^*$ transition form factors if the corresponding DA of these light mesons are worked out. Another potentially interesting application is to employ the two-pion DA [40] in studying $B \to \pi\pi l\nu$ decay. The first step in this direction was done in Ref. [41].

$D^*D\pi$-coupling, QCD sum rules vs experiment

Recently the total width of D^* meson was measured by CLEO collaboration [42]: $\Gamma_{tot}(D^*) = 96 \pm 4 \pm 22$ keV. This remarkable measurement yields the strong $D^*D\pi$ coupling $g_{D^*D\pi} = 17.9 \pm 0.3 \pm 1.9$ defined as in Ref. [43]. Among many theoretical predictions for this coupling obtained by various methods I would like to single out the LCSR prediction [43] updated in Ref. [44] by including the $O(\alpha_s)$ correction to the twist 2 term: $g_{D^*D\pi} = 10 \pm 3.5$. The sum rule is derived [43] from the correlator (3) employing the double dispersion relation. An estimate in the same ballpark is obtained from the QCD sum rules in the soft pion limit [43, 45]. Note that the theoretical uncertainty quoted above includes variation of all inputs within reasonable limits, therefore it is difficult to push the LCSR prediction above its upper limit $g_{D^*D\pi} = 13.5$ which is still 25% lower than the central value of the CLEO measurement. If the currently observed discrepancy between the LCSR prediction and experiment persists in future one might suspect that the simple quark-hadron duality ansatz which works in the one-variable dispersion relations is too crude for the double dispersion relation.

Let me make one parenthetical remark. It is often claimed that knowing the value of the $D^*D\pi$ coupling one fixes the effective scale-independent coupling in HQET defined as $\hat{g} = f_\pi g_{H^*H\pi}/2m_H$, $H = B, D$. However, in the charmed sector there are large $1/m_H$ corrections to the HQET limit. Indeed, as shown in Refs. [43, 46] where both $D^*D\pi$

and $B^*B\pi$ couplings are calculated from LCSR, they can be fitted to a single effective $\hat{g}$ only by adding a substantial $1/m_H$ correction: $g_{H^*H\pi} = 2m_H\hat{g}/f_\pi(1+\Delta/m_H)$, with $\Delta \sim 1$ GeV. Therefore, expressing $g_{D^*D\pi}$ in terms of $\hat{g}$ is not a straightforward procedure.

LIGHT-CONE SUM RULES FOR HADRONIC B DECAYS

The CP-violation studies are nowadays concentrated on the two-body hadronic B decays. In order to fully explore experimental data one needs reliable theoretical predictions on hadronic matrix elements of the operators O_i entering the effective weak Hamiltonian, $H_W = \frac{G_F}{\sqrt{2}}\sum_i \lambda_i^{CKM} c_i(\mu)O_i$. The solution of this tremendously difficult problem can only be achieved within approximate QCD methods, such as the recently developed QCD factorization [47].

Here I will shortly outline a new approach to this problem [48] which is based on LCSR and allows to calculate the hadronic matrix elements in the same framework as the $B \to \pi$ form factor. As a study case the matrix elements $\langle \pi^-\pi^+ \mid O_{1,2} \mid B\rangle_{Emission}$ of the current-current operators $O_{1,2}$ for $\bar{B}^0 \to \pi^+\pi^-$ in the emission topology are considered, where $O_1 = (\bar{d}\Gamma_\mu u)(\bar{u}\Gamma^\mu b)$ and O_2 is replaced by a combination of O_1 and the colour-octet operator $\widetilde{O}_1 = (\bar{d}\Gamma_\mu \frac{\lambda^a}{2}u)(\bar{u}\Gamma^\mu \frac{\lambda^a}{2}b)$.

As usual in the sum rule derivation, one constructs a suitable correlation function:

$$F_\alpha^{(O)}(p,q,k) = -\int d^4x e^{-i(p-q)x}\int d^4y e^{i(p-k)y}\langle 0 \mid T\{\bar{u}\gamma_\alpha\gamma_5 d(y)O(0)m_b\bar{b}i\gamma_5 d(x)\} \mid \pi^-(q)\rangle . \tag{10}$$

Here the effective operator $O = O_1$ or $\widetilde{O}_1$ is correlated with the currents interpolating B meson and pion. In the above, a fictive momentum k is attributed to the weak vertex to avoid certain technical difficulties in the dispersion relations. Furthermore, we put $p^2 = k^2 = 0$ and consider the kinematical region of large spacelike external momenta $\mid (p-k)^2 \mid \sim \mid (p-q)^2 \mid \sim \mid P^2 \mid \gg \Lambda_{QCD}^2$, where $P = p-k-q$. Due to the large b quark mass it is possible to apply the light-cone OPE to the correlator (10) in this region. The lowest-order diagram for the operator O_1 is shown in Fig. 2a. It factorizes into a simple light-quark loop and the vacuum-to-pion correlation function similar to Eq. (3). The OPE of the correlator (10) with the operator $\widetilde{O}_1$ starts from the diagrams containing a one-gluon nonfactorizable exchange: either a soft (low virtuality) gluon which is absorbed by the pion DA (Fig. 2b) or a hard gluon exchanged between the light-quark loop and the heavy-light part of the correlator. The latter effect corresponds to the $O(\alpha_s)$ two-loop diagrams which demand technically difficult calculation. In what follows we concentrate on the soft nonfactorizable effect represented by the diagrams in Fig. 2b. Their calculation involves twist 3 and 4 quark-antiquark-gluon DA of the pion. The key nonperturbative parameters determining these DA are the matrix elements $\langle 0 \mid \bar{u}\sigma_{\mu\nu}\gamma_5 g_s G_{\alpha\beta} d \mid \pi\rangle$ and $\langle 0 \mid g_s\bar{u}\tilde{G}_{\alpha\beta}\gamma^\mu d \mid \pi\rangle$ estimated from SVZ sum rules [49].

The procedure of the sum rule derivation from Eq. (10) is more complicated than in the $B \to \pi$ form factor case. It can be shortly summarized as follows:

1. The dispersion relation in the pion-current channel with the momentum $(p-k)$ is employed together with the quark-hadron duality approximation, allowing one to obtain

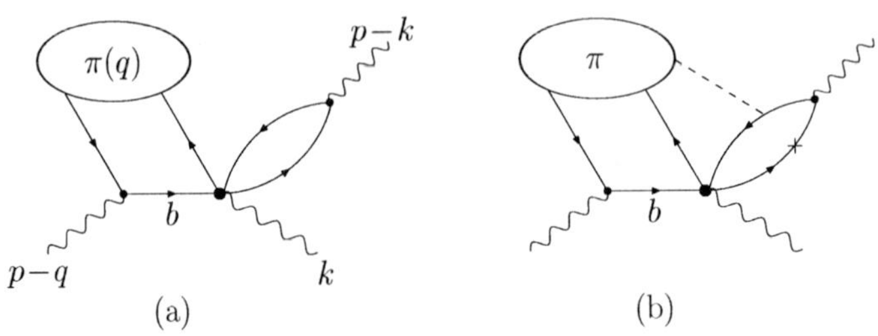

FIGURE 2. *Diagrams corresponding to the leading order of the correlator (10) for $O = O_1, \widetilde{O}_1$; the cross indicates the point of gluon emission in the second similar diagram.*

an analytic expression for the hadronic matrix element

$$\Pi^{(O)}_{\pi\pi}((p-q)^2, P^2) = i \int d^4x e^{-i(p-q)x} \langle \pi^-(p-k) \mid T\{O(0) m_b \bar{b} i\gamma_5 d\} \mid \pi^-(q)\rangle . \quad (11)$$

This matrix element resembles the pion form factor at large spacelike momentum transfer P^2 where, instead of a simple e.m. vertex, one has a more complicated short-distance part with a virtual b quark.

2. Analytic continuation of Eq. (11) in the variable P^2 to the large timelike $P^2 = m_B^2$ is performed. This procedure is analogous to the transition from large spacelike to large timelike momentum transfers for the pion e.m. form factor. Note that an imaginary part may emerge as a result of this continuation. It has to be interpreted as a strong final state interaction phase.

3. The dispersion relation for $\Pi^{(O)}_{\pi\pi}((p-q)^2, m_B^2)$ in the variable $(p-q)^2$ (in the B-meson channel) is written down and the duality ansatz for the higher B states is applied. As a final result, one obtains LCSR for the on-shell $\bar{B}_d \to \pi^+\pi^-$ matrix element of the operator O where the fictive momentum k vanishes due to $P^2 = m_B^2$. As usual, in order to suppress the higher states and to reduce the sensitivity to the duality approximation in both pion and B meson channels, two independent Borel transformations are performed in the variables $(p-k)^2$ and $(p-q)^2$, respectively. Note that all parameters entering the obtained sum rules are fixed either from the SVZ sum rules for two-point correlation functions or from LCSR for $f^+_{B\pi}$.

The resulting sum rule for the matrix element of O_1 in the leading order simply factorizes into a product of SVZ sum rule for f_π and the LCSR for $B \to \pi$ form factor: $\langle \pi^-\pi^+ \mid O_1 \mid B\rangle_E = i f_\pi f^+_{B\pi}(0)_{LCSR}\, m_B^2$, thereby reproducing the factorization approximation in the limit of the heavy quark mass. The LCSR for the matrix element of the colour-octet operator $\widetilde{O}_1$ calculated from the diagrams in Fig. 2b quantifies the soft nonfactorizable effect. Importantly, at $m_b \to \infty$ it is $1/m_b$ suppressed with respect to the factorizable part, in accordance with QCD factorization [47]. Numerically,

$$\frac{\langle \pi^-\pi^+ \mid \widetilde{O}_1 \mid B\rangle}{\langle \pi^-\pi^+ \mid O_1 \mid B\rangle} \equiv \frac{\lambda_E}{m_B}, \quad \lambda_E = 50 \div 150 \text{ MeV}, \quad (12)$$

that is, the soft nonfactorizable effect due to $\widetilde{O}_1$ is small and does not contain imaginary part. At the same time, the soft effect is as important as the small $O(\alpha_s)$ hard nonfactorizable effects calculated in the QCD factorization approach.

The $\bar{B}_d \to \pi^+\pi^-$ decay amplitude obtained from LCSR

$$\mathcal{A}(\bar{B}_d \to \pi^+\pi^-) = i\frac{G_F}{\sqrt{2}} V_{ub}V_{ud}^* f_\pi [f_{B\pi}^+(0)]_{LCSR}\, m_B^2 \Big\{ c_1(\mu) + \frac{c_2(\mu)}{3} + 2c_2(\mu)\Big(\frac{\lambda_E}{m_B} + O(\alpha_s)\Big)\Big\} + \ldots \quad (13)$$

has to be completed by calculating the $O(\alpha_s)$ nonfactorizable effects. After including the latter in the decay amplitude the scale μ dependence has to be partially canceled and the imaginary part will emerge at $O(\alpha_s)$. It is also important to calculate one by one the contributions of annihilation, penguin topologies for $O_{1,2}$ as well as the matrix elements of penguin operators O_n, $n \geq 3$ denoted by ellipses in the above. The LCSR approach can be generalized to other channels such as $B \to K\pi, KK, D\pi, J/\psi K$.

SUMMARY

The aim of this minireview was to demonstrate that employing QCD sum rules one is able to determine various heavy-flavour parameters starting from the most fundamental ones, the heavy quark masses, and ending with the most complicated ones, the hadronic matrix elements of nonleptonic B decays. The method is selfsufficient, that is, extracting a certain parameter from a sum rule, one uses the result in the other sum rules to calculate more complicated parameters. The following hierarchy can be traced:

$$m_b \to f_B \to f_{B\pi}^+ \to \langle \pi\pi \mid O_i \mid B \rangle \;.$$

Summarizing, QCD sum rules remain a reliable approximate approach well equipped to attack various topical problems in the physics of heavy flavours.

Note added:
After this meeting, two new QCD sum rule calculations of the heavy meson decay constants have been published, both including the $O(\alpha_s^2)$ correction [32] to the heavy-light correlator. The first analysis is done in HQET[50] and predicts $f_B = 206 \pm 20$ MeV, $f_D = 195 \pm 20$ MeV . The second one [51] uses $\overline{MS}$-mass of the b quark instead of the pole mass yielding $f_B = 197 \pm 23$ MeV and $f_{B_s} = 232 \pm 25$ MeV. Within uncertainties, both results [50, 51] agree with the intervals in Eq. (7).

Acknowledgements. I am grateful to P. Colangelo and G. Nardulli for an opportunity to participate in this fruitful and enjoyable workshop.

REFERENCES

1. M. A. Shifman, A. I. Vainshtein and V. I. Zakharov, Nucl. Phys. B **147**, 385, 448 (1979).
2. V. A. Novikov, L. B. Okun, M. A. Shifman, A. I. Vainshtein, M. B. Voloshin and V. I. Zakharov, Phys. Rept. **41**, 1 (1978).
3. L. J. Reinders, H. Rubinstein and S. Yazaki, Phys. Rept. **127**, 1 (1985).
4. D. J. Broadhurst, Phys. Lett. B **101**, 423 (1981);
   T.M. Aliev and V.L. Eletsky, Sov. J. Nucl. Phys. **38**, 936 (1983).
5. A. Khodjamirian and R. Ruckl, in *Heavy Flavors*, 2nd edition, eds., A.J. Buras and M. Lindner, World Scientific (1998), p. 345, hep-ph/9801443.
6. I. I. Balitsky, V. M. Braun and A. V. Kolesnichenko, Nucl. Phys. **B312** (1989) 509;
   V. M. Braun and I. E. Filyanov, Z. Phys. **C44** (1989) 157;
   V. L. Chernyak and I. R. Zhitnitsky, Nucl. Phys. **B345** (1990) 137.
7. V. M. Belyaev, A. Khodjamirian and R. Rückl, Z. Phys. **C60** (1993) 349;
   A. Khodjamirian, R. Rückl, S. Weinzierl and O. Yakovlev, Phys. Lett. **B410** (1997) 275;
   E. Bagan, P. Ball and V. M. Braun, Phys. Lett. **B417** (1998) 154;
   P. Ball, JHEP **9809** (1998) 005.
8. G. P. Lepage and S. J. Brodsky, Phys. Lett. **87B** (1979) 359; Phys. Rev. **D22** (1980) 2157. Phys. Lett. **B94** (1980) 245; Theor. Math. Phys. **42** (1980) 97;
   V. L. Chernyak and A. R. Zhitnitsky, JETP Lett. **25** (1977) 510; Sov. J. Nucl. Phys. **31** (1980) 544.
9. A. Khodjamirian, Eur. Phys. J. C **6** (1999) 477.
10. V. Braun and I. Halperin, Phys. Lett. B **328**, 457 (1994);
    V.M. Braun, A. Khodjamirian and M. Maul, Phys. Rev. **D61** (2000) 073004.
11. I. I. Balitsky and A. V. Yung, Phys. Lett. B **129**, 328 (1983);
    B. L. Ioffe and A. V. Smilga, *JETP Lett.* **37**, 298 (1983); Nucl. Phys. B **232**, 109 (1984).
12. P. Colangelo and A. Khodjamirian, hep-ph/0010175, in the *Boris Ioffe Festschrift 'At the Frontier of Particle Physics / Handbook of QCD'*, ed. M. Shifman (World Scientific, Singapore, 2001), p.1495.
13. M. Beneke, hep-ph/9911490.
14. C.A. Dominguez and N. Paver, Phys. Lett. B **293**, 197 (1992).
15. S. Narison, Phys. Lett. B **341**, 73 (1994).
16. M.A. Voloshin, Int. J. Mod. Phys. A **10**, 2865, (1995).
17. M. Jamin and A. Pich, Nucl. Phys. B Proc. Suppl. **74**, 300 (1999).
18. K. Melnikov and A. Yelkhovsky, Phys. Rev. D **59**, 114009 (1999).
19. A. Hoang, Phys. Rev. D **59**, 014039 (1999).
20. A.A. Penin and A.A. Pivovarov, Nucl. Phys. B **549**, 217 (1999).
21. M. Beneke and A. Signer, Phys. Lett. B **471**, 233 (1999).
22. A. Pineda and F.J. Yndurain, Phys. Rev. D **61**, 077505 (2000).
23. A. Pineda, JHEP **0106** (2001) 022.
24. S. Hashimoto, Nucl. Phys. B (Proc. Suppl.) **83**, 3 (2000).
25. K. G. Chetyrkin, J. H. Kuhn and M. Steinhauser, Nucl. Phys. B **505**, 40 (1997).
26. M. Eidemuller and M. Jamin, Phys. Lett. B **498** (2001) 203.
27. C.A. Dominguez, G.R. Gluckman and N. Paver, Phys. Lett. B **333**, 184 (1994).
28. J. Penarrocha and K. Schilcher, hep-ph/0105222.
29. L. Giusti (APE Coll.), Nucl. Phys. B (Proc. Suppl.) **63**, 167 (1998).
30. A. S. Kronfeld, Nucl. Phys. B (Proc. Suppl.) **63**, 311 (1998).
31. K. Hornbostel (NRQCD Coll.), Nucl. Phys. B (Proc. Suppl.) **73**, 339 (1999).
32. K. G. Chetyrkin and M. Steinhauser, hep-ph/0108017.
33. A. Khodjamirian, R. Ruckl, S. Weinzierl, C. W. Winhart and O. Yakovlev, Phys. Rev. **D62** (2000) 114002.
34. D. Becirevic and A. B. Kaidalov, Phys. Lett. B **478** (2000) 417.
35. J. P. Alexander *et al.* [CLEO Collaboration], Phys. Rev. Lett. **77** (1996) 5000.
36. Belle Collaboration, preprint BELLE-CONF-0110.
37. P. Ball, V. M. Braun, Phys. Rev. D **58**, 094016 (1998).
38. A. Khodjamirian, G. Stoll and D. Wyler, Phys. Lett. B **358**, 129 (1995);
39. A. Ali and V. M. Braun, Phys. Lett. B **359**, 223 (1995).

40. M. Diehl, T. Gousset, B. Pire and O. Teryaev, Phys. Rev. Lett. **81**, 1782 (1998);
M. V. Polyakov, Nucl. Phys. B **555**, 231 (1999).
41. M. Maul, Eur. Phys. J. C **21** (2001) 115.
42. S. Ahmed [CLEO Collaboration], hep-ex/0108013.
43. V. M. Belyaev, V. M. Braun, A. Khodjamirian and R. Rückl, Phys. Rev. **D51** (1995) 6177.
44. A. Khodjamirian, R. Ruckl, S. Weinzierl and O. Yakovlev, Phys. Lett. B **457** (1999) 245.
45. P. Colangelo, G. Nardulli, A. Deandrea, N. Di Bartolomeo, R. Gatto and F. Feruglio, Phys. Lett. B **339**, 151 (1994).
46. P. Colangelo and F. De Fazio, Eur. Phys. J. C **4** (1998) 503.
47. M. Beneke, G. Buchalla, M. Neubert and C. T. Sachrajda, Phys. Rev. Lett. **83** (1999) 1914; Nucl. Phys. **B591** (2000) 313.
48. A. Khodjamirian, Nucl. Phys. B **605** (2001) 558.
49. V. A. Novikov, M. A. Shifman, A. I. Vainshtein, M. B. Voloshin and V. I. Zakharov, Nucl. Phys. **B237** (1984) 525;
V. L. Chernyak, A. R. Zhitnitsky and I. R. Zhitnitsky, Sov. J. Nucl. Phys. **38** (1983) 645; Sov. J. Nucl. Phys. **41** (1985) 284.
50. A. A. Penin and M. Steinhauser, hep-ph/0108110.
51. M. Jamin and B. O. Lange, hep-ph/0108135.

Charming penguin in nonleptonic B decays

T.N. Pham

Centre de Physique Theorique, Centre National de la Recherche Scientifique, UMR 7644, Ecole Polytechnique, 91128 Palaiseau Cedex, France

Abstract. In the study of two-body charmless B decays as a mean of looking for direct CP-violation and measuring the CKM mixing parameters in the Standard Model, the short-distance penguin contribution with its absorptive part generated by charm quark loop seems capable of producing sufficient $B \to K\pi$ decays rates, as obtained in factorization and QCD-improved factorization models. However there are also long-distance charming penguin contributions which also give rise to a strong phase due to the rescattering $D^*D^* \to K\pi$. In this talk, I would like to discuss [19] a recent work on the long-distance charming penguin as a a different approach to the calculation of the penguin contributions in $B \to K\pi$ decays from charmed meson intermediate states. Using chiral effective Lagrangian for light and heavy mesons, corrected for hard pion and kaon momenta, we show that the charming-penguin contributions increase significantly the $B \to K\pi$ decays rates from its short-distance contributions, giving results in better agreement with experimental data.

INTRODUCTION

Recent measurements by the CLEO [1], Babar [2] and Belle [3] collaboration give consistent values for the $B \to K\pi$ branching ratios, which are respectively $(18.2^{+4.6}_{-4.0} \pm 1.6) \times 10^{-6}$, $(18.2^{+3.3+1.6}_{-3.0-2.0}) \times 10^{-6}$, $(13.7^{+5.7+1.9}_{-4.8-1.8}) \times 10^{-6}$ for $B^+ \to K^0\pi^+$ and $(17.2^{+2.5}_{-2.4} \pm 1.2) \times 10^{-6}$, $(16.7 \pm 1.6^{+1.2}_{-1.7}) \times 10^{-6}$, $(19.3^{+3.4+1.5}_{-3.2-0.6}) \times 10^{-6}$ for $B^0 \to K^+\pi^-$ decays. The short-distance contributions to $B \to K\pi$ decays as given by the penguin operators without charm quark loop in factorization model seem to produce the $B \to K\pi$ decays rates too small compared to the data [4]. A better agreement is obtained by including the so-called charming penguin contribution in the effective Wilson coefficients [5, 6, 7, 8, 9]. In this way an absorptive part of the decay amplitude is generated and the strong phase from this absorptive part can produce CP violation in $B \to K\pi$ decays [5, 10]. This approach seems to produce decay rates in agreement with data, at least qualitatively, as shown previously [9, 11, 12, 13] , where the charm quark loop contribution increases the effective Wilson coefficients of the strong penguin operators by about 30%, More recently charm quark effects computed by this method have been obtained in recent works dealing with the validity of factorization [14, 15, 16]. Another approach is to assume that the charm quark contributions are basically long-distance effects essentially due to rescattering processes such as, e.g. $B \to DD_s \to K\pi$. These contributions, first discussed in [17], have been more recently stressed by [4], where they are called charming penguin terms. The situation is similar to the $B_s \to \gamma\gamma$ decay for which the absorptive part obtained in [18] is comparable to the short-distance contribution. I would like to discuss here a recent work [19] on the charming penguin contributions in $B \to K\pi$ decays. As

CP602, *QCD@Work: International Workshop on Quantum Chromodynamics*
edited by P. Colangelo and G. Nardulli

details can be found in this reference, I will present only the main results of the work.

SHORT AND LONG DISTANCE WEAK MATRIX ELEMENT

In the standard model, effective Hamiltonian for non-leptonic B decays are given by

$$H_{\rm eff} = \frac{G_F}{\sqrt{2}}\left[V_{ub}^*V_{us}(c_1O_1^u + c_2O_2^u) + V_{cb}^*V_{cs}(c_1O_1^c + c_2O_2^c) - V_{tb}^*V_{ts}\left(\sum_{i=3}^{10} c_iO_i + c_gO_g\right)\right] \quad (1)$$

where c_i are the Wilson coefficients evaluated at the normalization scale $\mu = m_b$ [6, 8, 20, 21, 22] and next-to-leading QCD radiative corrections are included. O_1 and O_2 are the usual tree-level operators, O_i $(i = 3, ..., 10)$ are the penguin operators and O_g is the chromomagnetic gluon operator.

The $B \to K\pi$ decay amplitude $A_{K\pi}$ is given by

$$A_{K\pi} \;=\; < K(p_K)\pi(p_\pi)|iH_{\rm eff}|B(p_B) > \; . \quad (2)$$

In the factorization approximation, the above matrix element is evaluated at the tree-level as higher order QCD radiative corrections are already included in the effective Wilson coefficients and the charm quark operators O_1^c and O_2^c do not contribute. The short-distance part A_{SD} is obtained with $c_2 = 1.105$, $c_1 = -0.228$, $c_3 = 0.013$, $c_4 = -0.029$, $c_5 = 0.009$, $c_6 = -0.033$ [20]; $|V_{ub}| = 0.0038$, $V_{us} = 0.22$, $V_{tb} \simeq 1$, $V_{ts} = -0.040$ and $\gamma = -arg\,(V_{ub}) = 54.8^o$[23] and $F_0^{B\to\pi}(m_K^2) = 0.37$. We find

$$\begin{aligned} A_{SD}(B^+ \to K^0\pi^+) &= 2.43\times 10^{-8}\ \text{GeV} \\ A_{SD}(B^0 \to K^+\pi^-) &= (1.86 - i0.95)\times 10^{-8}\ \text{GeV}\,. \end{aligned} \quad (3)$$

As mentioned, the $B \to K\pi$ branching ratios obtained from Eq.(3) are too small compared with experiments. Instead of using perturbative QCD to treat the charm quark loop contributions, we now consider the one-particle D, D^* intermediate state contribution to the T-product of two charged weak currents corresponding to the local operators O_2^c. The matrix element of O_2^c is evaluated using a sum rule due to Wilson [24] . Following Wilson, consider now the short-distance limit of the T-product of two weak currents

$$T\left[J_{\mu N}(x)J_{\mu S}(0)\right] = B_1'(x)\sigma_m'(0) \quad (4)$$

where the contributions from the more singular, lower dimension operators have been taken out. $B_1'(x)$ is the coefficient of the local operator $\sigma_m'(0)$. Let $M_{AB}(q) = \int d^4x\exp(iq\cdot x) < A|J_{\mu S}(x)J_{\mu N}(0)|B >$ we have in momentum space,

$$\int^{q_{max}} M_{AB}(q)d^4q = B_1'(q_{max})\sigma_{AB}', \qquad B_1'(q_{max}) = \int^{q_{max}} B_1'(q)d^4q \quad (5)$$

If $B_1'(x)$ scales as $(x^2)^0$ as in QCD, and for q_{max} not too large, we obtain

$$\int^{q_{max}} M_{AB}(q)d^4q = \sigma_{AB}' \quad (6)$$

Eq.(6) thus gives us the matrix element of the local operators in terms of a Cottingham-like formula evaluated only up to a cut-off momentum q_{max} as the high momenta of the integral has already been factorized in the Wilson coefficients, as stressed in previous work [17, 25, 26]. It should be stressed here that in factorization model, the exchange term in the effective Hamiltonian is usually Fierz-reordered into a product of two color-singlet operators and then evaluated by vacuumm saturation. Actually, it can also be expressed in terms of an integral over the virtual momentum q which is the difference of the two quark momenta in the initial and final hadron. For example, the exchange term in the $K\pi$ transition is given as ($\psi(k,k-p)$ is the pion B-S wave function),

$$A(K^- \to \pi^-) = \int \frac{d^4k}{(2\pi)^4} \int \frac{d^4k'}{(2\pi)^4} \bar{\psi}(k,k-p)\, T_W(k,k-p;k',k'-p')\psi(k',k'-p') \quad (7)$$

Making a change of variable $k' = q+k$, we have

$$A(K^- \to \pi^-) = \int \frac{d^4q}{(2\pi)^4} T(p,q) \quad (8)$$

$$T(p,q) = \int \frac{d^4k}{(2\pi)^4} \bar{\psi}(k,k-p)\, T_W(k,k-p;k+q,k+q-p)\, \psi(k+q,k+q-p) \quad (9)$$

which is a higher twist contribution to the forward virtual scattering of the W boson with momentum q off a hadron. A similar expression can also be given for the transition $\Sigma \to p$ in hyperon nonleptonic decays. The above expression shows that nonleptonic weak matrix elements can be expressed as integral over the virtual W boson scattering amplitude. We have, for the long-distance part A_{LD}

$$\begin{aligned} A_{LD} &= A_{LD}(B^+ \to K^0\pi^+) = A_{LD}(B^0 \to K^+\pi^-) = \\ &= \frac{G_F}{\sqrt{2}} V_{cb}^* V_{cs} a_2 \int \frac{d^4q}{(2\pi)^4} \theta(q^2+\mu^2) T(q,p_B,p_K,p_\pi) \end{aligned} \quad (10)$$

where μ (or q_{max}) is a cut-off momentum separating long-distance and short-distance contribution. $T(q,p_B,p_K,p_\pi) = g^{\mu\nu} T_{\mu\nu}$, with

$$T_{\mu\nu} = i \int d^4x \exp(iq\cdot x) < K(p_K)\pi(p_\pi)|\mathrm{T}(J_\mu(x)J_\nu(0))|B(p_B) > \quad (11)$$

$J_\mu = \bar{b}\gamma_\mu(1-\gamma_5)c$ and $J_\nu = \bar{c}\gamma_\nu(1-\gamma_5)s$. To compute A_{LD} we saturate the $T_{\mu\nu}$ with the D, D^* intermediate states. This gives us the usual D,D^* pole term (Born term) for $T(q,p_B,p_K,p_\pi)$. To compute these pole terms, we use heavy quark effective theory and chiral effective lagrangian to obtain the $B \to D,D^*$ and $D \to K\pi$ and $D^* \to K\pi$ semi-leptonic decay form factors [27] which appear at each vertex of the pole diagrams. $<(D,\ D^*)|J^\mu|B>$ is parameterized in terms of the Isgur-Wise function and the matrix elements $< K\pi|J^\mu|D >$ and $< K\pi|J^\mu|D^* >$ are computed using Chiral Effective Lagrangian for semileptonic decays of heavy mesons to light pseudo-scalar mesons. We extrapolate the soft meson limit to higher momenta by using the full D^* propagator in the pole terms (a similar use of the full D^* propagator to go beyond the soft pion result

TABLE 1. Numerical values for the real part of A_{LD} in GeV for $\mu_\ell = 0.5-0.7$ GeV. First column refers to the D, the second is the D^* contribution.

μ_ℓ	D	D^*	**Total**
0.5	-4.66×10^{-9}	1.62×10^{-8}	1.15×10^{-8}
0.6	-7.77×10^{-9}	2.79×10^{-8}	2.01×10^{-8}
0.7	-1.19×10^{-8}	4.40×10^{-8}	3.21×10^{-8}

has also been given in [28]) . We also introduce a form factor in the strong $DD^*\pi$ coupling constant (a similar approach is used in semileptonic decays [29]). Including this effect, we obtain, for hard pion,

$$G_{D^*D\pi} = \frac{2m_D g}{f_\pi} F(|\vec{p}_\pi|) \,, \tag{12}$$

where $F(|\vec{p}_\pi|)$ is normalized by $F(0) = 1$ which corresponds to the soft pion limit. ($g \approx 0.4$). This form factor can be evaluated by using the constituent quark model which gives roughly, for $|\vec{p}_\pi| \simeq m_B/2$, $F(|\vec{p}_\pi|) = 0.065 \pm 0.035$.

Since the threshold for the D, D_s and D, D_s^* production is below the B meson mass, the D_s and D_s^* pole term for the $D, D^* \to K\pi$ form factors have an absorptive part. This pole term is in fact a rescattering term via the Cabibbo-allowed $B \to D, D_s^*$ decays followed by the strong annihilation process $D, D_s^* \to K\pi$ and can be obtained from the unitarity of the $B \to K\pi$ decay amplitude. We have

$$\begin{aligned}
\text{Disc}\, A_{LD} &= 2i\,\text{Im} A_{LD} = (-2\pi i)^2 \int \frac{d^4q}{(2\pi)^4} \delta_+(q^2 - m_{D_s}^2)\, \delta_+(p_{D^{(*)}}^2 - m_D^2) \times \\
&\times\; A(B \to D_s^{(*)} D^{(*)}) A(D_s^{(*)} D^{(*)} \to K\pi) = \\
&= -\frac{m_D}{16\pi^2 m_B} \sqrt{\omega^{*2} - 1} \int d\vec{n} A(B \to D_s^{(*)} D^{(*)}) A(D_s^{(*)} D^{(*)} \to K\pi)
\end{aligned} \tag{13}$$

With the $A(B \to D_s D)$, $A(B \to D_s^* D^*)$ given by factorization and $A(D_s D \to K\pi)$, $A(D_s^* D^* \to K\pi)$ by the t-channel D, D^* exchange pole terms which are proportional to $G^2_{D^*D\pi}$ and could be large due to the factor m_D^2. However the rescattering amplitudes $A(D_s^* D^* \to K\pi)$ etc. which are exclusive processes at high energy, should be suppressed. This is taken account by the suppression factor $F(|\vec{p}_\pi|)$ mentioned above. We find, for the absorptive part

$$\text{Im}\, \text{A}_{\text{LD}} = 2.34 \times 10^{-8}\,\text{GeV} \tag{14}$$

of which 1.45×10^{-8} GeV and 0.89×10^{-8} GeV are respectively the D, D_s and D^*, D_s^* contributions. To find the real part, we compute all Feynman diagrams obtained with the effective Lagrangian for the weak form factors and integrate over the virtual current momentum q up to a cut-off $\mu = m_b$. This includes the direct term and the pole terms which produce the absorptive part. It is possible to choose a cut-off momentum by a change of variable $q = p_B - p_{D^{(*)}}$ to the momentum ℓ defined by the formula

$$q = p_B - p_{D^{(*)}} \equiv (m_B - m_{D^{(*)}})v - \ell \,. \tag{15}$$

As discussed in [19], the chiral symmetry breaking scale is about 1 GeV and the mean charm quark momentum k for the on-shell D meson is about 300 MeV, the virtual momentum ℓ should be below 0.6 GeV, hence a cut-off $\mu_\ell \approx 0.6$ GeV . The real part is then given by a Cottingham formula as follows [27]

$$\begin{aligned} \mathrm{Re}\, A_{LD} &= \frac{i}{2\,(2\pi)^3}\frac{G_F}{\sqrt{2}}V_{cb}^* V_{cs}\, a_2 \int_0^{\mu_\ell^2} dL^2 \int_{-\sqrt{L^2}}^{+\sqrt{L^2}} dl_0 \sqrt{L^2 - l_0^2} \int_{-1}^{1} d\cos(\theta)\; i \\ &\times \left\{ \frac{j_D^\mu\, h_{D\mu}}{p_D^2 - m_D^2} + \frac{\Sigma_{pol}\, j_{D^*}^\mu\, h_{D^*\mu}}{p_{D^*}^2 - m_{D^*}^2} \right\}. \end{aligned} \tag{16}$$

in the above expression, the coupling constant g are corrected by the form factor $F(|\vec{p}_\pi|)$. The results for the real part are shown in Table 1 for $\mu_\ell = 0.5 - 0.7$ GeV. Our numerical results show that the long-distance charming penguin contributions to the decays $B \to K\pi$ are significant. These results agree qualitatively with a phenomenological analysis of these contributions given in [4]. In particular, we found that the absorptive part due to the D, D_s states is somewhat bigger than that from the D^*, D_s^* states, but of the same sign. The real part due to the D^*, D_s^* states is however $3-4$ times bigger and opposite in sign to the contributions from the D, D_s states. As shown in Table 1, the real part and absorptive part are of the same order of magnitude, at the 10^{-8} GeV level. The results for the branching ratios are

$$\begin{aligned} B(B^+ \to K^0\pi^+) &= (2.4^{+2.7}_{-1.9}) \times 10^{-5} \\ B(B^0 \to K^+\pi^-) &= (1.5^{+1.8}_{-1.3}) \times 10^{-5}\,. \end{aligned} \tag{17}$$

which agrees with the results from CLEO [1], Babar [2] and Belle [3] mentioned above. The inelastic FSI strong phase we get from the absorptive part will produce a CP violation in $B \to K\pi$ decays via the interference with the tree-level terms. We get, for the CP-asymmetry between $B^0 \to K^+\pi^-$ and $\bar{B}^0 \to K^-\pi^+$ decay rates : $A_{CP} = 0.21$ for $\gamma = 54.8^0$ which is comparable with recent results from CLEO [30].

CONCLUSION

In conclusion, we believe that the charmed resonance contributions we found seem to be capable of producing the charming-penguin terms suggested in [4] within theoretical errors. The strong phase generated by the real charm meson intermediate states would be the essential mechanism for direct CP violation in charmless B decays as suggested by [5, 10]. Though our estimate of the real part get uncertainties from the value of the cut-off momentum μ_ℓ due to various form factors, its strength is comparable with the short-distance part, though not as important as the long-distance contribution in $K \to \pi\pi$ decays.

ACKNOWLEDGMENTS

I would like to thank G. Nardulli and the organizers of the QCD@Work at Martina Franca for the warm hospitality extended to me at the Conference.

REFERENCES

1. CLEO Collaboration, D.Cronin-Hennessy *et al.*, Phys. Rev. Lett. **85**, 515 (2000).
2. Talks presented by T.J. Champion (BaBar Collaboration) at the XXXth International Conference on High-Energy Physics, Osaka, Japan, 27 July – 2 August, 2000, SLAC-PUB-8696, BABAR-PROC-00/13, hep-ex/0011018 ; A. Höcker, Proceedings of the 4th International Conference on *B* Physics and CP Violation, 13-19 February 2001, Ise-Shima, Japan, hep-ex/0104028.
3. Talks presented by P. Chang (Belle Collaboration) at the XXXth International Conference on High-Energy Physics, Osaka, Japan, 27 July – 2 August, 2000.
4. M. Ciuchini, E. Franco, G. Martinelli and L. Silvestrini, Nucl.Phys. **B 501**, 271 (1997).
5. J.-M. Gérard and W.S. Hou, Phys. Rev. Lett. **62**, 855 (1989); Phys. Rev. **D43**, 2909 (1991) .
6. R. Fleischer, Z. Phys. C **58**, 483 (1993) ;**62**, 81 (1994).
7. R. Fleischer and T. Mannel, Phys. Rev. D **57**, 2752 (1998).
8. N. G. Deshpande and X. G. He, Phys. Lett. **B336**, 471 (1994) .
9. A. Ali, G. Kramer, Cai–Dian Lü, Phys. Rev. D **58**, 094009. (1998).
10. M. Bander, A. Silverman and D. Soni, Phys. Rev. Lett. **43**, 242 (1979).
11. A. Deandrea, N. Di Bartolomeo, F. Feruglio, R. Gatto and G. Nardulli Phys. Lett B **320**, 170 (1993).
12. N. G. Deshpande, X. G. He, W. S. Hou, and S. Pakvasa, Phys. Rev. Lett. **82**, 2240 (1999).
13. C. Isola and T.N. Pham, Phys. Rev. D **62**, 094002 (2000).
14. M. Beneke, G. Buchalla, M. Neubert and C.T. Sachrajda, Phys. Rev. Lett. **83**, 1914 (1999) ; Contribution to ICHEP 2000, July 27 - August 2, Osaka, Japan, hep-ph/0007256.
15. D. Du, D. Yang, G. Zhu, Phys. Lett. **B488**, 46 (2000).
16. T. Muta, A. Sugamoto, Mao–Zhi Yang, Ya–Dong Yang, Phys.Rev. D **62**, 094020 (2000).
17. P. Colangelo, G. Nardulli, N. Paver and Riazuddin, Z.Phys. **C 45**, 575 (1990).
18. D. Choudhury and J. Ellis, Phys. Lett. **B 433**, 102 (1998).
19. C. Isola, M. Ladisa, G. Nardulli, T. N. Pham and P. Santorelli, Phys. Rev. **D64**, 014029 (2001).
20. A. Buras, M. Jamin. M. Lautenbacher, and P. Weisz, Nucl. Phys. **B400**, 37 (1993); A. Buras, M. Jamin, and M. Lautenbacher, ibid. **B400**, 75 (1993).
21. M. Ciuchini, E. Franco, G. Martinelli, L. Reina, and L. Silvestrini, Phys. Lett. **B 316**, 127 (1993); M. Ciuchini, E. Franco, G. Martinelli, and L. Reina, Nucl. Phys. **B 415**, 403 (1994).
22. G. Kramer, W. Palmer, and H. Simma, Nucl. Phys. **B428**, 77 (1994).
23. M. Ciuchini *et al.*, hep-ph/0012308.
24. K. G. Wilson, Phys. Rev. **179**, 1499 (1969).
25. G. Nardulli and Preparata, Phys. Lett. **B104**, 399 (1981)
G. Nardulli, G. Preparata and D. Rotondi, Phys.Rev. D **27**, 557 (1983).
26. T. N. Pham and D. G. Sutherland, Phys. Lett. **B135**, 209 (1984); Z. Phys. C **41** 327 (1988).
27. C. L. Y. Lee, M. Lu and M. B. Wise, Phys. Rev. **D 46**, 5040 (1992).
28. B. Bajc, S. Fajfer and R. J. Oakes, Phys.Rev. **D 53**, 4957 (1996) ; B. Bajc, S. Fajfer, R. J. Oakes and S. Prelovsek, Phys. Rev. **D 56**, 7207(1997); B. Bajc, S. Fajfer, R. J. Oakes and T. N. Pham, Phys. Rev. **D 58**, 054009 (1998) .
29. R. Casalbuoni, A. Deandrea, N. Di Bartolomeo, F. Feruglio, R. Gatto and G. Nardulli, Phys. Rep. **281**, 145 (1997) .
30. CLEO Collaboration, S. Chen et al., Phys. Rev. Lett. **85**, 525 (2000).

Direct CP Violation In Radiative B Decays

Tobias Hurth* and Thomas Mannel†

*CERN, Theory Division, CH 1211 Geneva 23, Switzerland
† Institut für Theoretische Teilchenphysik, Universität Karlsruhe, D–76128 Karlsruhe, Germany

Abstract. We discuss the role of the radiative B decays $B \to X_{s/d}\gamma$ in our search for new physics focusing on the issue of direct CP violation. We discuss in some detail a SM prediction for the CP asymmetries in inclusive $b \to s/d$ transitions, namely $\mid \Delta\mathcal{B}(B \to X_s\gamma) + \Delta\mathcal{B}(B \to X_d\gamma) \mid \sim 1 \cdot 10^{-9}$. Any measured value in significant deviation of this estimate would indicate new sources of CP violation beyond the CKM phase.

INTRODUCTION

It is well-known that radiative B decays like $B \to X_{s/d}\gamma$ play an important role in our search of new physics. These processes test the SM directly on the quantum level and, thus, are particularly sensitive to new physics (for a recent review see [1]). While the direct production of new (supersymmetric) particles is reserved for the planned hadronic machines such as the LHC at CERN, the indirect search of the B factories already implies significant restrictions for the parameter space of supersymmetric models and will thus lead to important clues for the direct search of supersymmetric particles. It is even possible that these rare processes lead to the first evidence of new physics by a significant deviation from the SM prediction, for example in the observables concerning direct CP violation, although it will then be difficult to identify in this way the new structures in detail. But also in the long run, after new physics has already been discovered, these decays will play an important role in analyzing in greater detail the underlying new dynamics.

One of the main difficulties in examining the observables in B physics is the influence of the strong interaction. As is well known, for matrix elements dominated by long-distance strong interactions there is no adequate quantitative solution available in quantum field theory. The resulting hadronic uncertainties restrict the opportunities in B physics significantly. If new physics does not show up in B physics through large deviations as recent experimental data indicates the focus on theoretically clean variables like inclusive radiative B decays is mandatory.

Within inclusive decays like $B \to X_{s/d}\gamma$ the long-distance strong interactions are less important and well under control due to the heavy mass expansion. The decay $B \to X_s\gamma$ was first observed by the CLEO collaboration [2]; these measurements have been refined [3, 4] and confirmed by other experiments [5, 6] (see also [7]). The present world average using the present data from BELLE, CLEO and ALEPH is

$$\mathcal{B}(B \to X_s\gamma) = (3.22 \pm 0.40) \times 10^{-4}. \tag{1}$$

CP602, *QCD@Work: International Workshop on Quantum Chromodynamics*
edited by P. Colangelo and G. Nardulli

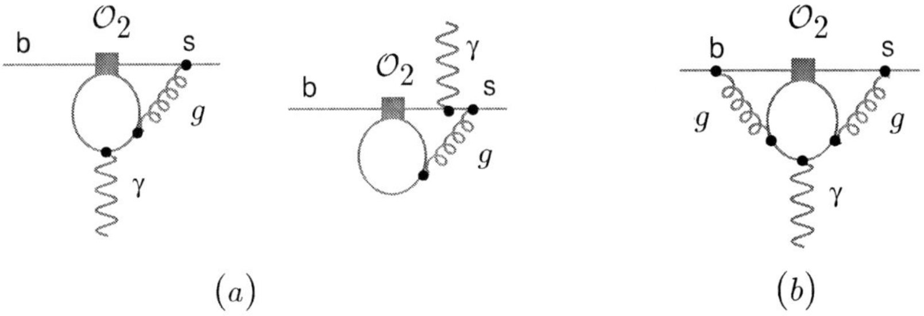

FIGURE 1. *a) Typical diagrams contributing to the matrix element of the operator O_2 at the NLL level. b) Typical diagram contributing to the NLL anomalous dimension matrix.*

The theoretical prediction of the Standard Model (SM) up to next-to-leading logarithmic (NLL) precision for the total decay rate of the $B \to X_s\gamma$ mode [8] is well in agreement with the experimental data. The theoretical NLL prediction for the $B \to X_s\gamma$ branching ratio [12] leads to

$$\mathcal{B}(B \to X_s\gamma) = (3.32 \pm 0.14 \pm 0.26) \times 10^{-4}, \tag{2}$$

where the first error represents the uncertainty regarding the scale dependences, while the second error is the uncertainty due to the input parameters. In the second error the uncertainty due to the parameter m_c/m_b is dominant.

This inclusive mode already allows for theoretically clean and rather strong constraints on the parameter space of various extensions of the SM [9, 10, 11]. While many phenomenolgical analyses of the inclusive $B \to X_{s/d}\gamma$ decays in supersymmetry are retricted to the assumption of minimal flavour violation including only CKM induced flavour change, a model-independent analysis also has to consider the generic new supersymmetric sources of flavour change due to the mixing in the squark mass matrix. In [11] new model-independent bounds on supersymmetric flavour-violating parameters were derived from $B \to X_s\gamma$. The importance of interference effects for the bounds on the parameters in the squark mass matrices within the unconstrained MSSM is explicitly demonstrated. In former analyses no correlations between the different sources of flavour violation were taken into account [10]. The new bounds are in general one order of magnitude weaker than the original bounds on the single off-diagonal element, which was derived in previous work [10] by neglecting any kind of interference effects.

Recently, quark mass effects within the decay $B \to X_s\gamma$ were further analysed [13], in particular the definitions of the quark masses m_c and m_b. The charm quark enters in specific NLL matrix elements (see Fig. 1) where the charm quark mass is dominantly off-shell. Therefore, the authors of [13] argue that the running charm mass should be chosen instead of the pole mass. The latter choice was used in all previous analyses [14, 8, 15, 16, 12].

$$m_c^{\text{pole}}/m_b^{\text{pole}} \quad \Rightarrow \quad m_c^{\overline{\text{MS}}}(\mu)/m_b^{\text{pole}}, \; \mu \in [m_c, m_b]. \tag{3}$$

Since these matrix elements start at NLL order only and, thus, the renormalization scheme for m_c and m_b is an NNLL issue, one should regard this choice as an educated guess of the NNLL corrections. The new choice is guided by the experience gained from many higher-order calculations in perturbation theory. Numerically, the shift from $m_c^{\rm pole}/m_b^{\rm pole} = 0.29 \pm 0.02$ to $m_c^{\overline{\rm MS}}(\mu)/m_b^{\rm pole} = 0.22 \pm 0.04$ is rather important and leads to a $+11\%$ shift of the central value of the $B \to X_s\gamma$ branching ratio. With their new choice of the charm mass their theoretical prediction for the branching ratio is [13]

$$\mathcal{B}(B \to X_s\gamma) = (3.73 \pm 0.30) \times 10^{-4}, \tag{4}$$

which induces a small difference between the theoretical and the experimental value.

However, because the choice of the renormalization scheme for m_c and m_b is a NNLL effect, their important observation should be reinterpreted into a larger error bar in $m_c^{\overline{\rm MS}}(\mu)/m_b^{\rm pole}$ which includes also the value of m_c^{pole}. A more conservative choice would then be $m_c^{\overline{\rm MS}}(\mu)/m_b^{\rm pole} = 0.22 \pm 0.07$ which deletes the significance of the perceived discrepancy.

DIRECT CP ASYMMETRY

The CP violation in the B system will yield an important independent test of the SM description of CP violation (see [17]). In particular, detailed measurements of CP asymmetries in rare B decays will be possible in the near future. Theoretical predictions for the *normalized* CP asymmetries of the inclusive channels (see [18, 19, 20]) within the SM lead to

$$\alpha_{CP}(B \to X_{s/d}\gamma) = \frac{\Gamma(\bar{B} \to X_{s/d}\gamma) - \Gamma(B \to X_{\bar{s}/\bar{d}}\gamma)}{\Gamma(\bar{B} \to X_{s/d}\gamma) + \Gamma(B \to X_{\bar{s}/\bar{d}}\gamma)} \tag{5}$$

$$\alpha_{CP}(B \to X_s\gamma) \approx 0.6\%, \qquad \alpha_{CP}(B \to B_d\gamma) \approx -16\% \tag{6}$$

when the best-fit values for the CKM parameters [21] are used. An analysis for the leptonic counterparts is presented in [22]. The normalized CP asymmetries may also be calculated for exclusive decays; however, these results are model-dependent. An example of such a calculation may be found in [23].

CLEO has already presented a measurement of the CP asymmetry in inclusive $b \to s\gamma$ decays, yielding [24]

$$\alpha_{CP}(B \to X_s\gamma) = (-0.079 \pm 0.108 \pm 0.022) \times (1.0 \pm 0.030)\,, \tag{7}$$

which indicates that very large effects are already excluded.

Supersymmetric predictions for the CP asymmetries in $B \to X_{s/d}\gamma$ depend strongly on what is assumed for the supersymmetry-breaking sector and are, thus, a rather model-dependent issue. The minimal supergravity model cannot account for large CP asymmetries beyond 2% because of the constraints coming from the electron and neutron electric dipole moments [25]. However, more general models allow for larger asymmetries, of the order of 10% or even larger [26, 19]. Recent studies of the $B \to X_d\gamma$ rate asymmetry in specific models led to asymmetries between -40% and $+40\%$ [27] or -45% and $+21\%$ [28]. In general, CP asymmetries may lead to clean evidence for new physics by a significant deviation from the SM prediction.

In [29] it was explicitly derived, that a bound on the *combined* asymmetries within the decays $b \to s\gamma$ and $b \to d\gamma$, as well as their leptonic counterparts is possible which is more suitable for the experimental settings. It provides a stringent test, if the CKM matrix is indeed the only source of CP violation. Using U-spin, which is the $SU(2)$ subgroup of flavour $SU(3)$ relating the s and the d quark and which is already a well-known tool in the context of nonleptonic decays [30, 31], one derives relations between the CP asymmetries of the exclusive channels $B^- \to K^{*-}\gamma$ and $B^- \to \rho^-\gamma$ and of the inclusive channels $B \to X_s\gamma$ and $B \to X_d\gamma$. One should make use of the U-spin symmetry only with respect to the strong interactions. Moreover, within *exclusive* final states, the vector mesons like the U-spin doublet (K^{*-}, ρ^-) are favoured as final states because these have masses much larger than the (current-quark) masses of any of the light quarks. Thus one expects, for the ground-state vector mesons, the U-spin symmetry to be quite accurate in spite of the nondegeneracy of m_d and m_s. Defining the rate asymmetries (not the *normalized* CP asymmetries) by

$$\Delta\Gamma(B^- \to V^-\gamma) = \Gamma(B^- \to V^-\gamma) - \Gamma(B^+ \to V^+\gamma) \tag{8}$$

one arrives at the following relation [29]:

$$\Delta\Gamma(B^- \to K^{*-}\gamma) + \Delta\Gamma(B^- \to \rho^-\gamma) = b_{exc}\Delta_{exc} \tag{9}$$

where the right-hand side is written as a product of a relative U-spin breaking b_{exc} and a typical size Δ_{exc} of the CP violating rate difference. In order to give an estimate of the right-hand side, one can use the model result from [23] for Δ_{exc},

$$\Delta_{exc} = 2.5 \times 10^{-7}\,\Gamma_B. \tag{10}$$

The relative breaking b_{exc} of U-spin can be estimated, e.g. from spectroscopy. This leads us to

$$\mid b_{exc} \mid = \frac{M_{K^*} - m_\rho}{\frac{1}{2}(M_{K^*} + m_\rho)} = 14\%. \tag{11}$$

Certainly, other estimates are also possible, such as a comparison of f_ρ and f_{K^*}. In this case one finds a very small U-spin breaking. Using the more conservative value for b_{exc}, which is also compatible with sum rule calculations of form factors (see [32]), one arrives at the standard-model prediction for the difference of branching ratios

$$\mid \Delta\mathcal{B}(B^- \to K^{*-}\gamma) + \Delta\mathcal{B}(B^- \to \rho^-\gamma) \mid \sim 4 \times 10^{-8} \tag{12}$$

Note that the right-hand side is model-dependent.

Quite recently, the U-spin breaking effects were also estimated in the QCD factorization approach [33]. Within this approach, it was shown that the U-spin breaking effect essentially scales with the differences of the two form factors, $(F_{K^*} - F_\rho)$. Using the formfactors from the QCD sum rule calculation in [34] and maximizing the CP asymmetries by a specific choice of the CKM angle γ, the authors of [33] obtain

$$\Delta\mathcal{B}(B^- \to K^{*-}\gamma) + \Delta\mathcal{B}(B^- \to \rho^-\gamma) \sim -3 \times 10^{-7}, \tag{13}$$

while for the separate asymmetries they obtain:

$$\Delta\mathcal{B}(B \to K^*\gamma) = -7\times10^{-7}, \qquad \Delta\mathcal{B}(B \to \rho\gamma) = 4\times10^{-7}. \tag{14}$$

This calculation explicitly shows the limitations of the relation (9) as a SM test.

The issue is much more attractive in the *inclusive* modes. Due to the $1/m_b$ expansion for the inclusive process, the leading contribution is the free b-quark decay. In particular, there is no sensitivity to the spectator quark and thus one arrives at the following relation for the CP rate asymmetries [29]:

$$\Delta\Gamma(B \to X_s\gamma) + \Delta\Gamma(B \to X_d\gamma) = b_{inc}\Delta_{inc}. \tag{15}$$

In this framework one relies on parton-hadron duality; so one can actually compute the breaking of U-spin by keeping a nonvanishing strange quark mass. The typical size of b_{inc} can be roughly estimated to be of the order of $|\, b_{inc}\,| \sim m_s^2/m_b^2 \sim 5\times10^{-4}$; $|\,\Delta_{inc}\,|$ is again the average of the moduli of the two CP rate asymmetries. These have been calculated (for vanishing strange quark mass), e.g. in [18] and thus one arrives at

$$|\,\Delta\mathcal{B}(B \to X_s\gamma) + \Delta\mathcal{B}(B \to X_d\gamma)\,| \sim 1\times10^{-9}. \tag{16}$$

Any measured value in significant deviation of (16) would be an indication of new sources of CP violation. Although only an estimate is given here, it should again be stressed that in the inclusive mode the right-hand side in (16) can be computed in a model-independent way with the help of the heavy mass expansion.

Going beyond leading order in the $1/m_b$ expansion the first subleading corrections are of order $1/m_b^2$ only. The $1/m_b^2$ corrections are induced by the imaginary part of the forward scattering amplitude $T(q)$:

$$T(q) = i\int d^4x < B\,|\,T\,O_7^+(x)\,O_7(0)\,|\,B > \exp(iqx) \tag{17}$$

where only the magnetic operator O_7 is taken into account. Using the operator product expansion for $T\,O_7^+(x)\,O_7(0)$ and heavy quark effective theory methods, the decay width $\Gamma(B \to X_s\gamma)$ reads [35, 36] (modulo higher terms in the $1/m_b$ expansion):

$$\begin{aligned} \Gamma^{(O_7,O_7)}_{B\to X_s\gamma} &= \frac{\alpha G_F^2 m_b^5}{32\pi^4}\,|\,V_{tb}V_{ts}\,|^2\, C_7^2(m_b)\left(1+\frac{\delta^{NP}}{m_b^2}\right), \\ \delta^{NP} &= \frac{1}{2}\lambda_1 - \frac{9}{2}\lambda_2, \end{aligned} \tag{18}$$

where λ_1 and λ_2 are the parameters for kinetic energy and the chromomagnetic energy. Using $\lambda_1 = -0.5\,\text{GeV}^2$ and $\lambda_2 = 0.12\,\text{GeV}^2$, one gets $\delta^{NP} \simeq -4\%$. Thus, the contributions are small and cancel in the sum of the rate asymmetries - in the limit of U-spin symmetry. The U-spin breaking effects in this contribution also induce an overall factor m_s^2/m_b^2 in addition - as one can read off from the explicit results for the $B \to X_s l^+ l^-$ case [37].

There are also nonperturbative corrections which scale with $1/m_c^2$ [38]. which are induced by the interference of the magnetic O_7 and the four-quark operator O_2. This

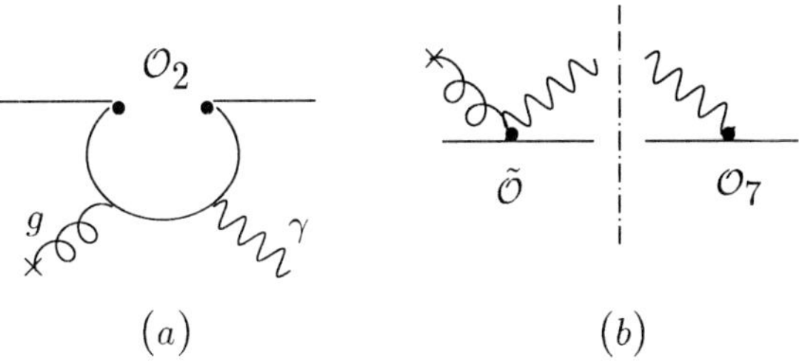

FIGURE 2. *a) Feynman diagram from which the operator $\tilde{O}$ arises. b) Relevant cut-diagram for the $(\mathcal{O}_2, \mathcal{O}_7)$-interference.*

effect is generated by the diagram in Fig. 2a (and by the one, not shown, where the gluon and the photon are interchanged); g is a soft gluon interacting with the charm quarks in the loop. Up to a characteristic Lorentz structure, this loop is given by the integral

$$\int_0^1 dx \int_0^{1-x} dy \frac{xy}{m_c^2 - k_g^2 x(1-x) - 2xyk_g k_\gamma} \quad . \tag{19}$$

As the gluon is soft, i.e. $k_g^2, k_g k_\gamma \approx \Lambda^{QCD} m_b/2 \ll m_c^2$, the integral can be expanded in k_g. The (formally) leading operator, denoted by $\tilde{O}$, is

$$\tilde{O} = \frac{G_F}{\sqrt{2}} V_{cb} V_{cs}^* C_2 \frac{eQ_c}{48\pi^2 m_c^2} \bar{s}\gamma_\mu(1-\gamma_5) g_s G_{\nu\lambda} b \varepsilon^{\mu\nu\rho\sigma} \partial^\lambda F_{\rho\sigma} \quad . \tag{20}$$

Then working out the cut diagram shown in Fig. 2b, one obtains the nonperturbative contribution $\Gamma^{(\tilde{O},\mathcal{O}_7)}_{B \to X_s \gamma}$ to the decay width, which is due to the $(\mathcal{O}_2, \mathcal{O}_7)$ interference. Normalizing this contribution by the LL partonic width, one obtains

$$\frac{\Gamma^{(\tilde{O},\mathcal{O}_7)}_{B \to X_s \gamma}}{\Gamma^{LL}_{b \to s\gamma}} = -\frac{1}{9} \frac{C_2}{C_7} \frac{\lambda_2}{m_c^2} \simeq +0.03 \quad . \tag{21}$$

This result corresponds to the leading term in an expansion in the parameter $t = k_g k_\gamma / 2m_c^2$. The expansion parameter is approximately $m_b \Lambda_{QCD}/2m_c^2 \approx 0.3$ (rather than Λ_{QCD}^2/m_c^2) and it is not a priori clear whether formally higher order terms in the m_c expansion are numerically suppressed. However, the explicit expansion of the complete vertex function, corresponding to Fig. 2a, shows that higher order terms are indeed suppressed, because the corresponding coefficients are small (see i.e. [39]).

Moreover, the operator $\tilde{O}$ does not contain any information on the strange mass, thus, also in these contributions one finds the same overall suppression factor from the U-spin breaking.

The corresponding long-distance contributions from up-quark loops are CKM suppressed in the $B \to X_s\gamma$ case, but this does not hold in the $B \to X_d\gamma$ case. Naively, one could expect that the corresponding contributions from up-quark loops scale with $1/m_u^2$. However, following the approach of [39], one easily shows that the general vertex function cannot be expanded in the parameter t in that case. However, the expansion in inverse powers of t is reasonable. The leading term in this expansion scales like $t^{-1} \sim m_u^2/k_g k_\gamma$ and therefore cancels the factor $1/m_u^2$ in the prefactor (see the analogous $1/m_c^2$ factor in (20)) and one gets a suppression factor $(\Lambda_{QCD}^2/m_u^2)\cdot(m_u^2/k_g k_\gamma)$ [39]. Thus, although the expansion in inverse powers in t induces nonlocal operators, one explicitly finds that the leading term scales with Λ_{QCD}/m_b [1].

This argument improves the discussion in [29] where the argument was given that vector dominance calculations in [40] show that the long-distance contributions from the up-quark loops to the decay rates are found to be rather small. However, that argument in [29] does not allow any statement about the effects of the U-spin breaking in constrast to the one given here.

Summing up, the analysis shows that the known nonperturbative contributions to (16) are under control and small and that this prediction provides a clean SM test, if generic new CP phases are present or not. Any significant deviation from the estimate (16) would be a strong hint to non-CKM contributions to CP violation.

Finally, we emphasize that an analogous prediction for the leptonic inclusive $B \to X_{s/d}\gamma$ modes is also possible taking into account some specific cuts on the invariant dilepton spectrum.

ACKNOWLEDGEMENTS

We are grateful to Gerhard Buchalla for discussions and to Gino Isidori for a careful reading of the manuscript.

REFERENCES

1. T. Hurth, hep-ph/0106050.
2. CLEO Collaboration, M. S. Alam *et al.*, Phys. Rev. Lett. **74**, 2885 (1995).
3. CLEO Collaboration, S. Ahmed *et al.*, hep-ex/9908022.
4. CLEO Collaboration, S. Chen *et al.*, hep-ex/0108032.
5. ALEPH Collaboration, R. Barate *et al.*, Phys. Lett. B **429**, 169 (1998).
6. BELLE Collaboration, Y. Ushiroda, hep-ex/0104045.
7. New results presented by H. Tajima for the BELLE Collaboration and by D. Cassel for the CLEO Collaboration can be found at the homepage of the XX International Symposium on Lepton and Photon Interactions at High Energies 23-28 July 2001, Rome Italy: http://www.lp01.infn.it/.
8. A. Ali, C. Greub, Zeit. f. Phys. **C60**, 433 (1993); N. Pott, Phys. Rev. **D54**, 938 (1996); C. Greub, T. Hurth, D. Wyler, Phys. Lett. **B380**, 385 (1996); Phys. Rev. **D54** 3350 (1996); K. Adel, Y.P. Yao, Phys. Rev. **D49** 4945 (1994); C. Greub, T. Hurth, Phys. Rev. **D56** 2934 (1997); K. Chetyrkin, M. Misiak, M. Münz, Phys. Lett. **B400** 206 (1997).
9. G. Degrassi, P. Gambino, G. F. Giudice, JHEP**0012**, 009 (2000); M. Carena, D. Garcia, U. Nierste, C. E. Wagner, Phys. Lett. B **499**, 141 (2001); W. de Boer, M. Huber, A. V. Gladyshev, D. I. Kazakov, hep-ph/0102163.

[1] Also in the exclusive case an explicit analysis within the QCD factorization approach leads to this suppression factor [33].

10. F. Gabbiani, E. Gabrielli, A. Masiero, L. Silvestrini, Nucl. Phys. B **477**, 321 (1996); F. Borzumati, C. Greub, T. Hurth, D. Wyler, Phys. Rev. D **62**, 075005 (2000); Nucl. Phys. Proc. Suppl. **86**, 503 (2000)
11. T. Besmer, C. Greub and T. Hurth, Nucl. Phys. B **609**, 359 (2001).
12. C. Greub and T. Hurth, Nucl. Phys. Proc. Suppl. **74**, 247 (1999) [hep-ph/9809468].
13. P. Gambino and M. Misiak, hep-ph/0104034.
14. C. Greub and T. Hurth, hep-ph/9608449. Talk given at DPF 96 Meeting, Minneapolis, MN, 10-15 Aug 1996, in *Minneapolis 1996, Particles and fields, vol. 2* 810-814.
15. M. Ciuchini, G. Degrassi, P. Gambino and G.F. Giudice, Nucl. Phys. **B527**, 21 (1998).
16. A.L. Kagan and M. Neubert, Eur. Phys. J. **C7**, 5 (1999).
17. T. Hurth *et al.*, J. Phys. G **G27**, 1277 (2001).
18. A. Ali, H. Asatrian and C. Greub, Phys. Lett. **B429**, 87 (1998).
19. A. L. Kagan and M. Neubert, Phys. Rev. D **58**, 094012 (1998).
20. K. Kiers, A. Soni and G. Wu, Phys. Rev. D **62**, 116004 (2000).
21. F. Caravaglios, F. Parodi, P. Roudeau and A. Stocchi, hep-ph/0002171.
22. A. Ali and G. Hiller, Eur. Phys. J. C **8**, 619 (1999).
23. C. Greub, H. Simma and D. Wyler, Nucl. Phys. B **434**, 39 (1995) [Erratum-ibid. B **444**, 447 (1995)].
24. T. E. Coan *et al.* [CLEO Collaboration], Phys.Rev.Lett. **86**, 5661 (2001).
25. T. Goto, Y. Y. Keum, T. Nihei, Y. Okada and Y. Shimizu, Phys. Lett. B **460**, 333 (1999).
26. M. Aoki, G. Cho and N. Oshimo, Nucl. Phys. B **554**, 50 (1999); C. Chua, X. He and W. Hou,Phys. Rev. D **60**, 014003 (1999); Y. G. Kim, P. Ko and J. S. Lee, Nucl. Phys. B **544**, 64 (1999); S. Baek and P. Ko, Phys. Rev. Lett. **83**, 488 (1999); L. Giusti, A. Romanino and A. Strumia, Nucl. Phys. B **550**, 3 (1999); E. J. Chun, K. Hwang and J. S. Lee, Phys. Rev. D **62**, 076006 (2000); D. Bailin and S. Khalil, Phys. Rev. Lett. **86**, 4227 (2001).
27. A. G. Akeroyd, Y. Y. Keum and S. Recksiegel, Phys. Lett. B **507**, 252 (2001).
28. H. H. Asatrian and H. M. Asatrian, Phys. Lett. B **460**, 148 (1999); H. H. Asatryan, H. M. Asatrian, G. K. Yeghiyan and G. K. Savvidy, hep-ph/0012085.
29. T. Hurth and T. Mannel, Phys. Lett. B **511**, 196 (2001).
30. R. Fleischer, Phys. Lett. B **459**, 306 (1999).
31. M. Gronau and J. L. Rosner, Phys. Lett. B **500**, 247 (2001); M. Gronau, Phys. Lett. B **492**, 297 (2000).
32. A. Ali, V. M. Braun and H. Simma, Z. Phys. C **63**, 437 (1994).
33. S. W. Bosch and G. Buchalla, hep-ph/0106081.
34. P. Ball and V. M. Braun, Phys. Rev. D **58**, 094016 (1998).
35. A. F. Falk, M. Luke and M. J. Savage, Phys. Rev. D **49**, 3367 (1994).
36. A. Ali, G. Hiller, L. T. Handoko and T. Morozumi, Phys. Rev. D **55**, 4105 (1997).
37. A. Ali and G. Hiller, Eur. Phys. J. C **8**, 619 (1999).
38. M.B. Voloshin, Phys. Lett. **B397**, 275 (1997); A. Khodjamirian, R. Ruckl, G. Stoll and D. Wyler, Phys. Lett. **B402**, 167 (1997); Z. Ligeti, L. Randall and M.B. Wise, Phys. Lett. **B402**, 178 (1997); A.K. Grant, A.G. Morgan, S. Nussinov and R.D. Peccei, Phys. Rev. **D56**, 3151 (1997).
39. G. Buchalla, G. Isidori and S. J. Rey, Nucl. Phys. B **511**, 594 (1998).
40. G. Ricciardi, Phys. Lett. B **355**, 313 (1995); N. G. Deshpande, X. He and J. Trampetic, Phys. Lett. B **367**, 362 (1996).

New Results on Charm Photoproduction at Fermilab

Sergio P. Ratti [1]

Dipartimento di Fisica Nucleare e Teorica dell' Universita' and Sezione INFN
via A. Bassi, 6 - I27100 Pavia (PV), Italy

Abstract. FOCUS is a high statistics photoproduction experiment at Fermilab. From among the many results that are being obtained, in the context of the CP violation issue, we measured the values of the $D^o\overline{D^o}$ lifetime mixing parameter $y_{CP}(=\Delta\Gamma/2\Gamma)$ and compared the double Cabibbo suppressed amplitude R_{WS} to the measurement of CLEO-II. We compared also the CP violation asymmetry parameter for several decays to other experiments. New lifetime measurements are presented for both mesons and baryons, together with a new *very preliminary* measurement of the Ω_c^o mass.

INTRODUCTION

FOCUS is a high statistics photoproduction experiment at Fermilab. It collected over 7 billion triggers enriched in charm events and fully reconstructed close to 1.5 million charm decays; it offers therefore a very high sensitivity to investigate rare phenomena in the charm sector and eventually probe New Physics.

Three are the topics addressed in the present paper:

a- the Standard Model predictions for D^o-$\overline{D^o}$ mixing and CP violation in charm decays are generally expected to be orders of magnitude below the sensitivities of current experiments[1].

Mixing occurs since the two weak eigenstates D^o and $\overline{D^o}$ are not mass eigenstates

---

[1] Coauthors: J.Link, M.Reyes, P.M.Yager (**UC DAVIS**); J.Anjos,I.Bediaga, C.Gobel, J.Magnin, A. Massafferri, J.M. de Miranda, I.M.Pepe, A.C. dos Reis, (**CPBF, Rio de Janeiro**); S.Carrillo, E.Casimiro, A.Sánchez-Hernández, C.Uribe, F.Vasquez (**CINVESTAV, México City**); L.Cinquini, J.P.Cumalat, B.O'Reilly, J.E.Ramirez, E.W.Vaandering (**CU Boulder**); J.N.Butler, H.W.K.Cheung, I.Gaines, P.H.Garbincius, L.A.Garren, E.Gottschalk, P.H.Kasper, A.E.Kreymer, R.Kuschke (**Fermilab**); S.Bianco, F.L.Fabbri, S.Sarwar, A.Zallo (**INFN Frascati**); C.Cawlfield, D.Y.Kim, A.Rahimi, J.Wiss (**UI Champaign**); R.Gardner, A.Kryemadhi (**IU Bloomington**); Y.S.Chung, J.S.Kang, B.R.Ko, J.W.Kwak, K.B.Lee, H.Park (**Korea University, Seoul**); G.Alimonti, M.Boschini, B.Caccianiga, P.D'Angelo, M.DiCorato, P.Dini, M.Giammarchi, P.Inzani, F.Leveraro, S.Malvezzi, D.Menasce, M.Mezzadri, L.Milazzo, L.Moroni, D.Pedrini, C.Pontoglio, F.Prelz, M.Rovere, S.Sala (**INFN and Milano**); T.F.Davenport III (**UNC Asheville**); L.Agostino, V.Arena, G.Boca, G.Bonomi, G.Gianini, G.Liguori, M.M.Merlo, D.Pantea, C.Riccardi, I.Segoni, P.Vitulo (**INFN and Pavia**); H.Hernandez, A.M.Lopez, H.Mendez, L.Mendez, E.Montiel, D.Olaya, A.Paris, J.Quinones, C.Rivera, W.Xiong, Y.Zhang (**Mayaguez, Puerto Rico**); J.R.Wilson (**USC Columbia**); K.Cho, T.Handler, R.Mitchell (**UT Knoxville**); D.Engh, M.Hosack, W.E.Johns, M.Nehring, P.D.Sheldon, K.Stenson, M.Webster (**Vanderbilt**); M.Sheaff (**Wisconsin, Madison**)

CP602, *QCD@Work: International Workshop on Quantum Chromodynamics*
edited by P. Colangelo and G. Nardulli

(or CP eigenstates in the limit of CP conservation). The mixing effects are typically parametrized by the two variables: $x = \frac{\Delta M}{\Gamma}$ and $y = \frac{\Delta\Gamma}{2\Gamma}$, referring respectively to the mass difference ΔM of the two mass eigenstates and to their width difference $\Delta\Gamma$, both scaled by the average decay width. The mixing effects are investigated experimentally by either studying D^o wrong sign decays or comparing the lifetimes of opposite CP final states.

CP violation asymmetries can be also investigated by looking for particle antiparticle asymmetries in several decay channels such as $D^+ \to K^-K^+\pi^+$, $D^0 \to K^-K^+$, $D^0 \to \pi^-\pi^+$. In single Cabibbo suppressed D decays, the penguin terms in the effective Hamiltonian may provide the different phases of the two weak amplitudes. The direct CP violating asymmetries for these decays are expected to be at most 10^{-3}[2].

b- the lifetimes of all charmed mesons would be the same in absence of W-exchange or W-annihilaition diagrams as well as final state interactions. The contribution of the fully leptonic decays to the total decay width is completely negligible, while the semileptonic widths of D^o and D^+ are equal within errors[3]. Thus, the fact that the two lifetimes are different[3], implies that the two hadronic widths are different. The W-exchange diagram, contributes to the D^o decay while the W-annihilation diagram contributes to the D_s^+ decay and the two contributions might not be identical.

c- the lifetimes of the charmed baryons are known at best with a 5% uncertainty and new measurements are needed. Here, final state interactions and spin effects contribute to specific jerarchies in their lifetimes. In addition, evidence for the double charmed baryon Ω_c^o is still somewaht weak and FOCUS is able now to provide at least a very preliminary value of its mass.

The analysis presented here is mostly based on the full data sample collected by the experiment during the Fermilab 1996-1997 fixed-target run with photons having an average energy of 180 GeV in a segmented BeO target, usingd an upgraded version of the E687 spectrometer[4]. The vertex detector is formed by 16 layers of silicon microstrips [four (x,y) interleaved with the target segments and 12 (x,y,u) downstream] which provide an excellent proper time resolution $\sigma_\tau \approx 30$ fs. The charged particle momenta are measured from their deflections by two analysis magnets of opposite polarity and the hits left into five stations of multiwire proportional chambers. Particle identification is provided by three multicell threshold Cerenkov counters, two electromagnetic calorimeters, one hadron calorimeter and two arrays of muon counters.

MIXING PARAMETERS

The mixing amplitude (squared) is easily written as:

$$| < \overline{D^o}|D^o(t) > |^2 \approx e^{-\Gamma_1 t}[1 + e^{\Delta\Gamma t} - 2e^{\frac{\Delta\Gamma t}{2}} cos(\Delta M t)] \tag{1}$$

being Γ the total decay width.

Defining $x = \frac{\Delta M}{\Gamma}$ and $y = \frac{\Delta\Gamma}{2\Gamma}$, in the case of $|x|, |y|$ small, one can write the mixing amplitude as:

$$A_{mix} \approx \frac{y + ix}{2} \Gamma t \, e^{-\frac{\Gamma t}{2}} \tag{2}$$

Integrating $|A_{mix}|^2$ over time, the rate for the mixing process is described by:

$$R_{mix} = \frac{\Gamma_{mix}}{\Gamma_{unmix}} = \frac{x^2 + y^2}{2} \qquad (3)$$

Upper limits on R_{mix} (e.g.: 95% confidence level) draw circles in the x,y plane. To measure R_{mix} from ΔM one needs accuracies in the mass difference out of reach (less than 100 μeV or so); to measure R_{mix} from $\Delta\Gamma$, one needs very good lifetime measurements.

Measurements of R_{mix} are performed by using semileptonic decays and looking at the *wrong sign* (hereafter WS) leptons. The particle-antiparticle ambiguity is solved by selecting the decays $D^{*\pm} \to D^o\pi^{\pm}$, able to discriminate what D^o has been produced. Contrary to neutral kaons, hadronic charm decays are complicated since doubly Cabibbo suppressed (hereafter DCS) channels come into play, adding a new term to the box diagram, as well as a new strong phase ϕ (relative to the Cabibbo favoured decay). The decay amplitude R_{WS} is then written as:

$$R_{WS}(t) = [R_{DCS} + y'\Gamma t\sqrt{R_{DCS}} + \frac{x'^2 + y'^2}{2}\frac{\Gamma^2 t^2}{2}]e^{\Gamma t} \qquad (4)$$

where $x' = xcos\delta + ysin\delta$ and $y' = ycos\delta - xsin\delta$. When the mixing effects are very small, the branching ratio R_{WS} of the *wrong sign* decay is close to that of the DCS channel.

LIFETIME MIXING

The y mixing parameter can be evaluated by directly measuring the lifetimes of pure CP states. Assuming CP conservation, the y parameter equals y_{CP}, the width asymmetry $\Delta\Gamma$ between the $CP = +1$ and $CP = -1$ states. The discovery of two CP eigenstates, say D_1^o and D_2^o similar to the K_1^o and K_2^o states, would be important "per se", independent of CP violation.

FOCUS has measured y_{CP} using the channel $D^o \to K^-K^+$ as $CP = +1$ state and the CP mixed state $D^o \to K^-\pi^+$. Assuming that $K^-\pi^+$ is an equal mixture of $CP = \pm 1$

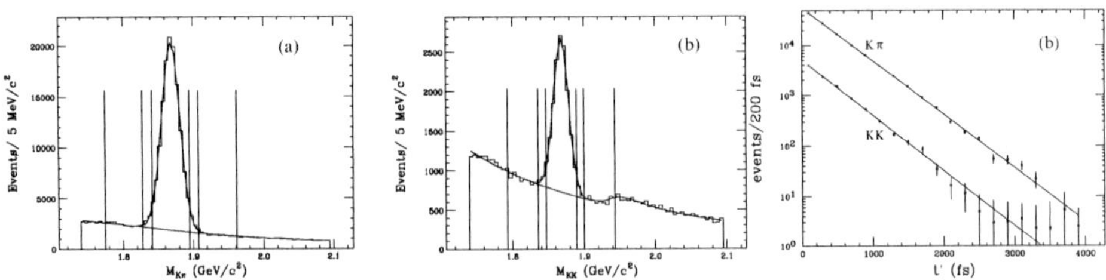

FIGURE 1. Mass and time distributions: a)- mass for 119738 $D^0 \to K^-\pi^+$ events; b)- mass for 10331 $D^0 \to K^-K^+$ events (b) Vertical lines show the selected signal and sideband regions for lifetime and y_{CP} fit; c)- reduced proper time distributions for $D^o \to K^-K^+$ and $D^o \to K^-\pi^+$ events. The distributions are background subtracted and include very small Monte Carlo corrections.

states, y_{CP} can be written in terms of the two lifetimes:

$$y_{CP} = \frac{\Gamma_{CP-even} - \Gamma_{CP-odd}}{\Gamma_{CP-even} + \Gamma_{CP-odd}} = \frac{\tau(D^o \to K\pi)}{\tau(D^o \to KK)} - 1 \quad (5)$$

To get a value of $\tau(D^o \to KK)$ with an error of $\approx 1\%$ one needs a sample of at least 10,000 events in the Cabibbo suppressed channel $D^o \to K^+K^-$. Fig. 1 shows the selected 119738 events for the decay $D^0 \to K^-\pi^+$ (fig. 1a) and the 10331 events for the decay $D^o \to K^-\pi^+$ (fig. 1b); fig. 1c shows instead, the 2 reduced proper time distributions, together with the binned-likelihood simultaneous fit using y_{CP} and $\tau(K^-\pi^+)$ as free parameters. The fit is performed over 20 bins, 200fs wide, and spans over more than $6\tau(K^-\pi^+)$. The values obtained are[5]: $y_{CP} = (3.42 \pm 1.39 \pm 0.74)$ and $\tau(K^-\pi^+) = (409.2 \pm 1.3)$ fs. We quote only the 0.3% statistical error as we must wait for values from other channels such as $D^o \to K^-\pi^-2\pi^+$ to estimate any systematic error.

The value of y_{CP} is bearly compatible with zero and definitely needs confirmation. The experiments at the B factories are clear candidates to provide a check of the result.

WRONG SIGN AND DOUBLY CABIBBO SUPPRESSED DECAY

Starting with a sample of over 200,000 D^* events, FOCUS tackled the measurement of the WS branching ratio i.e.: $D^o \to K^+\pi^-$ or $\overline{D^o} \to K^-\pi^+$ decays relative to the *right sign* (RS) decays. The sign of the D^* identifies unambiguously the C state of the neutral meson. Obviously, the D^* requirement reduces significantly the available statistics. The event selection depends upon the capacity of separating pions from kaons. Four different backgrounds are relevant: a- $D^o \to \pi^+\pi^-$; b- $D^o \to K^+K^-$; c- partially reconstructed D's and finally *doubly misidentified* $D^o \to K\pi$ decays. In particular, when the mass of the pion and the kaon are interchanged -leaving unchanged their charge- the final state produces a broad peak in the $\Delta M = M(D^*) - M(D^o)$ distribution at the charm mass difference. To deal with the different types of background we divided the $\Delta M vs M(K\pi)$ scatter plot into 80 ΔM bins, each 1 MeV wide, and fitted for each bin the RS and WR $K\pi$ mass distributions with the above backgrounds generated by Montecarlo[6].

The result of the overall fit is shown in fig. 2a. From the fit we get 149 ± 31 WS events compared to $36,760 \pm 195$ RS events. This provides a Branching Ratio:

$$R_{WS} = \frac{\Gamma(D^o \to K^+\pi^-)}{\Gamma(D^o \to K^-\pi^+)} = (0.405 \pm 0.085 \pm 0.025)\%. \quad (6)$$

The time evolution for the $D^o \to K^+\pi^-$ decay is given by equation (4). If mixing is negligible, the branching ratio R_{WS} provides a measurement of the DCS branching ratio, which is expected to be of order $tan^4\theta_c \approx 0.25\%$. If mixing is significant, R_{WS} depends upon the lifetime acceptance of our analysis. To investigate this point, in an adequate Monte Carlo sample of $D^0 \to K^-\pi^+$ decays, generated with a nominal lifetime, we re-weighted the accepted events by the ratio of the survival probability provided by its lifetime given by eq. (4) and the probability provided by the nominal lifetime. We then

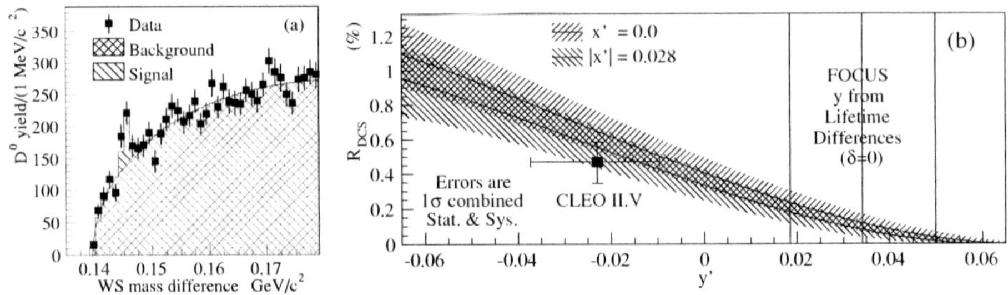

FIGURE 2. a) The $D^0 \to K^+\pi^-$ mass difference distribution with the signal and background contributions shown. b) R_{DCS} plotted as a function of y'. Contours are given for two values of x' covering the 95% CL of the CLEO result.

obtained R_{DCS} as a function of x' and y', which depend on our measurement for R_{WS} and on the average values $< t >$ and $< t^2 >$ obtained from our Monte Carlo. We determined two bands in the R_{DCS}-y' plane shown in fig. 2b. They correspond to the two values of x' which cover the CLEO 95% CL of $|x'| < 0.28$. For comparison, the CLEO ranges for R_{DCS} and y' are also shown. The two experiments are compatible, although FOCUS may be suggestive of a possible y_{CP} mixing different from zero.

CP VIOLATION

To tackle the CP violation issue, FOCUS selected the single Cabibbo suppressed decay modes: $D^+ \to K^-K^+\pi^+$, $D^o \to K^-K^+$, $D^o \to \pi^-\pi^+$ and their C conjugate states. The D^o flavour is tagged by the $D^{*\pm}$ decays. These decays are normalized to their allowed modes to correct for the difference in the D - $\overline{D}$ photoproduction rates.

For instance, for the single Cabibbo suppressed decay mode $D^0 \to K^-K^+$ and the normalizing mode $D^0 \to K^-\pi^+$ the normalization is written as:

$$\eta(D) = \frac{N(D^o \to K^-K^+)}{N(D^o \to K^-\pi^+)} \tag{7}$$

Here, $N(D^o \to K^-K^+)$ is the efficiency corrected number of candidate decays. The use of the η ratio has the additional benefit that most of the corrections due to efficiencies cancel out, therefore reducing the systematic uncertainties [2].

The CP asymmetries are then written as:

$$A_{CP} = \frac{\eta(D) - \eta(\overline{D})}{\eta(D) + \eta(\overline{D})} \tag{8}$$

[2] It is assumed that there is no measurable CP violation in the Cabibbo favoured decays.

TABLE 1. Measured CP asymmetries(10^{-2}) (see Ref.[8] and ref. therein).

Experiment	$D^+ \to K^-K^+\pi^+$	$D^o \to K^-K^+$	$D^o \to \pi^-\pi^+$
E687	-3.1 ± 6.8	$+2.4 \pm 8.4$	
CLEO II		$+8.0 \pm 6.1$	
E791	-1.4 ± 2.9	$-1.0 \pm 4.9 \pm 1.2$	$-4.9 \pm 7.8 \pm 3.0$
CLEO II[7]		$+0.05 \pm 2.18 \pm 0.84$	$+1.95 \pm 3.22 \pm 0.84$
FOCUS[8]	$+0.6 \pm 1.1 \pm 0.5$	$-0.1 \pm 2.2 \pm 1.5$	$+4.8 \pm 3.9 \pm 2.5$

In Table 1 the FOCUS results are compared to those from other experiments. The values are all consistent with zero. They altogether represent a substantial improvement over previous published limits.

LIFETIMES

E687[9] measured the lifetime of all charmed mesons and baryons and FOCUS plans to remeasure them with smaller errors. The present situation of the uncertainties in charmed lifetimes -as given by the 2000 edition of the Particle Data Book[3] is summarized in Table 2.

TABLE 2. Uncertainties in lifetime measurements.

Particle	$\sigma(\tau)/\tau$ *(stat)* %	$\sigma(\tau)/\tau$ *(syst)* %	Particle	$\sigma(\tau)/\tau$ *(stat)* %	$\sigma(\tau)/\tau$ *(syst)* %
D^+	1.4	1.0	Λ_c^+	5.9	3.7
D^o	1.0	1.0	Ξ_c^+	≈ 20	≈ 5
D_s^+	3.6	1.5	Ξ_c^o	≈ 25	≈ 5
			Ω_c^o	≈ 30	≈ 14

It will be hard to dominate the systematic errors in FOCUS to significantly improve the measurement of τ_{D^+} given by E687 with a 1.% systematic error. The analysis described in the lifetime mixing section provides a new measurement of τ_{D^o} with a statistics over hundred thousand events and a 0.3% statistical error.

Here we present still preliminary measurements of all the remaining lifetimes but $\tau_{\Omega_c^o}$.

The D_s^+ lifetime has been measured on 50% of the $D_s^+ \to \pi^+\phi$ FOCUS sample and provides a value: $\tau(D_s^+) = 506 \pm 8$ fs. This improves the present uncertainty by about a factor 3.

If the new world average for $\tau(D_s^+) = 499.9 \pm 6.1$ fs and the value of $\tau(D^o)$ given before are used, we calculate a ratio $\frac{\tau(D_s^+)}{\tau(D^o)} = 1.235 \pm 0.016$ which is a value almost 15σ away from unity, thus indicating that the W-exchange diagram and the W-annihilation diagram play different roles in the weak charm decays.

The lifetime values for charmed baryons whose uncertainties are reported in Table 2 can be considered not much more than estimates. Infact, only $\tau(\Lambda_c^+)$ has a statistical error of about 6%.

FOCUS has a preliminary remeasurement of $\tau(\Lambda_c^+)$ on 80% of the total sample: $\tau(\Lambda_c^+) = 204.5 \pm 3.4$ fs. This improves the uncertainty by more than a factor 3 and

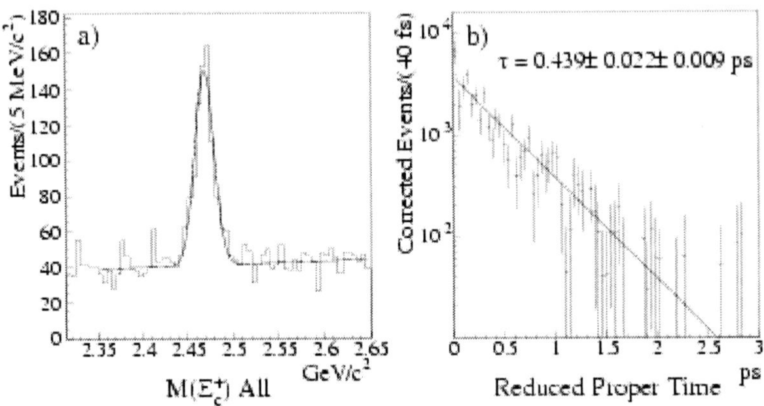

FIGURE 3. a- Cumulative mass distribution for the decays of the Ξ_c^+ baryon; b- reduced proper time distribution for the same decays.

sets as preliminary new world average a value $\tau(\Lambda_c^+) = (201.9 \pm 3.1)$ fs.

The lifetimes of the other charmed baryons will be dealt with in the next Section.

STRANGE-CHARMED BARYONS

In the charmed baryon sector, information on the strange-charmed particles is still rather scanty. FOCUS has remeasured the lifetimes of the Ξ baryons[3] and has in progress the remeasurement of the mass of the Ω_c^o particle.

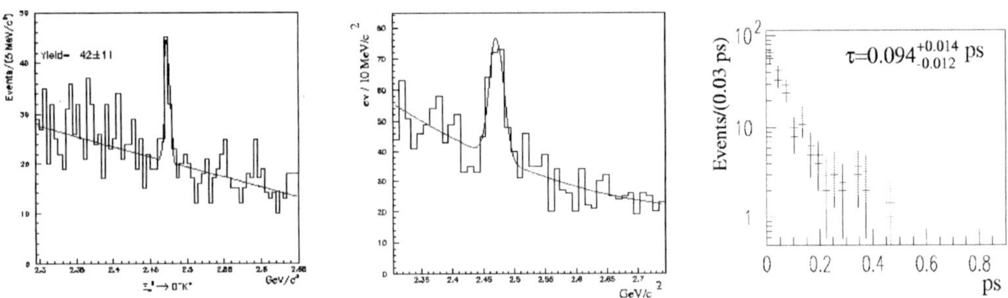

FIGURE 4. a- Mass distribution for the decay $\Xi_c^o \to \Omega^- K^+$; b- mass distribution for the decay $\Xi_c^o \to \Xi^- \pi^+$; c- reduced proper time distribution for the two samples.

[3] Preliminary data were presented at previous Conferences[10]a.

Collecting 300 events of the decay $\Xi_c^+ \to \Xi^- 2\pi^-$, 130 events of the decay $\Xi_c^+ \to \Sigma^+ K^- \pi^+$; 58 events of the decay $\Xi_c^+ \to \Lambda^o K^- 2\pi^+$ and 45 events of the Cabibbo suppressed decay $\Xi_c^+ \to pK^-\pi^+$ (for a total of 533 events shown in fig. 3a, we measured a lifetime[11] $\tau(\Xi_c^+) = (439 \pm 22 \pm 9)$ fs (see fig. 3b). This value improves the world sample[3] by more than a factor 2 and reduces the error by almost a factor 4.

The Ξ_c^o baryon has been observed in two channels: $\Xi_c^o \to \Omega^- K^+$ (42 events) and $\Xi_c^o \to \Xi^- \pi^+$ (117 events). The preliminary signals are shown in fig.s 4a,b respectively. The lifetime measurement on the total sample -reduced proper time distribution shown in fig. 4c- gives a still preliminary value for the lifetime: $\tau(\Xi_c^o) = 94^{+14}_{-12}$ fs, where the error is statistical only. Although the statistics improves by a factor 3 compared to the Particle Data Group, the error is still large and better measurements are to be expected.

Finally the Ω_c^o baryon issue is still open. The mass is still uncertain and the lifetime measured with very limited statistics[4].

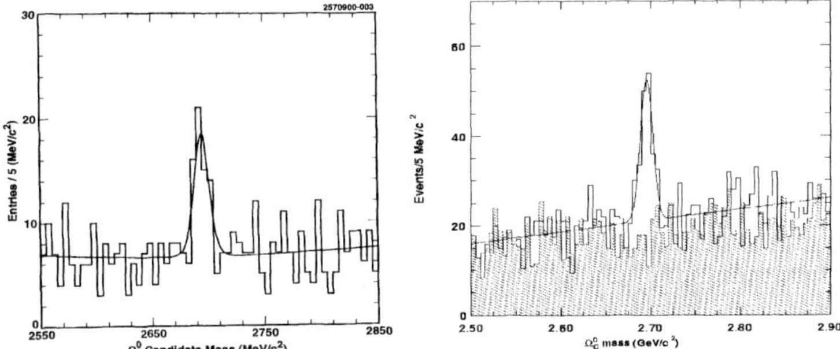

FIGURE 5. a- Cumulative mass distribution for the 5 Ω_c^o decay channels observed by CLEO; b- cumulative mass distribution for the 4 Ω_c^o decay channels observed by FOCUS.

Cleo[12] analyzed six different decay channels of the Ω_c^o, i.e.: $\Omega_c^o \to \Omega^-\pi^+$, $\Omega_c^o \to \Omega^-\pi^+\pi^o$, $\Omega_c^o \to \Omega^-\pi^- 2\pi^+$, $\Omega_c^o \to \Xi^- K^- 2\pi^+$, $\Omega_c^o \to \Xi^o K^-\pi^+$, getting a total of 40.4 ± 9.0 events but did not find evidence for the decay $\Omega_c^o \to \Sigma^+ 2K^-\pi^+$. The cumulative mass distribution is shown in fig. 5a.

FOCUS performed a very preliminarly analysis and selected the four charged decay channels (we did not tackle yet final states containg neutral pions), i.e.: $\Omega_c^o \to \Omega^-\pi^+$, $\Omega_c^o \to \Omega^-\pi^- 2\pi^+$, $\Omega_c^o \to \Xi^- K^- 2\pi^+$, $\Omega_c^o \to \Sigma^+ 2K^-\pi^+$, collecting over 100 events shown in fig. 5b. The very preliminary values for the mass and width found, as well as the yeld for each decay are collected and compared to previous experiments in Table 3

[4] It is indeed peculiar that WA89 never published a value for the mass of their 86 Ω_c^o events.

TABLE 3. Mass width and observed yield of the Ω_c^o **(very preliminary)**

FOCUS: Decay mode	Mass (Mev/c^2)	Width (Mev/c^2)	Yield
$\Omega_c^o \to \Omega^- \pi^+$	2696.0 ± 4.5	11.4 ± 4.1	22.5 ± 8.7
$\Omega_c^o \to \Omega^- 2\pi^- \pi^+$	2695.5 ± 2.0	7.8 ± 2.1	21.0 ± 7.2
$\Omega_c^o \to \Xi^- K^- 2\pi^+$	2605.5 ± 2.2	5.2 ± 1.9	42.3 ± 11.3
$\Omega_c^o \to \Sigma^+ 2K^- \pi^+$	2696.5 ± 2.5	6.4 ± 2.1	28.5 ± 9.5
cumulative histogram	2695.5 ± 2.2	7.8 ± 2.1	115.5 ± 17.6
Experiment			
E687 $\Omega_c^o \to \Omega^- \pi^+$	$2705.9 \pm 3.3 \pm 2.0$	=	10
E687 $\Omega_c^o \to \Sigma^+ 2K^- \pi^+$	$2699.9 \pm 1.5 \pm 2.5$	=	42
CLEO 5 decay modes	$2694.6 \pm 2.6 \pm 1.9$	=	40
FOCUS 4 decay modes	2695.9 ± 1.3	=	115

CONCLUSIONS AND ACKNOWLEDGMENTS

FOCUS obtained remarkable results on charm physiscs. In competition with Cleo and the B factory experiments BaBar and Belle, it will continue to provide valuable contributions to the understanding of both charmed mesons and baryons.

I wish to thank the organizers of the Workshop for the wonderfull hospitality.

Finally, this report wuold have never been written in time without the most valuable assistance of Cristina Riccardi and Paolo Vitulo.

REFERENCES

1. *Compilation of Standard Model predictions*: H.N.Nelson, hep-ex/9908021;
2. I.I.Bigi and A.I.Sanda, *"CP violation"*, Cambridge, UK: Univ. Pr., 382 (2000); F.Buccella *et al.*, Phys. Rev. **D51**, 3478 (1995).
3. C.Caso et al., Eur. Phys. Jour. **15**, 543-578 (2000);
4. P.L.Frabetti *et al.*, Nucl. Instr. Meth. **A320**, 519 (1992);
5. J.Link, *et al.*, Phys. Lett. **B485**, 62 (2000);
6. J.Link, *et al.*, Phys. Rev. Lett. **86**, 2955 (2001);
7. A.B.Smith for CLEO coll., hep-ex 0104008, (2001);
8. J.Link *et al.*, Phys. Lett. **B491**, 232 (2000);
9. a- P.L.Frabetti *et al.*, Phys. Lett. **B323**, 459 (1994); b- ibidem **B357**, 678 (1995); c- Phys. Rev. Lett. **70**, 1381; d- ibidem 1755, (1993); e- ibidem. **71**, 827 (1993);
10. S.P.Ratti *et al.* (FOCUS Coll.): a- Proc. Europ. Int. Conf. on H.E.P. -EPS-HEP1999 (Ed.s K. Huitu, H. Kurti-Suonio, J. Maalampi, I.o.P, 2001) p. 873; b- Proc. 30th Int. Conf. on H.E.P. (Ed.s C.S.Lim, T.Yamanaka, World Sci., 2001) p.381;
11. FOCUS Coll.:*High Statics Measurement of the Ξ_c^+ lifetime*, submitted for publication to Phis. Lett. **B** (2001);
12. CLEO-II Coll., *Obsercation of the Ω_c^o charmed baryon at CLEO*, CLNS 00-1695/CLEO 00-19 hep-ex/0010035v2 (jan 29th, 2001).

Dalitz decays of charmed mesons

Antimo Palano

INFN and University of Bari, Italy

Abstract. A review of Dalitz decays of charmed mesons is presented with particular emphasis to the aspects related to Light Meson Spectroscopy and to the solution of the scalar mesons puzzle.

Introduction

New generation experiments are providing large data sets for charm physics with statistics which allow to supersede most of previous measurements. The Dalitz plot analysis of three-body decays is a relatively new technique in development for charm physics studies.

This method of analysis is the most complete way of analyzing the data since it allows to measure both decay amplitudes and phases. The final state is the result of the interference of all intermediate states.

The significant results provided by these studies are:

- The accurate measurements of branching fractions.
- The study of Final State Interactions (FSI).
- The study of CP violation on rates and on the Dalitz plot.
- New inputs to several old unsolved problems in light meson spectroscopy, in particular to the scalar mesons puzzle.

Factorization models assume the decay weak amplitudes to be real. As a consequence of the FSI the observed amplitudes have a relative complex phase. An example of the FSI effects is given by the obscuring of the color suppression.

CP violation is expected to be small in charm decays. Two amplitudes with different phases are needed:

$$Ae^{i\delta_A} + Be^{i\delta_B}$$

In singly Cabibbo-suppressed dacays penguin terms may provide a weak phase, while Final State Interactions provide a strong phase shift. Under CP the weak phases change sign but the strong ones do not. Thus any difference in the Dalitz plot, between D and $\bar{D}$ would be an evidence for CP violation.

CP602, *QCD@Work: International Workshop on Quantum Chromodynamics*
edited by P. Colangelo and G. Nardulli

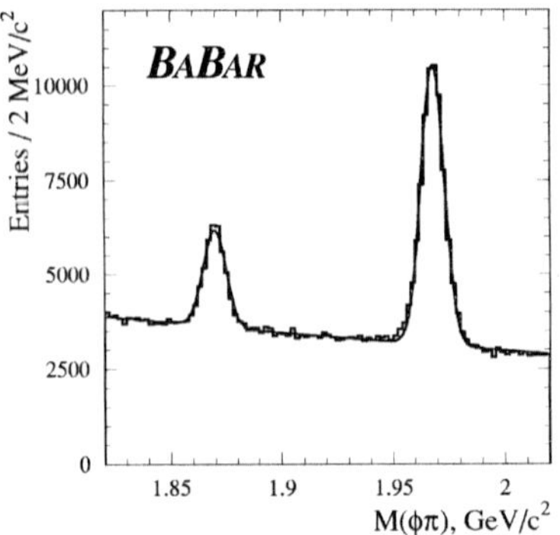

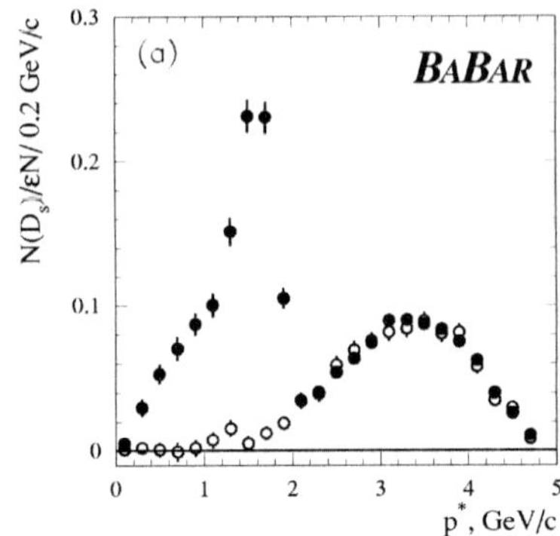

FIGURE 1. $\phi\pi$ mass spectrum and p^* distribution from BaBar. Filled points: data taken at the $\Upsilon(4S)$ energy, open points: normalized off resonance data.

Fixed target experiments.

The Fermilab E791 experiment has taken data during 1990-1991 using 500 GeV/c π^- beam obtaining 2.5×10^5 reconstructed charm.

The experiment FOCUS hosted at Fermilab as well, is the successor to E687 which took data in 1990-1991. The data have been collected during 1996-1997 using 170 GeV γ beam accumulating 10^6 reconstructed charm.

The technique employed in these experiments is to have good vertexing and good particle identification. The large Lorentz boost allows to have a good separation of the charm vertex.

e^+e^- colliders at the $\Upsilon(4S)$.

The experiment CLEOII at Cornell has published results using 9 fb^{-1} integrated luminosity.

BaBar at SLAC has accumulated 23 fb^{-1} at the end of 2000; Belle at KEK has accumulated 12 fb^{-1} in the same period.

In these experiments charmed mesons are obtained from continuum through a cut in their center of mass momentum p^* and tagged requiring that they come from a D^* decay:

$$e^+e^- \to D^* \quad X$$
$$\to D\pi$$

$$e^+e^- \to D_S^* \quad X$$
$$\to D_S\gamma$$

where $\pi = \pi^{\pm}, \pi^0$. Fig. 1 shows the $\phi\pi$ mass distribution and the p^* momentum spectrum for the $D_S \to \phi\pi$ decay mode as reconstructed by BaBar [1].

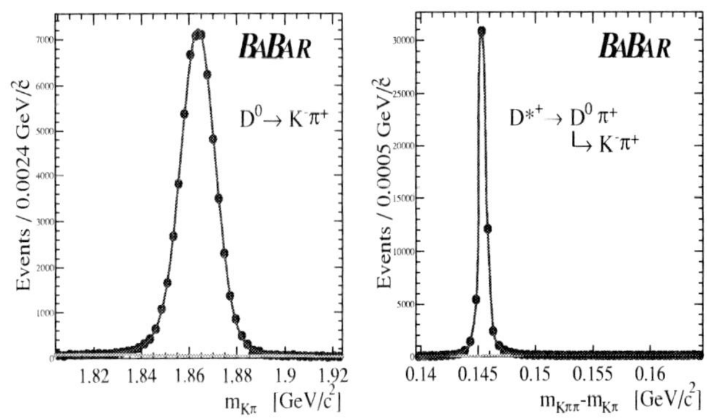

FIGURE 2. $K^-\pi^+$ mass spectrum from tagged D^0 and Δm for $D^{*+} \to \pi^+ D^0, D^0 \to K^-\pi^+$. Data are from BaBar.

Fig. 2 shows an example of a D^* tag from BaBar for the dacay $D^{*+} \to \pi^+ D^0$, $D^0 \to K^-\pi^+$[1].

Dalitz analysis technique.

Nearly all charmed mesons decays proceed through intermediate resonance production. The Dalitz plots distributions are fitted using a sum of interfering amplitudes.

$$| \sum c_i A_i e^{i\phi} |^2$$

Each amplitude A_i is the product of a Breit-Wigner BW(m) and a term describing the angular distributions $Z(\Omega)$ (for example: Zemach tensors).

$$A_i = BW_i(m) Z_i(\Omega)$$

Bare amplitudes are real ($\phi = 0$ or 180^0), the asymmetry can only be generated by FSI.

Fig. 3 shows an example of FSI from $D^+ \to K^+K^-\pi^+$ from FOCUS [2]. It is possible to observe a strong asymmetry between the two K^* lobes, indicating the presence of interferences among the final state amplitudes.

The puzzle of the scalar mesons.

The scalar mesons are still a puzzle in Light Meson Spectroscopy. The question is relevant, since the scalar glueball or 4-quark states may be hidden among the states observed up to now.

---

[1] Through all the paper, charge conjugation is implied.

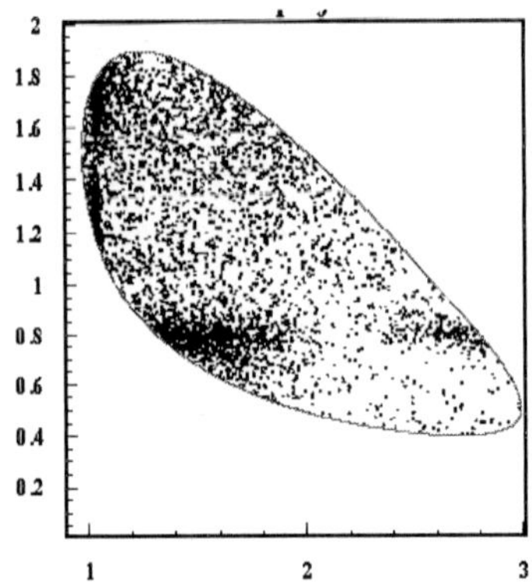

FIGURE 3. $D^+ \rightarrow K^+K^-\pi^+$ Dalitz plot from FOCUS. The plot is given in terms of $m^2(K^-\pi^+) vs. m^2(K^+K^-)$.

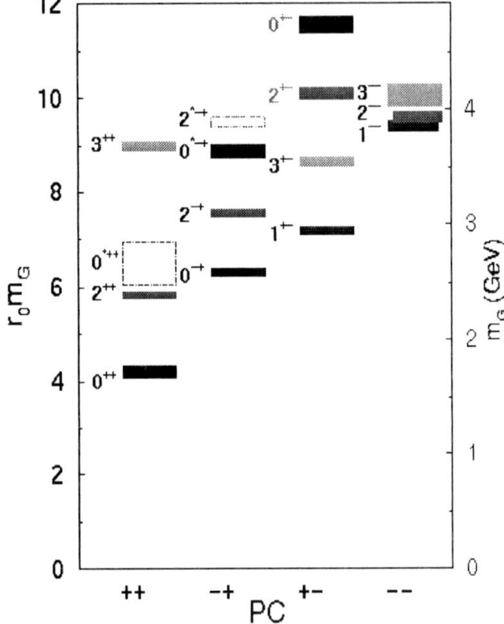

FIGURE 4. Glueballs spectrum from Lattice QCD.

Fig. 4 shows calculations from lattice QCD of the glueballs spectrum [3] where one can observe that the Scalar Glueball is expected at about 1700 MeV.

For a standard $q\bar{q}$ multiplet we expect 9 states, whereas the Particle Data Group lists 15 candidates below 1.8 GeV (see table 1). Among these resonances, $f_J(1710)$ appears with spin 0 or 2 in different experiments.

TABLE 1. The scalar mesons.

I=1/2	I=1	I=0
		$\sigma/f_0(400-1200)$
	$a_0(980)$	$f_0(980)$
		$f_0(1370)$
$K_0^*(1430)$	$a_0(1450)$	$f_0(1500)$
		$f_J(1710)$

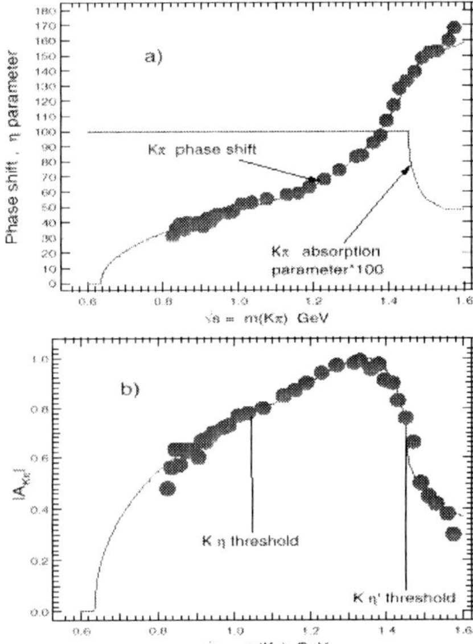

FIGURE 5. $K\pi$ phase shift and scalar amplitude from LASS.

Let us now review the new information coming from the analysis of charm decays.

The resonance $K_0^*(1430)$.

The actual parameters on this resonance in PDG are from LASS experiment at SLAC which studied the reaction:

$$K^-p \rightarrow K^-\pi^+n$$

using 11 GeV/c incident K (see fig. 5) [4]. This is a wide resonance, therefore its parameters are difficult to extract also because of the presence of an S-wave elastic background. The resulting parameters are:

$$m = 1.412 \quad GeV \qquad \Gamma = 294 \quad MeV$$

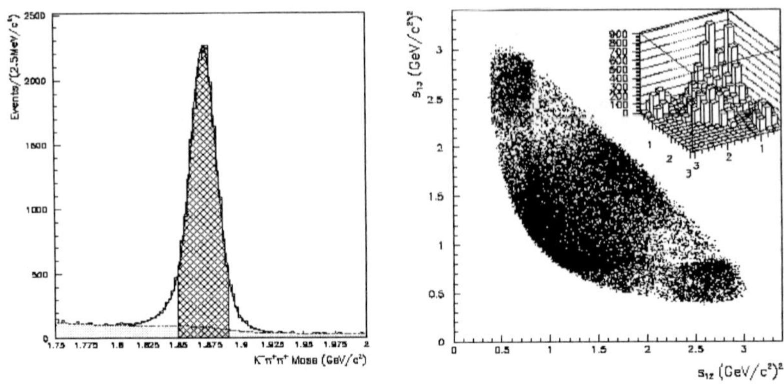

FIGURE 6. $D^+ \to K^-\pi^+\pi^+$ mass spectrum and Dalitz plot from E791.

Study of $D^+ \to K^-\pi^+\pi^+$.

This Dalitz plot has been analyzed by several experiments (E691, E687, E791) [5], [6], [7]. In contrast to all other charmed mesons decays, a large Non Resonant contribution is present in this decay. Fig. 6 shows the data from E791, with $\approx$ 23 000 events

The analysis of this decay mode requires the presence of strong interferences. The channel is dominated by $K^*(890)$ (13 %) and $K_0^*(1430)$ (34 %) with the need of a large Non Resonant contribution (104 %).

However this model does not fit the data well. A better fit is obtained including a new scalar meson $\kappa(800)$ with parameters:

$$m = 815 \pm 30 \quad MeV, \qquad \Gamma = 560 \pm 116 \quad MeV$$

In this scenario the Non Resonant contribution goes to (52 %) and that of κ to 21 % with a 180^0 relative phase.

The resonance $\sigma/f_0(400-1200)$.

Fig. 7 shows a sketch of the S-wave amplitude together with the $\pi\pi$ phase shift from different experiments [8]. The interpretation of these data is given in terms of the slowly moving phase attributed to a wide structure: the σ. The two other dips are attributed to other scalar mesons interfering with the background scalar amplitude: $f_0(980)$ and $f_0(1500)$

The σ state is subject to controversial interpretations. Minkowski and Ochs identify the σ with the scalar glueball (the *Red Dragon*) [9]. Klempt [10] and Pennington [11] on the other hand suggest the σ being a pole in the t-channel.

Two facts can help in discriminating between a resonance and a t-channel pole. A pole in the S-channel produces a resonant shape whose parameters do not depend on the

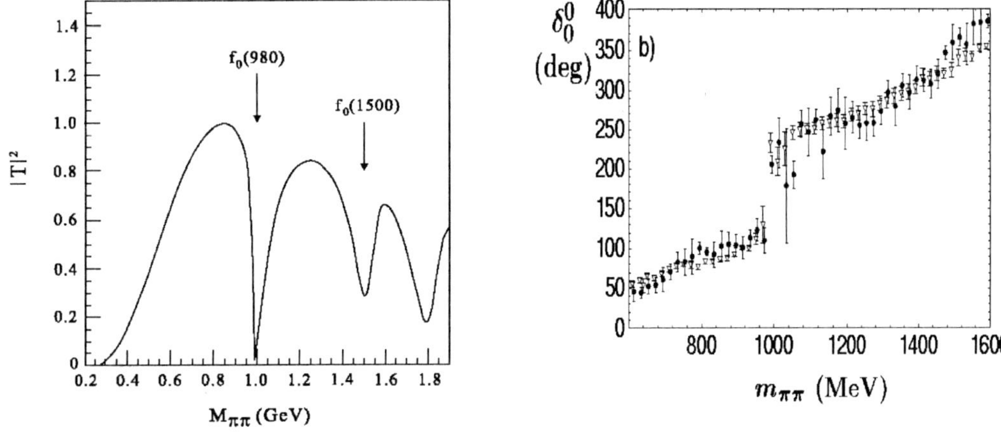

FIGURE 7. Sketch of the $\pi\pi$ amplitude and phase shift from different experiments.

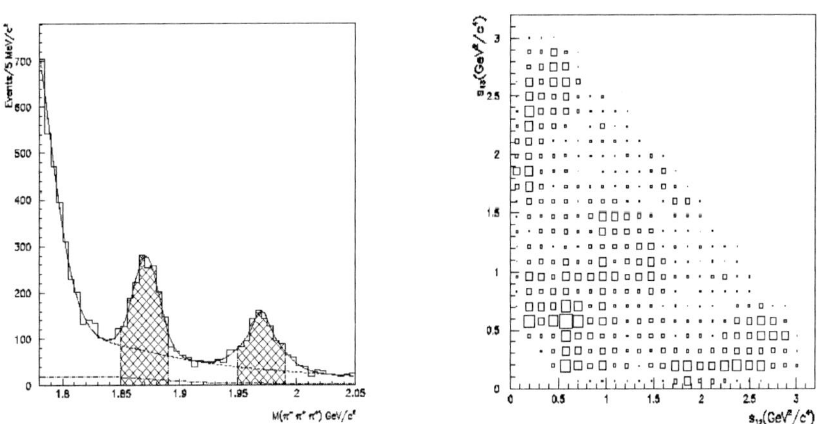

FIGURE 8. 3π mass spectrum and $D^+ \to \pi^+\pi^+\pi^-$ Dalitz plot from E791.

reaction. A pole in the t-channel gives rise to shapes which are strongly dependent from the physical process.

Study of $D^+ \to \pi^-\pi^+\pi^+$.

Fig. 8 shows the $\pi^+\pi^+\pi^-$ mass spectrum from E791 (1686 events in D^+ and 937 in D_S^+.) with a Signal/Background of 2/1 [12]. The symmetrized D^+ Dalitz plot is also shown. The $\pi^+\pi^-$ projection is presented in fig. 9 and shows evidence for $\rho(770)$ and

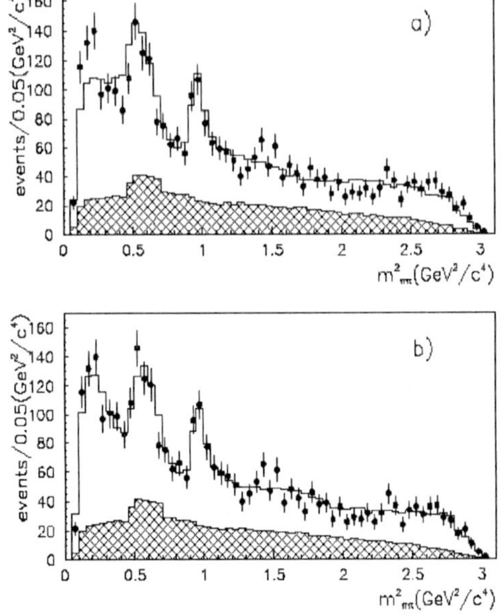

FIGURE 9. $\pi^+\pi^-$ projection from $D^+ \to \pi^+\pi^+\pi^-$. The two figures show the result from the fit with and without the insertion of the σ.The data are from E791.

$f_0(980)$. Also in this case a satisfacory fit is obtained only including an extra scalar resonance $\sigma(500)$ with parameters:

$$m = 478 \pm 24 \quad MeV, \qquad \Gamma = 324 \pm 41 \quad MeV$$

In this hypothesis the dacay $\sigma\pi$ accounts for nearly half (46 %) of D^+ decay and the $f_0(1370)$ contribution vanishes. Similar results have been obtained in a preliminary analysis by FOCUS.

The problem of $f_0(980)$ and $a_0(980)$.

These resonances are both considered as candidates for being 4-quark states due to their proximity to $\bar{K}K$ threshold. $f_0(980)$ appears as a peak or as a hole in different reactions: as a sharp drop in central production, as a peak in $J/\psi \to \phi\pi\pi$. The superposition of several experimental results is summarized in fig. 10.

$f_0(980)$ has been observed in the same experiment ($\pi^- p \to \pi^0\pi^0 n$ at 40 GeV/c from GAMS) to revert from a dip to a peak as a function of the 4-momentum transfer [13].

The $f_0(980)$ parameters are usually described by a coupled channel Breit-Wigner to $\pi^+\pi^-$, K^+K^- and $K_S^0 K_S^0$ final states [14].

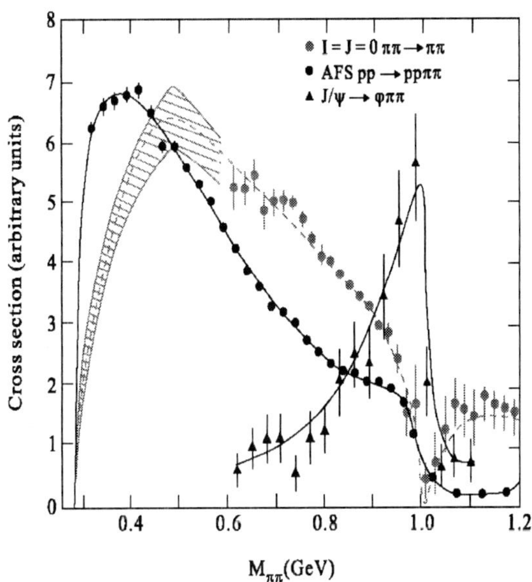

FIGURE 10. $\pi^+\pi^-$ mass spectra from different experiments.

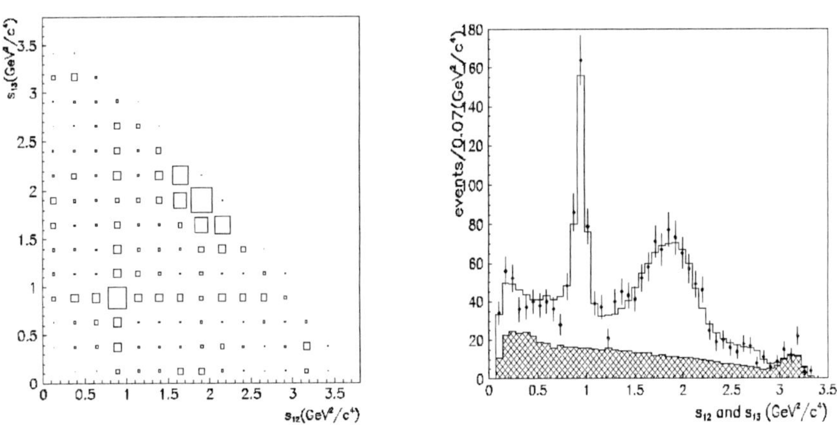

FIGURE 11. Dalitz plot from $D_S^+ \to \pi^+\pi^+\pi^-$ and $\pi^+\pi^-$ mass spetrum from E791.

Study of $D_S^+ \to \pi^-\pi^+\pi^+$

This decay mode has been studied by FOCUS (1300 events) and E791 (850 events). The Dalitz plot from E791 is shown in fig. 11 [15]. Both experiments observe a strong $f_0(980)$ appearing as a narrow peak (see fig. 11). They find the $f_0(980)$ parameters

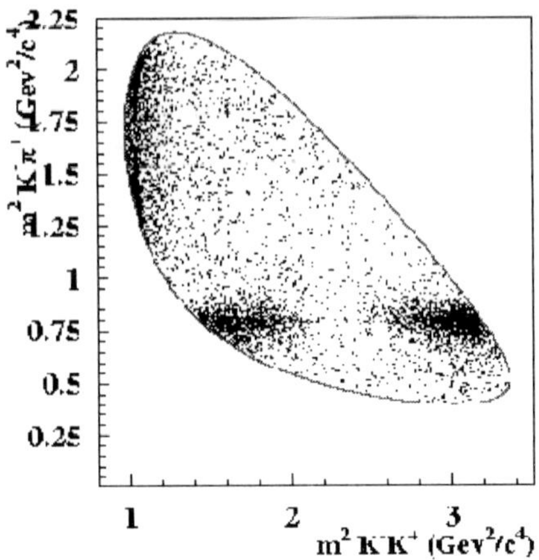

FIGURE 12. $D_S^+ \to K^+K^-\pi^+$ Dalitz plot from FOCUS.

insensitive to the $\bar{K}K$ coupling. Fitting with a standard Breit-Wigner, they obtain:

$$m = 975 \pm 3 \quad MeV \qquad \Gamma = 44 \pm 2 \quad MeV$$

They also estimate a large $f_0(980)$ contribution (57 %) that, on the other hand, would indicate a large $\bar{s}s$ content in this state.

Study of $D_S^+ \to K^-K^+\pi^+$.

The decay of $D_S^+ \to K^-K^+\pi^+$ has been studied by FOCUS (see fig. 12). The fit requires the presence of a relatively broad scalar meson in the threshold region which would suggest the presence of both $f_0(980)$ and $a_0(970)$.

Study of $D^0 \to K_S^0 K^- K^+$.

A 3D view of the Dalitz plot of the decay $D^0 \to K_S^0 K^- K^+$ from FOCUS is presented in fig. 13 [2]. Evidence is found for the decay $D^0 \to K^- a_0^+$. Together with a strong ϕ, a broad scalar in the threshold region suggest the presence of both $f_0(980)$ and/or $a_0(980)$ resonances.

Study of $D_S^+ \to K_S^0 K_S^0 \pi^+$.

The resonance $f_j(1710)$ has been measured with spin 0 or 2 in different experiments. It has been a candidate for being the tensor or scalar glueball. It has been observed in J/ψ decay, central production and $\gamma\gamma$ collisions.

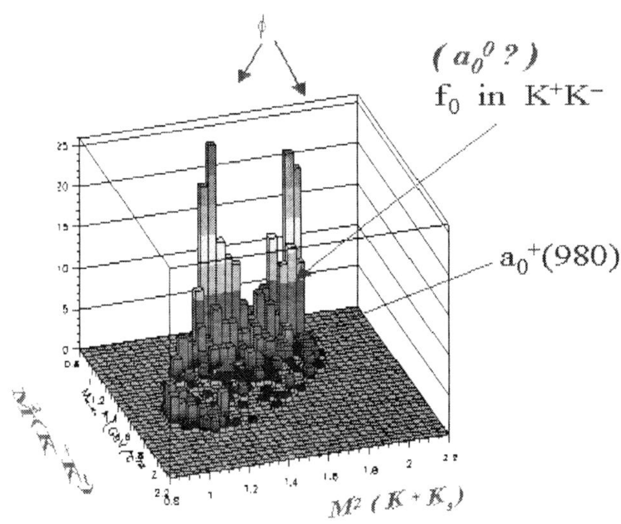

FIGURE 13. $D^0 \to K_S^0 K^+ K^-$ Dalitz plot from FOCUS.

The $D_S^+ \to K_S^0 K_S^0 \pi^+$ decay mode has been studied by BaBar. The channel has been isolated using the tag from $D_S^* \to D_S\gamma$ and p^* cuts. The $K_S^0 K_S^0 \pi^+$ mass spectrum is shown in fig. 14. The background subtracted Dalitz plot, togheter with its $K_S^0 K_S^0$ projection, are presented in fig. 15. The data show evidence for the decay $D_S \to f_j(1700)\pi$.

Study of $D^0 \to K^-\pi^+\pi^0$

The Decay $D^0 \to K^-\pi^+\pi^0$ has been studied by CLEO with 7000 events [16]. The Dalitz plot is presented in fig. 16. The Dalitz analysis shows strong $\rho^0(770)$ (79 %) and $K^*(890)$ (29 %) productions. There is no evidence for κ instead. They also measure a Dalitz plot asymmetry:

$$A_{CP} = -0.031 \pm 0.086$$

Conclusions.

A new chapter in particle physics has been opened: the high statistics Dalitz analysis of charmed mesons decays. These studies will provide information on the different diagrams which originate charm decays and eventually find signals of CP violation in

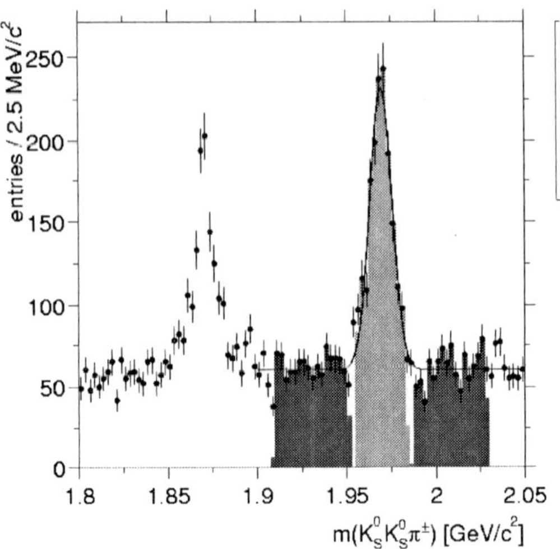

FIGURE 14. $K_S^0 K_S^0 \pi^+$ mass spectrum from BaBar.

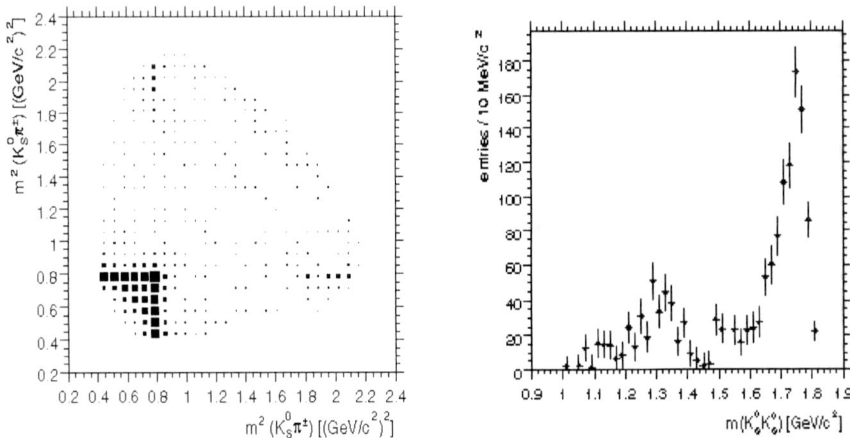

FIGURE 15. Background subtracted Dalitz plot of $D_S^+ \to K_S^0 K_S^0 \pi^+$ and $K_S^0 K_S^0$ projection. Data are from BaBar.

the charm sector. There is also the interesting possibility to solve several questions left open in light meson spectroscopy.

The near future will be dominated by B-factories and τ/charm factories.

Present available amount of data on Dalitz decays from fixed target and B-factories are the following.

- Cabibbo allowed 1-5: $\times 10^4$ events.

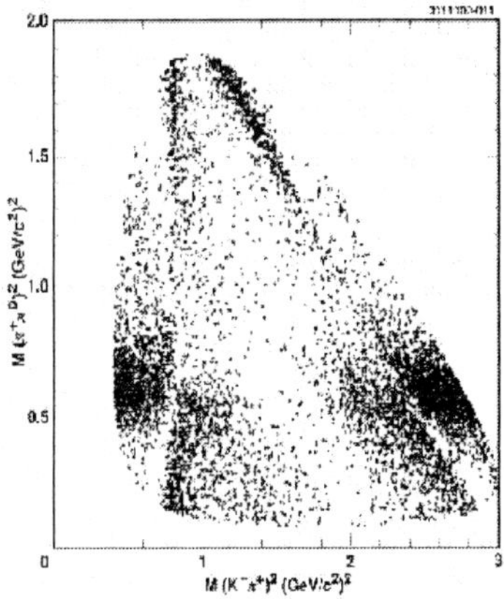

FIGURE 16. Dalitz plot of $D^0 \to K^- \pi^+ \pi^0$ from CLEO.

- Cabibbo suppressed: 1-10 $\times 10^3$ events.
- Doubly Cabibbo suppressed 50 - 300 events.

With the expected integrated luminosity from BaBar and Belle, in the next few years these yields could increase by a factor 20.

REFERENCES

1. The BaBar Collaboration, SLAC-PUB-8928, hep-ex/0107060.
2. G. Boca, Proceedings of Heavy Quarks at Fixed Target, Rio de Janeiro, Oct 9-12, 2000.
3. C.J. Morningstar and M. Peardon, Phys. Rev. **D60** (1999) 034509.
4. D. Aston et al., Nucl. Phys. **B296** (1988) 493.
5. J.C. Anjos et al., Phys. Rev. D **48** (1993) 56.
6. P.L. Frabetti et al., Phys. Lett. B **331** (1994) 217.
7. C. Göbel, hep-ex/0012009 5 Dec 2000.
8. R. Kaminski et al., Z. Phys. **C74** (1997) 79.
9. P. Minkowski and W. Ochs, Eur. Phys. J **C9** (1999) 283.
10. E. Klempt, hep-ex/0101031, 19 Jan 2001.
11. M.R. Pennington, hep-ph/9905241 13 May 1999.
12. E.M. Aitala et al., Phys.Rev.Lett. **86** (2001) 770.
13. D. Alde et al., Z. Phys. **C66** (1995) 375.
14. T.A. Armstrong et al., Z. Phys. **C51** (1991) 351.
15. E.M. Aitala et al., Phys.Rev.Lett. **86** (2001) 765.
16. S. Kopp, et al, Phys.Rev. **D63** (2001) 092001.

The $D^0 \to K^0\bar{K}^0$ decay beyond factorization [1]

J. O. Eeg*, S. Fajfer† and J. Zupan**

*Dept. of Physics, Univ. of Oslo, N-0316 Oslo, Norway
†Physics Department, University of Ljubljana, Jadranska 19, and Institut Jožef Stefan, Jamova 39, SI-1000 Ljubljana, Slovenia
**Institut Jožef Stefan, Jamova 39, SI-1000 Ljubljana, Slovenia

Abstract. The decay mode $D^0 \to K^0\bar{K}^0$ has no factorizable contribution. We calculate the nonfactorizable chiral loop contributions of order $O(p^3)$ and then we use a heavy-light type chiral quark model to calculate nonfactorizable tree level terms, also of order $O(p^3)$, proportional to the gluon condensate. Calculated chiral loops and the gluon condensate contributions are of the same order of magnitude as the experimental amplitude.

For nonleptonic decays of D mesons [1] - [10] as well as for K's and B's, the so called *factorization* hypothesis has been commonly used. The factorization hypothesis is known to fail badly for nonleptonic K decays [11, 12]. On the other hand, there are certain heavy hadron weak decays where factorization might apply. Recently, the understanding of factorization for exclusive nonleptonic decays of B mesons in terms of QCD in the heavy quark limit has been considerably improved [13]. Following [14] we discuss nonfactorizable terms for D decays, in particular for the decay mode $D^0 \to K^0\bar{K}^0$. In $D^0 \to K^0\bar{K}^0$, factorization misses completely, predicting a vanishing branching ratio, in contrast with the experimental situation. To see this, note that at tree level the $D^0 \to K^0\bar{K}^0$ decay might occur due to two annihilation diagrams [1] which could potentially create the $K^0\bar{K}^0$ state. However, they cancel each other by the GIM mechanism. Moreover, in factorization limit, the amplitude is proportional to

$$\langle K^0\bar{K}^0|V_\mu|0\rangle\langle 0|A^\mu|D^0\rangle \simeq (p_{K^0}-p_{\bar{K}^0})_\mu f_D p_D^\mu = 0\,. \qquad (1)$$

In many of the studies (e.g. [2, 3, 4, 5, 7]) this decay has been understood as a result of final state interactions (FSI) e.g. [2]. A recent investigation of the $D^0 \to K^0\bar{K}^0$ decay mode performed in [3] has focused on the s channel and the t channel one particle exchange contributions.

On the other hand it is well known that factorization does not work in nonleptonic K decays. Among many approaches the Chiral Quark Model (χQM) [16] was shown to be able to accommodate the intriguing $\Delta I = 1/2$ rule in $K \to \pi\pi$ decays, as well as CP violating parameters, by systematic involvement of the soft gluon emission forming gluon condensates and chiral loops at $O(p^4)$ order [12]. In the χQM [16] the light quarks (u,d,s) couple to the would-be Goldstone octet mesons (K,π,η) in a chiral invariant way,

[1] Talk given by S. Fajfer

CP602, *QCD@Work: International Workshop on Quantum Chromodynamics*
edited by P. Colangelo and G. Nardulli

such that all effects are in principle calculable in terms of physical quantities and a few model dependent parameters, namely the quark condensate, the gluon condensate and the constituent quark mass [12, 16, 17].

In the case of D meson decays one has to extend the ideas of the χQM to the sector involving a heavy quark (c) using the chiral symmetry of light degrees of freedom as well as heavy quark symmetry and Heavy Quark Effective Field Theory (HQEFT). Such ideas have already been presented in previous papers [18, 19, 20] and lead to the formulation of Heavy-Light Chiral Quark Models (HLχQM). In our formulation of the HLχQM Lagrangian, an unknown coupling constant appears in the term that couples the heavy meson to a heavy and a light quark [14]. Our strategy is to relate expressions involving this coupling to physical quantities, as it is done within the χQM [12]. We perform the bosonization by integrating out the light and heavy quarks and obtain a heavy quark symmetric chiral Lagrangian involving light and heavy mesons [21, 22, 23].

Because the $O(p)$ (factorizable) contribution is zero as seen in Eq. (1), we approach to the $D^0 \to K^0\bar{K}^0$ decay by calculating systematically $O(p^3)$ contributions. We do this by including first the nonfactorizable contributions coming from the chiral loops. These are based on the weak Lagrangian corresponding to the factorizable $O(p)$ terms for $D^0 \to \pi^+\pi^-$ and $D^0 \to K^+K^-$. Second, we consider the gluon condensate contributions, also of $O(p^3)$ within the χQM and HLχQM framework. The energy release in $D \to K\bar{K}$ is $p = 788$ MeV and hence p/Λ_χ (for $\Lambda_\chi \geq 1$ GeV), is close to unity. The next to leading $O(p^5)$ terms might be almost of the same order of magnitude compared to our $O(p^3)$ terms. However, we expect a weak suppression of the order p^2/Λ_χ^2. On the other hand, the inclusion of $O(p^5)$ order in this framework is not straightforward. Before doing loop calculations at that order, one has to find a reliable framework to include light resonances like ρ, K^*, $a_0(980)$, $f_0(975)$ etc. The poorly known scalar resonances would introduce a rather large uncertainty [3]. Right now, a consistent calculation of this or higher orders does not seem to be possible.

Note that we have omitted $1/m_Q$ terms in the framework of HQEFT.

The effective weak Lagrangian at quark level relevant for $D \to \pi\pi, K\bar{K}$ is

$$L_W = \widetilde{G}\left[c_A(Q_A - Q_C) + c_B(Q_B^{(s)} - Q_B^{(d)})\right], \tag{2}$$

where $\widetilde{G} = -2\sqrt{2}G_F V_{us}V_{cs}^*$, and

$$Q_A = (\bar{s}_L\gamma^\mu c_L)\,(\bar{u}_L\gamma_\mu s_L)\ , Q_C = (\bar{d}_L\gamma^\mu c_L)\,(\bar{u}_L\gamma_\mu d_L),$$
$$Q_B^{(q)} = (\bar{u}_L\gamma^\mu c_L)\,(\bar{q}_L\gamma_\mu q_L)\ , \qquad (q = s,d), \tag{3}$$

are quark operators.

Using Fierz transformations [14] one obtains operators $Q_A = Q_B^{(s)}/N_c + R_B^{(s)}$, $Q_C = Q_B^{(d)}/N_c + R_B^{(d)}$, $Q_B^{(s)} = Q_A/N_c + R_A$ and $Q_B^{(d)} = Q_C/N_c + R_C$, where the R's correspond to color exchange between two currents and are genuinely non-factorizable: $R_A = 2(\bar{s}_L\gamma^\mu t^a c_L)\ (\bar{u}_L t^a\gamma_\mu s_L)$, $R_C = 2(\bar{d}_L\gamma^\mu t^a c_L)\ (\bar{u}_L t^a\gamma_\mu d_L)$, $R_B(q) = 2(\bar{u}_L\gamma^\mu t^a c_L)(\bar{q}_L t^a\gamma_\mu q_L)\ ,\ (q = s,d)$. The operators can be written in terms of currents [14]. The factorization approach amounts to writing the currents in

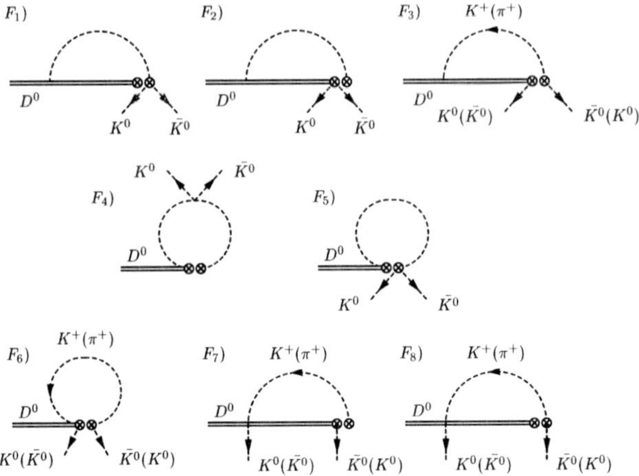

FIGURE 1. The diagrams which give nonzero amplitudes.

terms of hadron (in our case meson) fields only, so that the operator $Q_B^{(s)} - Q_B^{(d)}$ in the left equation is equal to the product of two meson currents. The color currents are then zero if hadronized (mesons are color singlet objects). There is also a replacement of the Wilson coefficients in the hadronized effective weak Lagrangian $c_{A,B} \rightarrow c_{A,B}(1+1/N_c)$. Combining heavy quark symmetry and chiral symmetry of the light sector, we can obtain the weak chiral Lagrangian for nonleptonic D meson decays due to factorizable terms. Then we can first use this to calculate nonfactorizable contributions due to chiral loops. Second, we can calculate the color currents' contribution using the gluon condensate within the framework of the HLχQM.

Treating the light pseudoscalar mesons as pseudo-Goldstone bosons one obtains the usual $O(p^2)$ chiral Lagrangian [14] and from this Lagrangian, we can deduce the light weak current to $O(p)$.

In the heavy meson sector interacting with light mesons we have used the lowest order $O(p)$ chiral Lagrangian (for details see [14]). From symmetry grounds, the heavy-light weak current is to $O(p^0)$ bosonized too and the unknown coupling α_H is related to the physical decay constant f_D.

In the factorization limit there are no contributions to $D^0 \rightarrow K^0\bar{K^0}$ at tree level. The observation of a partial decay width $B(D \rightarrow K^0\bar{K^0}) = (6.5 \pm 1.8) \times 10^{-4}$ on the other hand implies that we can expect sizable contributions at the one loop level. Calculations to one loop in the framework of combined chiral perturbation theory and HQEFT involves a construction of the most general effective Lagrangian that has the correct symmetry properties in order to make the renormalization work. Our calculatins were done in the strict $\overline{MS}$ renormalization scheme.

Writing down the most general one loop graphs with two outgoing Goldstone bosons (K^0 and $\bar{K^0}$) one arrives at 26 Feynman diagrams. The expressions for nonzero amplitudes corresponding to the graphs on Fig. 1 is given in [14].

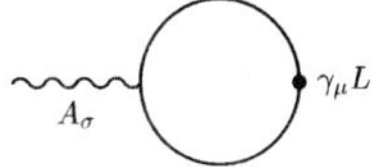

FIGURE 2. Feynman diagram for bosonization of left-handed current to order $O(p)$

The partial decay width for the decay $D^0 \to K^0\bar{K^0}$ is then

$$\Gamma_{D^0\to K^0\bar{K^0}} = \frac{1}{2\pi}\frac{G_F^2}{8m_D}c_A^2|V_{us}V_{cs}^*|^2\frac{|F|^2}{(8\pi^2)^2}p\,, \tag{4}$$

where $F = \sum_n F_n$ is the sum of the amplitudes corresponding to the graphs on Fig.1 and p is the $K^0(\bar{K^0})$ three - -momentum in the D^0 rest frame.

In numerical calculation we use the values of α_H, g and f (f is related to the π meson decay constant) obtained within the same framework [23, 29, 21, 27, 28]. The coupling g is extracted from existing experimental data on $D^* \to D\pi$ and $D^* \to D\gamma$ decays. The analysis in [23] includes chiral corrections at one loop order and yields $g = 0.27^{+0.04+0.05}_{-0.02-0.02}$, leaving the sign undetermined. Recently CLEO Collaboration has measured the D^{*+} decay width [25]. By combining this result with existing data on D^* decay widths [15], we obtain value $g = 0.57 \pm 0.08$. We present results for $g = 0.27$ and $g = 0.57$. The larger value seems to be in better agreement with the results coming from different approaches listed in [21]. We put everywhere $\mu = 1$ GeV $\simeq \Lambda_\chi$.

For the Wilson coefficients $c_{A,B}$ of (2) we use $c_A = 1.10 \pm 0.05$ and $c_B = -0.06 \pm 0.12$ [24], calculated at the scale $\mu = 1$ GeV with the number of colors $N_c = 3$. Due to the suppression of c_B in comparison with c_A, we do not include terms proportional to c_B.

TABLE 1. Table of the one chiral loop amplitudes (see Fig. 1), where $M = \sum_n M_n$. The second column shows the amplitudes calculated using $g = 0.27$ while the third column amplitudes have been calculated using $g = 0.57$.

	$M_i[\times 10^{-7}\text{GeV}]$ ($g = 0.27$)	$M_i[\times 10^{-7}\text{GeV}]$ ($g = 0.57$)
M_1	-0.42	-0.82
M_2	-0.31	-0.62
M_3	-0.62	-1.23
M_4	0.75 -2.54 i	0.70 -2.37 i
M_5	-0.81	-0.76
M_6	-0.61	-0.57
M_7	-0.99	-0.92
M_8	0.91	0.85
$\sum_i M_i$	-2.11 -2.54 i	-3.37 -2.37 i

The imaginary part of the amplitude comes from the F_4 graph, when the π's or the K's in the loops are on-shell. All other graphs contribute only to the real part of the amplitude. The imaginary part of the amplitude is scale and scheme independent within chiral perturbation theory. This amplitude is also obtained from unitarity, and is valid beyond the chiral loop expansion.

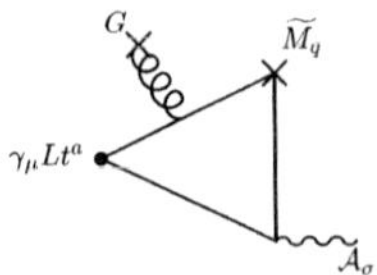

FIGURE 3. Diagram for bosonization of the color current to $O(p^3)$

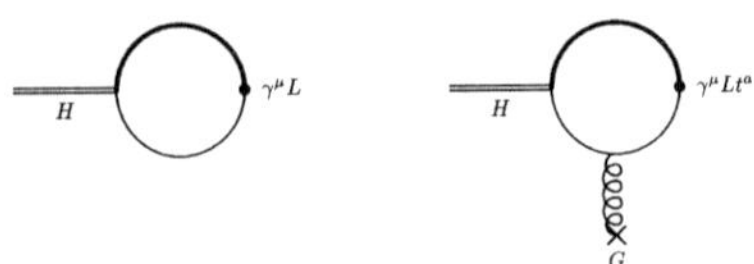

FIGURE 4. Diagrams representing bosonization of heavy-light weak current. The boldface line represents the heavy quark, the solid line the light quark.

In the effective weak Lagrangian there are, after Fierz transformations, terms that involve color currents. As mesons are color singlet objects, the product of color currents does not contribute at meson level in the factorization limit. However, at quark level they do contribute through the gluon condensate. In order to estimate this contribution we have to establish the connection between the underlying quark-gluon dynamics and the meson level picture. This is done through the use of a Heavy-Light Chiral Quark Model (HLχQM). Our starting point is the Lagrangian containing both quark and meson fields [14, 18, 19, 20, 26]. After bosonizing the light weak current at the order $O(p)$ and $O(p^3)$ (see Fig. 2 and Fig. 3) as in [14] one should bosonize the heavy - light weak current integrating out quark fields as presented on Fig. 4. The product of two external gluon fields (G in Fig. 3 and Fig. 4) is interpreted as the gluon condensate ($\langle G^2 \rangle$) [14]. This contribution is of the $O(p^3)$ order. In the language of chiral perturbation theory, the divergent part of the counterterm has the Lorentz and flavor structure of the effective Lagrangian given in eq. (47) in [14].

By taking into account various relations of the loop integrals we determine

$$M(D^0 \to K^0 \bar{K^0})_{\langle G^2 \rangle} \simeq 0.43 \times 10^{-7}\, GeV\ ; \tag{5}$$

which is of the same order of magnitude as the chiral loop contributions in Table 1. Adding both the chiral loops and the gluon condensate (5) contributions, we obtain the total amplitude to $O(p^3)$

$$g = 0.27; \quad M_{Th} = (-1.68 - 2.54\,i) \times 10^{-7}\, GeV \tag{6}$$
$$g = 0.57; \quad M_{Th} = (-2.94 - 2.37\,i) \times 10^{-7}\, GeV\ . \tag{7}$$

or in terms of branching ratio

$$g = 0.27; \quad B(D^0 \to K^0 \bar{K^0})_{Th} = (4.2 \pm 1.4) \times 10^{-4} \tag{8}$$
$$g = 0.57; \quad B(D^0 \to K^0 \bar{K^0})_{Th} = (6.5 \pm 1.7) \times 10^{-4} \tag{9}$$

where the estimated uncertainties reflect the uncertainties in the rest of input parameters. These results should be compared with experimental data [15] $B(D^0 \to K^0\bar{K}^0) = (6.5 \pm 1.8) \times 10^{-4}$.

We can summarize that we indicate the leading nonfactorizable contributions to $D^0 \to K^0\bar{K}^0$. Even though the use of chiral perturbation theory in this decay mode could be questioned, the calculated chiral loops can be considered as part of the final state interactions. In the treatment of the final state interactions the light pseudoscalar meson exchanges have to be present due to unitarity. Although, the next to leading $O(p^5)$ order terms might give sizable contributions to this decay, we have demonstrated that contributions due to the chiral loops and gluon condensates are of the same order of magnitude as the amplitude extracted from the experimental result.

REFERENCES

1. Bauer, M., Stech, B., and Wirbel, M., *Z. Phys. C* **34**, 103-117 (1987).
2. Pham, X. Y., *Phys. Lett. B* **193**, 331-341 (1987).
3. Dai, Y. S. et al., *Phys. Rev. D* **60**, 014014 (1999).
4. Lipkin, H., *Phys. Rev. Lett.* **44**, 710 -712 (1980).
5. Gerard, J.-M., Pastieau, J., Weyers, J., *Phys. Lett. B* **436**, 363-368 (1998).
6. Kamal. A. N., Santra, A. B., Uppal, T., Verma, R.C., *Phys. Rev. D* **53**, 2506- 2515 (1995).
7. Zenczykowski, P., *Phys. Lett. B* **460**, 390-396 (1999).
8. Terasaki, K., *Phys. Rev. D* 59, 114001-1-23 (1999).
9. Buras, A. J., Gerard, J.- M., Rückl, R., *Nucl. Phys. B* **268**, 16-59 (1986).
10. Buccella, F., Lusignoli, M., Miele, G., Pugliese A., and Santorelli, P., *Phys. Rev. D* **51**, 3478-3486 (1995), Buccella, F., M. Lusignoli, M., and Pugliese, A., *Phys. Lett. B* **379**, 249-256 (1996).
11. See for example: Buras, A.J., Jamin, M., and Lautenbacher, M. E., *Nucl. Phys. B* **408**, 209-285, (1993), *Phys. Lett. B* **389** , 749-756 (1996), Bertolini, S., Fabbrichesi, M., Eeg, J.O., *Rev. Mod. Phys.* **72**, 65-93 (2000), and references therein.
12. Bertolini, S., Fabbrichesi, M., Eeg, J.O., *Nucl. Phys. B* **449**, 197-228 (1995) V. Antonelli, V., S. Bertolini, S., Eeg, J.O., Fabbrichesi, M., and Lashin, E.I., *Nucl. Phys. B* **469** 143-180 (1996); Bertolini, S., Eeg, J.O., Fabbrichesi, M., and Lashin, E.I., *Nucl. Phys. B* **514**, 93-112 (1998).
13. Beneke, M., Buchalla,G., Neubert, M., and Sachrajda, C.T.,*Phys. Rev. Lett* **83**, 1914-1917 (1999).
14. Eeg. J.O., Fajfer, S., and Zupan, J., *Phys. Rev. D* **64**, 034010-1-13 (2001).
15. Review of Particle Physics, Groom, D.E., et al., *Eur. Phys. J. C* **151** (2000).
16. Pich, A., and de Rafael, E., *Nucl. Phys. B* **358**, 311-382 (1991).
17. Eeg, J. O., and Picek, I., *Phys. Lett. B* **301**, 423-429 (1993), Eeg, J. O., and Picek, I., *Phys. Lett. B* **323**, 193-200 (1994) 193, Bergan, A.E., and Eeg, J.O., *Phys. Lett. B* **390**, 420-426 (1997).
18. Bardeen, W.A., and Hill, C.T., *Phys. Rev.* D**49**, 409-425 (1994).
19. Ebert, D., Feldmann, T., Friedrich, R., and Reinhardt, H., Nucl. Phys. **B 434**, 619-646 (1995).
20. Deandrea, A., Di Bartelomeo, N., Gatto, R., Nardulli, G., and Polosa, A.D., Phys. Rev. **D 58**, 034004-1-17 (1998), A. D. Polosa, hep-ph/0004183.
21. Casalbuoni, R., et al., *Phys. Rep.* **281**, 145-238 (1997).
22. Wise, M. B., *Phys. Rev. D* **45**, R2188-2191 (1992).
23. Stewart, I.W., *Nucl. Phys. B* **529**, 62-80 (1998).
24. Buras, A.J., *Nucl. Phys. B* **434**, 606-618 (1995).
25. Coan, T.E., et al.(CLEO Collaboration), hep-ex/0102007.
26. Hiorth, A., and Eeg, J. O., in preparation.
27. Grinstein B., et al., *Nucl. Phys. B* **B 380**, 369-376 (1992).
28. Falk, A.F., Grinstein, B., *Nucl. Phys.B* **B 416**, 771-785 (1994).
29. Boyd C.G., and Grinstein, B., *Nucl. Phys. B* **442**, 205-227 (1995).

Final State Interaction for Non-Leptonic Exclusive Charm Decays

F. Buccella

Dipartimento di Scienze Fisiche, Università di Napoli,
INFN, Sezione di Napoli, Italy

Abstract. An approach to the PP and PV exclusive channel for non-leptonic charmed decays, where final state interaction and annihilation contributions play an important role and account for the large SU(3) violations found experimentally, is compared with the most recent and more precise experimental results.
The test is particularly successful for the Cabibbo first forbidden rates.

This talk describes a research, begun many years ago in Naples [1] and continued in collaboration with Maurizio Lusignoli and Alessandra Pugliese [2] [3] [4].
Non-leptonic decays of hadrons have always been a mine of information on their weak and strong interactions.
The $\Delta I = \frac{1}{2}$ selection rule for the decay of strange particles is still not understood, as it is shown from the present dispute [5] on the prediction for ε'.
I am not resisting from quoting very old papers, where:

$$A_S(\Sigma^+ \to n + \pi^+) = 0 \qquad (1)$$
$$A_P(\Sigma^- \to n + \pi^-) = 0$$

[6] [7], respectively, were predicted in excellent agreement with experiment.
Q.C.D. improves slightly the predictions for strange particles non-leptonic decays [8], by reducing the $\Delta I = \frac{3}{2}$ part of the effective lagrangian by the proper amount, but does not increase enough the $\Delta I = \frac{1}{2}$ part:

$$\bar{s}_L \gamma_\mu u_L \bar{u}_L \gamma^\mu d_L \to k_1 \bar{s}_L \gamma_\mu u_L \bar{u}_L \gamma^\mu d_L + k_2 \bar{s}_L \gamma_\mu d_L \bar{u}_L \gamma^\mu u_L \qquad (2)$$

with $k_1 + k_2 < 1 < k_1 - k_2$ and, at LO, $(k_1 + k_2)^2 (k_1 - k_2) = 1$.

A breakthrough seemed to arise from the discovery by the ITEP group of the penguin term [9]:

$$\bar{s}_L \gamma_\mu \lambda^a d_L \bar{q}^i \gamma^\mu \lambda^a q_i \qquad (3)$$

which contributes to the I=0 two pions final state for K decay and to the hyperon s-wave non-leptonic decay amplitudes in such a way, according to the authors, to reach agreement with experiment,but this claim is not universally accepted [10].
Before the experimental lifetimes of the pseudoscalar charmed mesons have been measured, Cabibbo and Maiani [11], made the spectator hypothesis, $c\,(\bar{q}) \to s + u + \bar{d}\,(\bar{q})$,

CP602, *QCD@Work: International Workshop on Quantum Chromodynamics*
edited by P. Colangelo and G. Nardulli

which implies an equal lifetime for them and for the charmed baryons.
This is contradicted by the experimental fact that:

$$\tau(D^0) \leq \tau(D_s^+) < \tau(D^+) \tag{4}$$

with $\frac{\tau(D^+)}{\tau(D^0)} = 2.4$. Since $D^+ = c\bar{d}$, this result has been related to a destructive interference between the $\bar{d}$ in the D^+ and the one produced in the decay [12].
Indeed the QCD correction to the bare non-leptonic weak hamiltonian gives a destructive interference for the $D^+ \to \bar{K^0}\pi^+$ decay amplitude within the factorization approximation:

$$(k_1+\xi k_2) < \bar{K^0}\, \bar{s}\gamma_\mu c\, D^+ >< \pi^+\, \bar{u}\gamma^\mu\gamma_5 d\, 0 > + (k_2+\xi k_1) < \pi^+\, \bar{u}\gamma_\mu c\, D^+ >< \bar{K^o}\, \bar{s}\gamma^\mu\gamma_5 d\, 0 > \tag{5}$$

The two products of matrix elements appearing in eq.(5) should be equal in the SU(3) limit and in that case their total contribution should be proportional to $(k_1+k_2)(1+\xi)$ and $k_1+k_2<1$ and ξ, which should be $\frac{1}{3}$ in perturbative QCD, to get a good fit should be ≈ 0 [13].
For D^0 (and D_s^+), there are annihilation contributions as [14]:

$$k_2 < 0\, \bar{u}\gamma_\mu\gamma_5 c\, D^0 >< \bar{K}\pi\ \bar{s}\gamma^\mu d\, 0 > \tag{6}$$

which vanish in the SU(3) limit (they are proportional to $m_s - m_d$).
There is an isospin sum rule for $D \to \bar{K}\pi$ decays:

$$A(D^0 \to \pi^+ K^-) = A(D^+ \to \pi^+\bar{K^0}) - \sqrt{2}A(D^0 \to \pi^0\bar{K^0}) \tag{7}$$

The values of the corresponding rates imply that the three complex numbers appearing in eq.(7) build a non-trivial triangle in the Gauss plane; therefore their phases are different, which implies a final state interaction, as it happens in $K \to \pi\pi$ decays.
A final state interaction does not change the prediction for the sum of the rates into states with the same strong interaction quantum numbers (parity, isospin, hypercharge), which for PP and PV final states agrees with experiment, if one takes the QCD values for k_1 and k_2 and $\xi = 0$ [1].
The amplitudes, modified by the final state interaction, are given by [1],[2]:

$$A^{[resc]}(D \to f_i) = A^{[n.r.]}(D \to f_i) + c_i(e^{i\delta_R} - 1)A^{[res]} \tag{8}$$

where:

$$A^{[res]} = \Sigma_i c_i A^{[n.r.]}(D \to f_i) \tag{9}$$

c_i is the coupling of f_i to the resonance and δ_R the phase shift.
After finding that a good description of data is obtained with a large value for the phase-shift corresponding to the $(0^+, Y = \pm 1, I = \frac{1}{2})$ state [1], one of us (G.Miele) realized that such a state had been experimentally found with mass and width well consistent with the value found by us for δ_R.
We assumed Gell-Mann Okubo mass splitting to find the position of the $(0^+, T = 1, Y = 0)$ resonance, which is relevant for the final state interaction of the states produced in the

Cabibbo allowed decays of D_s^+ and first forbidden of D^+, and SU(3) relationships for the PP couplings to the 0^+ resonances. Similarly we expect an octet of 0^- states to provide a final state interaction for D decays into PV and also for their mass and couplings we assume Gell-Mann Okubo splittings and SU(3) symmetry for the couplings.
In fact SU(3) symmetry implies many relationships for the non-leptonic weak amplitudes,which follow from the fact that the charmed mesons classify as a $\bar{3}$ under SU(3) (D_s^+,D^+ and D^0 are a doublet and a singlet under U spin, respectively) and in the four quark limit,which is a good approximation for the weak decays of charmed particles, if one does not want to consider CP violation, the $\Delta S = \Delta C$ (Cabibbo allowed), $\frac{1}{\sqrt{2}}$ $\Delta S = 0$ (Cabibbo first forbidden) and $\Delta S = -\Delta C$ (Cabibbo doubly forbidden) non-leptonic weak lagrangians form a U spin triplet [3]. So we have both relationships for rates proportional to the same power of $[\sin\theta_c]^2$ as well for differently allowed processes.
In particular for the Cabibbo first forbidden decays of $D^0 \to$ PP, one should have equal branching ratios for the K^+K^- and $\pi^+\pi^-$ channels and a selection rule forbidding the $K^0\bar{K^0}$ channel in strong disagreement with the experimental values: $.425 \pm .016\%$, $.152 \pm .009\%$ and $.065 \pm .018\%$, respectively.
It should also be:

$$\frac{Br(D^0 \to K^+\pi^-)}{[\tan\theta_c]^2} = Br(D^0 \to K^+K^-) = [\tan\theta_c]^2 Br(D^0 \to K^-\pi^+) \quad (10)$$

$$Br(D^+ \to \pi^+\pi^0) = \frac{[\tan\theta_c]^2}{2} Br(D^+ \to \bar{K^0}\pi^+) \quad (11)$$

to be compared with: $.368 \pm .100\%$, $.425 \pm .016\%$ and $.191 \pm .005\%$, $.25 \pm .07\%$ and $.038 \pm .004\%$, respectively.
Within our approach [3] these large violations are accounted mainly by the strong dependence of the masses of the 0^+ and 0^- resonant states from the mass splittings within the SU(3) octets.Also the SU(3) relationships between the D^+ and D_s^+ decay amplitudes are expected to be spoiled by the energy dependance of the final state interaction phases.
However the final state interaction, which does not change the total rates in the Cabibbo forbidden channels and the total contribution of the interference between the Cabibbo allowed and doubly forbidden channels, does not spoil the cancellation implied by SU(3) for the non-diagonal element of the 2x2 matrix for the width of the neutral charmed pseudoscalar mesons.
In fact, by considering the PP and PV contributions, we found [4]:

$$\Gamma_{1,2} = \Sigma_{f_i} A^*(D^0 \to f_i) A(\bar{D^0} \to f_i) \approx (35.2 - 33.7)\, 10^{-3}\, \Gamma_{D^0} = 1.5\, 10^{-3}\, \Gamma_{D^0} \quad (12)$$

To reproduce the data, which are becoming more precise, we had to assume also a non-resonant SU(3) invariant final state interaction phase in the PV channel, which transforms as a 27 multiplet [4].

A good test on the experimental agreement of our approach may be achieved by comparing our predictions [4] for the Cabibbo forbidden rates with the more recent and precise updated measurements: this is done in Table 1, where the predictions of [4] for the branching ratios (in %) are compared with the old and recent data.

Table 1

Initial particle	Final state	Old exp.	Our prediction	Recent exp.
$D^+ \to$	$\pi^+\pi^-$	0.25 ± 0.07	0.19	0.25 ± 0.07
	$\eta\pi^+$	0.75 ± 0.25	0.34	0.30 ± 0.06
	$\eta'\pi^+$	< 0.9	0.73	0.50 ± 0.10
	$\bar{K}^0K^+$	0.78 ± 0.17	0.81	0.74 ± 0.10
	$\rho^0\pi^+$	< 0.14	0.13	0.105 ± 0.03
	$\rho^+\eta$	< 1.2	0.013	< 0.43
	$\rho^+\eta'$	< 1.5	0.12	< 0.3
	$\omega\pi^+$	< 0.7	0.019	< 0.43
	$\phi\pi^+$	0.67 ± 0.08	0.61	0.61 ± 0.08
	$\bar{K}^0K^{*+}$	not measured	1.71	3.2 ± 1.5
	$\bar{K}^{*0}K^+$	0.51 ± 0.10	0.38	0.42 ± 0.05
$D^0 \to$	$\pi^0\pi^0$	0.088 ± 0.023	0.110	0.084 ± 0.022
	$\pi^+\pi^-$	0.159 ± 0.012	0.159	0.152 ± 0.009
	K^+K^-	0.454 ± 0.029	0.446	0.425 ± 0.016
	$\bar{K}^0K^0$	0.11 ± 0.04	0.098	0.065 ± 0.018
	$\phi\pi^0$	not measured	0.11	< 0.085
	$\phi\eta$	not measured	0.09	< 0.17
	$\bar{K}^0K^{*0}$	< 0.08	0.064	< 0.05
	$\bar{K}^{*0}K^0$	< 0.15	0.062	< 0.1
	K^-K^{*+}	$0.34 \pm .08$	0.43	0.34 ± 0.08
$D_s^+ \to$	$K^0\pi^+$	< 0.7	0.40	< 0.49
	$K^+\rho^0$	not measured	0.29	< 0.18
	$K^+\phi$	< 0.25	0.018	< 0.03

By looking at Table 1, it is satisfactory that, despite the great progress achieved for the precision of the measured data, the reduced χ^2 for all the Cabibbo first forbidden decays considered is still .9. For the Cabibbo allowed channels the larger precision reached by experiment brings to some disagreement for the decays of D^0 into neutral PP and PV final states and for the $D_s^+ \to \eta'\rho^+$ amplitude. While the smaller amplitude found for the last may be explained by our neglect of the glue component of η', some modification in the form of the final state interaction may bring to a better agreement for data, provided that the sum on the rates of all the channels reachable from $K^-\pi^+$ or $K^-\rho^+$, for which we have large non rescattered amplitudes, is predicted in good agreement with experiment in both cases.

We may conclude that our general approach has successfully overcome the challenging test of the comparison with more precise data, especially for the Cabibbo first forbidden sector, and that, if we shall be able to consider also the VV final states, which have branching ratios comparable with the ones considered here, we may face the appealing issues of CP violation and of the evaluation of the non-diagonal matrix element of

the width of the system $D^0 - \bar{D^0}$, for which the FOCUS experiment finds the value $(.0342 \pm .0139 \pm .0074) = \Gamma_{D^0}$ [15], larger than expected.
It would be also important to understand why data require a complete color screening, by demanding a vanishing value for ξ instead of $\frac{1}{3}$ expected within the perturbative QCD approach.

REFERENCES

1. F. Buccella, M. Forte, G. Miele and G. Ricciardi, *Z. Phys.* **C48**, 47 (1990).
2. F. Buccella, M. Lusignoli, G. Miele and A. Pugliese, *Z. Phys.* **C55**, 243 (1992).
3. F. Buccella, M. Lusignoli, G. Miele, A. Pugliese and P. Santorelli, *Phys. Rev.* **D51**, 3478 (1995).
4. F. Buccella, M. Lusignoli and A. Pugliese, *Phys. Lett.* **B379**, 249 (1996)
5. A.J. Buras, M. Jamin , M.E. Lautenbacher and P.H. Weisz *Nucl. Phys.* **B400**, 37 (1993)
   A.J. Buras, M. Jamin and M.E. Lautenbacher, *ibid.* 75;
   M. Ciuchini, E. Franco, G. Martinelli and L. Reina, *Nucl. Phys.* **B415**, 403 (1994)
   E. A. Paschos, hep-ph/9912230
   T. Hambye, G.O. Köhler, E.A. Paschos and P.H. Soldan, *Nucl. Phys.* **B564**, 391 (2000)
   S. Bertolini, J.O. Eeg and M. Fabbrichesi, hep-ph/0002234 and refs. therein.
6. G. Altarelli, F. Buccella and R. Gatto , *Phys. Lett.* **14**, 70 (1965)
7. E. Borchi, F. Buccella and R. Gatto , *Phys. Rev. Lett.* **14**, 507 (1965)
8. M. K. Gaillard and B. W. Lee , *Phys. Rev. Lett.* **33**, 108 (1974)
   G. Altarelli and L. Maiani , *Phys. Lett.* **B52**, 351 (1974)
9. M. A. Shifman, A. I. Vainshtein and V. I. Zacharov, *JETP Lett.* **22**, 55 (1975) and *Nucl. Phys.* **B120**, 316 (1977)
10. R. S. Chivukula, J. M. Flynn and H. Georgi , *Phys. Lett.* **B171**, 453 (1986).
11. N. Cabibbo and L. Maiani, *Phys. Lett.* **B73**, 418 (1978)
12. B. Guberina, S. Nussinov, R. D. Peccei and R. Ruckl, *Phys. Lett.* **B89**, 111 (1979)
13. M. Bauer and B. Stech, *Phys. Lett.* **B152**, 380 (1985)
    M. Bauer, B. Stech and M. Wirbel, *Z. Phys.* **C34**, 103 (1987)
14. A. N. Kamal, *Phys. Rev.* **D33**, 1344 (1986)
    M. Lusignoli and A. Pugliese, Rome preprint n.515 (1986)
15. J. M. Link et al. (FOCUS collaboration), *Phys. Lett.* **B485**, 62 (2000)

A quark loop model for heavy mesons

Aldo Deandrea

IPN, Université de Lyon I, 4 rue Enrico Fermi, F-69622 Villeurbanne Cedex, France

Abstract. I consider a model based on a quark–meson interaction Lagrangian. The transition amplitudes are evaluated by computing diagrams in which heavy and light mesons are attached to quark loops. The light chiral symmetry relations and the heavy quark spin-flavour symmetry dictated by the heavy quark effective theory are implemented. The model allows to compute the decay form factors and therefore can give predictions for the decay rates, the invariant mass spectra and the asymmetries.

INTRODUCTION

The increasing number of available data on heavy meson processes demands theoretical predictions for these processes to be compared with experiment. I consider a simple model, based on an effective constituent quark-meson Lagrangian containing both light and heavy degrees of freedom, constrained by the known symmetries of QCD in the limit $m_Q \to \infty$ and the light chiral symmetry relations. I write a Lagrangian at the meson-quark level [1]. This allows to deduce from a small number of parameters the heavy meson couplings and form factors, with a considerable reduction in the number of free parameters with respect to the Lagrangian written in terms of meson fields only [2].

The part of the quark-meson effective Lagrangian involving heavy and light quarks and heavy mesons is:

$$\begin{aligned} L_{h\ell} &= \bar{Q}_v iv\cdot\partial Q_v - \left(\bar{\chi}(\bar{H}+\bar{S}+i\bar{T}_\mu \frac{D^\mu}{\Lambda_\chi})Q_v + h.c.\right) \\ &+ \frac{1}{2G_3}\mathrm{Tr}[(\bar{H}+\bar{S})(H-S)] + \frac{1}{2G_4}\mathrm{Tr}[\bar{T}_\mu T^\mu] \end{aligned} \tag{1}$$

where Q_v is the effective heavy quark field, χ is the light quark field, G_3, G_4 are coupling constants and Λ_χ ($= 1$ GeV) is a dimensional parameter. The Lagrangian (1) is heavy spin and flavour symmetric. Note that the fields H and S have the same coupling constant. By putting these two coupling constants equal, one assumes that the effective quark-meson Lagrangian can be obtained from a four quark interaction of the NJL type [3].

The cut-off prescription is part of the dynamical information regarding QCD which is introduced in the model. The idea is to mimic the QCD behaviour in a simple and calculable way. In the infrared the model is not confining and its range of validity can not be extended below energies of the order of Λ_{QCD}. In practice one introduces an infrared cut-off μ, to take this into account.

CP602, *QCD@Work: International Workshop on Quantum Chromodynamics*
edited by P. Colangelo and G. Nardulli

Models related to the one discussed here, with different regularization prescriptions and different approaches are [4, 5]. The cut-off prescription used here is implemented via a proper time regularization. After continuation to the Euclidean it reads, for the light quark propagator:

$$\int d^4k_E \frac{1}{k_E^2+m^2} \to \int d^4k_E \int_{1/\mu^2}^{1/\Lambda^2} ds\, e^{-s(k_E^2+m^2)} \tag{2}$$

where μ and Λ are infrared and ultraviolet cut-offs.

The cut-off prescription is similar to the one used in [3], with $\Lambda = 1.25$ GeV; the numerical results are not strongly dependent on the value of Λ. The constituent mass m in the NJL models represents the order parameter discriminating between the phases of broken and unbroken chiral symmetry and can be fixed by solving a gap equation, which gives m as a function of the scale mass μ for given values of the other parameters. Here I take $m = 300$ MeV and $\mu = 300$ MeV.

HEAVY-TO-HEAVY FORM FACTORS

As an example of the quantities that can be analytically calculated in the model, one can examine the Isgur-Wise function ξ:

$$\langle D(v')|\bar{c}\gamma_\mu(1-\gamma_5)b|B(v)\rangle = \sqrt{M_B M_D}C_{cb}\,\xi(\omega)(v_\mu+v'_\mu) \tag{3}$$

where $\omega = v\cdot v'$ and C_{cb} contains logarithmic corrections depending on α_s; within the approximations used here, it can be put equal to 1. At leading order $\xi(1) = 1$. The same universal function ξ also parameterizes $B \to D^*$ semileptonic decay. One finds:

$$\xi(\omega) = Z_H\left[\frac{2}{1+\omega}I_3(\Delta_H) + \left(m+\frac{2\Delta_H}{1+\omega}\right)I_5(\Delta_H,\Delta_H,\omega)\right]\ . \tag{4}$$

where:

$$\begin{aligned} I_3(\Delta) &= -\frac{iN_c}{16\pi^4}\int^{\mathrm{reg}} \frac{d^4k}{(k^2-m^2)(v\cdot k+\Delta+i\varepsilon)} \\ &= \frac{N_c}{16\pi^{3/2}}\int_{1/\Lambda^2}^{1/\mu^2}\frac{ds}{s^{3/2}}\,e^{-s(m^2-\Delta^2)}\left(1+\mathrm{erf}(\Delta\sqrt{s})\right) \end{aligned} \tag{5}$$

$$\begin{aligned} I_5(\Delta_1,\Delta_2,\omega) &= \frac{iN_c}{16\pi^4}\int^{\mathrm{reg}} \frac{d^4k}{(k^2-m^2)(v\cdot k+\Delta_1+i\varepsilon)(v'\cdot k+\Delta_2+i\varepsilon)} \\ &= \int_0^1 dx \frac{1}{1+2x^2(1-\omega)+2x(\omega-1)}\times \\ &\quad \left[\frac{6}{16\pi^{3/2}}\int_{1/\Lambda^2}^{1/\mu^2} ds\,\sigma\, e^{-s(m^2-\sigma^2)}\, s^{-1/2}\left(1+\mathrm{erf}(\sigma\sqrt{s})\right)+\right. \\ &\quad \left.\frac{6}{16\pi^2}\int_{1/\Lambda^2}^{1/\mu^2} ds\, e^{-s(m^2-2\sigma^2)}\, s^{-1}\right] \end{aligned} \tag{6}$$

TABLE 1. Form factors and slopes. Δ_H in GeV.

Δ_H	$\xi(1)$	ρ^2_{IW}	$\tau_{1/2}(1)$	$\rho^2_{1/2}$	$\tau_{3/2}(1)$	$\rho^2_{3/2}$
0.3	1	0.72	0.08	0.8	0.48	1.4
0.4	1	0.87	0.09	1.1	0.56	2.3
0.5	1	1.14	0.09	2.7	0.67	3.0

In these equations

$$\Gamma(\alpha, x_0, x_1) = \int_{x_0}^{x_1} dt\, e^{-t}\, t^{\alpha-1} \tag{7}$$

is the generalized incomplete gamma function, erf is the error function and

$$\sigma(x, \Delta_1, \Delta_2, \omega) = \frac{\Delta_1\,(1-x) + \Delta_2 x}{\sqrt{1 + 2\,(\omega-1)\,x + 2\,(1-\omega)\,x^2}} \,. \tag{8}$$

One can compute in a similar way the form factors describing the semi-leptonic decays of a meson belonging to the fundamental negative parity multiplet H into the positive parity mesons in the S and T multiplets [1]. Examples of these decays are $B \to D^{**} l \nu$ where D^{**} can be either a S state or a T state. These decays are described by two form factors $\tau_{1/2}, \tau_{3/2}$ [6] which can be computed in the model by a loop calculation similar to the one used to obtain $\xi(\omega)$ [1, 8].

The numerical results for the form factors are in Table 1. The predictions for a few branching ratios calculated in the model are given in Table 2.

TABLE 2. Branching ratios (%) for semileptonic B decays. Theoretical predictions for three values of Δ_H and experimental results. Units of Δ_H in GeV.

Decay mode	$\Delta_H = 0.3$	$\Delta_H = 0.4$	$\Delta_H = 0.5$	Exp. [7]
$B^0 \to D \ell \nu$	3.0	2.7	2.2	2.10 ± 0.19
$B^0 \to D^* \ell \nu$	7.6	6.9	5.9	4.60 ± 0.27
$B^0 \to D_0 \ell \nu$	0.03	0.005	0.003	–
$B^0 \to D_1^{*\prime} \ell \nu$	0.03	0.008	0.0045	–
$B^0 \to D_1^* \ell \nu$	0.27	0.18	0.13	< 0.74
$B^0 \to D_2^* \ell \nu$	0.43	0.34	0.30	< 0.65

HEAVY-TO-LIGHT FORM FACTORS

The model allows to compute the B semileptonic decay form factor to π, ρ, etc. The form factors of B to a vector meson V consist of two kind of contributions. In the first one the current is directly attached to the loop of quarks. In the second, there is a intermediate state between the current and the $B\,V$ system [9]. For $B \to \pi$ form factors an extra contribution is also taken into account [10]. Results are in good agreement with available data. For $B \to \pi \ell \nu$ (using $V_{ub} = 0.0032$, $\tau_B = 1.5610^{-12}$ s):

$$B(\bar{B}^0 \to \pi^+ \ell \nu) = (1.1 \pm 0.5) \times 10^{-4} \,, \tag{9}$$

for $B \to \rho \ell \nu$:

$$B(\bar{B}^0 \to \rho^+ \ell \nu) = (2.5 \pm 0.8) \times 10^{-4}\,, \tag{10}$$

for $B \to a_1 \ell \nu$:

$$B(\bar{B}^0 \to a_1^+ \ell \nu) = (8.4 \pm 1.6) \times 10^{-4}\,. \tag{11}$$

In the limit of heavy mass for the initial meson and of large energy for the final one (LEET), the expressions of the form factors simplify and for $B \to V l \nu$, they reduce only to two independent functions [11]. The four-momentum of the heavy meson is written as $p = M_H v$ in terms of the mass and the velocity of the heavy meson. The four-momentum of the light vector meson is written as $p' = En$ where $E = v \cdot p'$ is the energy of the light meson and n is a four-vector defined by $v \cdot n = 1, n^2 = 0$. The relation between q^2 and E is:

$$q^2 = M_H^2 - 2M_H E + m_V^2 \tag{12}$$

The large energy limit is defined as :

$$\Lambda_{QCD}, m_V << M_H, E \tag{13}$$

keeping v and n fixed and m_V is the mass of the light vector meson. The relations between the form factors appearing in the LEET limit constitute a powerful theoretical cross–check of the formulas derived in the model. The result is as follows:

$$A_0(q^2) = \left(1 - \frac{m_V^2}{M_H E}\right) \zeta_{||}(M_H, E) + \frac{m_V}{M_H} \zeta_\perp(M_H, E) \tag{14}$$

$$A_1(q^2) = \frac{2E}{M_H + m_V} \zeta_\perp(M_H, E) \tag{15}$$

$$A_2(q^2) = \left(1 + \frac{m_V}{M_H}\right) \left[\zeta_\perp(M_H, E) - \frac{m_V}{E} \zeta_{||}(M_H, E)\right] \tag{16}$$

$$V(q^2) = \left(1 + \frac{m_V}{M_H}\right) \zeta_\perp(M_H, E). \tag{17}$$

The explicit expressions for $\zeta_{||}$ and $\zeta_\perp$ are [12]:

$$\zeta_{||}(M_H, E) = \frac{\sqrt{M_H Z_H}\, m_V^2}{2E f_V} \left[I_3\left(\frac{m_V}{2}\right) - I_3\left(-\frac{m_V}{2}\right) + 4\Delta_H m_V Z \right] \sim \frac{\sqrt{M_H}}{E} \tag{18}$$

$$\zeta_\perp(M_H, E) = \frac{\sqrt{M_H Z_H}\, m_V^2}{2E f_V} \left[I_3(\Delta_H) + m_V^2 Z\right] \sim \frac{\sqrt{M_H}}{E}\,, \tag{19}$$

where terms proportional to the constituent light quark mass m have been neglected. It is interesting to note that in LEET one can also relate the tensor form factor T_1, T_2 and T_3 to the semileptonic ones and to the $\zeta_\perp$ and $\zeta_{||}$ form factors of the LEET limit [11]:

$$T_1(q^2) = \zeta_\perp(M_H, E), \tag{20}$$

$$T_2(q^2) = \left(1 - \frac{q^2}{M_H^2 - m_V^2}\right)\zeta_\perp(M_H,E), \tag{21}$$

$$T_3(q^2) = \zeta_\perp(M_H,E) - \frac{m_V}{E}\left(1 - \frac{m_V^2}{M_H^2}\right)\zeta_{||}(M_H,E). \tag{22}$$

$\zeta_\perp$ and $\zeta_{||}$ obtained in this way agree with those of (18,19) [13]. Concerning the scaling properties of $\zeta_{||}$ and $\zeta_\perp$, the asymptotic E-dependence is not predicted by the large energy limit. As $E \sim M$ at $q^2 = 0$ the Feynman mechanism contribution to the form factors would indicate a $1/E^2$ behaviour rather than the $1/E$ found in the model. Note however that the E-dependence is not rigorously established in QCD.

CONCLUSIONS

Calculating directly from the QCD Lagrangian remains an extremely difficult task, in spite of the impressive success of lattice QCD calculations. A most promising approach is the one based on heavy meson effective Lagrangians, which incorporate the heavy quark symmetries and in addition the approximate chiral symmetry for light quarks. Although with increasing data such an approach is the best one beyond direct QCD calculations, a large number of parameters have to be fixed before obtaining predictions. An intermediate approach consists in using the effective Lagrangian at the level of mesons and constituent quarks plus few simple assumptions on the QCD dynamics. It allows to compute meson transition amplitudes by evaluating loops of heavy and light quarks. The model describes a number of essential features of heavy meson physics in a simple and compact way, in particular Isgur-Wise scaling in the heavy-to-heavy semileptonic decays and the large energy limit for the heavy-to-light ones.

ACKNOWLEDGMENTS

I would like to thank R. Gatto, G. Nardulli and A.D. Polosa for the fruitful scientific collaboration on which this work is based. Particular thanks go to the organizers of this workshop for inviting me and for the pleasant atmosphere in Martina Franca. Institut de Physique Nucléaire de Lyon (IPN Lyon) is UMR 5822.

REFERENCES

1. A. Deandrea, N. Di Bartolomeo, R. Gatto, G. Nardulli and A. D. Polosa, Phys. Rev. D **58**, 034004 (1998) [hep-ph/9802308]; for a review of the model see: A. D. Polosa, Riv. Nuovo Cim.**23** N11,1 (2000) [hep-ph/0004183].
2. R. Casalbuoni, A. Deandrea, N. Di Bartolomeo, R. Gatto, F. Feruglio and G. Nardulli, Phys. Lett. B **292** (1992) 371 [hep-ph/9209248]; Phys. Lett. B **294** (1992) 106 [hep-ph/9209247]; Phys. Lett. B **299** (1993) 139 [hep-ph/9211248]; Phys. Rept. **281** (1997) 145 [hep-ph/9605342].
3. D. Ebert, T. Feldmann, R. Friedrich and H. Reinhardt, Nucl. Phys. B **434**, 619 (1995) [hep-ph/9406220]; D. Ebert, T. Feldmann and H. Reinhardt, Phys. Lett. B **388**, 154 (1996) [hep-ph/9608223].
4. W. A. Bardeen and C. T. Hill, Phys. Rev. D **49** (1994) 409 [hep-ph/9304265].
5. B. Holdom and M. Sutherland, Phys. Rev. D **47** (1993) 5067 [hep-ph/9211226].
6. N. Isgur and M. B. Wise, Phys. Rev. D **43** (1991) 819.
7. D. E. Groom *et al.* [Particle Data Group Collaboration], Eur. Phys. J. C **15** (2000) 1.
8. A. Deandrea, R. Gatto, G. Nardulli and A. D. Polosa, JHEP **9902** (1999) 021 [hep-ph/9901266].
9. A. Deandrea, R. Gatto, G. Nardulli and A. D. Polosa, Phys. Rev. D **59** (1999) 074012 [hep-ph/9811259].
10. A. Deandrea, R. Gatto, G. Nardulli and A. D. Polosa, Phys. Rev. D **61** (2000) 017502 [hep-ph/9907225].
11. J. Charles, A. Le Yaouanc, L. Oliver, O. Pene and J. C. Raynal, Phys. Rev. D **60**, 014001 (1999) [hep-ph/9812358]; Phys. Lett. B **451**, 187 (1999) [hep-ph/9901378].
12. A. Deandrea, 34th Rencontres de Moriond, QCD and Hadronic interactions, Les Arcs, France, 20-27 Mar 1999, hep-ph/9905355.
13. A. Deandrea and A. D. Polosa, hep-ph/0105058, Phys. Rev. D in press.

Mass effects in the emission of gluons from heavy quarks at high energies

J. Fuster*, M.J. Costa* and P. Tortosa*

*IFIC Intitut de Física Corpuscular, Univ. València–CSIC, Apdo. 22085, E-46071 València, Spain

Abstract.
The effects in the emission of gluons due to the mass of the heavy quarks have clearly been observed by the experiments at LEP and SLC. The analyses of the data using theoretical corrections computed at Next-to-Leading Order have allowed to either test the flavour independence of the strong coupling constant with very high precision ($\sim$1%) or measure the b-quark mass at high energy, $\sqrt{s} \sim M_Z$. The results obtained by the various experiments, ALEPH, DELPHI, OPAL and SLD, agree well within errors. The systematic uncertainities limit present determinations though new methods and strategies are being developed to overcome the present bounds.

INTRODUCTION

The radiation of gluons by quarks obviously depends on the value of the strong coupling constant but also on the mass of the emitting quark. Massive quarks radiate less gluons than light quarks in a similar way to the suppression of bremsstrahlung for muons compared to electrons. At LEP/SLC energies, mass effects are usually negligible for inclusive obervables since they are suppressed by $m_b^2/m_Z^2 \leq 0.3\%$. But for more exclusive observables, like jet cross sections, the mass effects are considerably enhanced for small enough values of the jet resolucion parameter, y_c, as they go like $m_b^2/m_Z^2/y_c$.

Recently, Next-to-Leading Order (NLO) calculations taking mass effects into account have been performed for the ratios of b- over ℓ (uds)- events for some infrared-safe observables of event shape type and the 3-jet event rates [1, 2, 3, 4]. They can be used to either test α_s flavour independence, taking the b quark mass extracted from independent measurements performed at low energies, or, assuming α_s universality, measure the b quark mass at high energy, $\sqrt{s} \sim M_Z$.

In the Standard Model (SM) all fermion masses are free parameters and their origin, although linked to the spontaneous symetry breaking mechanism, is still an unresolved puzzle. Hence they are fundamental parameters whose values can only be extracted using experimental data. Masses of quarks are measured with less precision than those of charged leptons. This is because free quarks unlike charged leptons are not observed in nature. Therefore, one can only get indirect information about the values of the quark masses and their properties using dynamical relations which link these parameters to the observed quantities.

The perturbative pole mass, M_q, and the running mass, $m_q(\mu)$, are the two most commonly used quark mass definitions. The former is defined as the pole of the renormalized quark propagator. It is gauge and scheme independent but appears to be ambigous

CP602, *QCD@Work: International Workshop on Quantum Chromodynamics*
edited by P. Colangelo and G. Nardulli

because of non-perturbative renormalons [5]. The second definition corresponds to the renormalized mass in the $\overline{MS}$ scheme. It is a purely dynamical definition and depends on the energy scale of the process under study.

The measurements of $m_b(\mu)$ performed by the LEP and SLC collaborations [6, 7, 8, 9, 10, 11] at $\mu = M_Z$ and the value obtained at the production threshold, $\mu \sim M_\Upsilon/2$, extracted from the Υ bound states with QCD sum rules and QCD lattice calculations [12], constitute a set of two independent measurements at distant scales. This fact allows then to test the QCD or any other model prediction for the $m_b(\mu)$ quantity at the two different energy scales.

EXPERIMENTAL STRATEGY

The experimental method is based on measuring the observables for which the NLO calculations have been performed using samples of Z^0 hadronic decays. In order to select b-events different b-tagging techniques are used in each experiment, usually based on the lifetime impact parameter of the charged particles of the event. The working point is chosen so as to have small corrections and biases, having purities around 85%.

A sample of simulated events is used to correct the measured ratio for detector acceptance effects, kinematic biases introduced in the tagging procedure and the hadronization process. Different sources of systematic errors need to be estimated: imperfections of the detector and physics modelling, limited knowledge of the hadronization process and theoretical ambiguities coming from the uncalculated higher orders.

EXPERIMENTAL RESULTS

DELPHI

The DELPHI experiment at LEP considered the ratio of the normalized 3-jet rates for b to ℓ quarks, $R_3^{b\ell}$, with Durham and Cambridge as jet finder algorithms, and calculations using the running mass $m_b(M_Z)$ and the pole mass, M_b definitions. For Durham, the results obtained for $m_b(M_Z)$ and α_s^b/α_s^ℓ are [6]:

$$m_b(M_Z) = 2.67 \pm 0.25\ (\text{stat}) \pm 0.34\ (\text{had}) \pm 0.27\ (\text{th})\ \text{GeV}/\text{c}^2 \tag{1}$$

$$\alpha_s^b/\alpha_s^\ell = 1.007 \pm 0.010\ (\text{stat} + \text{had} + \text{th}) \tag{2}$$

and for Cambridge [13]:

$$m_b(M_Z) = 2.61 \pm 0.18\ (\text{stat})^{+0.45}_{-0.49}\ (\text{had}) \pm 0.18\ (\text{tag}) \pm 0.12\ (\text{th})\ \text{GeV}/\text{c}^2 \tag{3}$$

$$M_b = 4.2 \pm 0.6\ (\text{stat} + \text{had} + \text{th})\ \text{GeV}/\text{c}^2 \tag{4}$$

$$\alpha_s^b/\alpha_s^\ell = 1.005 \pm 0.012\ (\text{stat} + \text{had} + \text{th}) \tag{5}$$

The dominant contribution to the total error comes from the uncertainty on the hadronization model. Two fragmentation models, the string and cluster implemented

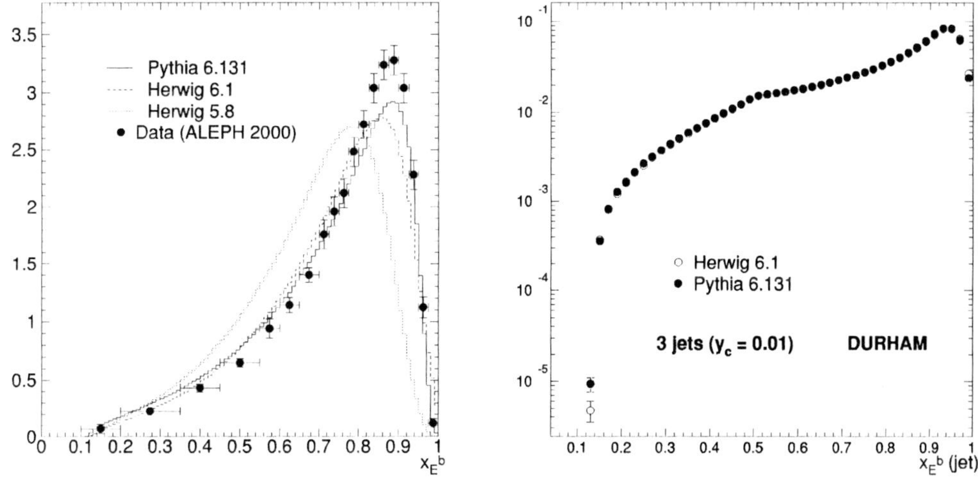

FIGURE 1. B-hadron and b-jet scaled energy distributions.

in Pythia and Herwig generators respectively, are used to correct the observable to parton level. Half of the difference between the two hadronization corrections is taken then as the hadronization model uncertainty, σ^{mod}.

In the present analysis, recent versions of the generators have been used, Pythia 6.131 and Herwig 6.1, both tuned to DELPHI data. One of the improvements achieved in Herwig is related with the b fragmentation, which has been made harder, being now closer to the Pythia prediction and to what has been measured (see figure 1 left).

The $R_3^{b\ell}$ observable and its fragmentation correction was found to depend on the B-hadron scaled energy, x_E^b. For a 3-jet event the b-jet scaled energy can defined $x_E^b(jet)$. This observable quantity is correlated with x_E^b (figure 1 right shows its distribution for the two generators). A region of $x_E^b(jet)$ where the two known fragmentation models behave similarly was selected and the hadronization model uncertainty was reduced from ± 0.006 to ± 0.001 for a $y_c = 0.02$ (see figure 2).

The way σ^{mod} is defined, it has two contributions: one coming from the difference between the fragmentation models, $\sigma^{frag-mod}$, and the other one from the difference between the decay models, $\sigma^{dec-mod}$, implemented in the two generators. $\sigma^{frag-mod}$ can be obtained by forcing all particles to be stable in the Monte Carlo. Figure 2 (right) indicates that $\sigma^{dec-mod}(y_c)$ is negligible for $y_c \geq 0.01$.

Two different tagging techniques have been used to select the b- and ℓ-quark initiated event samples and, in the case of the b sample, also to distinguish between the gluon and quark jets with a gluon jet purity $\sim 80\%$. The final result has been considered to be the mean value of the results obtained by the two methods.

For the Durham jet clustering algorithm, the $R_3^{b\ell}$ obtained at parton level as a function of y_c is compared in figure 3 (left) with the LO and NLO predictions written in

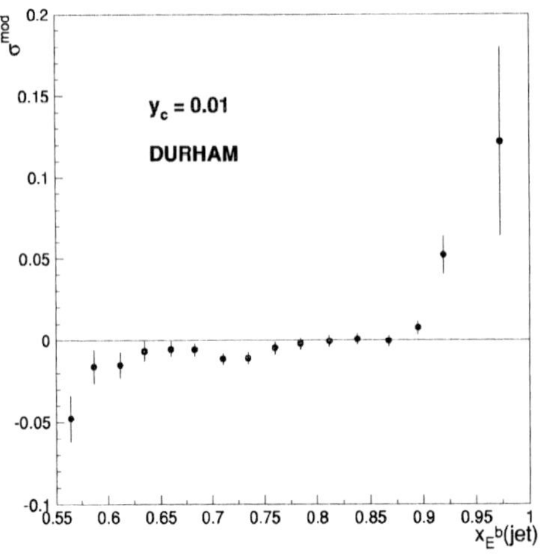

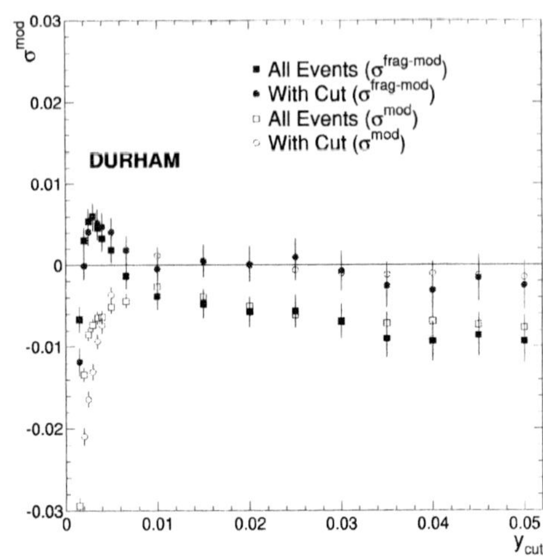

FIGURE 2. Hadronization model uncertainty as a function of the b-jet scaled energy (left) and as a function of the y_c (right).

terms of $m_b(M_Z) = 2.8\ GeV/c^2$ and in terms of $M_b = 4.6\ GeV/c^2$. This new result is compatible with the previous measurement (empty markers) but the error coming from the hadronization process has been reduced from ± 0.007 to ± 0.003 thanks to the cut performed on $x_E^b(jet)$.

SLD

The observable used by the SLD experiment at SLC and Brandenburg et al. is the $R_3^{b\ell}$ defined in [7]. Six different jet clustering algorithms have been used: four of the Jade family (E, E0, P and P0), Durham and Geneva.

NLO calculations with a $m_b(M_Z) = 3\ GeV/c^2$ have been used to obtain α_s^b/α_s^ℓ for each algorithm. A weighted average of the six results has been taken, considering 100% correlation among algorithms, to get the final values:

$$\begin{aligned} \alpha_s^b/\alpha_s^\ell &= 1.004 \pm 0.018\ (\text{stat})^{+0.026}_{-0.031}\ (\text{syst})^{+0.018}_{-0.029}\ (\text{th}) \\ \alpha_s^c/\alpha_s^\ell &= 1.036 \pm 0.043\ (\text{stat})^{+0.041}_{-0.045}\ (\text{syst})^{+0.020}_{-0.018}\ (\text{th}) \end{aligned} \tag{6}$$

where $R_3^{c\ell}$ has been measured to get α_s^c/α_s^ℓ.

Brandenburg et al. used the measured $R_3^{b\ell}$ to extract a value of $m_b(M_Z)$ for each algorithm. The obtained results show a large spread of values which could be due to the different higher order effects affecting each algorithm [14]. No strong argument to not use all six algorithms was found and a consistent value of $m_b(M_Z)$ was obtained

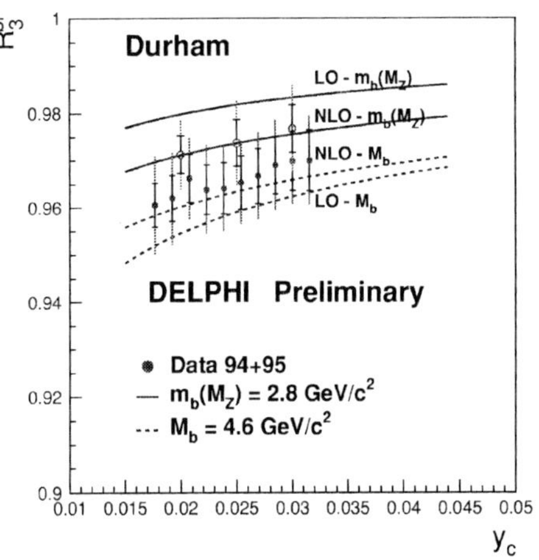

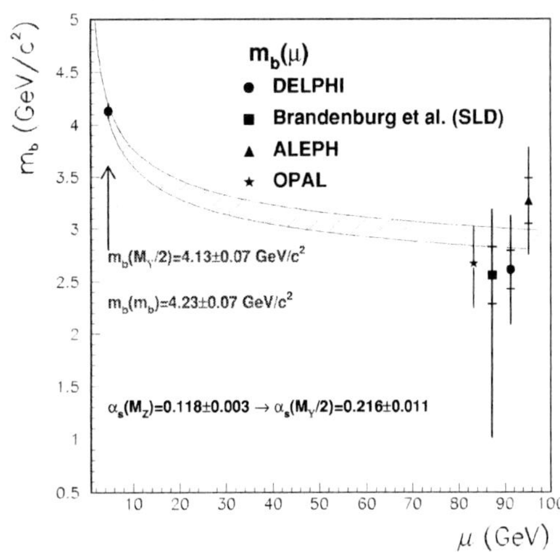

FIGURE 3. $R_3^{b\ell}$ with Durham (left). Measured value of $m_b(\mu)$ at different energy scales (right).

provided that there are additional sources of uncertainties of $\sim 0.5\ GeV/c^2$ [8]:

$$m_b(M_Z) = 2.56 \pm 0.27\ (\text{stat})^{+0.28}_{-0.38}\ (\text{syst})^{+0.49}_{-1.48}\ (\text{th})\ \text{GeV/c}^2 \tag{7}$$

ALEPH

The ALEPH experiment at LEP has taken as observables the 3-jet rate defined with the Durham algorithm at a $y_c = 0.02$ and the first and second moments of some event shape variables (T, C, y_3, B_T and B_W). The ratios from b tagged events to an inclusive sample were determined for the obtained observables.

The requirement to have small NLO and hadronization corrections compared with the mass effect, leaves only the 3-jet rate and the first moment of y_3 as observables. The smallest total uncertainty was obtained with y_3, which gave the results of [10]:

$$m_b(M_Z) = 3.27 \pm 0.22\ (\text{stat}) \pm 0.22\ (\text{syst}) \pm 0.38\ (\text{had}) \pm 0.16\ (\text{th})\ \text{GeV/c}^2 \tag{8}$$

$$M_b = 4.7 \pm 0.7\ (\text{stat} + \text{had} + \text{th})\ \text{GeV/c}^2 \tag{9}$$

$$\alpha_s^b/\alpha_s^\ell = 0.997 \pm 0.004\ (\text{stat}) \pm 0.004\ (\text{syst}) \pm 0.007\ (\text{had}) \pm 0.003\ (\text{th}) \tag{10}$$

OPAL

The OPAL experiment at LEP tested α_s universality by measuring some event shape observables (T, M_H, B_W, y_3 and C) for ℓ, c and b tagged samples and performing a

3 parameter fit to $O(\alpha_s^2)$ massive calculations to get simultaneosly α_s^{uds}, $\alpha_s^c/\alpha_s^{uds}$ and $\alpha_s^b/\alpha_s^{uds}$ [9]:

$$\alpha_s^c/\alpha_s^\ell = 0.997 \pm 0.038 \text{ (stat)} \pm 0.030 \text{ (syst)} + 0.012 \text{ (th)} \tag{11}$$
$$\alpha_s^b/\alpha_s^\ell = 0.993 \pm 0.008 \text{ (stat)} \pm 0.006 \text{ (syst)} \pm 0.011 \text{ (th)}$$

A very recent result has been obtained for $m_b(M_Z)$ by measuring the ratio of the normalized 3-jet rate for b to $udsc$ tagged events. The same six jet finder algorithms of section plus the Jade algorithm were used and a value for $m_b(M_Z)$ was obtained for each of them. By minimizing the χ^2 of the seven correlated determinations a single result was found to be [11]:

$$m_b(M_Z) = 2.67 \pm 0.03 \text{ (stat)}^{+0.29}_{-0.37} \text{ (syst)} \pm 0.19 \text{ (th) GeV/c}^2 \tag{12}$$

CONCLUSIONS

NLO calculations taking the mass effects into account are needed to describe the b mass effects observed at LEP and SLC energies and have allowed to test flavour independence of α_s at 1% precision and to measure the b quark mass at the M_Z scale. The four results obtained so far for $m_b(M_Z)$ are compatible with each other and are in good agreement with determinations at lower scales extrapolated to M_Z by using the evolution predicted by QCD (see figure 3 right). Some observables used in the analyses are affected by large theoretical corrections compared to the mass effect itself. In view of that, some analyses (DELPHI, ALEPH) do not include the results derived from these observables. The dominant error on all performed measurements is due to the uncertainty on the fragmentation process though new studies are being developed in order to reduce its impact. In fact, big improvements have been achieved in the Monte Carlo generators to reproduce the observed mass effects [15].

REFERENCES

1. G.Rodrigo et al., Phys. Rev. Lett. **79** (1997) 193.
2. W.Bernreuther et al., Phys. Rev. Lett. **79** (1997) 189.
3. P.Nason et al., Phys. Lett. **B407** (1997) 57.
4. M. Bilenkii et al., Phys. Rev. **D60** (1999) 114006.
5. G.Rodrigo, Nucl. Phys. Proc. Suppl. **54A** (1997) 60-66.
6. DELPHI Coll., P.Abreu et al., Phys. Lett. **B418** (1998) 430-442.
7. SLD Coll., K. Abe et al., Phys. Rev. **D59** (1999) 12002.
8. A. Brandenburg et al., Phys. lett. **B468** (1999) 168.
9. OPAL Coll., G. Abbiendi et al., Eur. Phys. Jour. **C11** (1999) 643.
10. ALEPH Coll., R. Barate et al., Eur. Phys. Jour. **C18** (2000) 1.
11. OPAL Coll., G.Abbiendi et al., CERN-EP-2001-034, submitted to Eur. Phys. Jour.
12. M. Jamin and A. Pich, Nucl. Phys. **B507**, 334 (1997);
    V. Giménez et al., JHEP 0003:018 (2000).
13. S. Cabrera, PhD Thesis (1999) (unpublished);
14. G. Rodrigo et al., Nucl. Phys. **B554** (1999), 257-297.
15. A. Ballestrero et al., Report of the QCD Working Group, hep-ph/0006259

Large order behavior in perturbation theory of the pole mass and the singlet static potential

Antonio Pineda

Institut für Theoretische Teilchenphysik, Universität Karlsruhe, D-76128 Karlsruhe, Germany

Abstract. We discuss upon recent progress in our knowledge of the large order behavior in perturbation theory of the pole mass and the singlet static potential. We also discuss about the renormalon subtracted scheme, a matching scheme between QCD and any effective field theory with heavy quarks where, besides the usual perturbative matching, the first renormalon in the Borel plane of the pole mass is subtracted.

MASS NORMALIZATION CONSTANT

In this paper, we review some results obtained in Ref. [1].

The on-shell (OS) or pole mass can be related to the $\overline{\rm MS}$ renormalized mass by the series

$$m_{\rm OS} = m_{\overline{\rm MS}} + \sum_{n=0}^{\infty} r_n \alpha_s^{n+1}, \tag{1}$$

where the normalization point $\nu = m_{\overline{\rm MS}}$ is understood for $m_{\overline{\rm MS}}$ and the first three coefficients r_0, r_1 and r_2 are known [2] ($\alpha_s = \alpha_s^{(n_l)}(\nu)$, where n_l is the number of light fermions). The pole mass is also known to be IR finite and scheme-independent at any finite order in α_s [3]. We then define the Borel transform

$$m_{\rm OS} = m_{\overline{\rm MS}} + \int\limits_0^\infty {\rm d}t\, e^{-t/\alpha_s} B[m_{\rm OS}](t), \qquad B[m_{\rm OS}](t) \equiv \sum_{n=0}^{\infty} r_n \frac{t^n}{n!}. \tag{2}$$

The behavior of the perturbative expansion of Eq. (1) at large orders is dictated by the closest singularity to the origin of its Borel transform, which happens to be located at $t = 2\pi/\beta_0$, where we define

$$\nu \frac{d\alpha_s}{d\nu} = -2\alpha_s \left\{ \beta_0 \frac{\alpha_s}{4\pi} + \beta_1 \left(\frac{\alpha_s}{4\pi}\right)^2 + \cdots \right\}.$$

Being more precise, the behavior of the Borel transform near the closest singularity at the origin reads (we define $u = \frac{\beta_0 t}{4\pi}$)

$$B[m_{\rm OS}](t(u)) = N_m \nu \frac{1}{(1-2u)^{1+b}} \left(1 + c_1(1-2u) + c_2(1-2u)^2 + \cdots\right) + (\text{analytic term}), \tag{3}$$

CP602, *QCD@Work: International Workshop on Quantum Chromodynamics*
edited by P. Colangelo and G. Nardulli

where by *analytic term*, we mean a piece that we expect it to be analytic up to the next renormalon ($u = 1$). This dictates the behavior of the perturbative expansion at large orders to be

$$r_n \overset{n\to\infty}{=} N_m \nu \left(\frac{\beta_0}{2\pi}\right)^n \frac{\Gamma(n+1+b)}{\Gamma(1+b)} \left(1 + \frac{b}{(n+b)} c_1 + \frac{b(b-1)}{(n+b)(n+b-1)} c_2 + \cdots\right). \quad (4)$$

The different b, c_1, c_2, etc ... can be obtained from the procedure used in [4] (see [4, 1] for the explicit expressions). We then use the idea of [5] and define the new function

$$\begin{aligned} D_m(u) &= \sum_{n=0}^{\infty} D_m^{(n)} u^n = (1-2u)^{1+b} B[m_{OS}](t(u)) \quad (5) \\ &= N_m \nu \left(1 + c_1(1-2u) + c_2(1-2u)^2 + \cdots\right) + (1-2u)^{1+b}(\text{analytic term}). \end{aligned}$$

This function is singular but bounded at the first IR renormalon. Therefore, we can expect to obtain an approximate determination of N_m if we know the first coefficients of the series in u and by using

$$N_m \nu = D_m(u = 1/2). \quad (6)$$

The first three coefficients: $D_m^{(0)}$, $D_m^{(1)}$ and $D_m^{(2)}$ are known in our case. In order the calculation to make sense, we choose $\nu \sim m$. For the specific choice $\nu = m$, we obtain (up to $O(u^3)$ with $u = 1/2$)

$$\begin{aligned} N_m &= 0.424413 + 0.137858 + 0.0127029 = 0.574974 \quad (n_f = 3) \quad (7) \\ &= 0.424413 + 0.127505 + 0.000360952 = 0.552279 \quad (n_f = 4) \\ &= 0.424413 + 0.119930 - 0.0207998 = 0.523543 \quad (n_f = 5). \end{aligned}$$

The convergence is surprisingly good. The scale dependence is also quite mild (see [1]).

By using Eq. (4), we can now go backwards and give some estimates for the r_n. They are displayed in Table 1. We can see that they go closer to the exact values of r_n when increasing n. This makes us feel confident that we are near the asymptotic regime dominated by the first IR renormalon and that for higher n our predictions will become an accurate estimate of the exact values. In fact, they are quite compatible with the results obtained by other methods like the large β_0 approximation (see Table 1).

We can now try to see how the large β_0 approximation works in the determination of N_m. In order to do so, we study the one chain approximation from which we obtain the value [6]

$$N_m^{(\text{large } \beta_0)} = \frac{C_f}{\pi} e^{\frac{5}{6}} = 0.976564. \quad (8)$$

By comparing with Eq. (7), we can see that it does not provide an accurate determination of N_m. This may seem to be in contradiction with the accurate values that the large β_0 approximation provides for the r_n (starting at $n = 2$) in Table 1. Lacking of any physical explanation for this fact, it may just be considered to be a numerical accident. In fact, the agreement between our determination and the large β_0 results does not hold at very high orders in the perturbative expansion, whereas we believe, on physical grounds since our approach incorporates the exact nature of the renormalon, that our determination should go closer to the exact result at high orders in perturbation theory. Nevertheless, the large β_0 approximation remains accurate up to relative high orders.

TABLE 1. Values of r_n for $\nu = m_{\overline{\rm MS}}$. Either the exact result, the estimate using Eq. (4), or the estimate using the large β_0 approximation [7].

$\tilde{r}_n = r_n/m_{\overline{\rm MS}}$	$\tilde{r}_0$	$\tilde{r}_1$	$\tilde{r}_2$	$\tilde{r}_3$	$\tilde{r}_4$
exact ($n_f = 3$)	0.424413	1.04556	3.75086	—	—
Eq. (4) ($n_f = 3$)	0.617148	0.977493	3.76832	18.6697	118.441
large β_0 ($n_f = 3$)	0.424413	1.42442	3.83641	17.1286	97.5872
exact ($n_f = 4$)	0.424413	0.940051	3.03854	—	—
Eq. (4) ($n_f = 4$)	0.645181	0.848362	3.03913	13.8151	80.5776
large β_0 ($n_f = 4$)	0.424413	1.31891	3.28911	13.5972	71.7295
exact ($n_f = 5$)	0.424413	0.834538	2.36832	—	—
Eq. (4) ($n_f = 5$)	0.706913	0.713994	2.36440	9.73117	51.5952
large β_0 ($n_f = 5$)	0.424413	1.21339	2.78390	10.5880	51.3865

STATIC SINGLET POTENTIAL NORMALIZATION CONSTANT

One can think of playing the same game with the singlet static potential in the situation where $\Lambda_{\rm QCD} \ll 1/r$. Its perturbative expansion reads

$$V_s^{(0)}(r;\nu_{us}) = \sum_{n=0}^{\infty} V_{s,n}^{(0)} \alpha_s^{n+1}. \tag{9}$$

The first three coefficients $V_{s,0}^{(0)}$, $V_{s,1}^{(0)}$ and $V_{s,2}^{(0)}$ are known [8]. At higher orders in perturbation theory the log dependence on the IR cutoff ν_{us} appears [9]. Nevertheless, these logs are not associated to the first IR renormalon (see [1]), so we will not consider them further in this section. We now use the observation that the first IR renormalon of the singlet static potential cancels with the renormalon of (twice) the pole mass. We can then read the asymptotic behavior of the static potential from the one of the pole mass and work analogously to the previous section. We define the Borel transform

$$V_s^{(0)} = \int_0^{\infty} dt\, e^{-t/\alpha_s} B[V_s^{(0)}](t), \qquad B[V_s^{(0)}](t) \equiv \sum_{n=0}^{\infty} V_{s,n}^{(0)} \frac{t^n}{n!}. \tag{10}$$

The closest singularity to the origen is located at $t = 2\pi/\beta_0$. This dictates the behavior of the perturbative expansion at large orders to be

$$V_{s,n}^{(0)} \overset{n\to\infty}{=} N_V \nu \left(\frac{\beta_0}{2\pi}\right)^n \frac{\Gamma(n+1+b)}{\Gamma(1+b)} \left(1 + \frac{b}{(n+b)} c_1 + \frac{b(b-1)}{(n+b)(n+b-1)} c_2 + \cdots\right), \tag{11}$$

and the Borel transform near the singularity reads

$$B[V_s^{(0)}](t(u)) = N_V \nu \frac{1}{(1-2u)^{1+b}} \left(1 + c_1(1-2u) + c_2(1-2u)^2 + \cdots\right) + (\text{analytic term}). \tag{12}$$

In this case, by *analytic term*, we mean an analytic function up to the next IR renormalon at $u = 3/2$.

As in the previous section, we define the new function

$$
\begin{aligned}
D_V(u) &= \sum_{n=0}^{\infty} D_V^{(n)} u^n = (1-2u)^{1+b} B[V_s^{(0)}](t(u)) \qquad (13)\\
&= N_V \nu \left(1 + c_1(1-2u) + c_2(1-2u)^2 + \cdots\right) + (1-2u)^{1+b}(\text{analytic term})
\end{aligned}
$$

and try to obtain an approximate determination of N_V by using the first three (known) coefficients of this series. By a discussion analogous to the one in the previous section, we fix $\nu = 1/r$. We obtain (up to $O(u^3)$ with $u = 1/2$)

$$
\begin{aligned}
N_V &= -1.33333 + 0.571943 - 0.345222 = -1.10661 \quad (n_f = 3) \qquad (14)\\
&= -1.33333 + 0.585401 - 0.329356 = -1.07729 \quad (n_f = 4)\\
&= -1.33333 + 0.586817 - 0.295238 = -1.04175 \quad (n_f = 5)\,.
\end{aligned}
$$

The convergence is not as good as in the previous section. Nevertheless, it is quite acceptable and, in this case, apparently, we have a sign alternating series. In fact, the scale dependence is quite mild (see [1]). Overall, up to small differences, the same picture than for N_m applies.

So far we have not made use of the fact that $2N_m + N_V = 0$. We use this equality as a check of the reliability of our calculation. We can see that the cancellation is quite dramatic. We obtain

$$
2\frac{2N_m + N_V}{2N_m - N_V} = \begin{cases} 0.038 & , n_f = 3 \\ 0.025 & , n_f = 4 \\ 0.005 & , n_f = 5. \end{cases}
$$

We can now obtain estimates for $V_{s,n}^{(0)}$ by using Eq. (11). They are displayed in Table 2. Note that in Table 2 no input from the static potential has been used since even N_V have been fixed by using the equality $2N_m = -N_V$. We can see that the exact results are reproduced fairly well (the same discussion than for the r_n determination applies). This makes us feel confident that we are near the asymptotic regime dominated by the first IR renormalon and that for higher n our predictions will become an accurate estimate of the exact results. The comparison with the values obtained with the large β_0 approximation would go (roughly) along the same lines than for the mass case, although the large β_0 results seem to be less accurate in this case (see Table 2).

In order to avoid large corrections from terms depending on ν_{us}, the predictions should be understood with $\nu_{us} = 1/r$.

RENORMALON SUBTRACTED SCHEME

In effective theories with heavy quarks, the inverse of the heavy quark mass becomes one of the expansion parameters (and matching coefficients). A natural choice in the past (within the infinitely many possible definitions of the mass) has been the pole mass because it is the natural definition in OS processes where the particles finally measured in

TABLE 2. Values of $V_{s,n}^{(0)}$ with $\nu = 1/r$. Either the exact result (when available), the estimate using Eq. (11), or the estimate using the large β_0 approximation [10].

$\tilde{V}_{s,n}^{(0)} = rV_{s,n}^{(0)}$	$\tilde{V}_{s,0}^{(0)}$	$\tilde{V}_{s,1}^{(0)}$	$\tilde{V}_{s,2}^{(0)}$	$\tilde{V}_{s,3}^{(0)}$	$\tilde{V}_{s,4}^{(0)}$
exact ($n_f = 3$)	-1.33333	-1.84512	-7.28304	—	—
Eq. (11) ($n_f = 3$)	-1.23430	-1.95499	-7.53665	-37.3395	-236.882
large β_0 ($n_f = 3$)	-1.33333	-2.69395	-7.69303	-34.0562	—
exact ($n_f = 4$)	-1.33333	-1.64557	-5.94978	—	—
Eq. (11) ($n_f = 4$)	-1.29036	-1.69672	-6.07826	-27.6301	-161.155
large β_0 ($n_f = 4$)	-1.33333	-2.49440	-6.59553	-27.0349	—
exact ($n_f = 5$)	-1.33333	-1.44602	-4.70095	—	—
Eq. (11) ($n_f = 5$)	-1.41383	-1.42799	-4.72881	-19.4623	-103.190
large β_0 ($n_f = 5$)	-1.33333	-2.29485	-5.58246	-21.0518	—

the detectors correspond to the fields in the Lagrangian (as in QED). Unfortunately, this is not the case in QCD and one reflection of this fact is that the pole mass suffers from renormalon singularities. Moreover, these renormalon singularities lie close together to the origin and perturbative calculations have gone very far for systems with heavy quarks. At the practical level, this has reflected in the worsening of the perturbative expansion in processes where the pole mass was used as an expansion parameter. It is then natural to try to define a new expansion parameter replacing the pole mass but still being an adequate definition for threshold problems. This idea is not new and has already been pursued in the literature, where several definitions have arisen [11]. We can not resist the tentation of trying our own definition. We believe that, having a different systematics than the other definitions, it could further help to estimate the errors in the more recent determinations of the $\overline{\text{MS}}$ quark mass. Our definition, as the definitions above, try to cancel the bad perturbative behavior associated to the renormalon. On the other hand, we would like to understand this problem within an effective field theory perspective. From this point of view what one is seeing is that the coefficients multiplying the (small) expansion parameters in the effective theory calculation are not of natural size (of $O(1)$). The natural answer to this problem is that we are not properly separating scales in our effective theory and some effects from small scales are incorporated in the matching coefficients. These small scales are dynamically generated in n-loop calculations (n being large) and are of $O(me^{-n})$ (we are having in mind a large β_0 evaluation) producing the bad (renormalon associated) perturbative behavior. In order to overcome this problem, we may think of doing the Borel transform. In that case, the renormalon singularities correspond to the non-analytic terms in $1-2u$. These terms also exist in the effective theory. Therefore, our procedure will be to subtract the pure renormalon contribution in the new mass definition, which we will call renormalon subtracted (RS) mass, $m_{\rm RS}$. We define the Borel transform of $m_{\rm RS}$ as follows

$$B[m_{\rm RS}] \equiv B[m_{\rm OS}] - N_m \nu_f \frac{1}{(1-2u)^{1+b}} \left(1 + c_1(1-2u) + c_2(1-2u)^2 + \cdots\right), \qquad (15)$$

where ν_f could be understood as a factorization scale between QCD and NRQCD (or HQET) and, at this stage, should be smaller than m. The expression for $m_{\rm RS}$ reads

$$m_{\rm RS}(\nu_f) = m_{\rm OS} - \sum_{n=0}^{\infty} N_m \nu_f \left(\frac{\beta_0}{2\pi}\right)^n \alpha_s^{n+1}(\nu_f) \sum_{k=0}^{\infty} c_k \frac{\Gamma(n+1+b-k)}{\Gamma(1+b-k)}, \qquad (16)$$

where $c_0 = 1$. We expect that with this renormalon free definition the coefficients multiplying the expansion parameters in the effective theory calculation will have a natural size and also the coefficients multiplying the powers of α_s in the perturbative expansion relating $m_{\rm RS}$ with $m_{\overline{\rm MS}}$. Therefore, we do not loose accuracy if we first obtain $m_{\rm RS}$ and later on we use the perturbative relation between $m_{\rm RS}$ and $m_{\overline{\rm MS}}$ in order to obtain the latter. Nevertheless, since we will work order by order in α_s in the relation between $m_{\rm RS}$ and $m_{\overline{\rm MS}}$, it is important to expand everything in terms of α_s, in particular $\alpha_s(\nu_f)$, in order to achieve the renormalon cancellation order by order in α_s. Then, the perturbative expansion in terms of the $\overline{\rm MS}$ mass reads

$$m_{\rm RS}(\nu_f) = m_{\overline{\rm MS}} + \sum_{n=0}^{\infty} r_n^{\rm RS} \alpha_s^{n+1}, \qquad (17)$$

where $r_n^{\rm RS} = r_n^{\rm RS}(m_{\overline{\rm MS}}, \nu, \nu_f)$. These $r_n^{\rm RS}$ are the ones expected to be of natural size (or at least not to be artificially enlarged by the first IR renormalon).

In Ref. [1], we have applied this scheme to potential NRQCD and HQET. For the former, by using the $\Upsilon(1S)$ mass, we have obtained a determination of the $\overline{\rm MS}$ bottom quark mass. For the latter, we have obtained a value of the charm mass by using the difference between the D and B meson mass. In both cases the convergence is significantly improved if compared with the analogous OS evaluations.

REFERENCES

1. A. Pineda, **JHEP06**, 022 (2001).
2. N. Gray, D.J. Broadhurst, W. Grafe and K. Schilcher, Z. Phys. **C48**, 673 (1990); K. Melnikov and T. van Ritbergen, Phys. Lett. **B482**, 99 (2000); K.G. Chetyrkin and M. Steinhauser, Nucl. Phys. **B573** 617 (2000).
3. A.S. Kronfeld, Phys. Rev. **D58**, 051501 (1998).
4. M. Beneke, Phys. Lett. **B344**, 341 (1995).
5. T. Lee, Phys. Rev. **D56**, 1091 (1997); Phys. Lett. **B462**, 1 (1999).
6. M. Beneke and V. M. Braun, Nucl. Phys. **B426**, 301 (1994).
7. M. Beneke and V.M. Braun, Phys. Lett. **B348**, 513 (1995); P. Ball, M. Beneke and V.M. Braun, Nucl. Phys. **B452**, 563 (1995).
8. W. Fischler, Nucl. Phys. **B129**, 157 (1977); Y. Schröder, Phys. Lett. **B447**, 321 (1999); M. Peter, Phys. Rev. Lett. **78**, 602 (1997).
9. N. Brambilla, A. Pineda, J. Soto and A. Vairo, Phys. Rev. **D60**, 091502 (1999); T. Appelquist, M. Dine and I.J. Muzinich, Phys. Rev. **D17**, 2074 (1978).
10. Y. Kiyo and Y. Sumino, Phys. Lett. **B496**, 83 (2000); A.H. Hoang, hep-ph/0008102.
11. I. Bigi, M. Shifman, N. Uraltsev and A. Vainshtein, Phys. Rev. **D50**, 2234 (1994); M. Beneke, Phys. Lett. **B434**, 115 (1998); A.H. Hoang, Z. Ligeti and A.V. Manohar, Phys. Rev. Lett. **82**, 277 (1999); O. Yakovlev and S. Groote, Phys. Rev. **D63**, 074012 (2001).

Renormalization of HQET at Three Loops

A. G. Grozin

Budker Institute of Nuclear Physics, Novosibirsk 630090, Russia

Abstract. Three-loop propagator diagrams in HQET can be reduced, using integration by parts, to 8 basis integrals: 5 trivial ones, two expressible via ${}_3F_2$, and one only known up to ε^0. Calculation of the heavy-quark propagator in HQET is considerably simplified by the non-abelian exponentiation theorem.

THREE-LOOP PROPAGATOR DIAGRAMS IN HQET

Heavy Quark Effective Theory (HQET) is an effective field theory approximating QCD for problems with a single heavy quark when characteristic momenta are much less than its mass m, see the textbook [1]. A method of calculation of two-loop propagator diagrams in HQET based on integration by parts [2] has been constructed in [3]. It has been extended to three loops in [4]. Methods of perturbative calculations in HQET are reviewed in [5].

All generic topologies of three-loop propagator diagrams in HQET are presented in Fig. 1. The corresponding Feynman integrals depend on indices of the lines — the degrees of the corresponding denominators. They can also contain powers of numerators which cannot be expressed via the denominators. If index of a line becomes zero, this line shrinks to a point. In some cases, the resulting diagram contains lower-loop propagator insertions, and is easy to calculate. An algorithm which reduces any three-loop propagator integral in HQET to a linear combination of 8 basis integrals (with coefficients being rational functions of space-time dimension d) has been constructed and implemented in [4]. The basis integrals are presented in Fig. 2; indices of all lines are equal to 1 here.

The first 5 basis integrals are trivial:

$$B_1 = I_1^3\,, \quad B_2 = I_1 I_2\,, \quad B_3 = I_3\,, \quad B_4 = I_3 \frac{I_2}{I_1^2}\,, \quad B_5 = I_3 \frac{G_2}{G_1^2}\,, \tag{1}$$

where

$$\begin{aligned} I_n &= \frac{\Gamma(1+2n\varepsilon)\Gamma^n(1-\varepsilon)}{(1-n(d-2))_{2n}}\,, \\ G_n &= \frac{1}{\left(n+1-n\frac{d}{2}\right)_n \left((n+1)\frac{d}{2}-2n-1\right)_n} \frac{\Gamma(1+n\varepsilon)\Gamma^{n+1}(1-\varepsilon)}{\Gamma(1-(n+1)\varepsilon)} \end{aligned} \tag{2}$$

are the n-loop sunset HQET and massless integrals.

CP602, *QCD@Work: International Workshop on Quantum Chromodynamics*
edited by P. Colangelo and G. Nardulli

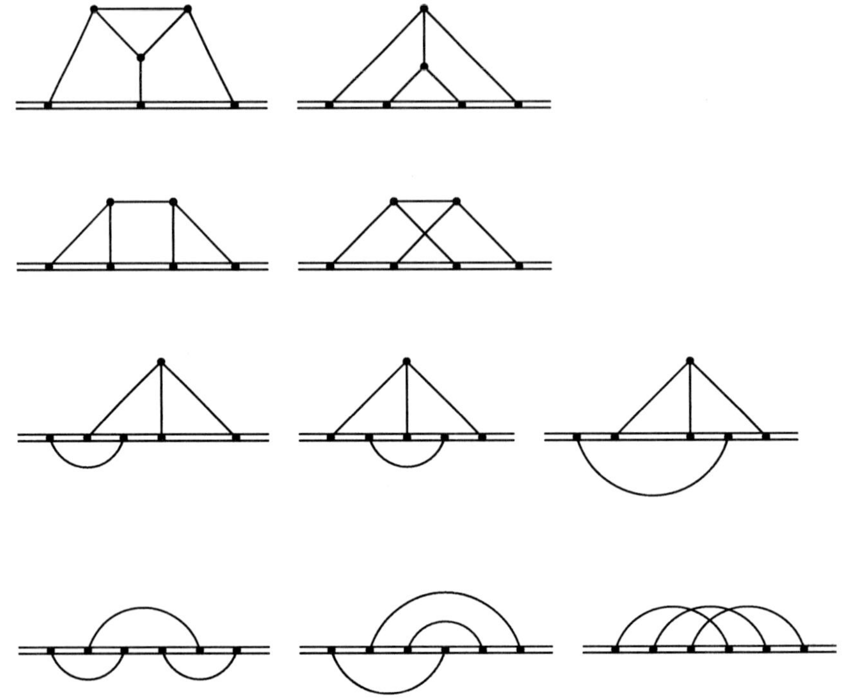

FIGURE 1. Topologies of three-loop propagator diagrams in HQET

The next two basis integrals are

$$B_6 = G_1 I(1,1,1,1,\varepsilon)\,, \quad B_7 = I_1 J(1,1,-1+2\varepsilon,1,1)\,. \tag{3}$$

We use recurrence relations for the two-loop integrals I and J to re-express them via convergent integrals (the first two diagrams in Fig. 3) $I(1,1,1,1,2+\varepsilon)$ and $J(1,1,2+2\varepsilon,1,1)$. At a non-integer n, the integral $I(1,1,1,1,n)$ has been calculated in [6], and $J(1,1,n,1,1)$ — in [4]. In particular,

$$\begin{aligned}
I(1,1,1,1,2+\varepsilon) &= \frac{8(3d-13)}{(d-4)(d-5)(d-6)(d-8)(2d-11)} \frac{\Gamma(1+6\varepsilon)\Gamma(1-2\varepsilon)\Gamma(1-\varepsilon)}{\Gamma(1+\varepsilon)} \\
&\times \left[{}_3F_2\left(\left. \begin{array}{c} 1,2-2\varepsilon,3+4\varepsilon \\ 3+\varepsilon,4+4\varepsilon \end{array} \right| 1 \right) \right. \\
&\left. - \frac{(d-6)(d-8)(2d-11)}{12(d-3)(d-4)(3d-13)} \frac{\Gamma^2(1+3\varepsilon)\Gamma(1-3\varepsilon)\Gamma(1+\varepsilon)}{\Gamma(1+6\varepsilon)\Gamma(1-2\varepsilon)} \right], \\
J(1,1,2+2\varepsilon,1,1) &= \frac{1}{3(d-4)(d-5)(d-6)(2d-9)} \frac{\Gamma(1+6\varepsilon)\Gamma^2(1-\varepsilon)}{\Gamma(1+2\varepsilon)} \\
&\times {}_3F_2\left(\left. \begin{array}{c} 1,2-2\varepsilon,1+4\varepsilon \\ 3+2\varepsilon,2+4\varepsilon \end{array} \right| 1 \right).
\end{aligned} \tag{4}$$

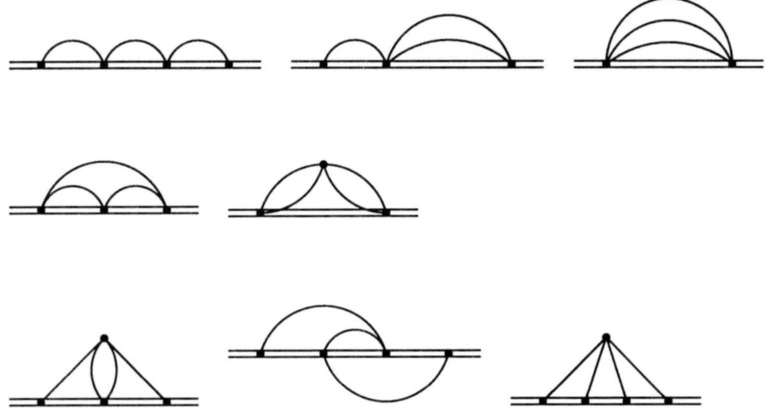

FIGURE 2. Basis integrals

Using the methods of [7], it is not difficult to obtain the expansions

$$
\begin{aligned}
&{}_3F_2\left(\begin{matrix}1,2-2\varepsilon,3+4\varepsilon\\3+\varepsilon,4+4\varepsilon\end{matrix}\middle|1\right)=3+\left(18\zeta_2-\frac{73}{2}\right)\varepsilon+3(-48\zeta_3-7\zeta_2+75)\varepsilon^2\\
&\quad+3(216\zeta_4+56\zeta_3+115\zeta_2-506)\varepsilon^3\\
&\quad+12(-375\zeta_5+90\zeta_2\zeta_3-63\zeta_4-230\zeta_3-135\zeta_2+789)\varepsilon^4+\cdots\\
&{}_3F_2\left(\begin{matrix}1,2-2\varepsilon,1+4\varepsilon\\3+2\varepsilon,2+4\varepsilon\end{matrix}\middle|1\right)=2+6(-2\zeta_2+3)\varepsilon+12(10\zeta_3-11\zeta_2+6)\varepsilon^2\\
&\quad+24(-28\zeta_4+55\zeta_3-27\zeta_2+9)\varepsilon^3\\
&\quad+48(94\zeta_5-16\zeta_2\zeta_3-154\zeta_4+135\zeta_3-45\zeta_2+12)\varepsilon^4+\cdots
\end{aligned}
\tag{5}
$$

They can be extended to ε^8, should it be needed.

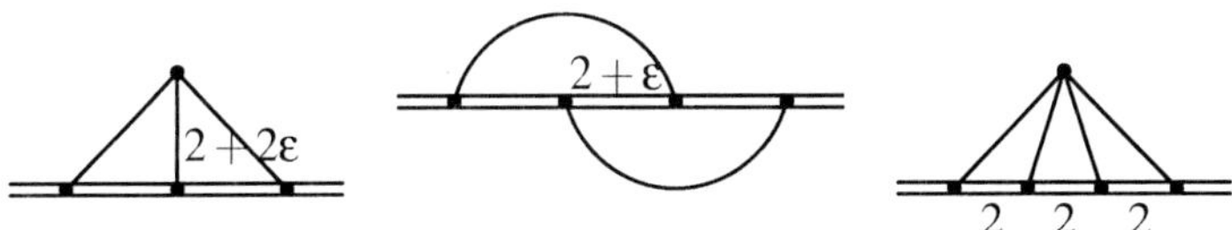

FIGURE 3. Convergent integrals related to $B_{6\ldots8}$

The most difficult basis integral $B_8 = J_c(1,1,1,1,1,1,1)$ can be re-expressed via a convergent integral $J_c(2,2,2,1,1,1,1)$ (the last diagram in Fig. 3). It can be calculated in 4-dimensional euclidean coordinate space in cylindrical coordinates. Taking first the integral in the radial coordinate of the light vertex, then the integral in its time, and then the integrals in the two times of heavy-light vertices, we obtain

$$J_c(2,2,2,1,1,1,1) = 2(\zeta_3 - 1) + O(\varepsilon)\,. \tag{6}$$

Due to the non-abelian exponentiation theorem [8], the unrenormalized HQET quark propagator in the coordinate space can be written as

$$S_0(t) = -i\theta(t)\exp\left[C_F\frac{g_0^2}{(4\pi)^{d/2}}\left(\frac{it}{2}\right)^{2\varepsilon}s+C_F\frac{g_0^4}{(4\pi)^d}\left(\frac{it}{2}\right)^{4\varepsilon}(C_A s_A+T_F n_l s_l)\right.$$
$$\left.+C_F\frac{g_0^6}{(4\pi)^{3d/2}}\left(\frac{it}{2}\right)^{6\varepsilon}\left(C_A^2 s_{AA}+C_F T_F n_l s_{lF}+C_A T_F n_l s_{lA}+(T_F n_l)^2 s_{ll}\right)+\cdots\right].(7)$$

Suppose we have calculated the one-loop correction to the HQET propagator in the coordinate space (Fig. 4*a*). Let's multiply this correction by itself (Fig. 5). We get an integral in t_1, t_2, t_1', t_2' with $0<t_1<t_2<t$, $0<t_1'<t_2'<t$. Ordering of primed and non-primed integration times can be arbitrary. The integration area is subdivided into 6 regions, corresponding to 6 diagrams in Fig. 5. If the colour factors of the diagrams Fig. 4*c*, *d* were the same as that of the one-particle-reducible diagram (Fig. 4*b*), i. e. equal to the square of the colour factor C_F of the one-loop diagram (Fig. 4*a*) (as in the abelian case), then the sum of the diagrams Fig. 4*b*, *c*, *d* would be equal to $\frac{1}{2}$ of the square of the one-loop correction (Fig. 4*a*), as given by the square of the first term in the expansion of the exponent (7). In the non-abelian case, however, the colour factor of Fig. 4*c* differs from C_F^2 by $-C_F C_A/2$, which is the colour factor of Fig. 4*e*. This is because when we reduce the colour factor of Fig. 4*c* to that of the reducible diagram Fig. 4*b*, we get an extra term with the commutator $[t^a,t^b]$, which has the colour structure of Fig. 4*e*. Therefore, we should include the contribution of Fig. 4*c* with $-C_F C_A/2$ instead of its full colour factor into the term s_A in the exponent (7). This part of the colour factor is called maximally non-abelian or colour-connected [8]. Of course, the diagram Fig. 4*e* also contributes to s_A; the diagram Fig. 4*f*, with the one-loop gluon self-energy correction, contributes to s_l (quark loop) and s_A (gluon and ghost loops).

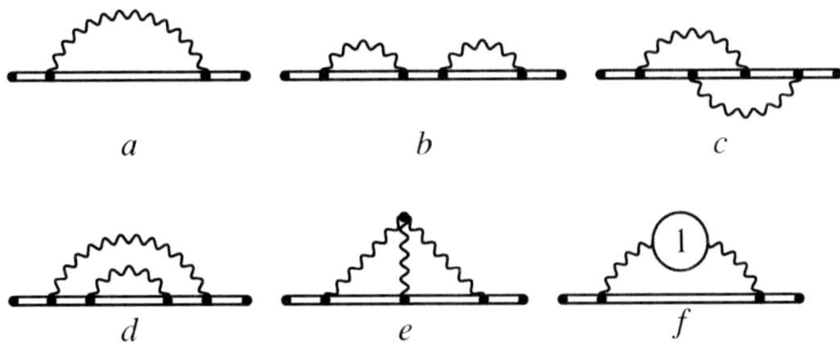

FIGURE 4. One- and two-loop diagrams for the HQET propagator

In the same way, we can consider the set of three-loop diagrams obtained by multiplying the corrections of Fig. 4*a* and *f*. We can imagine that this set is obtained from the one-particle-reducible diagram by allowing the gluon – heavy-quark vertices to slide along the heavy-quark line, crossing each other. These diagrams are said to contain two connected webs (or c-webs) [8]. Everything is already accounted for by the product of the one-loop correction and the (Fig. 4*f* part of) two-loop correction in the expansion of the exponent (7), except the contribution of Fig. 6*a* (and its mirror-symmetric), taken

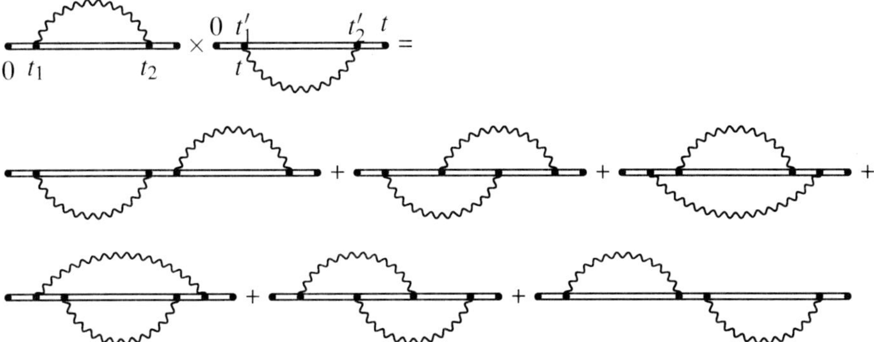

FIGURE 5. Exponentiation theorem

with the maximally non-abelian part of its colour factor. It contributes to the three-loop correction in the exponent. Similarly, out of all the diagrams with two connected webs Fig. 4*a* and *e*, only those of Fig. 6*b*, *c* (plus their mirror-symmetric ones), and *d* contribute to s_{AA}, with the maximally non-abelian part of their colour factors. This part appears, in the case of Fig. 6*b*, for example, when we commute t^a matrices to obtain the colour structure of the reducible diagram; it is identical to the colour factor of Fig. 6*h*, equal to $C_F C_A^2/4$.

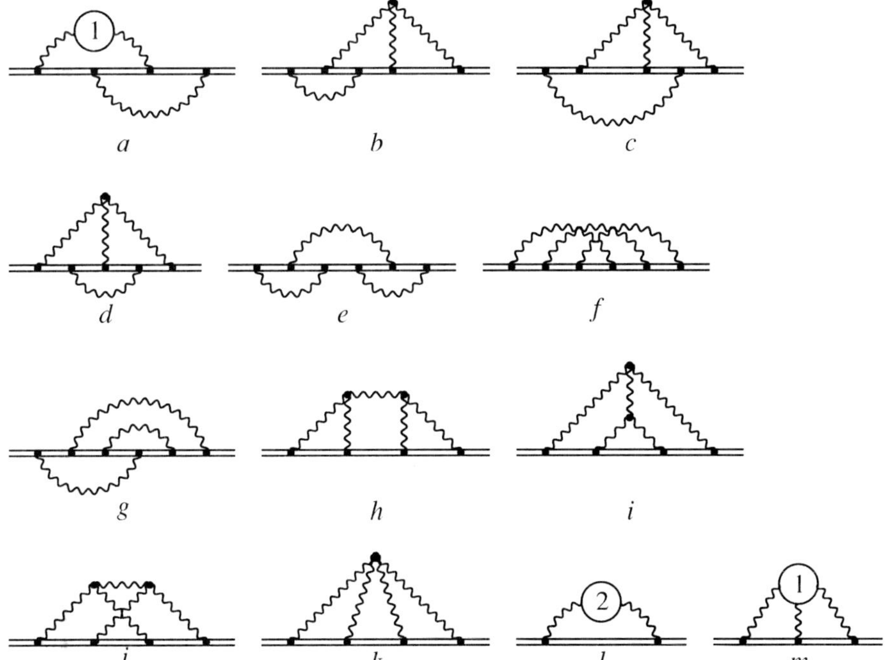

FIGURE 6. Three-loop diagrams contributing to the three-loop term in the exponent (7)

Now we are going to consider all the diagrams with three gluons, both ends of which are attached to the heavy-quark line. They are said to contain three c-webs, each of them is that of Fig. 4*a*. We decompose the colour factors of these 15 diagrams in the following way. We move the vertices along the heavy-quark line in such a way as to disentangle those c-webs. While doing so, we get extra terms from the commutators, having colour structures of the corresponding diagrams with the three-gluon vertex. These diagrams have fewer c-webs, which are more complicated. Finally, each colour factor can be expressed as a linear combination of three ones: C_F^3 (3 c-webs of Fig. 4*a*); $-C_F^2 C_A/2$ (2 c-webs of Fig. 4*a* and *e*); $C_F C_A^2/4$ (1 c-web of Fig. 6*h*). The first one occurs with the unit coefficient in all 15 colour factors. The sum of the corresponding contributions is just the term with the cube of the one-loop correction in the expansion of the exponent (7). The second colour structure occurs in the diagrams obtained by multiplying Fig. 4*a* and *c*; the sum of the corresponding contributions is contained in the product of the one-loop term and the two-loop one in the expansion of the exponent. We are left with the colour-connected parts of the colour factors (a single c-web contributions). They are present in the diagrams Fig. 6*e*, *f*, *g* (and its mirror-symmetric). They contribute to s_{AA} in (7).

We are left with the diagrams containing only a single connected web. Those of Fig. 6*h* and *i* have equal colour factors (this is evident if we close the quark line), they contribute to s_{AA}. The colour factor of Fig. 6*j* is zero. This becomes clear if we close the quark line and write a three-gluon vertex as the commutator (Fig. 7). The diagram with the four-gluon vertex (Fig. 6*k*) can be decomposed into three terms, with colour factors of Fig. 6*h*, *i*, *j*. The diagram Fig. 6*l* contains two-loop gluon self-energy corrections, including one-particle-reducible ones; it contributes to s_{AA}, s_{lF}, s_{lA}, s_{ll}. The diagram Fig. 6*m* contains one-loop corrections to the three-gluon vertex, including one-particle-reducible ones (i. e., one-loop self-energy corrections to each gluon propagator in Fig. 4*e*); it contributes to s_{AA}, s_{lA}.

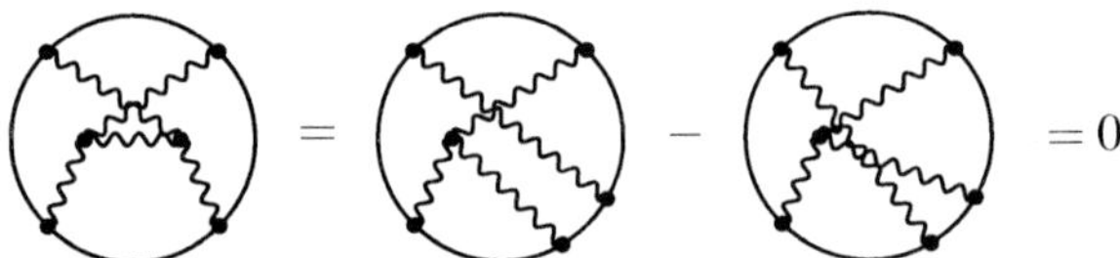

FIGURE 7. Vanishing colour factor

In order to find the heavy-quark field renormalization constant, we should re-express the unrenormalized propagator (7) in terms of the renormalized coupling $\alpha_s(\mu)$ and gauge-fixing parameter $a(\mu)$ instead of g_0^2 and a_0, and require that the renormalized propagator $S(t) = Z_Q^{-1} S_0(t)$ is finite in the limit $\varepsilon \to 0$. After this re-expression, $S_0(t)$ still has the exponential form with the same colour structures as in (7). Therefore, the renormalization constant can be written in the exponential form, too:

$$Z_Q = \exp\left[C_F \frac{\alpha_s}{4\pi} z + C_F \left(\frac{\alpha_s}{4\pi}\right)^2 (C_A z_A + T_F n_l z_l) \right.$$
$$\left. + C_F \left(\frac{\alpha_s}{4\pi}\right)^3 \left(C_A^2 z_{AA} + C_F T_F n_l z_{lF} + C_A T_F n_l z_{lA} + (T_F n_l)^2 z_{ll} \right) + \cdots \right]. \quad (8)$$

The coefficients z are obtained simply by singling out poles in ε from the corresponding terms in the exponent for $S_0(t)$ (expressed via the renormalized quantities).

The heavy-quark field anomalous dimension in HQET is defined as $\gamma_Q = d\log Z_Q/d\log\mu$. The μ-dependence comes from $\alpha_s(\mu)$ and $a(\mu)$ in Z_Q. Using $\beta(\alpha_s)$ and $\gamma_A(\alpha_s)$ for their differentiation, we obtain

$$\gamma_Q = 2(a-3)C_F\frac{\alpha_s}{4\pi} + \left[\left(\frac{a^2}{2}+4a-\frac{179}{6}\right)C_A + \frac{32}{3}T_F n_l\right]C_F\left(\frac{\alpha_s}{4\pi}\right)^2 \tag{9}$$
$$+\left[\left(\frac{5}{8}a^3+\frac{3}{4}\left(\zeta_3+\frac{13}{4}\right)a^2+\left(-8\zeta_4+6\zeta_3+\frac{271}{16}\right)a-24\zeta_4-\frac{123}{4}\zeta_3-\frac{23815}{216}\right)C_A^2\right.$$
$$\left.-6(16\zeta_3-17)C_F T_F n_l+\left(-\frac{17}{2}a+96\zeta_3+\frac{782}{27}\right)C_A T_F n_l+\frac{160}{27}(T_F n_l)^2\right]C_F\left(\frac{\alpha_s}{4\pi}\right)^3.$$

The two-loop term contains no C_F^2 part, and the three-loop one contains no C_F^3 and $C_F^2 C_A$ parts, as a consequence of the non-abelian exponentiation theorem. The two-loop term was obtained in [3]. The three-loop term has been obtained, by a completely different method, in [9]; our calculation confirms this result.

I am grateful to D. J. Broadhurst, K. G. Chetyrkin and T. van Ritbergen for useful discussions.

REFERENCES

1. A. V. Manohar and M. B. Wise, *Heavy Quark Physics*, Cambridge University Press (2000).
2. F. V. Tkachov, *Phys. Lett.* **B100** (1981) 65; K. G. Chetyrkin and F. V. Tkachov, *Nucl. Phys.* **B192** (1981) 159.
3. D. J. Broadhurst and A. G. Grozin, *Phys. Lett.* **B267** (1991) 105.
4. A. G. Grozin, *JHEP* **03** (2000) 013.
5. A. G. Grozin, hep-ph/0008300.
6. M. Beneke and V. M. Braun, *Nucl. Phys.* **B426** (1994) 301.
7. J. M. Borwein, D. M. Bradley and D. J. Broadhurst, *Electronic J. Combinatorics* **4(2)** (1997) R5, hep-th/9611004.
8. J. G. M. Gatheral, *Phys. Lett.* **B133** (1983) 90; J. Frenkel and J. C. Taylor, *Nucl. Phys.* **B246** (1984) 231.
9. K. Melnikov and T. van Ritbergen, *Nucl. Phys.* **B591** (2000) 515.

HOT AND DENSE QCD

From SPS to RHIC: Breaking the Barrier to the Quark-Gluon Plasma

Ulrich Heinz

Physics Department, The Ohio State University, Columbus, OH 43210, USA
Email: heinz@mps.ohio-state.edu

Abstract. After 15 years of heavy-ion collision experiments at the AGS and SPS, the recent turn-on of RHIC has initiated a new stage of quark-gluon plasma studies. I review the evidence for deconfined quark-gluon matter at SPS energies and the recent confirmation of some of the key ideas by the new RHIC data. Measurements of the elliptic flow at RHIC provide strong evidence for efficient thermalization during the very early partonic collision stage, resulting in a well-developed quark-gluon plasma with almost ideal fluid-dynamical collective behaviour and a lifetime of several fm/c.

I. THE QUARK-HADRON TRANSITION

Quantum Chromodynamics (QCD), the theory of strong interactions, predicts for strongly interacting bulk matter a phase transition from a gas of hadron resonances (HG) at low energy densities to a quark-gluon plasma (QGP) at high energy densities. The critical energy density ϵ_c is of the order of $1\,\mathrm{GeV/fm}^3$. It can be reached by heating matter at zero net baryon density to a temperature of about $T_c \approx 170\,\mathrm{MeV}$, or by compressing cold nuclear matter to baryon densities of about $\rho_c \sim 3-10\,\rho_0$ (where $\rho_0 = 0.15\,\mathrm{fm}^{-3}$ is the equilibrium density), or by combinations thereof. A simple version of the phase diagram is shown in Fig. 1 [1].

By colliding heavy ions at high energies one hopes to create hadronic matter at energy densities above ϵ_c. At lower energies (SIS @ 1 A GeV), the nuclei are stopped, compressed and moderately heated. At higher energies (AGS @ 10 A GeV and SPS @ 160 A GeV) one reaches higher temperatures, but the colliding nuclei are no longer completely stopped and the baryon chemical potential of the matter created at rest in the c.m.s. decreases. At the colliders RHIC ($\sqrt{s} = 200\,A$ GeV) and LHC ($\sqrt{s} = 5500\,A$ GeV) the baryon chemical potential is so small that one essentially simulates the nearly baryon-free hot hadronic matter of the early universe. If the matter thermalizes quickly at energy densities above ϵ_c, it will pass through the *quark-hadron phase transition* as the collision fireball expands and cools.

Along the temperature axis at $\mu_B = 0$ our knowledge of the QCD phase diagram is based on hard theory (lattice QCD) [2], but for nonzero baryon density we must still rely on models interpolating between low-density hadronic matter, described by low-energy effective theories, and high-density quark-gluon plasma, described by

CP602, *QCD@Work: International Workshop on Quantum Chromodynamics*
edited by P. Colangelo and G. Nardulli

perturbative QCD [1]. The uncertainties at high-baryon densities are thus relatively large (typically $\mathcal{O}(30-50\%)$). At zero baryon density, numerical simulations of QCD with 3 dynamical light quark flavors on the lattice are now available, and the systematic errors due to lattice discretization and continuum extrapolation are beginning to get small [2]. The critical temperature T_c for real-life QCD is predicted as $T_c \approx 170\,\text{MeV} \pm 10\%$ [2]. Near T_c the energy density in units of T^4 changes dramatically by more than a factor of 10 within a very narrow temperature interval. Above $T \simeq 1.2\,T_c$, ϵ/T^4 appears to settle at about 80% of the Stefan-Boltzmann value for an ideal gas of non-interacting quarks and gluons.

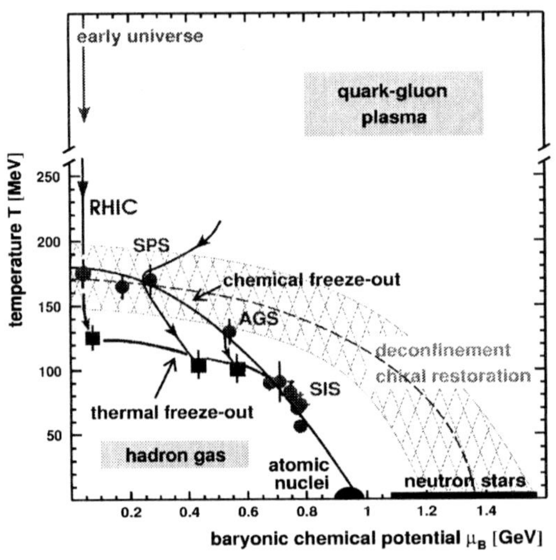

FIGURE 1. Sketch of the QCD phase diagram, temperature T vs. the chemical potential μ_B associated with net baryon density ρ_B. The cross-hatched region indicates the expected phase transition and its present theoretical uncertainty, the dashed line representing its most likely location. Lines with arrows indicate expansion trajectories of thermalized matter created in different environments. For a discussion of the chemical and thermal freeze-out lines and the location of the data points see text.

According to lattice QCD, only about $0.6\,\text{GeV/fm}^3$ of energy density are needed to make the transition to deconfined quark-gluon matter [2]. But it is very expensive to reach temperatures well above T_c: an initial temperature of 220 MeV, $\approx$30% above T_c, already requires an initial energy density $\epsilon \simeq 3.5\,\text{GeV/fm}^3$, about 6 times the critical value. This limits severely the reach of the CERN SPS into the QGP phase: only the region at and slightly above T_c can be probed. But with RHIC considerably higher energy densities have now become accessible so that we can penetrate deeper into the new phase. The situation will further improve with the beginning of the heavy-ion program at the LHC at CERN in 2007.

At T_c two phenomena occur simultaneously [3]: color confinement is broken, i.e. colored degrees of freedom can propagate over distances much larger than the size of a hadron, and the approximate chiral symmetry of QCD, which is spontaneously broken at low temperatures and densities, gets restored. Both effects significantly accelerate particle production: the liberation of gluons in large densities opens up new gluonic production channels, and the threshold for quark-antiquark pair production is lowered due to the restoration of the spontaneously broken chiral symmetry at low densities.

II. RECONSTRUCTING THE LITTLE BANG

As the two nuclei hit each other, a superposition of nucleon-nucleon (NN) collisions occurs. What is different from individual NN collisions is that (i) each nucleon scatters several times, and (ii) the partons liberated in different NN collisions rescatter with each other even before hadronization, as do the hadrons afterwards. Both change the particle production *per participating nucleon*, but only the rescattering processes can lead to a state of local thermal equilibrium, by redistributing the energy lost by the beams into the statistically most probable configuration. They cause thermodynamic pressure acting against the outside vacuum, and this makes the reaction zone expand collectively. *This is a genuine collective nuclear effect, with no analogue in elementary particle collisions.* The expansion dilutes the fireball below ϵ_c, at which point hadrons are formed from the quarks and gluons (hadronization). Further interactions among these hadrons cease once their average distance exceeds the range of the strong interactions: the hadrons "freeze out".

The strong interactions among the partons and hadrons before freeze-out wipe out much information about their original production processes. Extracting information about the hot and dense early collision stage thus requires to exploit features which are either established early and survive the rescattering and collective expansion or can be reliably back-extrapolated. Correspondingly one classifies the observables into two classes, *early* and *late* signatures.

The conceptually cleanest early signatures are the directly produced real and virtual photons since these re-interact weakly and escape directly from the fireball (virtual photons materialize as e^+e^- or $\mu^+\mu^-$ pairs). They are emitted throughout the expansion, but their production should be strongly weighted towards the hot and dense initial stages. Unfortunately, direct photons are rare, and the experimental background from hadronic decay photons after freeze-out is enormous.

Another early signature are hadrons containing charmed quarks. At the SPS, $c\bar{c}$ pairs can be only created in the primary NN collisions; secondary scatterings are already below the $c\bar{c}$ threshold. The latter can only redistribute them in phase-space, changing the relative amounts of mesons with hidden and open charm. It was shown that such a redistribution is much easier in the color-deconfined QGP phase than by reinteractions of charmed and other hadrons; in this way charm-redistribution becomes also an early signature. At SPS energies and below, charm production is a very rare process and only hidden charm mesons ($J/\psi, \psi'$) have been measured. The interpretation of what happens to them crucially depends on the assumption that no secondary charm is produced. At RHIC energies and above the production of additional charmed particles in secondary collisions becomes possible, and a consistent interpretation of charmonium data requires also the measurement of open charm. None of the present RHIC experiments can do that.

Hadrons made of u and d quarks can be easily produced and destroyed throughout the fireball expansion. Their abundances and spectra are *late signatures* which provide only *indirect* information about the early collision stages. But they are very numerous and can be measured very accurately. We'll use these *late* signals to reconstruct the Little Bang and then check the consistency of the resulting picture with the less detailed direct information from the *early* signals. Hadrons involving strange quarks play an intermediate role: $s\bar{s}$ pairs are easily produced in the dense color-deconfined, chirally almost symmetric QGP phase, but hadron interactions after hadronization leave their abundances essentially unaltered. Their yields thus reflect the situation reached at the quark-hadron transition point.

It was recently realized that in *non-central* collisions the observed collective flow pattern of *all* hadrons shows certain anisotropies ("elliptic flow") which are established very early in the collision, even before hadrons are formed, and are hardly changed in the late expansion stages. The elliptic flow of the emitted hadrons (including the abundant light ones) thus constitutes another *early collision signature.*

III. INITIAL CONDITIONS

We can estimate the initially *produced* energy density by dividing the measured total transverse energy E_T by an estimated initial reaction volume [4]

$$\epsilon_{\mathrm{Bj}}(\tau_0) = \frac{1}{\pi R^2} \frac{1}{\tau_0} \frac{dE_T}{dy} \,. \tag{1}$$

Here $\tau_0\, dy$ is an estimate for the length of a cylinder undergoing boost-invariant longitudinal expansion [4] from which particles with longitudinal momenta in a rapidity interval dy are emitted. Inserting for πR^2 the overlap area of two Pb nuclei colliding at zero impact parameter, choosing $\tau_0 = 1\,\mathrm{fm}/c$, and using $dE_T/dy(y = 0) \approx 400\,\mathrm{GeV}$ for central Pb+Pb collisions [5] gives

$$\epsilon_{\mathrm{Bj}}^{\mathrm{Pb+Pb}}(1\,\mathrm{fm}/c) = 3.2 \pm 0.3\,\mathrm{GeV}/\mathrm{fm}^3 \,. \tag{2}$$

Note that this is the average over the transverse plane; the value $\epsilon_0 = \epsilon(r{=}0)$ in the center is about twice as high. The analogous value extracted from $\sqrt{s} = 130\,A\,\mathrm{GeV}$ Au+Au collisions at RHIC [6] is 60% larger, and recent PHOBOS data from $\sqrt{s} = 200\,A\,\mathrm{GeV}$ show another increase by about 15% [7]. As QGP searchers we are clearly playing in the right ball-park: if the matter were already thermalized after $1\,\mathrm{fm}/c$ (for which there is evidence from elliptic flow measurements, see below), the temperature corresponding to the SPS value (2) would be $T_0 \simeq 210 - 220\,\mathrm{MeV}$. At RHIC, thermalization seems to happen even earlier, at around 0.5-0.6 fm/c [8], and the initial temperature may have been as high as 350 MeV.

IV. THERMAL FREEZE-OUT: AN EXPLODING FIREBALL

The measured hadron spectra contain two pieces of information: (i) Their normalizations, i.e. the *yields and abundance ratios*, provide the chemical composition of the fireball at the "chemical freeze-out" point where the hadron abundances freeze out; this gives information about the degree of chemical equilibration, see Sec. VI. (ii) The hadronic *momentum spectra* provide information about thermalization of the momentum distributions and collective flow. The latter is caused by thermodynamic pressure and thus reflects, in a time-integrated way, the equation of state of the fireball matter. We concentrate on the transverse flow since all of it is generated *during* the reaction.

In a thermal, collectively expanding system the shapes of *all* hadronic $m_\perp$-spectra ($m_\perp = \sqrt{m^2+p_\perp^2}$) can be characterized by just two numbers: the temperature T_f and the mean transverse flow velocity $\langle v_\perp \rangle$ at freeze-out. This is true if all hadrons decouple simultaneously, i.e. if their rescattering cross sections are similar. This can be checked experimentally: e.g., it was observed that the Ω [9] and J/ψ [10] show steeper slopes than expected from the systematics of the remaining hadrons, because they have smaller hadronic rescattering cross sections and thus decouple earlier [11]. For all other hadrons, a common parametrization by a single pair ($T_\mathrm{f}, \langle v_\perp \rangle$) works very well. The transverse flow velocity $\langle v_\perp \rangle$ manifests itself as a flattening of the $m_\perp$-spectra, by a mass-independent blueshift factor $T_\mathrm{slope} = T_\mathrm{f}\sqrt{(1+\langle v_\perp \rangle)/(1-\langle v_\perp \rangle)}$ in the relativistic domain $m_\perp > 2m_0$ and by a mass-dependent term $T_\mathrm{slope} = T_\mathrm{f} + \frac{1}{2}m_0 \langle v_\perp \rangle^2$ at small $p_\perp$. A roughly linear mass-dependence of the transverse slopes at low $p_\perp$ was observed at the SPS (see e.g. [9]). Fig. 2 shows that the RHIC data [12,13] follow the same systematics, with antiproton spectra being much flatter at low $p_\perp$ than the pion spectra, in quantitative agreement with hydrodynamic calculations which assume complete thermalization of the fireball after $\tau_0 = 0.6\,\mathrm{fm}/c$ [14].

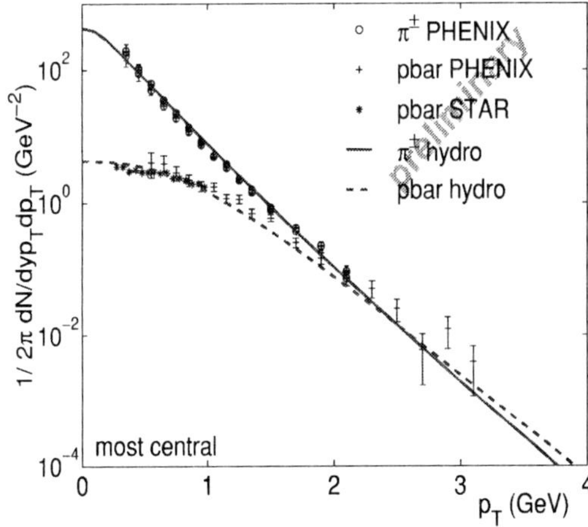

FIGURE 2. Preliminary transverse momentum spectra of charged pions and antiprotons from 130 A GeV Au+Au collisions at RHIC [12,13], compared with hydrodynamic predictions [14], corresponding to T_f = 128 MeV, μ_B = 70 MeV, and $\langle v_\perp \rangle$ = 0.6 c (private communication by P.F. Kolb whom I thank for preparing this figure).

Transverse flow also affects two-particle momentum correlations and gives, for example, rise to a characteristic $m_\perp$-dependence of the source sizes extracted from Bose-Einstein correlation measurements [15] or deuteron yields [16–18]. The most accurate separation of thermal and collective contributions to the final hadron momenta is obtained from a simultaneous analysis of single-particle spectra and two-particle correlations [19,20]. At the SPS this gives $T_\mathrm{f} \approx 100$ MeV and $\langle v_\perp \rangle \approx 0.55\,c$. The corresponding freeze-out energy density is only about 50 MeV/fm^3! For RHIC such a combined analysis is not yet available, but the spectral slopes alone (see Fig. 2) indicate a somewhat higher freeze-out temperature $T_{\rm f} \approx 125\,\text{MeV}$, with a 20% larger radial flow at RHIC than at the SPS at the same value of T.

V. THE MISSING RHO: WATCHING THERMALIZATION

If the Little Bang started at initial energy densities above $3\,\text{GeV/fm}^3$, but decoupled only at about $50\,\text{MeV/fm}^3$, how can we find out what happened in between? The ρ meson provides a first answer: it can decay into e^+e^- or $\mu^+\mu^-$ pairs which escape from the fireball without further interactions, and this ρ-decay clock ticks at a rate of $1.3\,\text{fm}/c$, the ρ lifetime. What I mean by this is that after one generation of ρ's has decayed, a second generation is created by resonant $\pi\pi$ scattering, which can again decay into dileptons, etc. The number of extra dileptons with the invariant mass of the ρ is thus a measure for the time in which the fireball consists of strongly interacting hadrons [21]. Obviously, ρ mesons do not exist before hadrons appear in the fireball, so they won't tell us anything about a possible initial QGP phase. But they still allow us to look *inside* the strongly interacting hadronic fireball at a later stage, still long before the hadrons decouple.

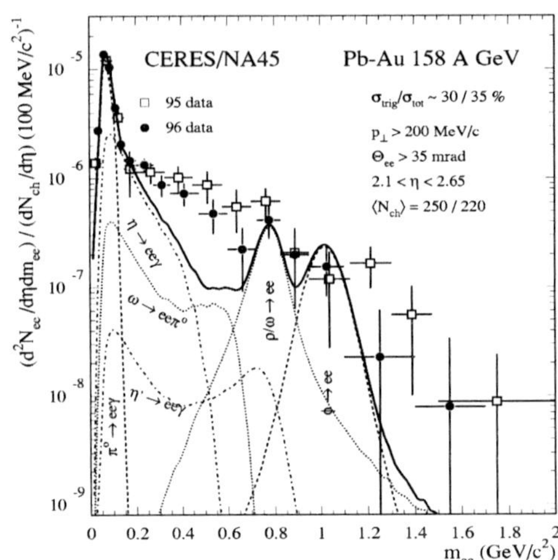

Figure 3. Invariant mass spectrum of e^+e^- pairs from $158\,A\,\text{GeV}/c$ Pb+Au collisions [22]. The solid line is the expected spectrum (the sum of the many shown contributions) from the decays of hadrons produced in pp and pA collisions (where it was experimentally checked [22]), properly scaled to the Pb+Au case. Two sets of data with different analyses are shown. Note that the ρ-peak reappears if only e^+e^- pairs with $p_\perp > 500\,\text{MeV}/c$ are selected [22]; such fast ρ's escape quickly from the fireball and are not as strongly affected by collision broadening.

However, when the CERES/NA45 collaboration looked at the e^+e^- spectrum in $158\,A\,\text{GeV}/c$ Pb+Au collisions (see Fig. 3) and couldn't find the ρ at all! They saw extra e^+e^- pairs in the mass region of the ρ and below (about 2.5–3 times as many as expected), but instead of a nice ρ-peak at $m_\rho = 770\,\text{MeV}$ one sees only a broad smear [22]. Many explanations of the CERES-effect have been proposed, but the simplest one consistent with the data (for a review see [23]) is *collision broadening*: there is strong rescattering of the pions, not only among each other, but also with the baryons in the hadronic resonance gas, and this modifies their spectral densities and, as a consequence, leads to a smearing of the ρ-resonance in the $\pi\pi$ scattering cross section.

This demonstrates that, after first being formed in the hadronization process, the pions (the most abundant species at the SPS) undergo intense rescattering before finally freezing out. And this again is the mechanism which allows the fireball to reach and maintain a state of approximate local thermal equilibrium, to build up thermodynamic pressure and to collectively explode, as seen from the above analysis of the freeze-out stage. That the dileptons from collision-broadened ρ's outnumber those from the decay of unmodified ρ's emitted at thermal freeze-out (which should show up as a normal ρ-peak) shows that the hadronic rescattering stage must have lasted several ρ lifetimes.

VI. SEEING THE QUARK-HADRON TRANSITION

In the rest of this talk I will concentrate on observables which were found to differ drastically in AA and NN collisions but which we now believe cannot be changed efficiently by hadronic rescattering during the time between hadronization and kinetic freeze-out. Observables for which this can be firmly established yield insights about where AA and NN collisions differ already *before or during* hadronization, irrespective whether or not the hadrons rescatter after being formed.

One possibility how the early stage of an AA collision may differ from that in NN collisions is the formation of a quark-gluon plasma. I therefore review a few key QGP predictions and check how they fare in comparison with the data. In the present Section I discuss *strangeness enhancement* as a QGP signature, returning to two further QGP predictions in the following two Sections.

Strangeness enhancement and chemical equilibration were among the earliest predicted QGP signatures [24]. The idea is simple: 1. Color deconfinement leads to high gluon density, fostering $s\bar{s}$ creation by gluon fusion. 2. Chiral symmetry restoration renders the s-quarks relatively light, and in the QGP they can be created without the need for additional light quarks to make a hadron; both effects lower the production threshold. Both should considerably reduce the time needed for strangeness saturation and chemical equilibration compared to hadronic rescattering processes. Since in NN and e^+e^- collisions strange hadron production is known to be suppressed relative to simple phase-space considerations [25], this should cause a relative strangeness enhancement in heavy-ion collisions.

Kinetic simulations based on known hadronic properties and interaction cross sections have shown that it is impossible to create a state of hadronic chemical equilibrium and a significant amount of strangeness enhancement out of a non-equilibrium initial state by purely hadronic rescattering [26]. If you want to get these features out, you have to put them in at the beginning of the simulation.

There may be many different ways of doing so, but the most efficient way of creating a state of (relative or absolute) hadronic chemical equilibrium may be provided by the hadronization process itself: If before hadronization the quarks and gluons are uncorrelated (such as in a QGP), then the most likely outcome of the non-perturbative hadronization process is a statistical occupation of the hadronic phase-space, a state of maximum entropy. If (as predicted for the QGP

[24]) the number of $s\bar{s}$-pairs is enhanced *before* the onset of hadronization, or by the fragmentation of gluons *during* hadronization, their statistical distribution over the available hadronic channels will naturally lead to an apparent hadronic chemical equilibrium state (with the corresponding enhancement of, say, the $\bar{\Omega}$) *even if none of the hadrons ever scattered with each other after being formed.*

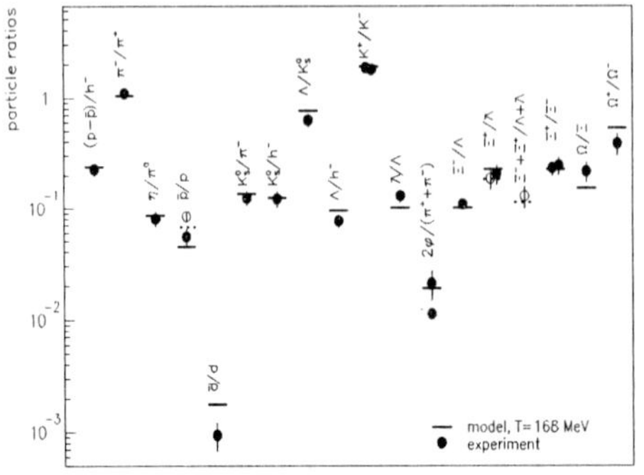

FIGURE 4. A compilation of measured particle ratios from 158 A GeV/c Pb+Pb collisions, compared with a hadron resonance gas in complete chemical equilibrium (full strangeness saturation) at $T_{\rm chem}$ = 168 MeV and $\mu_{\rm B}$ = 266 MeV [27].

Such a state of "apparent" or "pre-established" chemical equilibrium is indeed seen in the experiments: Fig. 4 shows 18 hadronic particle ratios from 158 A GeV Pb+Pb collisions, compared with a hadronic chemical equilibrium state at $T_{\rm chem}$ = 168 MeV and $\mu_{\rm B}$ = 266 MeV [27]. A similar analysis of recent RHIC data [28] gives $T_{\rm chem} \simeq 175$ MeV and $\mu_{\rm B} \simeq 50$ MeV. The ratio of strange to non-strange quarks in the final hadrons is about a factor 2 higher than in *pp* collisions [29], and triply strange Ω baryons are enhanced relative to pBe collisions by a factor 17 [30]! This and the value of $T_{\rm chem}$ are interesting: $T_{\rm chem}$ characterizes the energy density at which hadronization occurs (about 0.5 GeV/fm^3) and coincides with the critical temperature for color deconfinement from lattice QCD. If the hadrons were formed by hadronization at the critical energy density ϵ_c and their abundances froze out at $T_{\rm chem} \simeq 170$ MeV, there was no time to achieve this equilibrium configuration by hadronic rescattering; the hadrons must have been "born" into chemical equilibrium [31,32]. Since subsequent hadronic rescattering is inefficient in changing the hadronic abundances [26], we can thus use them to measure T_c.

VII. J/ψ SUPPRESSION AND COLOR DECONFINEMENT

What is the nature of the state from which the hadrons emerge in this manner? This question brings us to the second key QGP prediction [33]: the high gluon density resulting from color deconfinement should Debye-screen the color interaction between a c and a $\bar{c}$ quark produced during the initial nuclear impact, thus preventing them from binding into charmonium states (J/ψ, χ_c, ψ'). Instead, they would eventually find light quark partners to make hadrons with open charm. The result should be a suppression of charmonium production in heavy-ion collisions, and the screening mechanism should lead to a specific suppression pattern which, as a function of the achieved energy density, first affects the loosely bound ψ' and χ_c states and then the strongly bound J/ψ ground state [34].

Nuclear J/ψ suppression was indeed found at the SPS by the NA38/NA50 Collaboration. Fig. 5 shows that, as a function of collision centrality, measured by the produced transverse energy E_T and in the left panel translated into an energy density at $\tau = 1\,\mathrm{fm}/c$ using a generalization of Eq. (1), the yield of J/ψ mesons (identified by their $\mu^+\mu^-$ decay) is suppressed "anomalously" (i.e. below expectations). For a discussion of "normal" J/ψ suppression I must refer to Refs. [34,35]; it is well-defined and carefully experimentally tested and represented by the horizontal line in the left panel of Fig. 5. The observed deviation from this normal suppression is in qualitative agreement with the QGP prediction; in particular, the weakly bound ψ' (not shown) already suffers anomalous suppression in central S+U collisions while the strongly bound J/ψ shows it only in semicentral and central PbPb collisions. The observed pattern is not yet understood in all details [34,36]; however, it definitely cannot be reproduced by final state rescattering of the J/ψ with the dense hadronic environment after hadronization, represented by four independent calculations shown as lines in the right panel of Fig. 5.

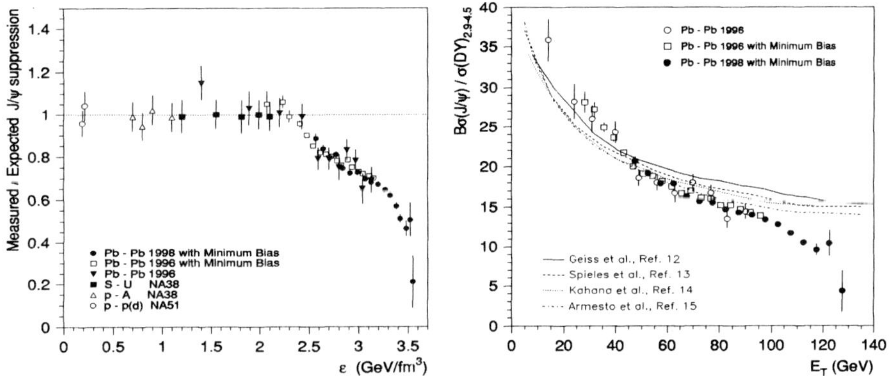

FIGURE 5. Left: "Anomalous J/ψ suppression" as a function of initial energy density [35]. Right: The ratio (J/ψ)/Drell-Yan production, as a function of the measured transverse energy E_T, compared to hadronic comover models [35].

VIII. THERMAL ELECTROMAGNETIC RADIATION

The third (and earliest) key prediction for QGP formation is thermal radiation of (real and virtual) photons from the thermalized quarks in the QGP [37]. In spite of the difficulties arising from large backgrounds, experiments at the SPS have searched for such radiation. In both the real [38] and virtual photon ($\mu^+\mu^-$ [39]) channels enhancements over hadronic decay backgrounds were reported. However, no unambiguous connection with thermal radiation has been established, in part because the predicted signal for the latter [40] is marginal at the SPS, given the accuracy of the experiments. To see the plasma "shine" one needs the higher temperatures and longer plasma lifetimes which can be reached at RHIC.

IX. ELLIPTIC FLOW: AN EARLY HADRONIC SIGNATURE

In non-central heavy-ion collisions the initial overlap region is spatially deformed into an almond shape in the transverse plane. Nevertheless, at each point $\boldsymbol{r}$ the initial transverse momentum distribution is isotropic. If the produced matter expands without further interaction ("free streaming"), the $p_\perp$ distribution remains isotropic while the spatial deformation eventually disappears (the almond looks more spherical as it grows). On the other hand, if the produced matter thermalizes quickly, pressure builds up inside, and the spatial deformation results in anisotropic pressure gradients. These generate stronger collective flow in the shorter direction (i.e. into the reaction plane) than in the longer one (i.e. out of the reaction plane) [41], and the $p_\perp$-distribution becomes anisotropic. This phenomenon is called elliptic flow, measured by the second coefficient v_2 of an azimuthal Fourier decomposition of the $p_\perp$-spectrum [42]. Elliptic flow lets the almond grow faster along its short than along its long direction, thereby reducing its deformation. When the spatial deformation and the accompanying anisotropic pressure gradients vanish, v_2 stops growing. The higher the initial energy density, the longer the total fireball lifetime until hadronization and freeze-out, and the earlier this saturation occurs in the expansion history. This makes v_2 an *early signature* [43] (the more so the higher the initial energy density) which is carried by the abundant soft final state hadrons.

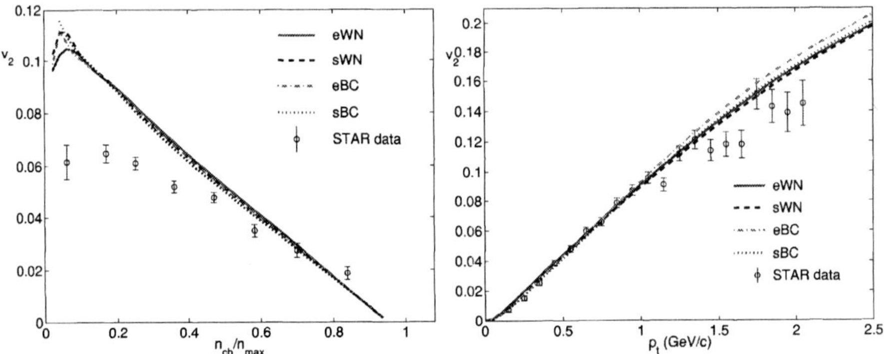

FIGURE 6. $p_\perp$ integrated elliptic flow coefficient v_2 as function of collision centrality (left) and differential elliptic flow $v_2(p_\perp)$ for minimum bias collisions (right) for 130 A GeV Au+Au collisions at RHIC [46]. The curves are from hydrodynamic collisions with different initial density profiles [47].

Kinetic simulations [44] show that the momentum-space response to the spatial deformation grows monotonically with the interaction strength among the constituents of the produced matter. For a fixed spatial deformation, the largest v_2 response results from a system which is coupled so strongly that it thermalizes "instantaneously" on fluid dynamic time scales. This maximal response can thus be computed by solving the equations of ideal (non-viscous) hydrodynamics [41,45]. Fig. 6 shows that the RHIC data exhaust this upper limit for semi-central Au+Au collisions and low $p_\perp$, indicating very rapid thermalization on a time scale of 0.5 fm/c [45]. For large impact parameters $b > 7\,\text{fm}$ and $p_\perp > 1.5 - 2\,\text{GeV}/c$ the measured v_2 stays below the hydrodynamic prediction, reflecting less efficient thermalization in small reaction volumes and at large transverse momenta. The hydrodynamic calculations show that the observed agreement between theory and data requires thermalization at energy densities well above ϵ_c. It thus becomes difficult to avoid the conclusion that the prehadronic state formed in Au+Au collisions at RHIC is indeed a quark-gluon plasma.

X. CONCLUSIONS

Relativistic heavy-ion collisions at the CERN SPS and RHIC have taken us into new and unprecedented regions of energy density, almost two orders above that of cold nuclear matter and well above the critical value for quark deconfinement. We have strong direct experimental evidence for a large degree of thermalization and strong collective behaviour setting in at a very early collision stage. Elliptic flow data show that at RHIC most of the anisotropic flow pattern is established before the energy density has dropped below ϵ_c. This requires thermalization in the quark-gluon phase. Even at the SPS, the observed patterns of chemical equilibrium freeze-out at T_c (now confirmed at RHIC), strangeness enhancement and J/ψ suppression are fully consistent with predictions based on hypothetical QGP creation.

At the SPS the search for electromagnetic radiation directly emitted from the QGP phase did not provide convincing answers. The higher initial temperatures and longer plasma lifetimes at RHIC should make that task a bit easier, but we may still have to wait a while to see results. First tentative evidence for jet quenching at RHIC (see [48]) indicates that with this process we may have found another method of directly probing the early prehadronic collisions stage. The era of detailed studies of the properties of the QGP has now begun.

REFERENCES

1. For a more detailed discussion see Rajagopal, K., these proceedings.
2. See Karsch, F., these proceedings.
3. Karsch, F., and Laermann, E., *Phys. Rev. D* **50**, 6954-6962 (1994).
4. Bjorken, J.D., *Phys. Rev. D* **27**, 140-151 (1983).
5. Alber, T., et al. (NA49 Collaboration), *Phys. Rev. Lett.* **75**, 3814-3817 (1995); Aggarwal, M., et al. (WA98 Collaboration), *Nucl. Phys.* **A610**, 200c-212c (1996); and *Eur. Phys. J. C* **18**, 651-663 (2001).
6. Adcox, K., et al. (PHENIX Collaboration), *Phys. Rev. Lett.* **87**, 052301 (2001).
7. Back, B.B., et al. (PHOBOS Collaboration), nucl-ex/0108009.
8. Kolb, P., et al., *Phys. Lett. B* **500**, 232-240 (2001).
9. Antinori, F., et al. (WA97 Collaboration), *Eur. Phys. J. C* **14**, 633-641 (2000).
10. Abreu, M.C., et al. (NA50 Collaboration), *Phys. Lett. B* **499**, 85-96 (2001).
11. van Hecke, H., Sorge, H., and Xu, N., *Phys. Rev. Lett.* **81**, 5764-5767 (1998).
12. Velkovska, J., et al. (PHENIX Collaboration), nucl-ex/0105012.
13. Harris, J., et al. (STAR Collaboration), Proceedings "Quark Matter 2001", Nucl. Phys. **A**, in press.

14. Huovinen, P., et al., *Phys. Lett. B* **503**, 58-64 (2001).
15. Chapman, S., Nix, J.R., and Heinz, U., *Phys. Rev. C* **52**, 2694-2703 (1995).
16. R. Scheibl and U. Heinz, Phys. Rev. C 59 (1999) 1585.
17. Murray, M., et al. (NA44 Collaboration), *Nucl. Phys.* **A661**, 456c-459c (1999).
18. Ambrosini, G., et al. (NA52 Collaboration), *New J. Phys.* **1**, 22 (1999).
19. Appelshäuser, H., et al. (NA49 Collaboration), *Eur. Phys. J. C* **2**, 661-670 (1998).
20. Tomášik, B., Wiedemann, U.A., and Heinz, U., nucl-th/9907096.
21. Heinz, U., and Lee, K.S., *Phys. Lett. B* **259**, 162-168 (1991).
22. Lenkeit, B., et al. (CERES Collaboration), *Nucl. Phys.* **A661**, 23c-32c (1999).
23. Rapp, R., and Wambach, J., *Adv. Nucl. Phys.* **25**, 1 (2000).
24. Rafelski, J., and Müller, B., *Phys. Rev. Lett.* **48**, 1066-1069 (1982); Koch, P., Müller, B., and Rafelski, J., *Phys. Rep.* **142**, 167-262 (1986).
25. Becattini, F., *Z. Phys. C* **69**, 485-492 (1996); Becattini, F., and Heinz, U., *ibid.* **76**, 269-286 (1997).
26. For a review see Heinz, U., *Nucl. Phys.* **A661**, 140c-149c (1999).
27. Braun-Munzinger, P., Heppe, I., and Stachel, J., *Phys. Lett. B* **465**, 15-20 (1999).
28. Braun-Munzinger, P., Magestro, D., Redlich, K., and Stachel, J., hep-ph/0105229.
29. Becattini, F., Gaździcki, M., and Sollfrank, J., *Eur. Phys. J. C* **5**, 143-153 (1998).
30. Lietava, R., et al. (WA97 Collaboration), *J. Phys. G* **25**, 181-188 (1999).
31. Heinz, U., *Nucl. Phys.* **A638**, 357c-364c (1998); *J. Phys. G* **25**, 263-274 (1999).
32. Stock, R., *Prog. Part. Nucl. Phys.* **42**, 295-309 (1999); *Phys. Lett. B* **456**, 277-282 (1999).
33. Matsui, T., and Satz, H., *Phys. Lett. B* **178** 416 (1986).
34. For a recent review see Satz, H., *Rep. Prog. Phys.* **63**, 1511-1574 (2000).
35. Abreu, M.C., et al. (NA50 Collaboration), *Phys. Lett. B* **477**, 28-36 (2000).
36. Blaizot, J.-P., Dinh, P.M., and Ollitrault, J.-Y., *Phys. Rev. Lett.* **85**, 4012-4015 (2000).
37. Shuryak,,E.V., *Phys. Lett.* **78B**, 150 (1978); Kajantie, K., and Miettinen, H.I., *Z. Phys. C* **9**, 341 (1981).
38. Aggarwal, M.M., et al. (WA98 Collaboration), *Phys. Rev. Lett.* **85**, 3595-3599 (2000).
39. Abreu, M.C., et al., (NA38 and NA50 Collaborations), *Eur. Phys. J. C* **14**, 443-455 (2000).
40. Gallmeister, K., Kämpfer, B., and Pavlenko, O.P., *Phys. Rev. C* **62**, 057901 (2000).
41. Ollitrault, J.-Y., *Phys. Rev. D* **46**, 229-245 (1992).
42. Voloshin, S.A., and Zhang, Y., *Z. Phys. C* **70**, 665-672 (1996).
43. Sorge, H., *Phys. Rev. Lett.* **78**, 2309-2312 (1997); *ibid.* **82**, 2048-2051 (1999).
44. Zhang, B., Gyulassy, M., and Ko, C.M., *Phys. Lett. B* **455**, 45-48 (1999); Molnar, D., and Gyulassy, M., nucl-th/0104073.
45. Kolb, P.F., Sollfrank, J., and Heinz, U., *Phys. Rev. C* **62**, 054909 (2000); Kolb, P.F., et al., *Phys. Lett. B* **500**, 232-240 (2001); Huovinen, P., et al., *ibid.* **503**, 58-64 (2001).
46. Ackermann, K.H., et al. (STAR Collaboration), *Phys. Rev. Lett.* **86**, 402-407 (2001).
47. Kolb, P.F., et al., *Nucl. Phys.* **A696**, 175-193 (2001).
48. Ullrich, T., these proceedings.

Charmonia States Suppression and Transverse Momentum Distribution in Pb-Pb Collisions at the CERN SPS

A.B.Kurepin

*for the NA50 Collaboration**

Abstract. We present the results of experiment NA50 on the production of charmonia states in Pb-Pb collisions at 158 GeV/c per nucleon. For the most peripheral collisions the experimental data agree with ordinary nuclear absorption as deduced from proton and light nuclei induced collisions. For central collisions, we observe an anomalous suppression which significantly departs from normal nuclear absorption and which could result from Debye color screening in a deconfined quark-gluon phase. The dependence of the transverse momentum distributions on the centrality of the Pb-Pb collision is also investigated.

INTRODUCTION

The data on the suppression of charmonia states in lead-lead collisions obtained by the NA50 collaboration at 158 GeV/c per nucleon [1] have attracted considerable attention in the search for experimental evidence of formation of deconfined matter.

The set of results obtained by both experiments NA38 and NA50 shows two different regimes for J/ψ absorption in nuclear matter. Ordinary nuclear absorption due to charmonium-nucleon interactions explains the production in proton and light ion induced reactions. On the other hand, for the more central Pb-Pb collisions, a significantly stronger absorption is observed which could be the indication of a new physics at the subnucleon level. It could be naturally explained by Debye color screening in a deconfined quark-gluon phase [2] which could account, at least qualitatively, for the peculiar pattern of the J/psi production rate as a function of energy density (see Fig.1). Other attempts to describe the observed pattern and based on the interaction of the charmonium state with the comovers produced in the collision lead to a significantly poorer agreement with the data [1].

The study of the transverse momentum distributions of the charmonia states, recently published [3], could shed some extra light on the underlying suppression mechanism.

DATA COLLECTION AND ANALYSIS

The data were collected with a thoroidal magnetic field muon spectrometer used to detect and measure the muon pair originating from the charmonium decay. The centrality of the collision was characterized by an electromagnetic calorimeter measuring

CP602, *QCD@Work: International Workshop on Quantum Chromodynamics*
edited by P. Colangelo and G. Nardulli

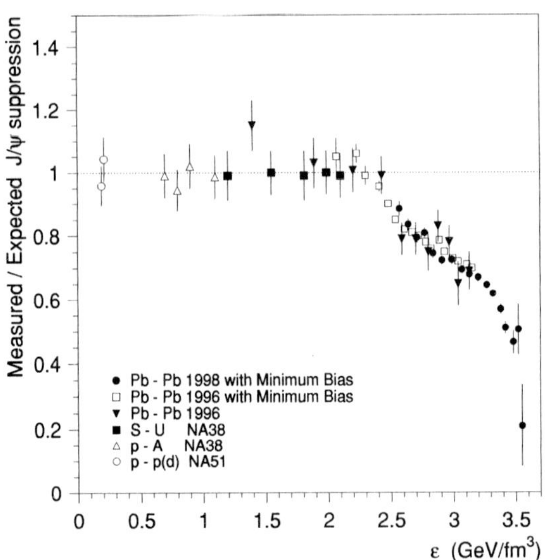

FIGURE 1. Measured J/ψ production yield, normalized to the yields expected assuming that the only source of suppression is the ordinary absorption by the nuclear medium. The data are shown as a function of the energy density reached in the several collision systems. Reprinted from [1] Copyright 2000, with permission from Elsevier Science.

the neutral transverse energy produced in the reaction, a "zero degree calorimeter" measuring the energy of the projectile spectator nuclei and a charged particle multiplicity detector. An active target system with seven lead subtargets was used to increase the yield of events. The dimuon background originated from uncorrelated pion and kaon decays was estimated and subtracted using the data of like-sign muon pairs. The final number of J/ψ events was about 190000 in 1996 and 50000 in 1995.

Muon pairs originating from the Drell-Yan process and from the semileptonic decays of pairs of charmed D and $\bar{D}$ mesons produce a physical invariant mass continuum which is estimated from a fit to the dimuon mass distribution. The raw muon pair p_T distributions of the charmonium resonaces is corrected using the fitted yields and the experimental p_T distributions for dimuon masses below and above the J/ψ and ψ′ resonances. For the J/ψ, the systematic uncertainty of the $\langle p_T^2 \rangle$ value due to the subtraction of the mass continuum amounts to only 2%. Details of the analysis method can be found in the recent publication [3].

DISCUSSION

The p_T distributions determined for the J/ψ, ψ′ and Drell-Yan mass intervals and integrated over the impact parameters have similar structure with a broad maximum near 1 GeV/c and decreasing by three orders of magnitude at about 4 GeV/c.

However the dependence of $\langle p_T^2 \rangle$ on the centrality of the collision is different for charmonia states and for the Drell-Yan process as can be seen on Fig.2.

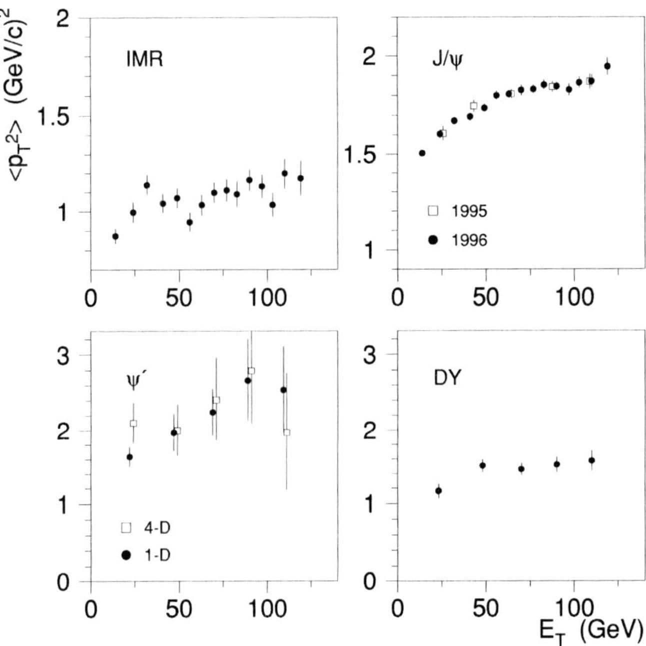

FIGURE 2. $\langle p_T^2 \rangle$ as a function of the transverse energy for several muon pair mass intervals. For the J/ψ the 5 open squares correspond to the 1995 data sample. The error bars are only statistical. For the ψ ′, both the 1-D and 4-D deconvolution results are shown. Reprinted from [3] Copyright 2001, with permission from Elsevier Science.

For J/ψ production, the $\langle p_T^2 \rangle$ value first increases and then tends to flatten for more central collisions. The same trend is seen for ψ′, however with much worse statistical accuracy. For the Drell-Yan process, $\langle p_T^2 \rangle$ is almost independent on the centrality of the collision, as it could be assumed for the hard process based on nucleon-nucleon interaction. The most interesting feature is a considerably higher value of $\langle p_T^2 \rangle$ for ψ′ than for J/ψ.

The nice agreement between the results obtained from the 1995 and 1996 data for the J/ψ suggests that the target thickness does not induce any significant effect on $\langle p_T^2 \rangle$. On the other hand, systematic effects due to the method used for correcting smearing and acceptance due to the apparatus are small, as deduced from the comparison, shown for the ψ′ on Fig.2, of the results obtained from two different methods.

In Fig.3 we plot the ratios $R_i(p_T)$ between the p_T distributions obtained in the E_T bin i and the same distribution obtained in the first E_T bin. The normalization to the Drell-Yan process was used as in the case of the production cross section analysis [1]. It is clearly seen that the suppression is stronger for low p_T .

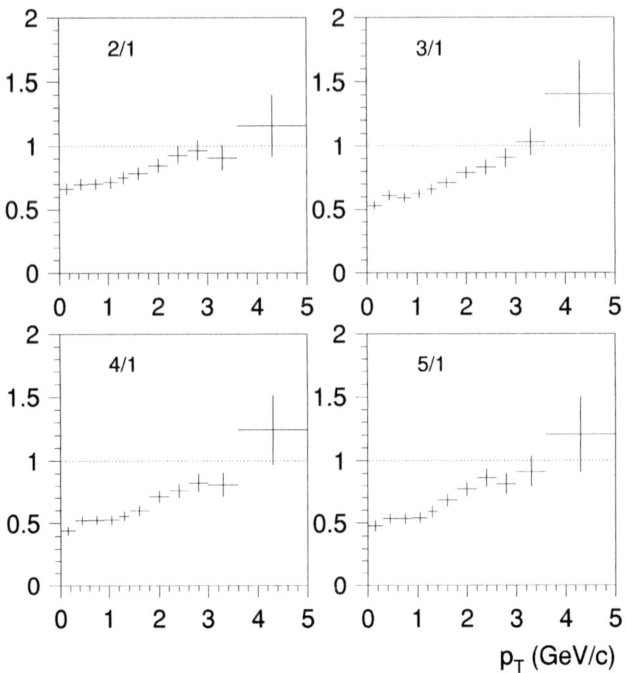

FIGURE 3. Ratios between the J/ψ p_T distributions in the E_T bin i (i = 2, 3, 4, 5) and in the first E_T bin, $R_i(p_T)$. Reprinted from [3] Copyright 2001, with permission from Elsevier Science.

The transverse mass distribution was fitted with the exponential function $M_T \cdot exp(-(M_T - M)/T)$, and T-values were determined for different E_T intervals. Fig.4 shows an increase of T from peripheral to semi-peripheral collisions followed by a plateau when centrality increases further up.

We can note that for J/ψ production the T-values for very central events are higher than it was predicted for the phase transition to quark-gluon plasma in lattice calculations [4]. A small increase of T for the most central events cannot be confirmed statistically. The estimate of the T-value for the most central ψ′ events is of the order of 180 MeV.

CONCLUSIONS

The observation by experiment NA50 of a two step pattern behaviour for the production of J/ψ as a function of centrality could be interpreted as a result of the Debye color screening in a quark-gluon plasma formed during the collision.

The study of the transverse momentum as a function of the centrality of collision reveals additional information. The interpretation of effects like larger $\langle p_T^2 \rangle$ for

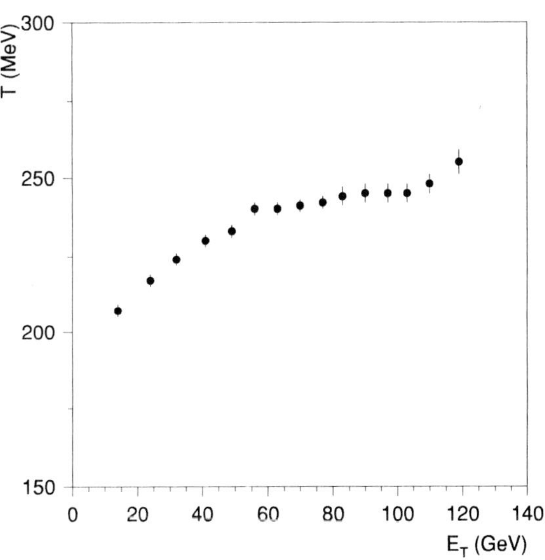

FIGURE 4. Inverse slope parameter, T, of the J/ψ transverse mass distributions, plotted as a function of the transverse energy. Reprinted from [3] Copyright 2001, with permission from Elsevier Science.

ψ′ production, larger J/ψ suppression for low p_T, large temperature for more central collisions still needs to be worked out. A combined theoretical analysis of charmonia states suppression and transverse momentum distributions is also needed.

This work was partially supported by the Russian Fondation for Fundamental Research, grant 99-02-16003.

REFERENCES

1. Abreu, M. C. et al.,na50 Collaboration, *Phys. Lett.* B **477**, 28 (2000)
2. Matsui, T. and Satz, H., *Phys. Lett.* B **178**, 416 (1986)
3. Abreu, M. C. et al., na50 Collaboration, *Phys. Lett.* B **499**, 85 (2001)
4. Boyd, G. et al., *Nucl. Phys.* B **469**, 419 (1996)

*NA50 Collaboration:

M.C. Abreu[6,a], B. Alessandro[10], C. Alexa[3], R. Arnaldi[10], M. Atayan[12], C. Baglin[1], A. Baldit[2], M. Bedjidian[11], S. Beolè[10], V. Boldea[3], P. Bordalo[6,b], S.R. Borenstein[9,c], G. Borges[6] A. Bussière[1], L. Capelli[11], J. Castor[2], C. Castanier[2], B. Chaurand[9], B. Cheynis[11], E. Chiavassa[10], C. Cicalò[4], T. Claudino[6], M.P. Comets[8], N. Constans[9], S. Constantinescu[3], P. Cortese[10,d], J. Cruz[6], N. De Marco[10], A. De Falco[4], G. Dellacasa[10,d], A. Devaux[2], S. Dita[3], O. Drapier[11], B. Espagnon[2], J. Fargeix[2], P. Force[2], M. Gallio[10], Y.K. Gavrilov[7], C. Gerschel[8], P. Giubellino[10], M.B. Golubeva[7], M. Gonin[9], A.A. Grigorian[12], S. Grigorian[12], J.Y. Grossiord[11], F.F. Guber[7], A. Guichard[11], H. Gulkanyan[12], R. Hakobyan[12], R. Haroutunian[11], M. Idzik[10,e], D. Jouan[8], T.L. Karavitcheva[7], L. Kluberg[9], A.B. Kurepin[7], Y. Le Bornec[8], C. Lourenço[5], P. Macciotta[4], M. Mac Cormick[8], A. Marzari-Chiesa[10], M. Masera[10], A. Masoni[4], M. Monteno[10], A. Musso[10], P. Petiau[9], A. Piccotti[10], J.R. Pizzi[11], W. Prado da Silva[10,f], F. Prino[10], G. Puddu[4], C. Quintans[6], S. Ramos[6,b], L. Ramello[10,d], P. Rato Mendes[6], L. Riccati[10], A. Romana[9], H. Santos[6], P. Saturnini[2], E. Scalas[10,d], E. Scomparin[10] S. Serci[4], R. Shahoyan[6,g], F. Sigaudo[10], S. Silva[6], M. Sitta[10,d], P. Sonderegger[5,b], X. Tarrago[8], N.S. Topilskaya[7], G.L. Usai[4], E. Vercellin[10], L. Villatte[8], N. Willis[8].

[1] LAPP, CNRS-IN2P3, Annecy-le-Vieux, France.
[2] LPC, Univ. Blaise Pascal and CNRS-IN2P3, Aubière, France.
[3] IFA, Bucharest, Romania.
[4] Università di Cagliari/INFN, Cagliari, Italy.
[5] CERN, Geneva, Switzerland.
[6] LIP, Lisbon, Portugal.
[7] INR, Moscow, Russia.
[8] IPN, Univ. de Paris-Sud and CNRS-IN2P3, Orsay, France.
[9] LPNHE, Ecole Polytechnique and CNRS-IN2P3, Palaiseau, France.
[10] Università di Torino/INFN, Torino, Italy.
[11] IPN, Univ. Claude Bernard Lyon-I and CNRS-IN2P3, Villeurbanne, France.
[12] YerPhI, Yerevan, Armenia.

a) also at UCEH, Universidade de Algarve, Faro, Portugal
b) also at IST, Universidade Técnica de Lisboa, Lisbon, Portugal
c) on leave of absence from York College CUNY
d) Universitá del Piemonte Orientale, Alessandria and INFN-Torino, Italy
e) also at Faculty of Physics and Nuclear Techniques, Academy of Mining and Metallurgy, Cracow, Poland
f) now at UERJ, Rio de Janeiro, Brazil
g) on leave of absence of YerPhI, Yerevan, Armenia

Results on Hyperon Production from CERN NA57 Experiment

V. Manzari on behalf of the NA57 Collaboration

Istituto Nazionale di Fisica Nucleare, Sez. di Bari, via Amendola 173, I-70126, Bari, Italy

F. Antinori[k], A. Badalà[f], R. Barbera[f], A. Bhasin[d], I.J. Bloodworth[d], G.E. Bruno[a], S.A. Bull[d], R. Caliandro[a], M. Campbell[g], W. Carena[g], N. Carrer[g], R.F. Clarke[d], A.P. de Haas[r], P.C. de Rijke[r], D. Di Bari[a], S. Di Liberto[n], R. Divia[g], D. Elia[a], D. Evans[d], K. Fanebust[b], F. Fayazzadeh[j], J. Fedorišin[i], G.A. Feofilov[p], R.A. Fini[a], J. Ftáčnik[e], B. Ghidini[a], G. Grella[o], H. Helstrup[c], M. Henriquez[j], A.K. Holme[j], A. Jacholkowski[a], G.T. Jones[d], P. Jovanovic[d], A. Jusko[h], R. Kamermans[r], J.B. Kinson[d], K. Knudson[g], A.A. Kolozhvari[p], V. Kondratiev[p], I. Králik[h], A. Kravcakova[i], P. Kuijer[r], V. Lenti[a], R. Lietava[e], G. Lovhøiden[j], M. Lupták[h], V. Manzari[a], G. Martinská[i], M.A. Mazzoni[n], F. Meddi[n], A. Michalon[q], M. Morando[k], D. Muigg[r], E. Nappi[a], F. Navach[a], P.I. Norman[d], A. Palmeri[f], G.S. Pappalardo[f], B. Pastirčák[h], J. Pišút[e], N. Pišútová[e], F. Posa[a], E. Quercigh[k], F. Riggi[f], D. Röhrich[b], G. Romano[o], K. Šafařík[g], L. Šándor[h], E. Schillings[r], G. Segato[k], M. Sené[l], R. Sené[l], W. Snoeys[g], F. Soramel[k], P. Staroba[m], T.A. Toulina[p], R. Turrisi[k], T.S. Tveter[j], J. Urbán[i], F. Valiev[p], A. van den Brink[r], P. van de Ven[r], P. Vande Vyvre[g], N. van Eijndhoven[r], J. van Hunen[g], A. Vascotto[g], T. Vik[j], O. Villalobos Baillie[d], L. Vinogradov[p], T. Virgili[o], M.F. Votruba[d], J. Vrláková[i] and P. Závada[m]

[a] Dipartimento IA di Fisica dell'Università e del Politecnico di Bari and INFN, Bari, Italy
[b] Fysisk Institutt, Universitetet i Bergen, Bergen, Norway
[c] Hogskolen i Bergen, Bergen, Norway
[d] University of Birmingham, Birmingham, UK
[e] Comenius University, Bratislava, Slovakia
[f] University of Catania and INFN, Catania, Italy
[g] CERN, European Laboratory for Particle Physics, Geneva, Switzerland
[h] Institute of Experimental Physics, Slovak Academy of Science, Košice, Slovakia
[i] P.J. Šafařík University, Košice, Slovakia
[j] Fysisk Institutt, Universitetet i Oslo, Oslo, Norway
[k] University of Padua and INFN, Padua, Italy
[l] Collège de France, Paris, France
[m] Institute of Physics, Prague, Czech Republic
[n] University "La Sapienza" and INFN, Rome, Italy
[o] Dipartimento di Scienze Fisiche "E.R. Caianiello" dell'Università and INFN, Salerno, Italy
[p] State University of St. Petersburg, St. Petersburg, Russia
[q] Institut de Recherches Subatomique, IN2P3/ULP, Strasbourg, France
[r] Utrecht University and NIKHEF, Utrecht, The Netherlands

Abstract. NA57 has been specifically designed to study the onset of strange baryon and antibaryon enhancements in Pb-Pb with respect to p-Be collisions, first observed by the WA97 experiment at 160 A GeV/*c* beam momentum. The aim is to look for the dependence of these enhancements on the interaction volume, as measured by the number of wounded nucleons, and the energy per incoming nucleon. In NA57 the centrality range goes down to a lower limit of about 50 wounded nucleons, compared with about 100 in WA97. Data have been collected both at 160 and at 40 A GeV/*c*, while WA97 collected data only at the top SPS beam momentum of 160 A GeV/*c*. In this contribution we recall the main features of the NA57 experiment and we present the first results on Ξ^- and $\overline{\Xi}^+$ hyperon production in Pb-Pb collisions at 160 A GeV/c.

CP602, *QCD@Work: International Workshop on Quantum Chromodynamics*
edited by P. Colangelo and G. Nardulli

INTRODUCTION

The WA97 experiment at CERN SPS has measured an enhancement of strange and multi-strange baryon and anti-baryon yields at midrapidity when going from p-Be to central Pb-Pb collisions at 160 A GeV/*c* [1].
Fig. 1 shows the WA97 yields per wounded nucleon per unit of rapidity at mid rapidity relative to p-Be for negative particles, Λ, Ξ^-, Ω^- and their antiparticles, as functions of the average number of wounded nucleons for p-Be, p-Pb and Pb-Pb collisions. The Pb-Pb data are divided in four centrality classes according to the number of wounded nucleon. Wounded nucleons are those nucleons both from projectile and target which undergo at least one primary inelastic scattering with another nucleon. All yields are extrapolated to a common phase space window, covering full p_T and one unit of rapidity centred at midrapidity.

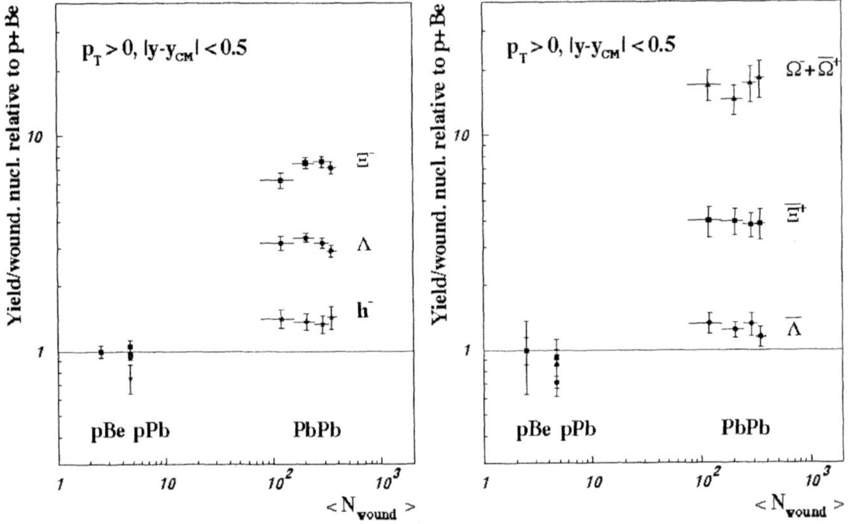

FIGURE 1. WA97 particle yields per unit of rapidity at central rapidity per wounded nucleon relative to p-Be yields, as functions of the number of wounded nucleons. The line refers to a yield proportional to the number of participants crossing the p-Be common point.

The horizontal bars of Pb-Pb points (yields) represent the FWHM of the distribution of the number of nucleons taking part to the collision in the four multiplicity classes.
Particles are divided in two groups: those with at least one valence quark in common with the nucleon (left) and those with no valence quark in common with the nucleon (right). We have kept them separate since it is empirically known that they may exhibit different production features, e.g.: Λ and $\bar{\Lambda}$ have different rapidity distributions both in p-S and S-S data [2].
From Fig. 1 one can observe that:

- while the yields per wouded nucleon have similar values in p-Be and p-Pb collisions for all the particles under study, they are found to be enhanced when going from p-A to Pb-Pb collisions. The horizontal line at 1 points out where Pb-Pb yields would have been in case of no enhancement;

- the enhancement is progressively stronger for the hyperons of higher strangeness content ($\Omega > \Xi > \Lambda$), up to a factor about 15 for $\Omega^- + \overline{\Omega}^+$;
- the yields per wounded nucleon appear to be rather constant within the centrality range covered by WA97, i.e. $\langle N_{wound} \rangle > 100$.

All these features cannot be explained by any current conventional hadronic microscopic model [3], while they nicely fulfil the predictions of a Quark Gluon Plasma phase transition in Pb-Pb collisions[4].

The picture emerging from the above results has led most of the WA97 Collaboration to carry on the study in order to search for an onset of the observed strangeness enhancement effect. In order to address the question of where the deconfinement sets in, a new experiment at the SPS, NA57, has been specifically designed to: a) extend the centrality range of Pb-Pb collisions at 160 A GeV/*c*; b) measure the pattern of multi-strange hyperon production as a function of the beam energy.

THE NA57 EXPERIMENT

The NA57 experiment has been designed for the systematic measurement of strange and multi-strange hyperon and anti-hyperon decays at central rapidity and medium p_T in A-A and p-A collisions. The NA57 layout, very similar to WA97, is outlined in Fig. 2, and has been described in details in[5,6].

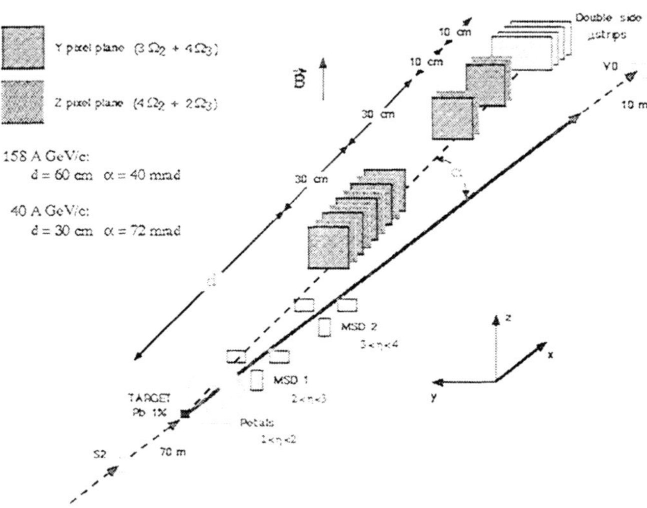

FIGURE 2. A schematicview of the NA57 layout (not to scale).

Tracks are reconstructed in a telescope tracker made of silicon pixel detectors, placed inside a 1.4 T magnetic field. Strange and multi-strange particles are identified by reconstructing their weak decays into final states containing only charged particles, e.g.:

$$\Lambda \rightarrow p+\pi^- \qquad + \text{c.c.}$$

$$\Xi^- \rightarrow \Lambda+\pi^- \qquad + \text{c.c.}$$
$$\quad \hookrightarrow p+\pi^-$$

$$\Omega^- \rightarrow \Lambda+\text{K}^- \qquad + \text{c.c.}$$
$$\quad \hookrightarrow p+\pi^-$$

The telescope is inclined above the beam line in order to accept particles produced in the target at central rapidity and medium transverse momentum: the full acceptance coverage is about one unit of rapidity around midrapidity and is the same for particles and anti-particles [3].

The centrality of the Pb-Pb collisions is determined by analyzing the charged particle multiplicity measured by the multiplicity strip detector (MSD) in the pseudorapidity interval $2<\eta<4$. The scintillator petals, placed 10 cm downstream the target, provide a signal to trigger on the centrality of the collision.

The sample of data collected by NA57 are reported in Table 1.

Table 1. Samples of data collected by NA57

System	Beam momentum	Sample size (trigger x 10^6)	Data taking year
Pb-Pb	160 A GeV/*c*	230	1998
		230	2000
Pb-Pb	40 A GeV/*c*	260	1999
p-Be	40 GeV/*c*	60	1999
		110	2001

The 160 A GeV/*c* Pb-Pb data will provide information on the hyperon production at lower centrality with respect to WA97.

The 40 A GeV/*c* Pb-Pb sample of data allows to study the strange baryon and anti-baryon production at lower center of mass energy: for instance, a significantly lower enhancement at 40 A GeV/*c* than at 160 A GeV/*c* would suggest the onset of deconfinement to be in between the two energies. The p-Be reference data at 40 GeV/c have been taken in the same condition for comparison with lead data at the same energy.

DATA ANALYSIS

Pb-Pb events are fully reconstructed and hyperon signals are extracted on the basis of kinematical selections with the same method used in WA97 [7].

Fig. 3 shows Ξ^- and $\overline{\Xi}^+$ signals from the whole Pb-Pb data sample at 40 A GeV/c and the relative y-p_T acceptance window.

Fig. 4 shows the Ξ^- and $\overline{\Xi}^+$ signals from the whole Pb-Pb data sample at 160 A GeV/c from the 1998 run and the relative $y-p_T$ acceptance window.

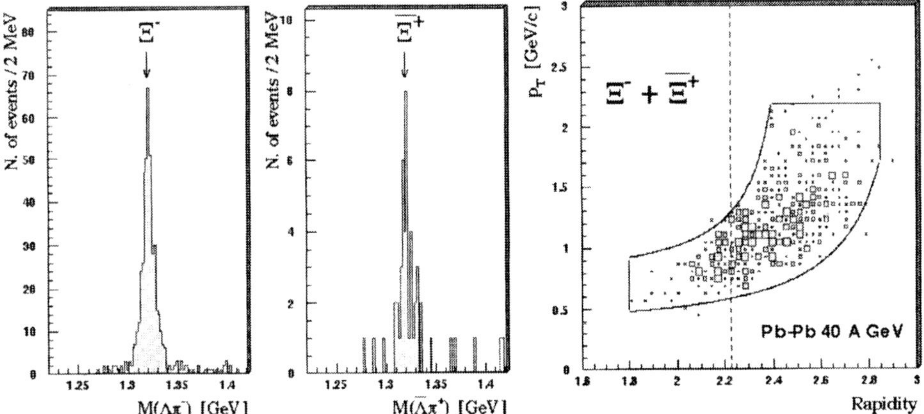

FIGURE 3. Ξ^- and $\overline{\Xi}^+$ signals for Pb-Pb collisions at 40 A GeV/c (left) and the corresponding y-p_T acceptance window.

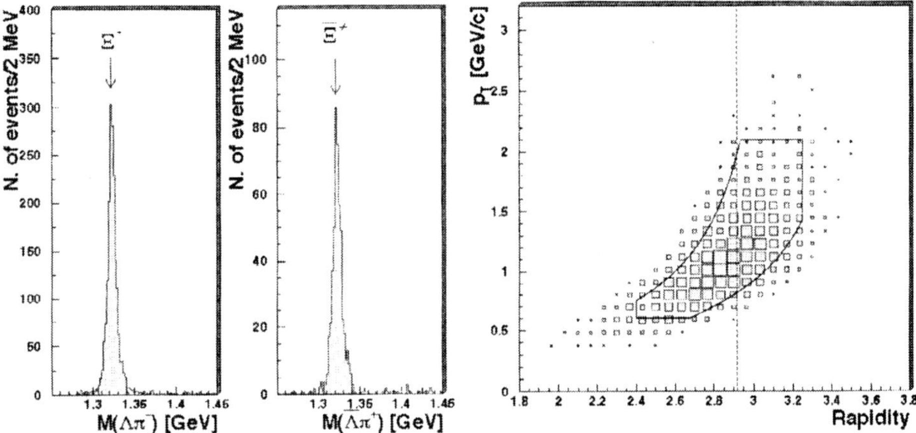

FIGURE 4. Ξ^- and $\overline{\Xi}^+$ signals for Pb-Pb collisions at 160 A GeV/c (left) and the corresponding y-p_T acceptance window.

In both figures the central rapidity is highlighted with a dashed line on the acceptance window. Due to the symmetry of the colliding system data can be reflected around midrapidity. The mass spectra for Ξ^- and $\overline{\Xi}^+$ are centered at the nominal value and show very low background: the FWHM is about 15 MeV at 40 A GeV/c and 10 MeV at 160 A GeV/c. Compared with the 160 A GeV/c data, one can observe that the production cross sections for hyperons and in particular anti-hyperons are significantly lower at 40 A GeV/c.

Since the enhancements observed by WA97 increase with the strangeness content of the particle, we started the analysis from cascades. In Pb-Pb events at 160 A GeV/c clear Ξ and Ω signals are visible in the correlation of the $\Lambda\pi$ and ΛK invariant masses, after preliminary analysis cuts, as shown in Fig. 5.

The Ξ^- and $\overline{\Xi}^+$ yields with the available statistics allows a study of the enhancement dependence on the collision centrality down to $N_{wound} \approx 50$. When the data sample from year 2000 data taking will be available for the analysis, the same kind of study will be feasible for the more rare Ω signal.

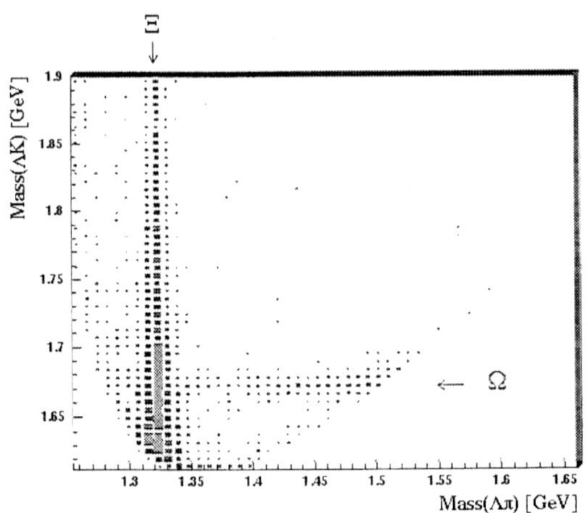

FIGURE 5. Scatter plot of ΛK versus $\Lambda\pi$ invariant mass for Pb-Pb collisions at 160 A GeV/c.

Data are corrected for geometrical acceptance, detector efficiency and reconstruction efficiency: a weight is calculated for each observed particle by a MonteCarlo simulation based on GEANT, as in WA97 [7].

The particle yields on the whole transverse momentum (p_T) range are calculated by integrating the transverse mass ($m_T = \sqrt{m^2 + p_T^2}$) distribution over one unit of rapidity and extrapolating to $p_T = 0$, according to the following formula:

$$Yield = \int_m^{\infty} dm_T \int_{y_{cm}-0.5}^{y_{cm}+0.5} dy \frac{d^2N(m_T, y)}{dm_T dy} \tag{1}$$

where the (m_T, y) distributions of identified hyperons are parametrized as

$$\frac{d^2N(m_T, y)}{dm_T dy} = f(y) m_T \exp\left(-\frac{m_T}{T}\right) \tag{2}$$

In equation (2) the rapidity function $f(y)$ is assumed to be constant in the acceptance region of the experiment and the inverse slope parameter T is evaluated by a maximum likelihood fit [8].

Ξ AND ANTI-Ξ YIELDS AT 160 A GEV/C

As mentioned in the Introduction, NA57 aims to study the enhancement as a function of the interaction volume, as measured by the number of nucleons taking part in the collision, i.e. wounded nucleons.

In order to enlarge the centrality range covered by WA97 to more peripheral collisions, special efforts were made in NA57 to reduce background sources. The NA57 charged particle multiplicity distribution measured in the pseudorapidity range $2<\eta<4$ is shown in Fig. 6. About the most central 60% of the total inelastic cross section was selected in NA57, to be compared with 40% in WA97. The drop in the multiplicity distribution at very low values is the effect of the centrality trigger applied to collect data, which suppresses low multiplicity events.
The multiplicity distribution is described in the framework of the Wounded Nucleon Model (WNM) [9]: the number of wounded nucleons is estimated from the trigger cross section using a Glauber model [10]. The Pb-Pb data sample is divided in 5 classes of multiplicity and the average number of wounded nucleons is determined for each class. The four most central classes (I to IV) are the same used in WA97 analysis, while the extended centrality range of NA57 allows to define one more peripheral class 0. The distribution of wounded nucleons for the five multiplicity classes is also shown in Fig. 6: for the new class 0 the average is 62.

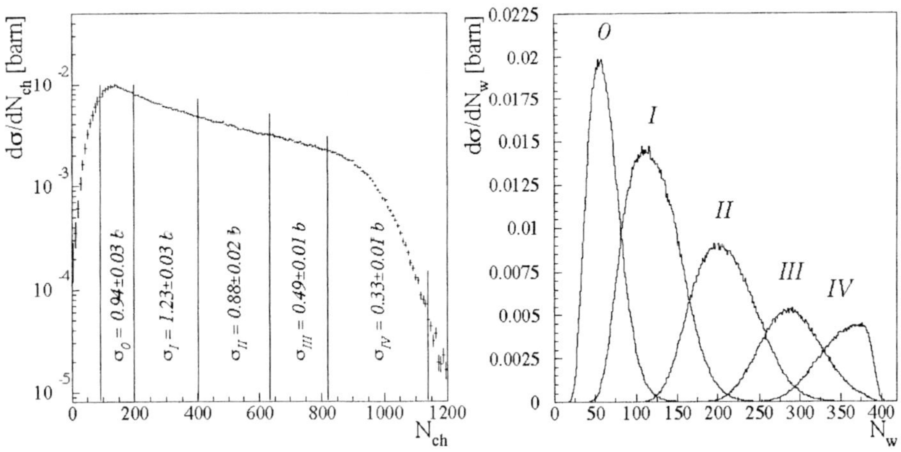

FIGURE 6. NA57 charged particle multiplicity distribution for Pb-Pb collisions at 160 A GeV/*c* (left) and wounded nucleon distribution for the five centrality classes (right). The cross section corresponding to each centrality class is also shown.

The NA57 Ξ^- and $\overline{\Xi}^+$ yields per participant in the five centrality classes are shown in Fig. 7, with the corresponding WA97 points for the four most central classes. In the common centrality range, the NA57 yields agree within 20-30% with WA97 yields, the latter being smaller. The reasons for this difference, which appears to be a systematic effect, are currently being investigated. From Fig. 7 a drop of Ξ^- and $\overline{\Xi}^+$ absolute yields, as measured by NA57, is visible in the new most peripheral bin. In particular, $\overline{\Xi}^+$ yield per wounded nucleons drops by a factor 2.6 when going from $N_{wound} = 121$ to $N_{wound} = 62$, i.e. from class I to class 0, corresponding to a 3.5 σ effect. The measured sudden reduction of yields as a function of centrality is not an artifact of our acceptance and efficiency correction procedure, since a similar drop is already shown by uncorrected data.

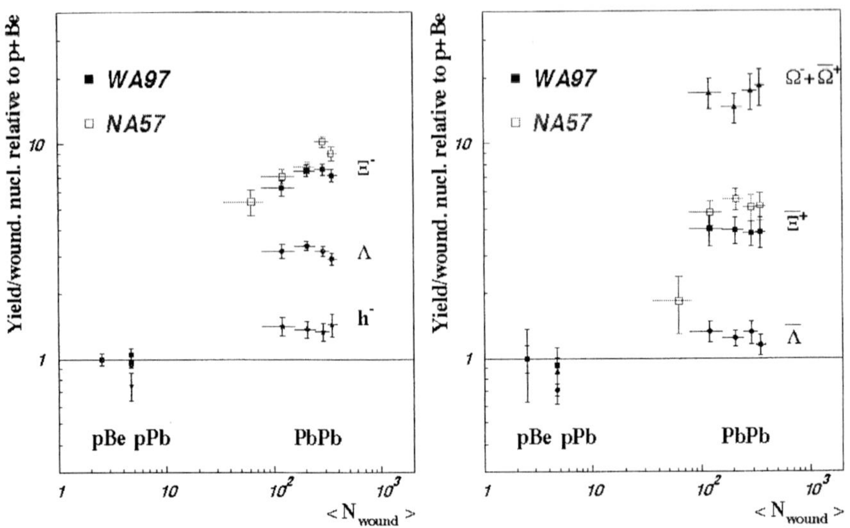

FIGURE 7. Ξ^- and $\overline{\Xi}^+$ yields per wounded nucleon relative to p-Be (same as in Figure 2) from NA57 data (open squares) are superimposed to the WA97 results (closed symbols).

CONCLUSIONS AND OUTLOOK

The NA57 has collected Pb-Pb and p-Be data at 40 A GeV/*c* beam momentum as well as Pb-Pb data at 160 A Gev/*c*. The first results on Ξ^- and $\overline{\Xi}^+$ yields at 160 A GeV/*c* from NA57 provide an indication of an onset of the enhancement between 50 and 100 nucleons taking part in the collision. The ongoing data analysis will provide information on the full pattern of strange baryon and antibaryon enhancements, which might show evidence of the QGP phase transition.

REFERENCES

1. R.A. Fini et al., *J. Phys. G.: Nucl. Part. Phys.* 27, 375-381 (2001).
   F. Antinori et al., *Nucl. Phys.* A661, 130c-139c (1999).
2. T. Alber et al., *Z. Phys.* C46, 195 (1994).
3. U. Heinz, *Nucl. Phys.* A661, 140c-149c (1999).
   F. Antinori et al., Eur. Phys. J. C11, 79-88 (1999).
4. J. Rafelski and B. Müller, *Phys. Rev. Lett.* 48, 1066 (1982).
5. V. Manzari et al., *J. Phys. G.: Nucl.. Part. Phys.* 27, 383-390 (2001).
6. V. Manzari et al., *Nucl. Phys.* A661, 716c-720c (1999).
7. I. Kralik et al., *Nucl. Phys.* A638, 115c-124c (1998).
8. F. Antinori et al., *Eur.. Phys. J.* C14, 633-641 (2000).
9. C.Y. Wong, *Introduction to High–Energy Heavy–Ion Collisions (World Scientific Publishing, Singapore 1994) 251, and references therein.*
10. N. Carrer et al., *J. Phys. G.: Nucl.. Part. Phys.* 27, 391-396 (2001).

The First Year At RHIC

T. S. Ullrich

Brookhaven National Laboratory, Upton, New York 11973

Abstract. The field of relativistic heavy ion physics is entering a new regime with the startup of the Relativistic Heavy Ion Collider (RHIC). RHIC commenced operation for physics in the Summer of 2000 in a short but successful run delivering the highest center-of-mass energies ever reached in the laboratory. In this talk I give a brief overview of RHIC and its experiments, and discuss some of the many physics results obtained so far.

INTRODUCTION

The aim of high energy heavy-ion physics is the study of strongly interacting matter at extreme energy densities. Statistical QCD [1] predicts that, at sufficiently high density, there will be a transition from hadronic matter to a plasma of deconfined quarks and gluons (QGP) [2] – a transition which in the early universe took place in the inverse direction some 10^{-5} s after the Big Bang and which might still play a role today in the core of collapsing neutron stars [3].

The study of the phase diagram of nuclear matter, using methods and concepts from both nuclear and high-energy physics, constitutes a new and interdisciplinary approach to investigate matter and its interactions. It is of interest to explore and test QCD on its natural scale (Λ_{QCD}) and to address the fundamental questions of confinement and chiral-symmetry breaking. Moreover, it is of general relevance in understanding the dynamic nature of phase transitions involving elementary quantum fields, as the QCD phase transition is the only one accessible to laboratory experiments.

Already the early exploratory programs at the Brookhaven National Laboratory Alternating Gradient Synchrotron (AGS) and the CERN Super Proton Synchrotron (SPS) have established the feasibility of high energy ion-ion experiments with their abundant particle production. They have shown that high energy densities of about 2.5 GeV/fm^3 can indeed be obtained in these reactions and have produced evidence for the onset of new collective phenomena. Some characteristic features of these reactions (e.g. momentum distributions and particle ratios) are already surprisingly close to the ones expected for a macroscopic system of hadronic matter heated to a temperature of close to or above the critical temperature in the initial phase of the collision [4].

Collisions of heavy ions at RHIC are expected to exceed the energy densities required for this transition. With the start of operations of the Relativistic Heavy Ion Collider (RHIC)[5] in June 2000 the field has entered a new realm of heavy ion collider physics, where perturbative QCD (pQCD) effects become important. Hard scattering processes are expected as well as increased energy and particle densities. A new round of collider experiments at RHIC is aimed at identifying the deconfinement phase transition and the

CP602, *QCD@Work: International Workshop on Quantum Chromodynamics*
edited by P. Colangelo and G. Nardulli

effects of chiral symmetry restoration, and characterizing the properties of each [6].

In this paper, results will be reported from the inaugural run at RHIC, which was just completed nine months prior to this Workshop. This paper will be divided into two main sections. Following the introduction, a brief overview of the RHIC collider and experiments will be presented. The remainder of this paper will focus on the first physics results from the Summer 2000 run including results on global observables, elliptic flow, two-particle interferometry, and preliminary results on high-pt phenomena. A concluding statement will be made on the physics that is anticipated from RHIC in the future.

THE RELATIVISTIC HEAVY ION COLLIDER AND ITS FIRST RUN

The first physics run at the Relativistic Heavy Ion Collider (RHIC) took place in the Summer of 2000. The RHIC accelerator-collider complex at Brookhaven is displayed in a schematic diagram in Fig. 1. During the Summer 2000 run, Au beams were accelerated from the Tandem Van de Graaff accelerator through a transfer line into the AGS Booster synchrotron and then into the AGS before injection into RHIC. RHIC was designed to accelerate and collide ions from protons up to the heaviest nuclei over a range of energies, up to 250 GeV for protons and 100 A GeV for Au nuclei. Beam energies during this first run were kept to a moderate 65 A GeV. RHIC attained its goal of ten percent of design luminosity by the end of its first run at the collision center-of-mass energy of $\sqrt{s_{NN}} = 130$ GeV.

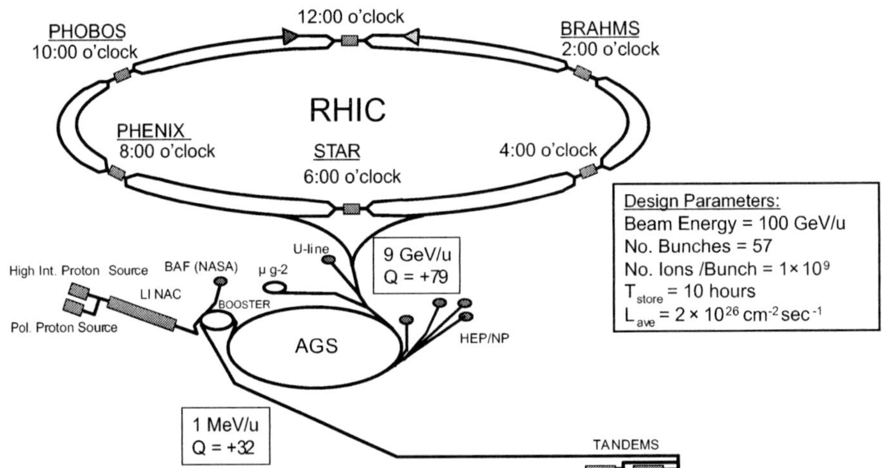

FIGURE 1. The Relativistic Heavy Ion Collider (RHIC) accelerator complex at Brookhaven National Laboratory. Nuclear beams are accelerated from the tandem Van de Graaff, through the transfer line into the AGS Booster and AGS prior to injection into RHIC. Details of the characteristics of proton and Au beams are also indicated after acceleration in each phase.

RHIC EXPERIMENTS

There are currently four experiments at RHIC. These experiments have various approaches to study the deconfinement phase transition to the quark gluon plasma. The STAR experiment [7] concentrates on measurements of hadron production over a large solid angle in order to measure single- and multi-particle spectra and to study global observables on an event-by-event basis. The PHENIX experiment [8] focuses on measurements of lepton and photon production and has the capability of measuring hadrons in a limited range of azimuth and pseudorapidity. The two smaller experiments BRAHMS (a forward and midrapidity hadron spectrometer) [9] and PHOBOS (a compact multiparticle spectrometer) [10] focus on single- and multi-particle spectra. The collaborations, which have constructed these detector systems and which will exploit their physics capabilities, consist of approximately 900 scientists from over 80 institutions internationally. In addition to colliding heavy ion beams, RHIC will collide polarized protons to study the spin content of the proton [11]. STAR and PHENIX are actively involved in the spin physics program planned for RHIC.

The **BR**oad **R**ange **H**adron **M**agnetic **S**pectrometer BRAHMS experiment is designed to measure and identify charged hadrons ($\pi^{\pm}$, $K^{\pm}$, $\overset{(-)}{p}$) over a wide range of rapidity and transverse momentum for all beams and energies available at RHIC. Because the conditions and thus the detector requirements at midrapidity and forward angles are different, the experiment uses two movable spectrometers for the two regions.

As shown in Fig. 2, there is a midrapidity spectrometer to cover the pseudorapidity range $0 \leq \eta \leq 1.3$ and a forward spectrometer to cover $1.3 \leq \eta \leq 4.0$. The latter employs four dipole magnets, three time projection chambers (TPC), and drift chambers. Particle identification is achieved with time-of-flight hodoscopes, a threshold Cherenkov counter, and one ring-imaging Cherenkov counter (RICH). The solid angle acceptance of the forward arm is 0.8 mstr. The midrapidity spectrometer has been designed for charged particle measurements for $p \leq 5\mathrm{GeV}/c$. The spectrometer has two TPCs for tracking, a magnet for momentum measurement, and a time-of-flight wall and segmented gas Cherenkov counter (GASC) for particle identification. It has a solid angle acceptance of 7 mstr. A set of beam counters and a silicon multiplicity array provide the experiment with trigger information and vertex determination.

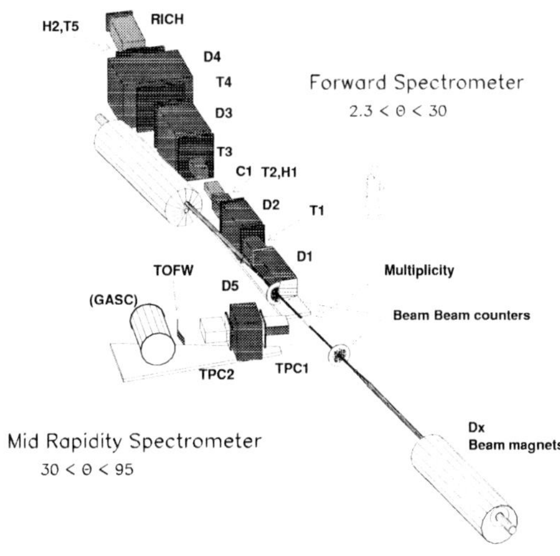

FIGURE 2 : Layout of the BRAHMS detector.

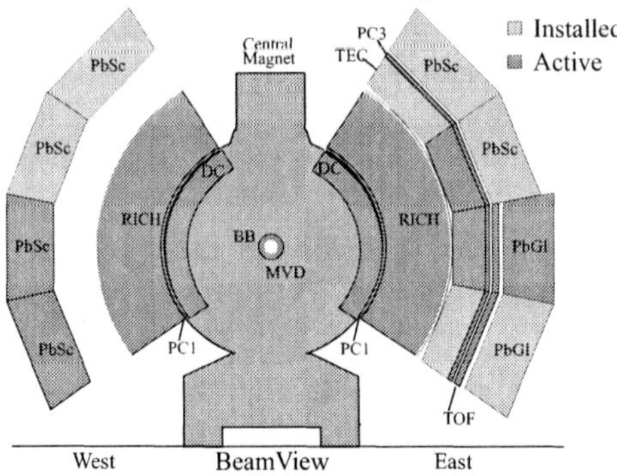

FIGURE 3 : Installed and active detectors for the RHIC Run-1 configuration of the PHENIX experiment.

The PHENIX detector consists of three spectrometers: two muon spectrometers covering the full azimuth for $1.1 < |\eta| < 2.4$ and a central spectrometer consisting of two arms each subtending 90° in azimuth and with $|\eta| < 0.35$. A central magnet provides an axial field, while each muon spectrometer contains a magnet that produces a roughly radial field. The central arms contain three tracking sub-systems: pad chambers (PC), drift chambers (DC) and time-expansion chambers (TEC); two forms of electromagnetic calorimetry (PbSc and PbGl); a time-of-flight hodoscope (TOF) and ring imaging Cerenkov counter (RICH). These sub-systems, together with a set of beam-beam counters (BBC) located in the region $3 < |\eta| < 3.9$, provide superb hadron and electron identification over a broad range of transverse momentum. The muon spectrometers use cathode strip chambers in three stations for tracking (muTr), and five layers of Iarocci tubes interleaved with iron absorber for muon identification (muID). For the first physics run of RHIC in the summer of 2000, the portions of the PHENIX detector shown in Fig. 3 were instrumented. Elements of all sub-systems, with the exception of the muTr, were in place and read out. All other sub-systems were instrumented in fractions ranging from 25% to 100% of their ultimate aperture and were used in the physics results. A total of approximately 5M events was recorded at $\sqrt{s_{NN}} = 130$ GeV.

The PHOBOS detector is designed to detect as many of the produced particles as possible and to allow a momentum measurement down to very low $p_{\perp}$. The setup consists of two parts: a multiplicity detector covering almost the entire pseudorapidity range of the produced particles and a two arm spectrometer at mid-rapidity. Figure 4 shows the detector, including the spectrometer arms, the multiplicity and vertex array, and the lower half of the magnet.

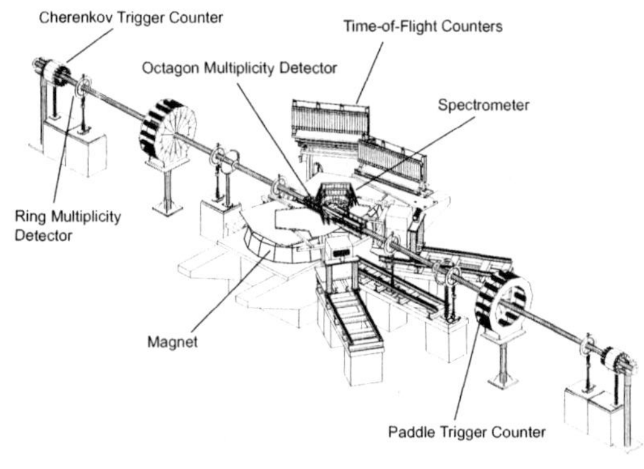

FIGURE 4 : PHOBOS detector setup for the 2000 running period.

One aspect of the design is that all detectors are produced using a common technology, namely as silicon pad or strip detectors. The multiplicity detector covers the range $-5.4 < \eta < 5.4$, measuring total charged multiplicity $dN_{ch}/d\eta$ over almost the entire phase space. For approximately 1% of the produced particles, information on momentum and particle identification will be provided by a two arm spectrometer located on either side of the interaction volume (only one arm was installed for the 2000 run). Each arm covers about 0.4 rad in azimuth and one unit of pseudorapidity in the range $0 < \eta < 2$, depending on the interaction vertex, allowing the measurement of $p_\perp$ down to 40 MeV/c. Both detectors are capable of handling the 600 Hz minimum bias rate expected for all collisions at the nominal luminosity.

The **S**olenoidal **T**racker **A**t **R**HIC (STAR) is a large acceptance detector capable of tracking charged particles and measuring their momenta in the expected high multiplicity environment. It is also designed for the measurement and correlations of global observables on an event-by-event basis and the study of hard parton scattering processes. The layout of the STAR experiment is shown in Fig. 5. The initial configuration of STAR in 2000 consists of a large time projection chamber (TPC) covering $|\eta| < 2$, a ring imaging Cherenkov detector covering $|\eta| < 0.3$ and $\Delta\phi = 0.1\pi$, and trigger detectors inside a solenoidal magnet with 0.25 T magnetic field. The solenoid provides a uniform magnetic field of maximum strength 0.5 T for tracking, momentum analysis and particle identification via ionization energy loss measurements in the TPC. Measurements in the TPC were carried out at mid-rapidity with full azimuthal coverage ($\Delta\phi = 2\pi$) and symmetry. A total of 1M minimum bias and 1M central events were recorded during the summer run 2000.

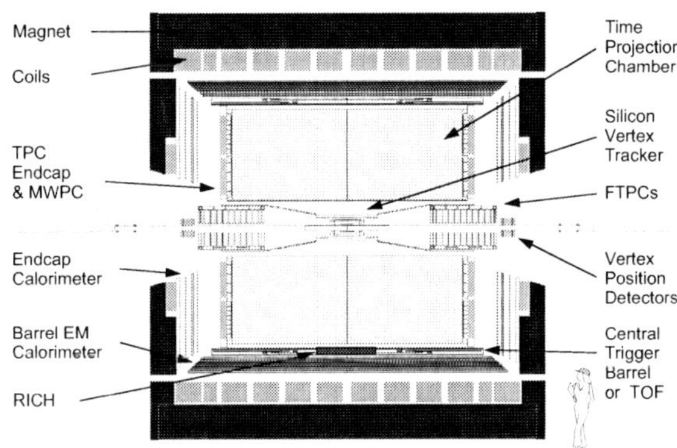

FIGURE 5 : Schematic view of the STAR detector.

Additional tracking detectors will be added for the run in 2001. These are a silicon vertex tracker (SVT) covering $|\eta| < 1$ and two Forward TPCs (FTPC) covering $2.5 < |\eta| < 4$. The electromagnetic calorimeter (EMC) will reach approximately 20% of its eventual $-1 < \eta < 2$ and $\Delta\phi = 2\pi$ coverage and will allow the measurement of high transverse momentum photons and particles. The endcap EMC will be constructed and installed over the next 2 – 3 years.

PHYSICS RESULTS

Charged Particle Multiplicity and Transverse Energy

The multiplicity of charged particles produced in heavy ion collisions arises from a variety of physics processes. In addition to the expected soft processes seen at lower energies, hard processes, nuclear shadowing, and rescattering all play a role [12].

The measurement of $(dN_{ch}/d\eta|_{\eta=0})/(\frac{1}{2}N_{part})$ as a function of energy [13] depicted in Fig. 6, was the first published result from RHIC. It shows that 70% more particles per participant pair are produced than at the SPS and 40% more than the extrapolation from $p\bar{p}$ (also shown) predicts at $\sqrt{s_{\mathrm{NN}}} = 130$ GeV. This is strong evidence that particle production is not simply due to independent NN interactions. Instead, whatever process amplifies the production at SPS energies relative to pp collisions is even stronger at RHIC. All RHIC experiments obtain very similar results. Also shown is the predicted energy dependence from the HIJING model [14] and an early EKRT calculation, a model based on parton saturation [15]. By integrating the measured $dN_{ch}/d\eta$ distributions obtained by PHOBOS, the total number of charged particles is $N_{ch} = 4200 \pm 470$.

A very surprising finding of PHENIX is that the average transverse energy per charged-particle appears to be constant over the full range of centrality ($\langle E_{\perp}\rangle/\langle N_{ch}\rangle \approx 0.8$ GeV) [16]. Even more surprising is that this number might be the same as at SPS energies as measured by WA98 and NA49. This seems to contradict the observations of STAR that $\langle p_{\perp}\rangle$ increases by 20% from SPS to RHIC energies [17].

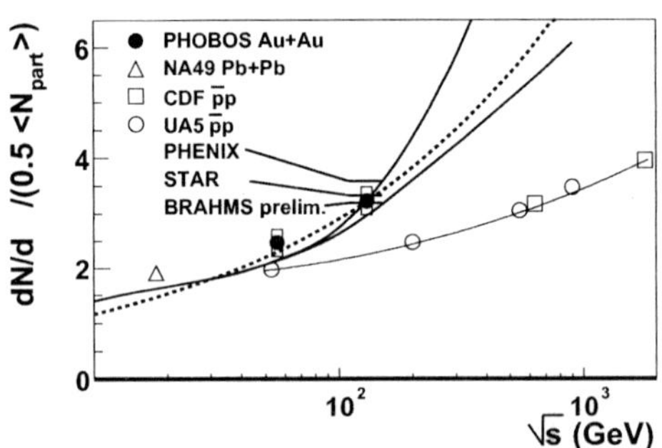

FIGURE 6 : Dependence of $dN_{ch}/d\eta/0.5N_{part}$ for $|\eta| < 1$ on the center-of-mass energy with results from all four RHIC experiments at $\sqrt{s_{NN}}$=130 GeV (from [13]).

One of the pressing questions concerns the energy density achieved in RHIC collisions. Using the Bjorken formula [18] at SPS one obtains an initial energy density of $\varepsilon_{BJ} \approx 3.2$ GeV/fm^3 [19]. Applying the same reasoning to the RHIC experiments would give a 70% increase. However, while the measurement of E_T seems straightforward, the Bjorken formula has theoretical uncertainties which do not decrease with beam energy. Further theoretical investigations are warranted in order to reduce the ambiguity in the concept of initial energy density.

Elliptic Flow

One of the most striking results from the initial round of RHIC results is the magnitude of the elliptical flow reported by the STAR experiment [20], which approaches the levels predicted by hydrodynamic calculations [21]. Since this initial result, the other experiments have confirmed this data and new information is now available about other aspects of this phenomenon [22].

The elliptic flow observable, which is sensitive to the early evolution of the system, is the anisotropic emission of particles "in" or "out" of the reaction plane defined for non-central collisions by the beam direction (z-axis) and the impact parameter direction (x-axis). Elliptic flow is usually characterized in terms of particle momenta by $v_2 = \langle (p_x^2 - p_y^2)/(p_x^2 + p_y^2)\rangle$, the second harmonic Fourier coefficient in the azimuthal distribution of particles with respect to the reaction plane.

Elliptic flow has its origin in the spatial anisotropy of the system when it is created in a non-central collision, and in particle rescatterings in the evolving system which convert the spatial anisotropy to momentum anisotropy. Being dependent on rescattering, the spatial anisotropy in general decreases with system expansion, thus quenching this effect and making elliptic flow particularly sensitive to the early stages of the system evolution [23]. Hydrodynamic models, which are based on the assumption of complete local thermalization, predict the strongest signals [21, 23, 24].

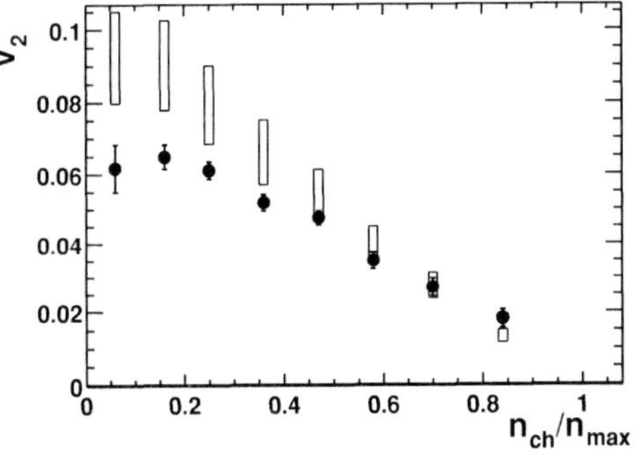

FIGURE 7 : Elliptic flow (solid points) for charged particles as a function of centrality defined as n_{ch}/n_{max}. The open rectangles show a range of values expected for v_2 in the hydrodynamic limit.

Figure 7 shows v_2 for charged particles as a function of centrality of the collision averaged over transverse momentum at midrapidity. v_2 reaches values of about 6% for relatively peripheral collisions, a value which is more than 50% larger than at the SPS [25], indicating stronger early-time thermalization at this RHIC energy. Also shown is the range of hydrodynamic predictions [23] for this energy. The data values for the lower multiplicities could indicate incomplete thermalization during the early time when elliptic flow is generated. On the other hand, for the most central collisions, comparison of the data with hydrodynamic calculations suggests that early-time thermalization may be nearly complete.

The differential anisotropic flow is a function of η and $p_\perp$. Figure 8 shows v_2 for charged particles as a function of $p_\perp$ for a minimum bias trigger. Mathematically, the v_2 value at $p_t = 0$, as well as its first derivative, must be zero, but it is interesting that v_2 appears to rise almost linearly with $p_\perp$ starting from relatively low values of $p_\perp$. This is consistent with a stronger "in-plane" hydrodynamic expansion of the system than the average radial expansion.

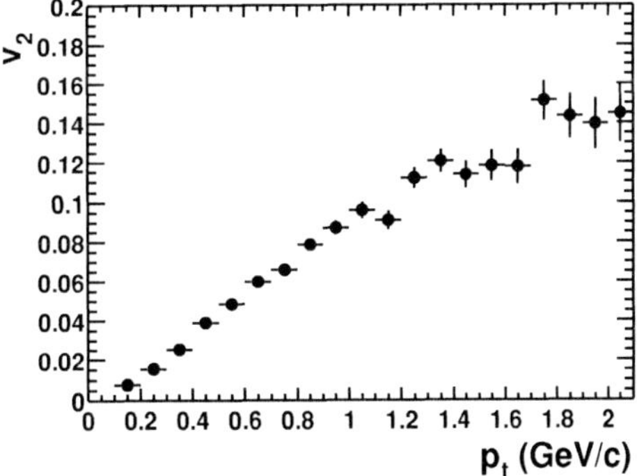

FIGURE 8 : v_2 as a function of $p_\perp$, as measured by STAR in $\sqrt{s_{NN}}$ = 130 GeV Au+Au collisions.

Comparing to estimates based on transport cascade models, one finds that elliptic flow is underpredicted by a factor of more than 2. Hydrodynamic calculations [23] for RHIC energies overpredict elliptic flow by about 20-50%. This is just the reverse of the situation at the SPS where cascade models gave a reasonable description of the data and hydrodynamic calculations were more than a factor of two too high. Also in contrast to lower collision energies, the observed shape of the centrality dependence of the elliptic flow is similar to hydrodynamic calculations and thus consistent with significant thermalization.

Two-Particle Interferometry (HBT)

The study of small relative momentum correlations, a technique also known as HBT [26] interferometry, is one of the most powerful tools to study complicated space-time dynamics of heavy ion collisions [27]. It provides crucial information which helps to improve our understanding of the reaction mechanisms and to constrain theoretical models of the heavy ion collisions. Interpretation of the extracted HBT parameters in terms of source sizes and lifetime is more or less straightforward for the case of chaotic static sources. In the case of expanding sources with strong space-momentum correlations (due to flow, etc.) the situation is more difficult, but the concept of length of homogeneity [28] provides a useful framework for the interpretation of data.

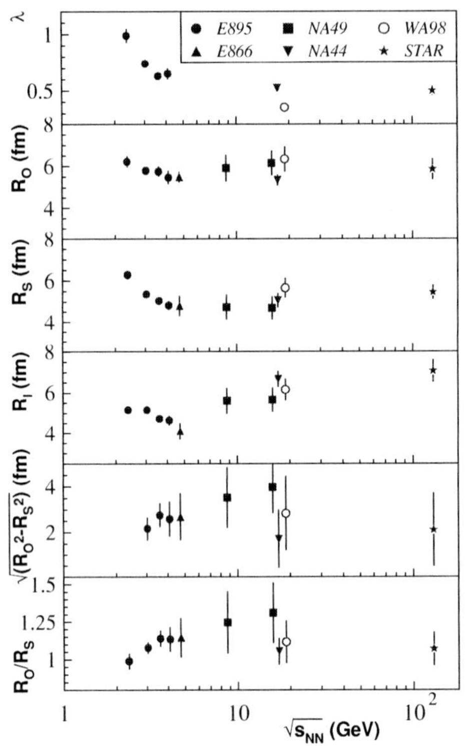

FIGURE 9 : Compilation of results on two-particle correlation (HBT) parameters from measurements using central collisions of Au + Au at the BNL-AGS, Pb + Pb at the CERN-SPS and Au + Au data from the STAR experiment at RHIC. Plotted are the coherence parameter λ, R_{out}, R_{side}, and $R_{longitudinal}$.

The dependence of the pion-emitting source parameters on the transverse momentum of the particle pairs (K_T) and on centrality was measured with high statistics in STAR [29]. In this study a multi-dimensional analysis is made using the standard Pratt-Bertsch decomposition [30] into outward, sideward, and longitudinal momentum differences and radius parameters. The data are analyzed in the longitudinally co-moving source frame, in which the total longitudinal momentum of the pair (collinear with the colliding beams) is zero.

As expected, larger sizes of the pion-emitting source are found for the more central (*i.e.* decreasing impact parameter) events, which in turn have higher pion multiplicities. This source size is observed to decrease with increasing transverse momentum of the pion pair. This dependence is similar to what has been observed at lower energies and is understood to be an effect of collective transverse flow. Shown in Fig. 9 is the coherence parameter λ and the radius parameters R_{out}, R_{side}, and R_{long} obtained in the analysis. Also shown are values of these parameters extracted from similar analyses at lower energies. All analyses are for low transverse momentum (∼ 170 MeV/c) negative pion pairs at midrapidity for central collisions of Au + Au or Pb + Pb. From Fig. 9 the values of λ, R_{out}, R_{side}, and R_{long} extend smoothly from the dependence at lower energies and do not reflect significant changes in the source from those observed at the CERN SPS energy. One of the biggest surprises is that the anomalously large source sizes or source lifetimes predicted for a long-lived mixed phase [31] have not been observed in this study. Preliminary results of the HBT analysis by the PHENIX Collaboration [32] agree with the STAR results within error bars.

One of the big puzzles, however, is the magnitude and the tranverse momentum (K_T) dependence of the ratio of R_{out}/R_{side} which contradicts *all* model predictions [31, 33]. These model calculations predict the ratio to be greater than unity due to system lifetime effects which cause R_{out} to be larger than R_{side}. They also predict that the ratio increases with K_T. Such an increase seems to be a generic feature of the models based on the Bjorken-type, boost-invariant expansion scenario. Hence, it was surprising to see that the experimentally observed ratio is less than unity and is decreasing as a function of K_T. Currently, it is far from clear what kind of scenario can lead to such a puzzling K_T dependence.

Particle Production at High Transverse Momentum

Out of all the impressive amount of new physics results from RHIC, the first evidence for "jet" quenching is certainly the finding that has been discussed most. The idea, proposed about 10 years ago [34], is quite simple. In the initial stage of the collision, quarks or gluons can scatter with high momentum transfer. The scattered partons, though fast, sense the hot and dense phase as the fireball evolves, losing a significant fraction of their momentum by induced gluon bremsstrahlung before escaping. The final fragmentation of the partons into jets of hadrons is then modified relative to the case in free space, exhibiting reduced transverse momenta of associated hadrons. In this sense the name "jet" quenching is somewhat unfortunate since it is not the "jet" that is quenched but the leading hadrons, *i.e.*, the fragmentation function is modified compared to that in elementary collisions.

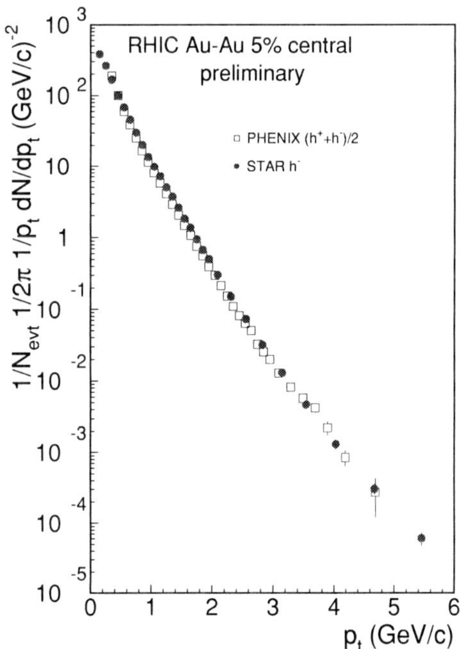

FIGURE 10 : Inclusive cross sections for the production of negatively charged hadrons from STAR and all charged hadrons from PHENIX for the most central 5% of the collisions. The data are independently normalized.

Both STAR [35] and PHENIX [36] have reported preliminary results on significantly reduced inclusive hadron cross sections at high $p_\perp$. Figure 10 shows inclusive $p_\perp$ distributions for negatively charged hadrons from STAR and all charged hadrons from PHENIX. The agreement of the data over a range of 7 orders of magnitude is impressive, illustrating the high level of analysis quality which these preliminary data have already achieved.

To demonstrate jet quenching, one needs a comparison basis. This is provided by high energy pp and $p\bar{p}$ data from CDF and UA1 (see [37]), assuming that all high $p_\perp$ particle production in AA (as in $pp/p\bar{p}$) results from binary hard collisions. A nuclear modification factor R_{AA} can then be defined as $R_{AA}(p_T) = (d\sigma_{AA}/dydp_\perp^2)/(\langle N_{binary}\rangle d\sigma_{pp}/dydp_\perp^2)$ [38], where the average number of binary collisions $\langle N_{binary}\rangle$ is obtained from the inelastic cross sections and the nuclear overlap integral. Results on R_{AA} as a function of $p_\perp$ for the two data sets of Fig. 10 are shown in Fig. 11. Values < 1 are expected for low $p_\perp$, since the cross sections in this region should scale with the number of participating nucleons rather than with the number of binary collisions. However, the high $p_\perp$ expectation of 1 for simple binary collision scaling is never reached, not to speak about the SPS level of 2 (due to the Cronin effect). Instead, a plateau is found at $R_{AA} = 0.6$–0.8, followed by a decrease at still higher $p_\perp$. This is further supported by the normalization of the central collision data to peripheral collisions rather than to pp [37].

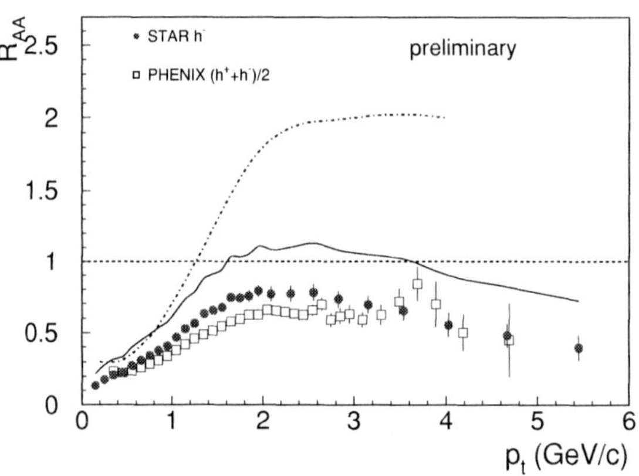

FIGURE 11 : Data from Fig. 10 normalized to nucleon-nucleon data. The thin line is the upper limit of the systematical uncertainty. The dash-dotted line corresponds to the average of the SPS data (from [37]).

Theoretically, the size of the observed effect can be accounted for by requiring an average energy loss of 0.25 GeV/fm for the scattered partons [39]. It should be clear that this is only a phenomenological value averaged over the evolution history of the fireball. It does not separately reveal the specific energy loss of the partons and the characteristics and density of the medium the partons penetrate.

In non-central collisions, the total parton propagation length should depend on the azimuthal direction. It is therefore conceivable that jet quenching would also show up as a specific azimuthal anisotropy of hadron spectra at large $p_\perp$, deviating from the low $p_\perp$ pattern. The suitable quantity to measure azimuthal anisotropy is v_2, the second Fourier coefficient of the azimuthal particle distribution relative to the reaction plane, usually called elliptic flow. Indeed, the STAR collaboration observes a flattening of v_2 for $p_\perp > 2\text{GeV}/c$. Wether this, however, is due to the expected decrease of elliptic flow for very high-pt particles or due to the onset of quenching is unresolved.

All in all, this "jet" quenching has given a hint for quark matter formation already after the first round of experiments. Of course, there are still uncertainties as to the normalization procedure, the influence of flow, and the role of the modification of the partonic structure functions (shadowing) which must be addressed.

SUMMARY

Compiling a summary of RHIC results from the first year is a challenge. The amount of results from the four experiments is overwhelming and their interpretation is still the subject of ongoing theoretical studies. There are many interesting results for which I had no room to discuss: strangeness, identified hadron spectra, particle ratios, and the ongoing efforts studying the question of thermalization. In the coming year the two large RHIC experiments will complete their detectors thus allowing to address many more physics topics. The first year and the huge amount of interesting results is only a first glimpse of what to expect from RHIC. Many of the data analyzed so far seem to confirm the overall picture that has emerged from previous studies at lower energies. A few observations, however, may force us already to review certain aspects of the standard picture of high energy nuclear collisions. There is evidence that quark and gluon degrees of freedom play a more important role at RHIC than at SPS and AGS and indeed QCD is being more directly involved in the interpretation of these data.

ACKNOWLEDGMENTS

I am grateful to many colleagues from all RHIC experiments for critical discussion, in particular I wish to thank: M. Baker, A. Drees, J. Harris, T. Ludlam, S. Panitkin, P. Steinberg, and F. Videbaek.

REFERENCES

1. F. Karsch *et al.*, Nucl. Phys. B605, 579-599 (2001).
2. See for example: J.W. Harris and B. Müller, Annu. Rev. Part. Sci. 46, 71-107 (1996) and J.-P. Blaizot, Nucl. Phys. A **661**, 3c (1999).
3. G. Baym, Nucl. Phys. **A590**, 233c-248c (1995)
4. U. Heinz, these proceedings.
5. Conceptual Design of the Relativistic Heavy Ion Collider (Report BNL 52195, 1989).
6. see Nucl. Phys. A661 (1999) and references therein.
7. Conceptual Design Report for the Solenoidal Tracker At RHIC, The STAR Collaboration, PUB-5347 (1992); J.W. Harris *et al.*, Nucl. Phys. A 566, 277c (1994).
8. PHENIX Experiment at RHIC - Preliminary Conceptual Design Report, PHENIX Collaboration Report (1992).
9. Interim Design Report for the BRAHMS Experiment at RHIC, BNL Report (1994).
10. RHIC Letter of Intent to Study Very Low pt Phenomena at RHIC, PHOBOS Collaboration (1991).
11. Proposal on Spin Physics Using the RHIC Polarized Collider, RHIC Spin Collaboration (1992).
12. D. Kharzeev and M. Nardi, Phys. Lett. B **507**, 121 (2001).
13. B. B. Back *et al.*, Phys. Rev. Lett. **85**, 3100 (2000).
14. X. Wang and M. Gyulassy, nucl-th/0008014.
15. K. J. Eskola, K. Kajantie, P. V. Ruuskanen and K. Tuominen, Nucl. Phys. **B570**, 379 (2000).
16. K. Adcox *et al.*, nucl-ex/0104015 (2001).
17. C. Adler *et al.*, Phys. Rev. Lett. 87, 112303 (2001).
18. J. D. Bjorken, Phys. Rev. D **27**, 140 (1983).
19. S. Margetis *et al.* Phys. Rev. Lett. **75**, 3814 (1995).
20. K. H. Ackermann *et al.*, Phys. Rev. Lett. **86**, 402 (2001).
21. J.-Y. Ollitrault, Phys. Rev. D **46**, 229 (1992).

22. C. Adler *et al.*, nucl-ex/0105011
23. P. Kolb, J. Sollfrank, U. Heinz, hep-ph/0006129.
24. D. Teaney and E.V. Shuryak, Phys. Rev. Lett. **83**, 4951 (1999).
25. A.M. Poskanzer and S.A. Voloshin for the NA49 Collaboration, Nucl. Phys. **A661**, 341c (1999).
26. R. Hanbury Brown and R.Q. Twiss, Phil. Mag. **45** (1954) 663.
27. U.A. Wiedemann and U. Heinz, Rhys. Rept. **319** (1999) 145.,
28. A. Makhlin and Y. Sinyukov, Z. Phys. **C39** (1998) 69.
29. C. Adler *et al.*, Phys. Rev. Lett. 87, 082301 (2001).
30. S. Pratt, Phys. Rev. D **33**, 1314 (1986); G. Bertsch, M. Gong and M. Tohyama, Phys. Rev. C **37**, 1896 (1988); and G. Bertsch, Nucl. Phys. A **498**, 151c (1989).
31. D. H. Rischke, Nucl. Phys. A610 (1996) 88c; D.H. Rischke and M. Gyulassy, Nucl. Phys. A608 (1996) 479.
32. S.C. Johnson, nucl-ex/0104020
33. S. Soff, S.A. Bass and A. Dumitru, nucl-th/0012085.
34. M. Gyulassi and M. Plümer, Phys. Lett. B243 (1990) 432; X.N. Wang and M. Gyulassi, Phys. Rev. D44 (1991) 3501 and Phys. Rev. Lett. 68 (1992) 1480.
35. J. Dunlop, in proceedings of the Quark Matter Conference 2001 to be published in Nucl. Phys A (2001).
36. J.Velkovska, nucl-ex/0105012
37. A. Drees, nucl-ex/0105019
38. E. Wang and X.N. Wang, nucl-th/0104031.
39. X.N. Wang, Phys. Rev. C61, 64910 (2000).

Identified Charged Particle Production In Au+Au Interactions At STAR-RHIC

D. Cozza* for the STAR Collaboration[1] and the STAR-RICH Collaboration[2]

*Dipartimento Interateneo di Fisica and INFN, Bari, Italy

Abstract. The study of the production of charged particles at high transverse momenta plays an important role in the understanding of the interaction mechanism of high-E_t jets with the dense nuclear matter.

A prototype of a RICH (Ring Imaging CHerenkov) detector developed in the framework of the CERN-ALICE experiment has been fully integrated in the STAR experiment at RHIC (BNL). It allows identification of primary charged particles (namely π, K and p) extending the region of acceptance of the experiment to medium-high transverse momenta (p_t < 2.5 GeV/c for π and K, p_t < 5 GeV/c for protons).

Methods of pattern recognition and identification of charged particles using the RICH will be presented. Preliminary results on identified charged particle ratios at midrapidity and 1.5 < p_t < 2.5 GeV/c will be also shown.

INTRODUCTION

The study of the relativistic heavy ion collisions at the RHIC energies, where hard scattering of quarks and gluons plays an important role in the collision process, will provide a more complete understanding of the properties of the dense partonic matter. At such energies, a new regime characterized by the appearance of semi-hard processes (called "minijets") is reached. Such collisions could, therefore, be described in terms of perturbative QCD (pQCD), allowing a cleaner comparison between nucleus-nucleus and nucleon-nucleon interactions.

At RHIC energies minijets have been estimated to contribute for a half of the total transverse energy achieved in central heavy ion collisions[1]. Although unresolvable experimentally as single jets, they will contribute significantly, as in pp and $\bar{p}p$ interactions, to the particle spectra in the region of the high transverse momenta.

In particular, proton and antiproton production at high p_t will represent a useful probe in the understanding of the mechanism of the jet quenching in dense matter[2].

---

[1] For a complete author list see: J.W. Harris, Proceedings of *"15th International Conference on Ultrarelativistic Nucleus-Nucleus Collisions"*, Stony Brook, NY, USA, 15 - 20 Jan 2001

[2] Member Institutions Bari/CERN ALICE-HMPID, Yale. For a complete author list see: B. Lasiuk, Proceedings of *"15th International Conference on Ultrarelativistic Nucleus- Nucleus Collisions"*, Stony Brook, NY, USA, 15 - 20 Jan 2001

CP602, *QCD@Work: International Workshop on Quantum Chromodynamics*
edited by P. Colangelo and G. Nardulli

At high p_t, while the gluon jets contribute in equal way to the proton and antiproton production, the quark jets will produce an asymmetry in the proton-antiproton production, because of the contribution of the valence quarks. Due to their larger color charge, glouns interact with the medium more strongly than quarks do, losing more energy. The final result is a suppression at high p_t in nuclear interactions of antiproton relative to proton yields. Therefore one expectes, in the nucleus-nucleus collision at RHIC, that $\bar{p}/p$ ratio decreases as p_t increases.

The capability to identify charged particles in the STAR experiment has been extended to high transverse momenta by using a RICH detector[3]. A prototype of a proximity-focusing RICH detector, developed in the framework of the ALICE experiment[4], has been fully integrated in the STAR experiment, allowing to identify π and K with $p_t < 2.5$ GeV/c and protons with $p_t < 5$ GeV/c. Details on the detector can be found in [5].

The STAR-RICH detector covers 2% of the TPC acceptance and has been installed in the central rapidity region to cover $|\eta|<0.3$ and $|\Delta\phi|<20°$ for particle coming from the main interaction vertex at the center of the TPC.

PARTICLE IDENTIFICATION WITH THE STAR-RICH

The identification of charged particles with the STAR-RICH detector has been performed on central Au+Au events at $\sqrt{s_{NN}} = 130$ GeV collected during the first year of running of RHIC. This study is based on 300k central triggers which represents 1/3 of the total statistics accumulated.

Charged tracks reconstructed in the TPC were accepted for futher analysis if they came from a primary vertex located within $|Zvert| < 50$ cm of the TPC center, where the z axis is along the beam direction. Only tracks with $|\eta|<0.15$ were selected in order to have a good acceptance in the detector.

The determination of the Cherenkov angle requires an accurate knowledge of the momentum of the particles and the incident angles on the detector, given by the tracking device (TPC), and a powerful pattern recognition of the photons created in the RICH radiator and detected by the photocathode.

The association between the extrapolated track from the TPC ($p_t>1.5$ GeV/c) and the pattern in the RICH requires the projection of the TPC track to the RICH to fall less than 5 mm from a minimum ionizing cluster in the RICH.. The minimum ionizing cluster are distiguishable from photoelectron signals by their larger amplitude[5]. Fake associations are avoided by selecting clusters with a charge Q > 120 ADC channels (1 ADC = 0.17fC).

After having calculated, for each candidate photon cluster in the event, the Cherenkov angle by tracing back to the radiator up to the most probable emission point along the track trajectory, an algorithm for the recognition of Cherenkov patterns based on the Hough Transform is applied. More details can be found in [6].

Fig.1 shows an offline display of a central event, where the reconstructed rings associated to tracks are also shown.

The reconstructed Cherenkov angle as a function of the track momentum measured in the TPC is shown in fig.2. Clear bands of events, with low background, around the predicted curves for π, K and p are visible.

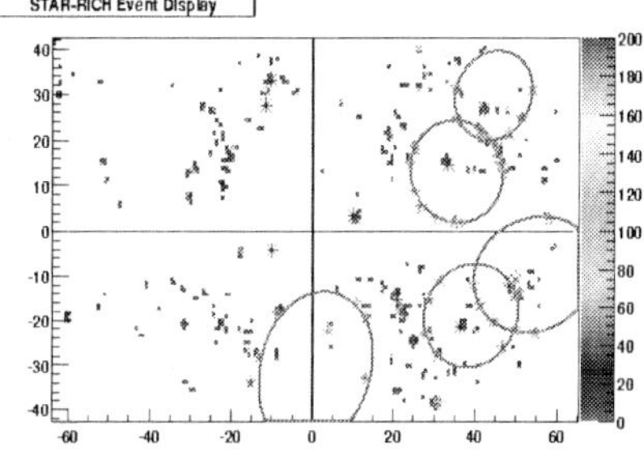

Figure 1. An offline event display of a central event in the RICH. Reconstructed rings with relative photon clusters (star) are shown for associated tracks.

Figure 2. Reconstructed Cherenkov angle in the RICH as a function of the track momentum in the TPC. Solid lines indicate the predicted curves for π, κ and p.

The inclusive measurements of the π, K and p yields (and their relative antiparticles) can be performed by fitting the distribution of the reconstructed Cherenkov angle in different p_t intervals. Fig.3 is an example where the three peaks corresponding to π^-, K^- and $\bar{p}$ are clear with negligible background. The fit has been performed using three gaussians, the total number of entries being the only constraint to the fit.

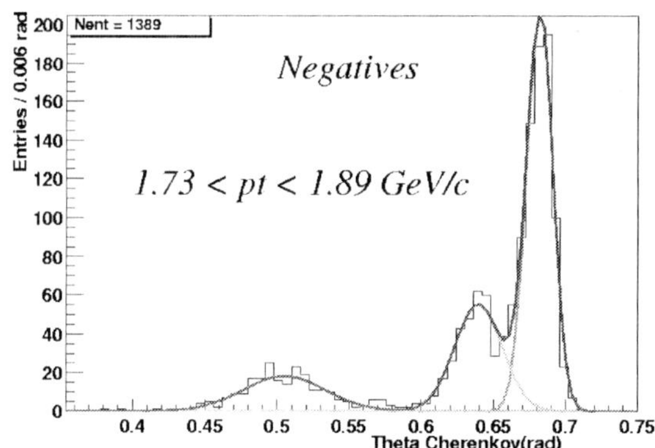

Figure 3. Reconstructed Cherenkov angle in the RICH in a range of restricted transverse momentum measured by TPC. A fit with three gaussians (solid line) has been performed (χ^2/NDF=1.8).

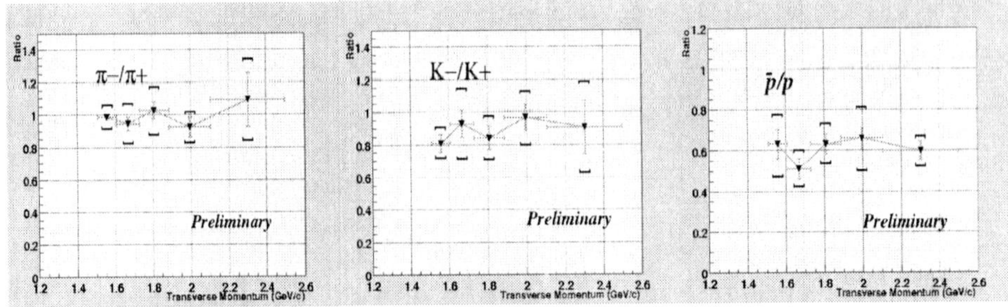

Figure 4. Ratios π^-/π^+ (left), K^-/K^+ (central) and $\bar{p}/p$ (right) as a function of p_t. The data are not corrected for acceptance and identification efficiency. Systematic errors are also shown.

The measured π^-/π^+ , K^-/K^+ and $\bar{p}/p$ ratios at central rapidity as a function of the transverse momentum are shown in fig.4. They are not corrected for geometrical acceptance and identification efficiency, although corrections are expected to cancel in the ratios. The π^-/π^+ , K^-/K^+ ratios at $p_t = 2.5$ GeV/c have been measured with a 2σ π/K separation, while the $\bar{p}/p$ ratio with more than 4σ K/p separation.

The π^-/π^+ , K^-/K^+ and $\bar{p}/p$ ratios are almost flat in this p_t range; in particular $\bar{p}/p$ is in agreement with the results from PHENIX[7].

CONCLUSIONS

A RICH detector, developed in the framework of the ALICE Collaboration, has been successfully integrated in the STAR experiment. The identification of charged particle is extended to $p_t < 2.5$ GeV/c for π and K and $p_t < 5$ GeV/c for protons. π^-/π^+, K^-/K^+ and $\bar{p}p$ have been measured in the transverse momentum range $1.5 < p_t < 2.5$ GeV/c. The RICH detector is already operating at $\sqrt{s_{NN}}$ = 200 GeV since June 2001 and therefore higher statistics will be available for addressing the question of partonic energy loss in dense matter in detail.

REFERENCES

1. K. Kajante, P.W.Landskoff, J. L.Lindfors, Phys. Rev. Lett. **59**, 2517(1987); K.J.Eskola, K.Kajantie and J.Lindfors, Nucl. Phys. B323, 37(1989); G.Calucci, D.Treleani, Phys. Rev. D 41, 3367 (1990).
2. X. N. Wang , hep-ph/9804357 (1998).
3. STAR-RICH Collaboration, "*Proposal for a Ring Imaging Cherenkov Detector in STAR*", YRHI 98-022.
4. F. Piuz et al., Nucl. Instr. and Meth. A433 (1999) 178-189.
5. ALICE Collaboration, "*ALICE Technical Design and Report: Detector for High Momentum PID*", CERN/LHCC 98-19.
6. D. Cozza et al., ALICE/RIC/98-39.
7. H. Ohnishi et al., PHENIX Collaboration, Proceedings of "*15th International Conference on Ultrarelativistic Nucleus- Nucleus Collisions*", Stony Brook, NY, USA, 15 - 20 Jan 2001.

Thermodynamics of 2 and 3 flavour QCD

F. Karsch

Department of Physics, Bielefeld University, D-33615 Bielefeld, Germany

Abstract. We will discuss recent results on the thermodynamics of QCD in the presence of light dynamical quark degrees of freedom. In particular, we will concentrate on an analysis of the flavour and quark mass dependence of the QCD phase diagram, the equation of state and the transition temperature. Moreover, we present recent results on the heavy quark free energy.

INTRODUCTION

Lattice calculations have provided a rather detailed picture of the thermodynamics of gluonic matter. We know that a phase transition exists, which separates a confining low temperature phase (glueball gas) from a deconfined gluon gas phase at high temperature. The transition between these two phases is first order, as originally predicted by Svetitsky and Yaffe [1]. Although the low temperature phase only consists of rather heavy glueballs, the lightest of which has a mass of about 1.8 GeV [2], the phase transition temperature turns out to be rather small, $T_c = 0.637(5)\sqrt{\sigma} \simeq 270$ MeV. This suggests that T_c does not sensitively depend on the mass of the lightest excitations in the confining phase. In fact, it agrees quite well with values extracted from resonance gas models based on the excitation spectrum of a gluonic string [3]. Bulk thermodynamic observables like the energy density and pressure change rapidly at T_c and asymptotically approach the ideal gas limit. Their rapid rise signals the liberation of many new light degrees of freedom at T_c. However, even at temperatures a few times T_c one still observes significant deviations from the asymptotic ideal gas behaviour. This suggests that also in the high temperature phase non-perturbative effects play an important role, which give rise to large screening masses and quasi-particle excitations.

Gluonic matter is described by an $SU(3)$ gauge theory, which is obtained as the infinite quark mass limit of QCD. In this limit the entire fermionic sector of QCD decouples and does not contribute to the thermodynamics; quarks only serve as static sources (quenched QCD). This allows, for instance, a study of thermal modifications (screening) of the forces between external charges, which shows that the linear confining potential weakens with increasing temperature; the string tension decreases and vanishes at T_c.

Some of these basic aspects of gluonic thermodynamics clearly will change drastically in the presence of light dynamical quark degrees of freedom. The asymptotic high temperature limit for bulk thermodynamic observables like the energy density or pressure rises with increasing number of light degrees of freedom. Absolute confinement is lost in the presence of quarks with finite mass already at low temperatures; the heavy quark free energy will no longer diverge when one tries to separate a static quark anti-quark pair. Moreover, the order of the QCD phase transition and even its very existence

CP602, *QCD@Work: International Workshop on Quantum Chromodynamics*
edited by P. Colangelo and G. Nardulli

will crucially depend on the number of light degrees of freedom and their masses.

During recent years these basic qualitative changes in the thermodynamics of QCD, which result from the presence of light quark degrees of freedom, have been observed in lattice calculations [4, 5]. We will discuss here the current status of their quantitative analysis. This will make clear that unlike in the quenched sector of QCD we did not yet reach a similarly detailed quantitative understanding of the relevant parameters that control the critical behaviour of QCD with a realistic spectrum of quark masses. However, thermodynamic calculations on the lattice steadily improve. This partly is due to the rapid development of computer technology. However equally important has been and still is the development of improved discretization schemes, *i.e.* improved actions. This is of particular relevance for thermodynamic calculations which are not only sensitive to the long distance physics at the phase transition but also probe properties at short distances in calculations of e.g. the energy density or the heavy quark potential [6].

In the following we will discuss some recent results on the flavour and quark mass dependence of QCD thermodynamics. We will not go into any details on the lattice formulation of QCD at finite temperature, which have been discussed elsewhere [7].

THE QCD PHASE DIAGRAM

The quark mass and flavour dependence of the QCD phase transition at finite temperature and vanishing baryon number density has been studied extensively in lattice calculations. The basic qualitative and quantitative features expected from universality arguments [1, 8] on the one hand and phenomenological considerations on the other hand have been reproduced by these calculations. The transition is found to be first order in the limit of infinitely heavy quarks as well as in the chiral limit of 3-flavour QCD. In the case of 2-flavour QCD the transition is found to be continuous. The current status of the analysis of universal properties in the chiral limit, however, is not really satisfying [9]. The demonstration that for 2-flavour QCD the transition belongs to the universality class of 3-d, O(4) symmetric spin models still is ambiguous.

The regions of first order transitions are separated from a broad crossover region by lines of second order phase transitions. These lines are expected to belong to the universality class of the 3-d Ising model [10], which is quite remarkable as the global $Z(2)$ symmetry which gets restored at these transitions is not a symmetry of the QCD Lagrangian. This makes it obvious that the bare couplings appearing in the QCD Lagrangian cannot be the relevant scaling fields that control the critical behaviour at these lines of second order transitions. It therefore also is not obvious a priori what is the correct order parameter for these transitions. The energy-like and magnetization-like operators of the effective Hamiltonian that controls the critical behaviour at this critical point will be linear combinations of the basic fields appearing in the QCD Lagrangian and the relevant temperature-like and ordering-field like couplings will be linear combinations of the bare parameters of the QCD Lagrangian, *i.e.* the gauge coupling $\beta \equiv 6/g^2$ and the quark masses m_q.

Understanding the critical behaviour and the relevant scaling fields in the vicinity of the *chiral critical line* at small values of the quark masses is, of course, interesting

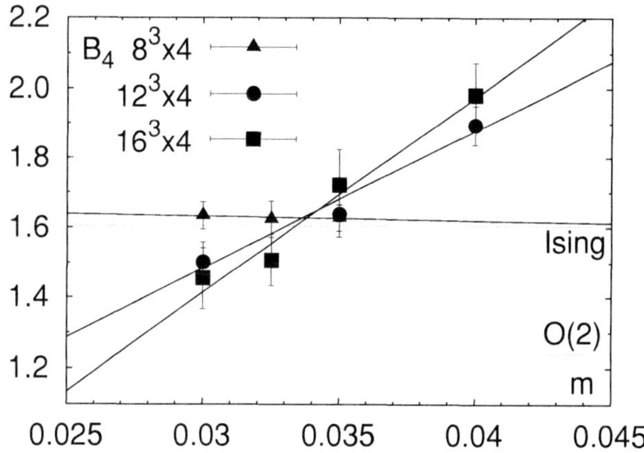

FIGURE 1. The cumulant B_4 versus the bare quark mass calculated in 3-flavour QCD on different size lattices with a standard staggered fermion action [12].

in its own rights. However, it eventually will also be of importance for a discussion of the physics in the vicinity of the critical endpoint which is expected to exist in the temperature-density phase diagram [11]. In an extended phase diagram, which also includes the dependence on the chemical potential, these critical points lie in the same critical surface of second order transitions.

The critical point separating the first order from the crossover region has recently been studied in some detail for the case of three degenerate quark mass flavours [12]. Through an analysis of cumulants of the chiral condensate, $\langle\bar{\psi}\psi\rangle$, as well as joint probability distributions for the chiral condensate and the gluonic part of the QCD action it could be shown that the transition indeed belongs to the universality class of the 3-d Ising model and that the corresponding order parameter can be constructed as a linear combination of the chiral condensate and the gluonic action. In Figure 1 we show the result of a calculation of the fourth cumulant of the chiral condensate (Binder cumulant),

$$B_4 = \frac{\langle(\bar{\psi}\psi)^4\rangle}{\langle(\bar{\psi}\psi)^2\rangle^2} \quad . \tag{1}$$

The cumulant has been calculated for different values of the quark mass at the pseudo-critical couplings. Up to finite volume corrections the cumulants obtained on different size lattices will cross at the critical quark mass corresponding to the second order phase transition point. The value of B_4 at this point is universal and unambiguously identifies the transition as an Ising-like transition.

In order to judge whether QCD with a realistic spectrum of light u,d quarks and a heavier strange quark lies in the first order or crossover region of the phase diagram it is necessary to determine the location of the critical line quantitatively. The calculations performed so far with 3 degenerate quark masses and different fermion actions [12] show that the critical mass parameter, e.g. the value of the pseudo-scalar meson mass at the chiral critical point, is rather sensitive to cut-off effects. While the calculation performed with the standard staggered fermion action led to a critical value $m_{ps} \simeq 300$ MeV, calculations with an improved staggered fermion action gave $m_{ps} \simeq 200$ MeV. These calculations thus need further confirmation through studies closer to the continuum limit. Nonetheless, the current estimates consistently yield rather small values for the pseudo-scalar meson mass at the critical endpoint in 3-flavour QCD, which makes it quite unlikely that the physical point in the phase diagram, corresponding to a realistic quark mass spectrum with two light u,d, and a heavier strange quark, would lie in the first order region. This would also be in agreement with an earlier estimate of the Columbia group [13].

Our current understanding of the QCD phase diagram of 3-flavour QCD at vanishing baryon number density is summarized in Figure 2.

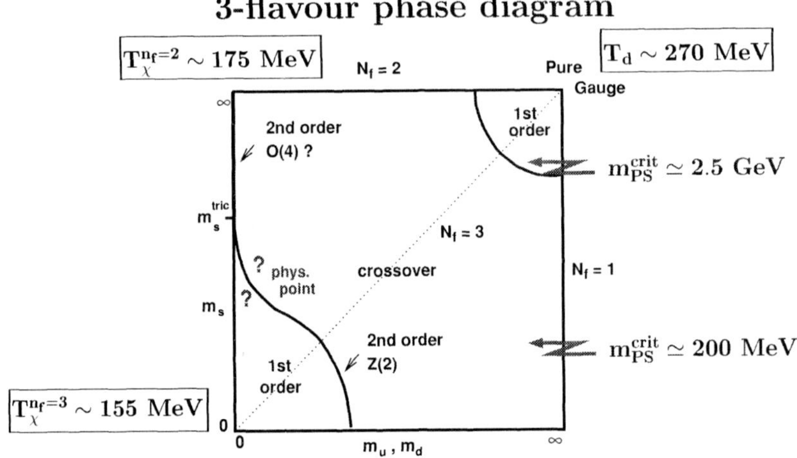

FIGURE 2. The QCD phase diagram of 3-flavour QCD with degenerate (u,d)-quark masses and a strange quark mass m_s.

THE TRANSITION TEMPERATURE

It should be clear from the previous discussion of the phase diagram that for a large range of quark masses the transition to the high temperature plasma phase is not a phase transition, *i.e.* the transition does not correspond to singularities in the QCD partition function. Nonetheless, also for these quark masses the transition occurs in a narrow temperature interval and thus is well localized. This transition region is characterized by peaks in susceptibilities, e.g. the chiral susceptibility χ_m or the Polyakov-loop susceptibility χ_L,

$$\chi_m = \frac{\partial}{\partial m_q}\langle\bar{\psi}\psi\rangle \quad , \quad \chi_L = N_\sigma^3\left(\langle L^2\rangle - \langle L\rangle^2\right) \quad . \qquad (2)$$

Here

$$L = \frac{1}{N_\sigma^3} \sum_{\vec{x}} \mathrm{Tr}\, L_{\vec{x}} \tag{3}$$

denotes the spatial average over Polyakov-loops[1], $L_{\vec{x}}$, defined at the spatial sites $\vec{x}$ of a lattice of size $N_\sigma^3 \times N_\tau$. A calculation of the pseudo-critical temperature, $T_c = 1/N_\tau\, a(\beta_{pc})$ still requires the determination of the lattice spacing $a(\beta_{pc})$ at the pseudo-critical couplings β_{pc} which in turn are defined through the location of the susceptibility peaks. The lattice spacing can be determined through the calculation of an independent physical observable. In order to quote T_c in physical units, *i.e.* MeV, we need a physical observable which does not crucially depend on the values of the quark masses. For physical values of the quark masses it would, of course, be most appropriate to use a hadron mass, e.g. the rho-meson mass, to set the physical scale for T_c. The hadron masses, however, are themselves strongly dependent on the quark mass values; their masses diverge in the infinite quark mass limit. Nonetheless, we can assign a physical value to the transition temperature in this limit. In quenched QCD the natural scale for T_c is the square root of the string tension, $\sqrt{\sigma}$. In fact, the string tension as well as quenched hadron masses[2] seem to describe the experimentally known QCD spectrum reasonably well already in the infinite quark mass limit. These quantities thus seem to have a weak quark mass dependence and are suitable to set the scale for T_c at arbitrary values of m_q. This situation is illustrated in Figure 3.

On the left hand side of Figure 3 we show the transition temperature in units of the vector meson mass m_V versus the ratio m_{ps}/m_V, which in the chiral limit is proportional to the square root of the quark mass. Here the drop in T_c/m_V observed with increasing quark mass mainly is due to the quark mass dependence of m_V and does not reflect the quark mass dependence of T_c. This figure shows, however, that calculations based on different discretization schemes for the gauge and fermion actions do yield consistent results for T_c and its dependence on m_q. In the chiral limit one finds for the critical temperature in 2 and 3-flavour QCD

$$\underline{2-\text{flavour QCD}:} \quad T_c = \begin{cases} (171 \pm 4)\ \text{MeV}, & \text{clover-improved Wilson fermions [14]} \\ (173 \pm 8)\ \text{MeV}, & \text{improved staggered fermions [15]} \end{cases}$$

$$\underline{3-\text{flavour QCD}:} \quad T_c = (154 \pm 8)\ \text{MeV}, \quad \text{improved staggered fermions [15]}$$

Here m_ρ has been used to set the scale for T_c. We note that all the results presented in Figure 3 have been obtained on lattices with a rather small temporal extent ($N_\tau = 4$). The lattice spacing is therefore still quite large in the vicinity of T_c, *i.e.* $a \simeq 0.3$ fm. Moreover,

---

[1] The Polyakov-loop is a purely time-like Wilson loop which is closed due to the periodicity of the lattice in temporal direction. Its definition as well as further details on the lattice formulation of QCD thermodynamics may, for instance, be found in [7].

[2] In the calculation of *quenched hadron masses* the contribution of dynamical light sea quarks is suppressed. Only the valence quark contributions and interactions with the gluonic vacuum are taken into account.

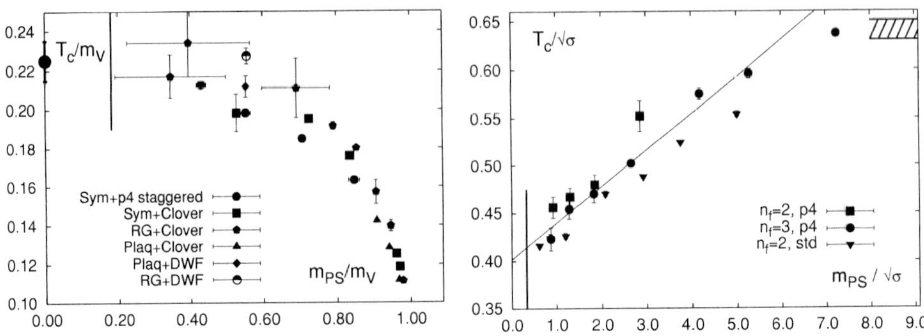

FIGURE 3. Transition temperatures in units of m_V (left) and in units of $\sqrt{\sigma}$ (right). The left hand figure shows results for 2-flavour QCD obtained with different gauge and fermion actions. The right hand figure shows the transition temperature in 2 (filled squares) and 3 (circles) flavour QCD obtained with improved staggered fermions (p4-action). Also shown are results for 2-flavour QCD obtained with the standard staggered fermion action (triangles). The dashed band indicates the uncertainty on $T_c/\sqrt{\sigma}$ in the quenched limit. The straight line is the fit given in Eq. 4. The vertical lines correspond to physical values of the hadron masses and a string tension of 425 MeV.

there are uncertainties involved in the ansatz used to extrapolate to the chiral limit. We estimate that the systematic error on the value of T_c/m_ρ still is of similar magnitude as the purely statistical error quoted above. The critical temperature thus is at present only known with an error of about 10%.

On the right hand side of Figure 3 we give T_c in units of $\sqrt{\sigma}$. This reflects the expected behaviour; with increasing values of the quark mass the hadrons become heavier and it becomes more difficult to create a dense hadronic system that could undergo a transition to a quark-gluon plasma phase; the transition temperature thus increases. It is, however, striking that the quark mass dependence of T_c is so weak. The straight line shown in Figure 3 is a fit to the 3-flavour data, which gave

$$(T_c/\sqrt{\sigma})_x = (T_c/\sqrt{\sigma})_0 + 0.04(1)\,x \quad \text{with} \quad x = m_{ps}/\sqrt{\sigma} \quad . \tag{4}$$

This suggests that also heavier resonances, which are little affected by changes of the quark mass, play an important role for building up the critical density needed to trigger the QCD phase transition. Moreover, we note that for pseudo-scalar masses larger than about $6\sqrt{\sigma} \simeq 2.5$ GeV the transition temperature agrees with the value found in the pure gauge theory, $T_c \simeq 0.637\sqrt{\sigma} \simeq 270$ MeV. In this heavy quark mass regime all the hadronic states are heavier than typical glueball masses and thus decouple from the thermodynamics, which only is controlled by gluonic degrees of freedom. In this heavy mass regime the transition also changes again from a crossover to a first order phase transition.

THE QCD EQUATION OF STATE

The temperature dependence of bulk thermodynamic observables like the energy density (ε) and pressure (p) has been analyzed in detail in the pure gauge sector. These calculations show that ε/T^4 as well as p/T^4 rapidly increase above T_c. However, even at $T \simeq 4\,T_c$ the Stefan-Boltzmann limit is not yet reached [16]. Deviations stay at the level of 15%, which is too much to be understood in terms of perturbative corrections [17]. Even at these high temperatures non-perturbative effects seem to play an important role for the thermodynamics. This finds further support from studies of the heavy quark free energy [18], the spatial string tension [19] as well as the gluon propagator in fixed gauges [20]. The analysis of the temperature dependence of these observables also suggests that the temperature dependent running coupling constant remains large in the plasma phase and electric and magnetic screening masses differ significantly from perturbative results even at temperatures which are several orders of magnitude larger than T_c [21]. In view of this it may even be surprising that models with weakly interacting quasi-particles [22] as well as calculations based on self-consistent resummation schemes [23] do provide a quite satisfactory description of bulk thermodynamics down to temperatures a few times T_c.

At least at high temperature, where the energy density and pressure are expected to approach the free gas limit, the QCD equation of state will strongly depend on the number of light partonic degrees of freedom. Already for two massless quark flavours the Stefan-Boltzmann constant increases by more than a factor two relative to the pure gauge theory. For n_f-flavour QCD one has,

$$\frac{\varepsilon_{SB}}{T^4} = \frac{3p_{SB}}{T^4} = \left(16 + \frac{21}{2} n_f\right) \frac{\pi^2}{30} \quad . \tag{5}$$

As the influence of non-zero quark masses on bulk thermodynamic observables will be exponentially suppressed at high temperature, Eq. 5 also gives the dominant high temperature behaviour for massive quark even when the masses are of the order of the (critical) temperature. In fact, this has been observed in lattice calculations with fairly large quark masses [24].

The overall pattern of the temperature dependence of bulk thermodynamic observables in QCD with 2 and 3 light quark flavours is very similar to the case of the pure gauge theory. With increasing number of light degrees of freedom the critical temperature shifts to smaller values and the asymptotic high temperature limit for p/T^4 becomes larger. This is seen in the left hand part of Figure 4. However, after rescaling the thermodynamic observables with the corresponding Stefan-Boltzmann constants they look quite alike in units of T/T_c. Of course, they still differ in details. In particular, it is apparent from the right hand part of Figure 4 that at T_c the rescaled pressure of QCD with light quarks is significantly larger than in the pure gauge theory. This also is the case for the energy density which is found to be $\varepsilon_c/T_c^4 \simeq 6$ in QCD with light quarks [24, 25], while it is only $\varepsilon_c/T_c^4 \simeq 1$ in the SU(3) gauge theory[3]. Much of this factor 6 difference in

---

[3] The energy density is discontinuous in the case of the pure gauge theory. The latent heat is found to be $\Delta\varepsilon/T_c^4 = 1.40(9)$ [26] with $\varepsilon/T_c^4 \simeq 2$ in the high temperature phase.

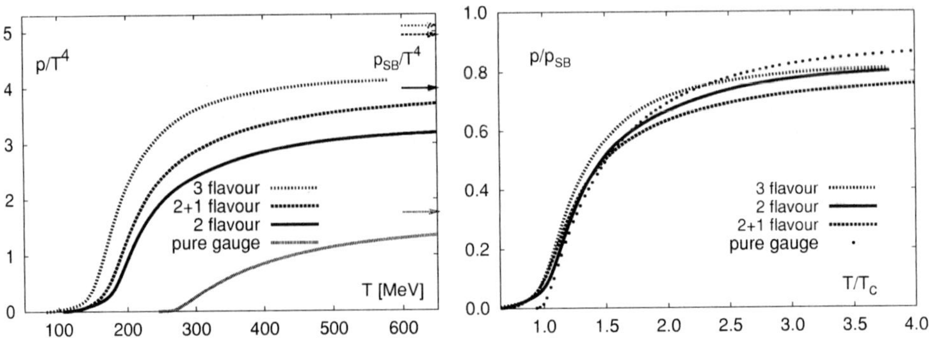

FIGURE 4. The pressure in QCD with $n_f = 0$, 2 and 3 light quarks as well as two light and a heavier (strange) quark. For $n_f \neq 0$ calculations have been performed on a $N_\tau = 4$ lattice using improved gauge and staggered fermion actions [15]. In the case of the SU(3) pure gauge theory the continuum extrapolated result is shown [16]. Arrows indicate the ideal gas pressure p_{SB} as given in Eq. 5.

ε_c/T_c^4, however, seems to arise from the difference in T_c between QCD with light quarks and the purely gluonic theory. The critical energy densities turn out to be quite similar. Unfortunately, the current error on T_c, which is about 10%, amplifies in the calculation of the energy density, which makes ε_c still badly determined in lattice calculations,

$$\varepsilon_c = (0.3 - 1.3)\,\mathrm{GeV/fm}^3 \quad . \tag{6}$$

THE HEAVY QUARK FREE ENERGY

The confining properties of the thermal heat bath generated by quarks and gluons can be probed by analyzing the response of the medium to the insertion of static sources. Static quark sources are described by the Polyakov-loop $L_{\vec{x}}$. In the pure gauge limit the expectation value of its spatial average, $\langle L \rangle$ (Eq. 3), is an order parameter for the deconfinement transition,

$$\langle L \rangle \begin{cases} = 0\,, & T < T_c \\ > 0\,, & T > T_c \end{cases} \quad . \tag{7}$$

This reflects the long distance behaviour of the correlation function for static quark anti-quark sources,

$$\mathrm{e}^{-F(r,T)/T} = \langle \mathrm{Tr}L_0\, \mathrm{Tr}L_{\vec{x}}^\dagger \rangle \quad , \quad r \equiv |\vec{x}| \quad , \tag{8}$$

which approaches the cluster value $|\langle L \rangle|^2$ in the limit of infinite separation between the quark anti-quark pair. In the confined phase this correlation function vanishes for $r \to \infty$ which signals that the free energy needed to separate the two sources is infinite. This is no longer the case if dynamical quarks with a finite mass are contributing to the thermal heat bath. The static sources can then be screened by quarks and anti-quarks present in the heat bath and even in the zero temperature limit this becomes possible through the generation of $q\bar{q}$-pairs from the vacuum. The free energy needed to separate the static sources thus will stay finite at all temperatures.

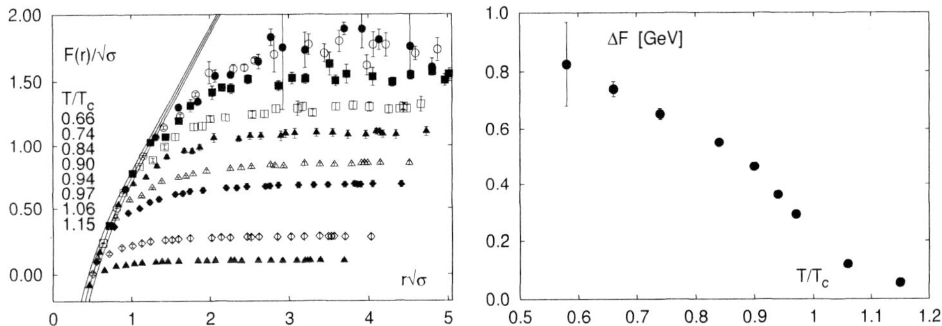

FIGURE 5. The heavy quark free energy in units of the square root of the string tension versus the quark anti-quark separation (left) and the estimate for the dissociation energy defined in Eq. 9 (right). The band shows the normalized Cornell potential $V(r) = -\alpha/r + \sigma r$ with $\alpha = 0.25 \pm 0.05$. The lattice data have been normalized to this potential at the shortest distance available, *i.e.* at $rT = 1/4$.

In Figure 5 we show the free energy of a static quark anti-quark pair calculated in 3-flavour QCD at various temperatures and for different separation of the static sources. At low temperatures, $T \lesssim 0.7\, T_c$, the heavy quark free energy coincides with the confining Cornell-type potential, $V(r) = -\alpha/r + \sigma r$ with $\alpha = 0.25 \pm 0.05$, up to distances $r \simeq 1.5/\sqrt{\sigma} \simeq 0.7$ fm. With increasing temperature $F(r,T)$, however, gets screened earlier and at T_c it starts to deviate from the zero temperature potential already at distances $r \simeq 0.3$ fm. Moreover, it becomes much easier to separate the heavy quark sources. As an estimate for the dissociation energy we show in the right hand part of Figure 5 the difference in free energy of a $q\bar{q}$-pair at infinite separation and a $q\bar{q}$-pair at distance $r_q = \sqrt{\alpha/\sigma}$,

$$\Delta F \equiv \lim_{r\to\infty} F(r,T) - F(\sqrt{\alpha/\sigma}) \quad . \tag{9}$$

The rapid decrease of ΔF close to T_c clearly will have consequences for the formation and existence of heavy quark bound states not only in the high temperature phase but also in the vicinity of T_c. Already at $T \simeq 0.9\, T_c$ the free energy difference for a heavy quark pair separated by a distance similar to the J/ψ radius ($r_\psi \sim 0.2$ fm) and a $q\bar{q}$-pair at infinite separation is only 500 MeV, which is compatible with the average thermal energy of a gluon ($\sim 3\, T_c$). The $c\bar{c}$-bound states are thus expected to dissolve already close to T_c, maybe even below T_c [27].

ACKNOWLEDGMENTS

The work has been supported by the TMR network ERBFMRX-CT-970122 and by the DFG under grant FOR 339/1-2.

REFERENCES

1. B. Svetitsky and L.G. Yaffe, Nucl. Phys. B210 [FS6], 423 (1982).
2. H. Wittig, Int. J. Mod. Phys. A12 (1997) 4477.
3. R. D. Pisarski and O. Alvarez, Phys. Rev. D26 (1982) 3735.
4. F. Karsch, Nucl. Phys B (Proc. Suppl.) 83-84 (2000) 14.
5. S. Ejiri, Nucl. Phys B (Proc. Suppl.) 94 (2001) 19.
6. F. Karsch, Nucl. Phys. Proc. Suppl.) 60A (1998) 169.
7. F. Karsch, *Lattice QCD at high temperature and density*, hep-lat/0106019.
8. R.D. Pisarski and F. Wilczek, Phys. Rev. D29 (1984) 338.
9. for a recent discussion and further references see Ref. 5.
10. S. Gavin, A. Gocksch and R.D. Pisarski, Phys. Rev. D49 (1994) 3079.
11. M. Stephanov, K. Rajagopal and E.Shuryak, Phys. Rev. Lett. 81 (1998) 4816.
12. F. Karsch, E. Laermann and Ch. Schmidt, *The Chiral Critical Point in 3-Flavour QCD*, to be published in Phys. Lett. B, hep-lat/0107020.
13. F.R. Brown et al., F.R. Brown, F.P. Butler, H. Chen, N.H. Christ, Zhi-hua Dong, W. Schaffer, L.I. Unger and A. Vaccarino, Phys. Rev. Lett. 65 (1990) 2491.
14. A. Ali Khan et al. (CP-PACS), Phys. Rev. D63 (2001) 034502.
15. F. Karsch, A. Peikert and E. Laermann, Nucl. Phys. B605 (2001) 579.
16. G. Boyd, J. Engels, F. Karsch, E. Laermann, C. Legeland, M. Lütgemeier and B. Petersson, Phys. Rev. Lett. 75 (1995) 4169 and Nucl. Phys. B469 (1996) 419.
17. P. Arnold and C. Zhai, Phys. Rev. D51 (1995) 1918;
    C. Zhai and B. Kastening, Phys. Rev. D52 (1995) 7232.
18. O. Kaczmarek, F. Karsch, E. Laermann and M. Lütgemeier, Phys. Rev. D62 (2000) 034021.
19. F. Karsch, E. Laermann and M. Lütgemeier, Phys. Lett. B346 (1995) 94.
20. A. Cucchieri, F. Karsch and P. Petreczky, Phys. Rev. D64 (2001) 036001.
21. K. Kajantie, M. Laine, J. Peisa, A. Rajantie, K. Rummukainen and M. Shaposhnikov, Phys. Rev. Lett. 79 (1997) 3130.
22. A. Peshier, B. Kämpfer, O.P. Pavlenko, and G. Soff, Phys. Rev. D54 (1996) 2399;
    P. Lévai and U. Heinz, Phys. Rev. C57 (1998) 1879;
    R.A. Schneider and W. Weise, *On the quasiparticle description of lattice QCD thermodynamics*, hep-ph/0105242.
23. J. P. Blaizot, E. Iancu and A. Rebhan, Phys. Rev. D63 (2001) 065003 and references therein.
24. A. Ali Khan et al. (CP-PACS), *Equation of state in finite-temperature QCD with two flavors of improved Wilson quark*, hep-lat/0103028.
25. F. Karsch, *Lattice results on QCD thermodynamics*, hep-ph/0103314 .
26. B. Beinlich, F. Karsch, A. Peikert, Phys. Lett. B390 (1997) 268.
27. S. Digal, P. Petreczky and H. Satz, Phys. Lett. B514 (2001) 57.

Strangeness production in a statistical effective model of hadronisation

F. Becattini*, G. Pettini

University of Florence and INFN Sezione di Firenze, Largo E. Fermi 2, I-50125, Firenze, Italy

Abstract. We suppose that overall strangeness production in both high energy elementary and heavy ion collisions can be described within the framework of an equilibrium statistical model in which the effective degrees of freedom are constituent quarks as used in effective lagrangian models. In this picture, the excess of relative strangeness production in heavy ion collisions with respect to elementary particle collisions arises from the unbalance between initial non-strange matter and antimatter and from the exact colour and flavour quantum number conservation over different finite volumes. The comparison with the data and the possible sources of model dependence are discussed.

INTRODUCTION: STRANGENESS PRODUCTION AND STATISTICAL HADRONISATION

Recent observations have shown that hadron multiplicities in both e^+e^- and hadronic high energy collisions agree very well with a statistical-thermal ansatz [1]. This finding has been interpreted in terms of a multi-cluster hadronisation process in which each cluster fills its relevant multi-hadronic phase space in a pure statistical fashion, at a critical value of energy density [1, 2]. Within this framework, temperature and other thermodynamical quantities have an essential statistical meaning which does not imply the existence of a thermalisation process at hadronic level through multiple collisions; rather, hadronisation itself yields a statistically equilibrated hadronic population. One of the main features of this approach is the very low number of free parameters required. Under suitable assumptions about cluster masses and charges fluctuations at fixed volumes [1, 2], there are essentially two free parameters, namely the sum V of the volumes of the clusters, and the temperature T. Yet, in order to reproduce the yields of strange particles, the model has to be supplemented with one more phenomenological parameter, γ_S, which suppresses the production of particles containing n strange quarks by a factor γ_S^n [1]. From fits to the available data, γ_S turns out to be < 1 in all examined collisions and strongly dependent on the initial colliding systems [1, 2, 3] so that full strangeness chemical equilibrium is never observed.

A further insight in strangeness production is achieved by calculating the ratio between newly produced valence strange quarks and u, d quarks, the so-called Wroblewski factor $\lambda_S = 2\langle s\bar{s}\rangle/(\langle u\bar{u}\rangle + \langle d\bar{d}\rangle)$, which is in fact fairly constant in all kinds of elemen-

[1] very recently [2] a new parametrisation of extra strangeness suppression has been proposed which is equivalent to that with γ_S at very high multiplicity

CP602, *QCD@Work: International Workshop on Quantum Chromodynamics*
edited by P. Colangelo and G. Nardulli

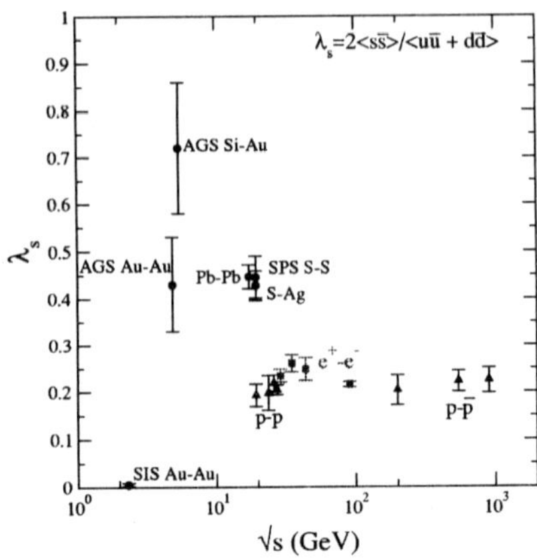

FIGURE 1. λ_S as a function of centre-of-mass energy in several kinds of collisions (from ref. [3])

tary collisions (EC) over a centre-of-mass energy range spanning about two orders of magnitude, while it shows a non-trivial behaviour in heavy ion collisions (HIC) and it is as twice as large at $\sqrt{s} \approx 20$ GeV [3] (see Fig. 1). It should be mentioned that this ratio is calculated by using the fitted primary (i.e. directly emitted from the hadronising source) hadron multiplicities before hadronic decays take place, which are not measurable. Thus, this parameter depend in fact on the fitted parameters T, V and γ_S and, in principle, is a model dependent quantity; still, the model accurately reproduces all measured multiplicities and the model dependence decreases as the number of measured final multiplicities increases, so that the estimated λ_S is expected to be reasonably close to its actual value. The explanation of the strangeness production in high energy collisions, and especially λ_S, is a major goal for the understanding of hadronisation and of the possible QGP formation in HIC.

In the following, we will try to account for this behaviour by resorting to a statistical description in terms of consituent quarks rather than hadrons, or, more specifically, using microscopic models containing quarks as fundamental degrees of freedom. As full QCD cannot be handled, we use effective models (EM) at finite temperature, which are commonly employed to investigate QCD phase transition at high temperatures and low baryon chemical potential, that is our region of interest. Particularly, we will refer to low energy models with four-fermion interactions such as the Nambu-Jona-Lasinio model [4, 5] and to what is sometimes referred to, in literature, as *ladder*-QCD [6].

THE MODEL

In using effective models with quarks as fundamental degrees of freedom to calculate relative flavour production, essentially the same physical scheme of the statistical hadronisation model (SHM) both in HIC and EC is kept. This means that the formation of a set of hadronising clusters endowed with mass, volume and flavour quantum charges is assumed, in which every allowed quantum state is equally likely. Also the assumption of suitable maximum-disorder fluctuations of cluster flavour charges and masses for fixed volumes, enabling the introduction of a global temperature [2], is retained. Three further assumptions are introduced:

1. Each single cluster is a colour singlet
2. The produced s quarks, or at least the ratio s/u, survive in the hadronic phase
3. The temperature T and the chemical potentials (in the grand-canonical framework) fitted in the SHM with hadron multiplicities are interpreted as the critical values for deconfinement and (approximate) chiral symmetry restoration

The physical process of single cluster hadronisation, in both EC and HIC is envisaged as an evolution towards a chaotic quantum state which finally leads to statistical multi-hadronic phase space population by coalescence of produced constituent quarks. Within this picture, the lack of complete strangeness chemical equilibrium at hadron level is the effect of a complete strangeness chemical equilibrium at constituent quark level. Whilst in HIC an *early* thermalisation and colour deconfinement over a large volume is expected, with consequent formation of relatively *large* colour singlet clusters, in EC the chaotic behaviour of quantum dynamics is supposed to set in at a *late* stage when the colour preconfinement mechanism has already brought about the formation of *small* colour singlet clusters. These distinctive features of hadronisation in HIC with respect to hadronisation in EC imply a difference of relative strange quark production which is, hopefully (as we have assumed in (2)), reflected into final hadrons. Moreover, the existence of a characteristic single cluster volume in EC independent of colliding system and centre-of-mass energy is argued to be the main responsible for the constancy of λ_S because of the strange quark suppression entailed by the colour singlet constraint over a small spacial region (canonical colour suppression). The assumption (3) is certainly the strongest one, as it implies that hadronisation itself is assumed to be a critical process and, thence, hadronisation temperature is to be identified with the critical QCD temperature; this is suggested by the observed constancy of fitted temperature for many collisions [2] and by its value $T \simeq 160$ MeV close to the calculated lattice value [7].

As has been mentioned, effective lagrangian models are a useful tool to deal with quark degrees of freedom at finite temperature. In fact, none of them can account for colour confinement but many of them embody chiral symmetry breaking (χSB) and its restoration (χSR), which is expected to occur at the same critical point [7]. The predictions of EM about temperature dependence of several physical quantities may strongly vary, nevertheless they also show striking common features. Specifically, the phase diagram for χSR exhibits a tricritical point in the chiral limit $m_q \to 0$ separating second order from first order phase transitions [6] and low-μ and high-T region is second order. Moreover, the expression for the number of quarks for the ith flavour generally stems from the one-loop term and, in the mean-field approximation of four-fermion models, it can be derived from a free Dirac Hamiltonian with constituent quark masses replacing current masses, so that in the grand-canonical ensemble:

$$\langle n_i \rangle = \frac{N_c V}{\pi^2} \int_0^\Lambda \mathrm{d}p \, \frac{p^2}{\exp[\sqrt{p^2 + M_i^2}/T - \mu_i/T] + 1} \tag{1}$$

where $N_c = 3$ and μ_i are the relevant chemical potentials. In Eq. (1) Λ is an UV cutoff which is needed in EM with four-fermion interactions as these models are not renormalisable (630 MeV in ref. [5]). In *ladder*-QCD models essentially the same expressions holds below Λ whereas at higher momenta the one-loop Hamiltonian has to be modified. However, this modification usually implies a minor change of the ratio s/u since the largest contribution to $\langle n_i \rangle$ comes from the integration region $p < \Lambda$. Finally, commonly to most EM, the u and d quark constituent masses $M_{u,d}$ steeply decrease to a value close to the current mass within a small interval of temperature (which is fairly identified with the critical region) whereas the strange quark constituent mass M_s decreases much more slowly and in fact, in the critical region, it has a value still much higher than current mass value.

In order to obtain quantitative predictions for EC within a given effective model the canonical expressions of quark numbers are needed, hence we have to implement exact colour and flavour conservation over finite volumes. This task can be accomplished by means of well known methods based on group theory [8]. In our case the involved symmetry group is $G = \mathrm{SU}(3)_c \times \mathrm{U}(1)_u \times \mathrm{U}(1)_d \times \mathrm{U}(1)_s$ and physical states to be counted in the partition function of a single cluster should be projected onto the irreducible invariant 1-dimensional subspace associated with the colour singlet representation with given initial flavour numbers. The overall number of quarks for a given flavour i can be obtained by taking the derivative of the overall partition function (i.e. the partition function of the multi-cluster system) with respect to fictitious fugacities λ_i:

$$\langle n_i \rangle = \left. \frac{\partial \log Z(\lambda_i)}{\partial \lambda_i} \right|_{\lambda_i = 1} \tag{2}$$

where the overall partition function reads (the proof is given in ref. [9]):

$$Z(\lambda_1,\lambda_2,\lambda_3) = \left[\prod_{i=1}^{3} \int_{-\pi}^{\pi} \frac{d\phi_i}{2\pi} \exp[iN_i\phi_i] \right] \left\{ \int d\mu(\theta_1,\theta_2) \exp\left[\sum_{i=1}^{3} \frac{2V_c}{(2\pi)^3} \right.\right.$$
$$\left.\left. \times \sum_{n=1}^{\infty} \frac{(-1)^{n+1}}{n} \chi_{1,0}(n\theta_1, n\theta_2) \, e^{in\phi_i} \lambda_i^n \int_0^{\Lambda} d^3p \exp[-\sqrt{p^2 + M_i^2}/T] + \text{c.c.} \right] \right\}^{V/V_c} \tag{3}$$

where $d\mu$ is the normalised $\mathrm{SU}(3)_c$ group measure [8] and $\chi_{1,0}$ is the character of the fundamental $\mathrm{SU}(3)_c$ representation. Two different volumes show up in Eq. (3) owing to the fact that colour singlet constraint applies to each single cluster (with volume V_c) whereas the flavour constraint (equal to that of initial state) applies to the system of clusters overall; indeed V is meant to be the sum of all clusters proper volumes.

ANALYSIS AND RESULTS

In our numerical analysis we essentially determine λ_S on the basis of Eqs. (2), (3). Generally speaking, λ_S depends on the value of constituent quark masses at a fixed temperature T, on the cutoff Λ and on the two volumes V and V_c. According to statement

(3) in previous section, the temperature T has been set equal to the value fitted within SHM, which is about 160 MeV in EC and in Pb–Pb collisions [3].

As far as HIC are concerned, λ_S calculation can be performed in the grand-canonical ensemble on the basis of Eq. (1) because of the very large involved overall volumes V. Colour canonical suppression (see below) sets in only at very small V_c volumes, which are excluded because of the assumption of colour deconfinement. Thereby, since λ_S is a ratio of particle numbers, the dependence on volumes vanishes and, by setting $\Lambda \to \infty$ in Eq. (1), it is found that one needs $M_s \sim 500$ MeV and $M_u < 100$ MeV to match the fitted λ_S value in Pb–Pb ($\simeq 0.4$) at $T = 160$ MeV. This means that an eligible effective model should have a critical region (at very low baryon chemical potential) around 160 MeV with constituent u and d masses essentially dropped from their value at $T = 0$. These features can be precisely found in one version (case 2) of the NJL model in ref. [5] which has thus been used to make a quantitative comparison with the data. In particular, this model has an UV cutoff $\Lambda = 630$ MeV, $M_{u,d}(T = 160\ \text{MeV}, \mu = 0) = 64$ MeV and $M_s(T = 160\ \text{MeV}, \mu = 0) = 449$ MeV. The agreement between calculated and fitted λ_S values in Pb–Pb is very good as it can be seen in Table 1.

TABLE 1. Comparison between λ_S fitted with SHM [3] and the calculated λ_S in high energy heavy ion collisions by using the central fitted values of T and $\mu_q \simeq \mu_B/3$ [3, 10]

Collision	T (MeV)	μ_q (MeV)	λ_S	λ_S (calc.)
Pb–Pb SPS	158.1 ± 3.2	79.3 ± 4.3	0.447 ± 0.025	0.455
Au–Au RHIC	165 ± 7	13.7 ± 1.7		0.335

The effect of volume finitiness on λ_S (*canonical suppression*) is mainly relevant to EC and has been accurately studied by enforcing colour and flavour constraints either separately or simultaneously. Generally speaking, the requirement of an exact conservation of some global quantity (such as colour or charge) suppresses heavier particles more than lighter ones with respect to the grand-canonical limit because, at finite T, less energy can be spent to compensate the unbalance created by one particle generation. Indeed, for a given volume $V = V_c$, the canonical suppression of s quarks with respect to u, d quarks is expected, and has been found, to be predominantly determined by the net zero strangeness constraint rather than by colour singlet constraint as colour can be compensated by the generation of two light u, d quarks instead of one heavier $\bar{s}$ quark. Unlike the number of quarks, the masses of constituent quarks have been determined by minimising the free energy in the grand-canonical limit, i.e. neglecting the effect of finite volume, at $T = 160$ MeV and $\mu_i = 0$.

The effect of combined colour and flavour conservation over different volumes on λ_S is shown in Fig. 2 as a function of V and V_c for colliding systems such as e^+e^- (flavour neutral) or pp, along with conservatively estimated ranges of λ_S determined with the multiplicity fits within the SHM (see Fig. 1). The colour canonical suppression of λ_S clearly shows up for single-cluster volumes below $\approx 15\ \text{fm}^3$. Apparently, a V_c between 5 and 10 fm^3 can account for the observed constant λ_S value in e^+e^- collisions whereas the predicted value in pp is too high. In fact, the relative strangeness enhancement due to the presence of initial u and d quarks in the colliding protons which inhibits the creation of $u\bar{u}$, $d\bar{d}$ pairs, seems to prevail over the relative strangeness suppression entailed by colour and $S = 0$ constraint.

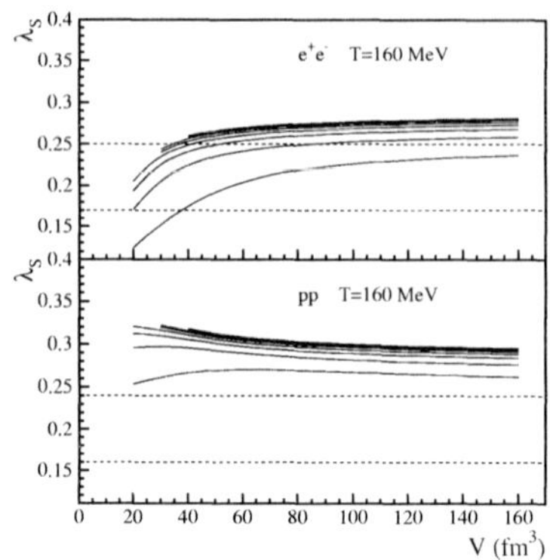

FIGURE 2. Calculated λ_S in e^+e^- and pp collisions at T=160 MeV within the NJL model [5] as a function of the total volume V for single cluster volume V_c varying from 5 fm^3 (lowest curves) to 40 fm^3 in steps of 5 fm^3. The horizontal bands are the ranges of fitted λ_S values in the SHM (see Fig. 1).

CONCLUSIONS

We have studied strangeness production within a statistical model of hadronisation by using effective models with constituent quarks. The basic idea is that full statistical equilibrium is achieved at the level of quark degrees of freedom in both elementary and heavy ion collisions. Besides the effect of different initial light flavour content and density, the smaller relative strangeness production in EC with respect to HIC is supposed to be related to the smaller sytem size and to colour confinement over small distances. A quantitative study in this regard gives a satisfactory agreement with the data in heavy ion and e^+e^- collisions but a significant disagreement in pp, which might be cured by taking more complex assumptions about quantum numbers distribution among the hadronising clusters.

REFERENCES

1. F. Becattini, Z. Phys. **C69** (1996) 485; F. Becattini and U. Heinz, Z. Phys. **C76** (1997) 269.
2. F. Becattini, L. Bellucci, G. Passaleva, Nucl. Phys. Proc. Suppl. **92** (2001) 137; F. Becattini, G. Passaleva, in preparation.
3. F. Becattini, M. Gazdzicki, J. Sollfrank, Eur. Phys. J. **C5** (1998) 143; F. Becattini et al., Phys. Rev. **C64** (2001) 024901.
4. Y. Nambu, G. Jona-Lasinio, Phys. Rev. **122** (1961) 345; **124** (1961) 246.
5. T. Hatsuda and T. Kunihiro, Phys. Rep. **247** (1994) 221.
6. A. Barducci, R. Casalbuoni, R. Gatto and G. Pettini, Phys. Rev. **D49**, (1994) 426 and references therein.
7. F. Karsch, this conference.
8. K. Redlich and L. Turko, Z. Phys. **C5**, (1980) 201; M.I. Gorenstein et al., Phys. Lett. **B123** (1983) 437; G. Auberson et al., J. Math. Phys. **27** (6) (1986) 1658.
9. F. Becattini and G. Pettini, in preparation.
10. W. Florkowski, W. Broniowski and M. Michalec, nucl-th/0106009.

Crystalline Color Superconductivity

Krishna Rajagopal

Center for Theoretical Physics, Massachusetts Institute of Technology, Cambridge, MA, USA 02139. E-mail: krishna@ctp.mit.edu

Abstract. We give an introduction crystalline color superconductivity, arguing that it is likely to occur wherever quark matter in which color-flavor locking does not occur is found. We survey the properties of this form of quark matter, and argue that its presence in a compact star may result in pulsar glitches, and thus in observable consequences. However, elucidation of this proposal requires an understanding of the crystal structure, which is not yet in hand.

INTRODUCTION

At asymptotic densities, the ground state of QCD with three quarks with equal masses is expected to be the color-flavor locked (CFL) phase [1, 2, 3]. This phase features a condensate of Cooper pairs of quarks which includes *ud*, *us*, and *ds* pairs. Quarks of all colors and all flavors participate in the pairing, and all excitations with quark quantum numbers are gapped.

The CFL phase persists for unequal quark masses, so long as the differences are not too large [4, 5]. It is very likely the ground state for real QCD, assumed to be in equilibrium with respect to the weak interactions, over a substantial range of densities. In this phase, chiral symmetry is broken via the locking of left-flavor and right-flavor symmetries to color [1]. Terms of order m_s^4 in the effective Lagrangian for the resulting pseudo-Goldstone bosons [6] may rotate the CFL condensate in the K^0 direction [7], resulting in further pseudo-Goldstone bosons [8]. Throughout the range of parameters over which the CFL phase exists as a bulk (and therefore electrically neutral) phase, it consists of equal numbers of u, d and s quarks and is therefore electrically neutral in the absence of any electrons [9]. The equality of the three quark number densities is enforced in the CFL phase by the fact that this equality maximizes the pairing energy associated with the formation of *ud*, *us*, and *ds* Cooper pairs. This equality is enforced even though the strange quark, with mass m_s, is heavier than the light quarks. If higher order effects do in fact introduce an additional K^0 condensate, the conclusion that the CFL phase is electrically neutral in the absence of electrons remains, because K^0 mesons are neutral.

If one imagines increasing m_s (or, more physically, decreasing μ) color-flavor locking is maintained until a transition to a state of quark matter in which some quarks remain ungapped. This "unlocking transition", which must be first order [4, 5], occurs when $m_s^2 \approx 4\mu\Delta_0$ [9, 10]. In this expression, Δ_0 is the BCS pairing gap, estimated in both models and asymptotic analyses to be of order tens to 100 MeV [3]. The strange quark mass m_s is a density-dependent effective mass. For $\mu \sim 400 - 500$ MeV, corresponding

CP602, *QCD@Work: International Workshop on Quantum Chromodynamics*
edited by P. Colangelo and G. Nardulli

to quark matter at densities which may arise at the center of compact stars, m_s is certainly significantly larger than the current quark mass, and its value is not known. In drawing the QCD phase diagram, therefore, there are two possibilities. As a function of decreasing μ, one possibility is a first order phase transition directly from color-flavor locked quark matter to hadronic matter, as explored in Ref. [10]. The second possibility is an unlocking transition [4, 5] to quark matter in which not all quarks participate in the dominant pairing, followed only at a lower μ by a transition to hadronic matter. We assume the second possibility here, and explore its consequences.

In quark matter in which CFL pairing involving all quarks does not occur, it is likely that up and down quarks continue to pair. In this 2SC phase, which was the earliest color superconducting phase to be studied [11], the attractive channel involves the formation of Cooper pairs which are antisymmetric in both color and flavor, yielding a condensate with color (greek indices) and flavor (latin indices) structure $\langle q^\alpha_a q^\beta_b \rangle \sim \varepsilon_{ab}\varepsilon^{\alpha\beta 3}$. This condensate leaves five quarks unpaired: up and down quarks of the third color, and strange quarks of all three colors. Because the BCS pairing scheme leaves ungapped quarks with differing Fermi momenta, it is natural to ask whether there is some generalization of the pairing ansatz, beyond BCS, in which pairing between two species of quarks persists even once their Fermi momenta differ. Crystalline color superconductivity is the answer to this question.

THE CRYSTALLINE COLOR SUPERCONDUCTING STATE

To date, crystalline color superconductivity has only been studied in the simplified model context in which one considers only pairing between massless up and down quarks whose Fermi momenta we attempt to push apart by turning on a chemical potential difference [14, 15, 16, 17], rather than considering CFL pairing in the presence of quark mass differences. That is, we introduce

$$\begin{aligned} \mu_u &= \mu - \delta\mu \\ \mu_d &= \mu + \delta\mu\,. \end{aligned} \qquad (1)$$

If $\delta\mu$ is nonzero but less than some $\delta\mu_1$, the ground state is precisely that obtained for $\delta\mu = 0$ [18, 19, 14]. In this 2SC state, red and green up and down quarks pair, yielding four quasiparticles with superconducting gap Δ_0 [11]. Furthermore, the number density of red and green up quarks is the same as that of red and green down quarks. As long as $\delta\mu$ is not too large, this BCS state remains unchanged (and favored) because maintaining equal number densities, and thus coincident Fermi surfaces, maximizes the pairing and hence the gain in interaction energy. As $\delta\mu$ is increased, the BCS state remains the ground state of the system only as long as its negative interaction energy offsets the large positive free energy cost associated with forcing the Fermi seas to deviate from their normal state distributions. In the weak coupling limit, in which $\Delta_0/\mu \ll 1$, the BCS state persists for $\delta\mu < \delta\mu_1 = \Delta_0/\sqrt{2}$ [18, 14]. These conclusions are the same (as long as $\Delta_0/\mu \ll 1$) whether the interaction between quarks is modeled as a point-like four-fermion interaction or is approximated by single-gluon exchange [3]. The loss of BCS pairing at $\delta\mu = \delta\mu_1$ is the analogue in this toy model of the unlocking transition.

If $\delta\mu$ is too large, no pairing between species is possible. The transition between the BCS and unpaired states as $\delta\mu$ increases past $\delta\mu_1$ has been studied in electron superconductors [18] and QCD superconductors [4, 5, 19] assuming that no other state intervenes. However, there is good reason to think that another state can occur. This is the "LOFF" state, first explored by Larkin and Ovchinnikov[12] and Fulde and Ferrell[13] in the context of electron superconductivity in the presence of magnetic impurities. They found that near the unpairing transition, in a range $\delta\mu_1 < \delta\mu < \delta\mu_2$, it is favorable to form a state in which the Cooper pairs have nonzero momentum. This generalization of the pairing ansatz (beyond BCS ansätze in which only quarks with momenta which add to zero pair) is favored because it gives rise to a region of phase space where each of the two quarks in a pair can be close to its Fermi surface, and such pairs can be created at low cost in free energy. Condensates of this sort spontaneously break translational and rotational invariance, leading to gaps which vary periodically in a crystalline pattern. If in some shell within the quark matter core of a neutron star (or within a strange quark star) the quark number densities are such that crystalline color superconductivity arises, rotational vortices may be pinned in this shell, making it a locus for glitch phenomena [14].

In Ref. [14], we have evaluated the width of the window $\delta\mu_1 < \delta\mu < \delta\mu_2$ for which crystalline color superconductivity occurs, upon making the simplifying assumption that quarks interact via a four-fermion interaction with the quantum numbers of single gluon exchange.

In the LOFF state, each Cooper pair carries momentum $2\mathbf{q}$ with $|\mathbf{q}| \approx 1.2\delta\mu$. The condensate and gap parameter vary in space with wavelength $\pi/|\mathbf{q}|$. Although the magnitude $|\mathbf{q}|$ is determined energetically, as we sketch below, the direction $\hat{\mathbf{q}}$ is chosen spontaneously. The LOFF state is characterized by a gap parameter Δ and a diquark condensate, but not by an energy gap in the dispersion relation: we obtain the quasiparticle dispersion relations[14] and find that they vary with the direction of the momentum, yielding gaps that vary from zero up to a maximum of Δ. The condensate is dominated by the regions in momentum space in which a quark pair with total momentum $2\mathbf{q}$ has both members of the pair within $\sim \Delta$ of their respective Fermi surfaces. These regions form circular bands on the two Fermi surfaces. Choosing a single, fixed, $\mathbf{q}$ means that only one circular band on each Fermi surface participates in the pairing. In all work published to date, we assume that all Cooper pairs make the same choice of direction $\hat{\mathbf{q}}$. Making this ansatz corresponds to choosing a single circular band on each Fermi surface. In position space, it corresponds to a condensate which varies in space like

$$\langle\psi(\mathbf{x})\psi(\mathbf{x})\rangle \propto \Delta e^{2i\mathbf{q}\cdot\mathbf{x}} \,. \tag{2}$$

This ansatz is certainly *not* the best choice. If a single plane wave is favored, why not two? That is, if one choice of $\hat{\mathbf{q}}$ is favored, why not add a second $\mathbf{q}$, with the same $|\mathbf{q}|$ but a different $\hat{\mathbf{q}}$? If two are favored, why not three? This question, namely the determination of the favored crystal structure of the crystalline color superconductor phase, remains open. Note, however, that if we find a region $\delta\mu_1 < \delta\mu < \delta\mu_2$ in which the simple LOFF ansatz, with a single $\hat{\mathbf{q}}$, is favored over the BCS state and over no pairing, then the LOFF state with whatever crystal structure turns out to be optimal must be favored in *at least* this region. Note also that even the single $\hat{\mathbf{q}}$ ansatz, which we use henceforth, breaks translational and rotational invariance spontaneously. The resulting phonon has been analyzed in considerable detail in Ref. [16]. It will be very interesting to use these methods to analyze the phonons in more complicated crystal structures.

Having simplified the interaction by making it pointlike, and simplifed the ansatz by assuming the condensate varies like a plane wave, in Ref. [14] we give the ansatz for the LOFF wave function, and by variation obtain a gap equation which allows us to solve for the gap parameter Δ, the free energy and the values of the diquark condensates which characterize the LOFF state at a given $\delta\mu$ and $|\mathbf{q}|$. We then vary $|\mathbf{q}|$, to find the preferred (lowest free energy) LOFF state at a given $\delta\mu$, and compare the free energy of the LOFF state to that of the BCS state with which it competes.[1]

Crystalline color superconductivity is favored for $\delta\mu_1 < \delta\mu < \delta\mu_2$. As $\delta\mu$ increases, one finds a first order phase transition from the ordinary BCS phase to the crystalline color superconducting phase at $\delta\mu = \delta\mu_1$ and then a second order phase transition at $\delta\mu = \delta\mu_2$ at which Δ decreases to zero. Because the condensation energy in the LOFF phase is much smaller than that of the BCS condensate at $\delta\mu = 0$, the value of $\delta\mu_1$ is almost identical to that at which the naive unpairing transition from the BCS state to the state with no pairing would occur if one ignored the possibility of a LOFF phase, namely $\delta\mu_1 = \Delta_0/\sqrt{2}$. For all practical purposes, therefore, the LOFF gap equation is not required in order to determine $\delta\mu_1$. The LOFF gap equation is used to determine $\delta\mu_2$ and the properties of the crystalline color superconducting phase [14].

We find that the LOFF gap parameter decreases from $0.23\Delta_0$ at $\delta\mu = \delta\mu_1$ to zero at $\delta\mu = \delta\mu_2$ [14]. The critical temperature above which the LOFF state melts is $T_c = \Delta\sqrt{3/2\pi^2}$ [20, 15], which means that, except for very close to $\delta\mu_2$, T_c will be much higher than typical neutron star temperatures.

In the limit of a weak four-fermion interaction, $G \to 0$, the crystalline color superconductivity window is bounded by $\delta\mu_1 = \Delta_0/\sqrt{2}$ and $\delta\mu_2 = 0.754\Delta_0$, as first demonstrated in Refs. [12, 13]. These results have been extended beyond the $G \to 0$ limit in Ref. [14]. Note that the BCS gap Δ_0 increases monotonically with G. We may therefore use Δ_0 to parametrize the strength of the interaction G. This proves convenient because, as we have seen, both $\delta\mu_1$ and $\delta\mu_2$ are given simply in terms of the physical quantity Δ_0. (Writing them in terms of the model-dependent parameters G and Λ requires unwieldy expressions.) As first recognized by Refs. [12, 13], at any fixed $\delta\mu$ the LOFF phase only occurs when the interaction strength (that is, Δ_0) lies in a specified window. LOFF pairing does not survive the weak-coupling ($\Delta_0 \to 0$) limit at fixed $\delta\mu$, because in this limit the width of the band on the Fermi surface in which pairing occurs goes to zero. On the other hand, if one takes $\delta\mu$ and Δ_0 both to zero while keeping $\delta\mu/\Delta_0$ fixed and in the appropriate range, LOFF pairing persists down to arbitrarily weak coupling [12, 13, 14].

---

[1] Our model Hamiltonian has two parameters, the four-fermion coupling G and a cutoff Λ. We often use the value of Δ_0, the BCS gap obtained at $\delta\mu = 0$, to describe the strength of the interaction: small Δ_0 corresponds to small G. When we wish to study the dependence on the cutoff, we vary Λ while at the same time varying the coupling G such that Δ_0 is kept fixed. We find that the relation between other physical quantities and Δ_0 is reasonably insensitive to variation of Λ.

OPENING THE CRYSTALLINE COLOR SUPERCONDUCTIVITY WINDOW

The first part of this section is more technical than the rest of this paper. Some readers may wish to skip to the text after Eqs. (15,16).

The variational derivation of the gap equation for the crystalline color superconducting phase is somewhat cumbersome [14]. One constructs a variational ansatz in which only quarks within a "pairing region" are allowed to pair, minimizes the free energy with respect to all variational parameters (two per mode in momentum, color, flavor and spin space), and obtains a self-consistency relation which may then be solved to obtain Δ. The intricacy arises from the fact that the definition of the boundary of the pairing region involves Δ itself. In Ref. [15], we provide a diagrammatic rederivation in which one simply makes an ansatz for the quantum numbers of the condensate and then "turns a field-theoretical crank" and sees this intricate result emerge. We then use the diagrammatic derivation to analyze the LOFF phase at nonzero temperature, obtaining the result for T_c given in the previous section. The diagrammatic formalism also allows us to go beyond a point-like interaction, and treat the exchange of a propagating gluon [17]. Before presenting the results of this generalization of the interaction, we first sketch the diagrammatic derivation of the gap equation.

In the crystalline color superconducting phase, the condensate contains pairs of u and d quarks with momenta such that the total momentum of each Cooper pair is given by $2\mathbf{q}$, with the direction of $\mathbf{q}$ chosen spontaneously. As noted above, wherever there is an instability towards (2), we expect the true ground state to be a crystalline condensate which varies in space like a sum of several such plane waves with the same $|\mathbf{q}|$.

In order to describe pairing between u quarks with momentum $\mathbf{p}+\mathbf{q}$ and d quarks with momentum $-\mathbf{p}+\mathbf{q}$, we must use a modified Nambu-Gorkov spinor defined as

$$\Psi(p,q) = \begin{pmatrix} \psi_u(p+q) \\ \psi_d(p-q) \\ \bar{\psi}_d^T(-p+q) \\ \bar{\psi}_u^T(-p-q) \end{pmatrix} . \tag{3}$$

Note that by q we mean the four-vector $(0,\mathbf{q})$. The Cooper pairs have nonzero total momentum, and the ground state condensate (2) is static. The momentum dependence of (3) is motivated by the fact that in the presence of a crystalline color superconducting condensate, anomalous propagation does not only mean picking up or losing two quarks from the condensate. It also means picking up or losing momentum $2\mathbf{q}$. The basis (3) has been chosen so that the inverse fermion propagator in the crystalline color superconducting phase is diagonal in p-space and is given by

$$S^{-1}(p,q) = \begin{bmatrix} \not{p}+\not{q}+\mu_u\gamma_0 & 0 & -\bar{\Delta}(p,-q) & 0 \\ 0 & \not{p}-\not{q}+\mu_d\gamma_0 & 0 & \bar{\Delta}(p,q) \\ -\Delta(p,-q) & 0 & (\not{p}-\not{q}-\mu_d\gamma_0)^T & 0 \\ 0 & \Delta(p,q) & 0 & (\not{p}+\not{q}-\mu_u\gamma_0)^T \end{bmatrix} , \tag{4}$$

where $\bar{\Delta} = \gamma_0 \Delta^\dagger \gamma_0$ and, here, Δ is a matrix proportional to $C\gamma_5 \varepsilon^{\alpha\beta 3}$. Note that the condensate is explicitly antisymmetric in flavor. $2\mathbf{p}$ is the relative momentum of the quarks in a given pair and is different for different pairs. In the gap equation below, we shall integrate over p_0 and $\mathbf{p}$. As desired, the off-diagonal blocks describe anomalous propagation in the presence of a condensate of diquarks with momentum $2\mathbf{q}$.

We obtain the gap equation by solving the one-loop Schwinger-Dyson equation given by

$$S^{-1}(k,q) - S_0^{-1}(k,q) = -g^2 \int \frac{d^4p}{(2\pi)^4} \Gamma^A_\mu S(p,q) \Gamma^B_\nu D^{\mu\nu}_{AB}(k-p), \tag{5}$$

where $D^{\mu\nu}_{AB} = D^{\mu\nu}\delta_{AB}$ is the gluon propagator, S is the full quark propagator, whose inverse is given by (4), and S_0 is the fermion propagator in the absence of interaction, given by S with $\Delta = 0$. The vertices are defined as follows:

$$\Gamma^A_\mu = \begin{pmatrix} \gamma_\mu \lambda^A/2 & 0 & 0 & 0 \\ 0 & \gamma_\mu \lambda^A/2 & 0 & 0 \\ 0 & 0 & -(\gamma_\mu \lambda^A/2)^T & 0 \\ 0 & 0 & 0 & -(\gamma_\mu \lambda^A/2)^T \end{pmatrix}. \tag{6}$$

In Refs. [14, 15], we introduce a point-like interaction by replacing $g^2 D^{\mu\nu}$ by $g^{\mu\nu}$ times a constant G. After some algebra (essentially the determination of S given S^{-1} specified above), and upon suitable projection, the Schwinger-Dyson equation (5) reduces to a gap equation for the gap parameter Δ given (in Euclidean space) by

$$\Delta = 2G \int \frac{d^4p}{(2\pi)^4} \frac{4\Delta w}{w^2 - 4\left[(|\mathbf{p}|^2 - (ip_0 + \delta\mu)^2)(\mu^2 - |\mathbf{q}|^2) + (\mathbf{p}\cdot\mathbf{q} + \mu(ip_0 + \delta\mu))^2\right]} \tag{7}$$

where $w = |\mathbf{p}|^2 - |\mathbf{q}|^2 - (ip_0 + \delta\mu)^2 + \mu^2 + \Delta^2$. We show in Ref. [15] that upon neglecting the (small) contributions of antiparticle pairing, this gap equation simplifies to

$$\Delta = 2G \int \frac{d^4p}{(2\pi)^4} \frac{2\Delta \sin^2\frac{\beta}{2}}{(p_0 - iE_1(\mathbf{p}))(p_0 + iE_2(\mathbf{p}))} \tag{8}$$

where $E_{1,2}(\mathbf{p})$ are defined as in Ref. [14]:

$$\begin{aligned} E_1(\mathbf{p}) = &\ +\delta\mu + \tfrac{1}{2}(|\mathbf{p}+\mathbf{q}| - |\mathbf{p}-\mathbf{q}|) + \tfrac{1}{2}\sqrt{(|\mathbf{p}+\mathbf{q}| + |\mathbf{p}-\mathbf{q}| - 2\mu)^2 + 4\Delta^2 \sin^2\tfrac{\beta}{2}} \\ E_2(\mathbf{p}) = &\ -\delta\mu - \tfrac{1}{2}(|\mathbf{p}+\mathbf{q}| - |\mathbf{p}-\mathbf{q}|) + \tfrac{1}{2}\sqrt{(|\mathbf{p}+\mathbf{q}| + |\mathbf{p}-\mathbf{q}| - 2\mu)^2 + 4\Delta^2 \sin^2\tfrac{\beta}{2}} \end{aligned} \tag{9}$$

and β is defined as the angle between the up quark momentum $\mathbf{q}+\mathbf{p}$ and the down quark momentum $\mathbf{q}-\mathbf{p}$. Upon doing the p_0 integration, we obtain the gap equation derived variationally in Ref. [14]:

$$\begin{aligned} 1 &= 2G \int_{\mathbf{p}\in P} \frac{d^3p}{(2\pi)^3} \frac{2\sin^2\frac{\beta}{2}}{E_1(\mathbf{p}) + E_2(\mathbf{p})} \\ &= 2G \int_{\mathbf{p}\in P} \frac{d^3p}{(2\pi)^3} \frac{2\sin^2\frac{\beta}{2}}{\sqrt{(|\mathbf{p}+\mathbf{q}| + |\mathbf{p}-\mathbf{q}| - 2\mu)^2 + 4\Delta^2 \sin^2\frac{\beta}{2}}} \end{aligned} \tag{10}$$

where the "pairing region" P in $\mathbf{p}$-space is given by

$$P = \{\mathbf{p} \mid E_1(\mathbf{p}) > 0 \text{ and } E_2(\mathbf{p}) > 0\} . \tag{11}$$

Thus, an exercise in residue calculus has reproduced the blocking regions, excluding from the gap equation those regions in momentum space where $E_1(\mathbf{p})$ or $E_2(\mathbf{p})$ is negative. Note that because $E_1(\mathbf{p}) + E_2(\mathbf{p}) \geq 0$, as can be seen from the definitions (9), there is no value of $\mathbf{p}$ for which both E_1 and E_2 are negative. Note also that the gap equation is dominated by those regions in momentum space where $E_1(\mathbf{p}) + E_2(\mathbf{p})$ is as small as possible, where the integrand in (10) is of order $1/\Delta$. These values of $\mathbf{p}$ are such that both members of a LOFF pair have momenta close to (within $\sim \Delta$ of) their respective Fermi surfaces. That is, $|\mathbf{p}+\mathbf{q}|$ is within Δ of μ_u and $|-\mathbf{p}+\mathbf{q}|$ is within Δ of μ_d. The results described in the previous section all follow from analysis of the gap equation (10) [12, 13, 20, 14].

In Ref. [17], we begin with the Schwinger-Dyson equation (5) but this time keep the gluon propagator. That is, we analyze the crystalline color superconducting phase upon assuming that quarks interact by the exchange of a medium-modified gluon, as is quantitatively valid at asymptotically high densities. The medium-modified gluon propagator is given by

$$D_{\mu\nu}(p) = \frac{P^T_{\mu\nu}}{p^2 - G(p)} + \frac{P^L_{\mu\nu}}{p^2 - F(p)} - \xi \frac{p_\mu p_\nu}{p^4} , \tag{12}$$

where ξ is the gauge parameter, $G(p)$ and $F(p)$ are functions of p_0 and $|\mathbf{p}|$, and the projectors $P^{T,L}_{\mu\nu}$ are defined as follows:

$$P^T_{ij} = \delta_{ij} - \hat{p}_i \hat{p}_j, \; P^T_{00} = P^T_{0i} = 0, \; P^L_{\mu\nu} = -g_{\mu\nu} + \frac{p_\mu p_\nu}{p^2} - P^T_{\mu\nu}. \tag{13}$$

The functions F and G describe the effects of the medium on the gluon propagator. If we neglect the Meissner effect (that is, if we neglect the modification of $F(p)$ and $G(p)$ due to the gap Δ in the fermion propagator) then $F(p)$ describes Thomas-Fermi screening and $G(p)$ describes Landau damping and they are given in the hard dense loop (HDL) approximation by [21]

$$\begin{aligned} F(p) &= m^2 \frac{p^2}{|\mathbf{p}|^2} \left(1 - \frac{ip_0}{|\mathbf{p}|} Q_0\left(\frac{ip_0}{|\mathbf{p}|}\right)\right), \qquad \text{with } Q_0(x) = \frac{1}{2}\log\left(\frac{x+1}{x-1}\right), \\ G(p) &= \frac{1}{2} m^2 \frac{ip_0}{|\mathbf{p}|}\left[\left(1 - \left(\frac{ip_0}{|\mathbf{p}|}\right)^2\right) Q_0\left(\frac{ip_0}{|\mathbf{p}|}\right) + \frac{ip_0}{|\mathbf{p}|}\right] , \end{aligned} \tag{14}$$

where $m^2 = g^2\mu^2/\pi^2$ is the Debye mass for $N_f = 2$. The further modification to the gluon propagator due to the Meissner effect in spatially uniform color superconducting phases has been the subject of much recent work [22], but the Meissner effect in the crystalline color superconducting phase has not yet been analyzed. Fortunately, in our calculation of $\delta\mu_2$ we we shall only need to study the crystalline color superconducting phase in the limit in which $\Delta \to 0$, and in this limit the expressions (12) and (14) are valid.

Upon neglecting the (small) contributions from antiparticle pairing, the gap equation becomes

$$\Delta(k_0) = \frac{-ig^2}{3\sin^2\frac{\beta(k,k)}{2}} \int \frac{d^4p}{(2\pi)^4} \frac{\Delta(p_0)}{(p_0+E_1)(p_0-E_2)} \times \left[\frac{C_F}{(k-p)^2 - F(k-p)} + \frac{C_G}{(k-p)^2 - G(k-p)} + \frac{C_\xi \xi}{(k-p)^2} \right], \quad (15)$$

where

$$\begin{aligned}
C_F &= \cos^2\frac{\beta(k,p)}{2}\cos^2\frac{\beta(p,k)}{2} - \cos^2\frac{\beta(k,-p)}{2}\cos^2\frac{\beta(-p,k)}{2} - \sin^2\frac{\beta(k,k)}{2}\sin^2\frac{\beta(p,p)}{2}, \\
C_G &= \frac{\cos\beta(k,-p)\cos\beta(-p,k) - \cos\beta(k,p)\cos\beta(p,k)}{2} - 2\sin^2\frac{\beta(k,k)}{2}\sin^2\frac{\beta(p,p)}{2} \\
&\quad - \cos\alpha(k,p)\left(\cos\alpha(p,k)\sin^2\frac{\beta(k,-p)}{2} + \cos\alpha(-p,-k)\sin^2\frac{\beta(k,p)}{2}\right) \\
&\quad - \cos\alpha(-k,-p)\left(\cos\alpha(p,k)\sin^2\frac{\beta(p,k)}{2} + \cos\alpha(-p,-k)\sin^2\frac{\beta(-p,k)}{2}\right), \\
C_\xi &= \sin^2\frac{\beta(k,p)}{2}\sin^2\frac{\beta(p,k)}{2} - \sin^2\frac{\beta(k,k)}{2}\sin^2\frac{\beta(p,p)}{2} - \sin^2\frac{\beta(k,-p)}{2}\sin^2\frac{\beta(-p,k)}{2} \\
&\quad + \cos\alpha(k,p)\left(\cos\alpha(p,k)\sin^2\frac{\beta(k,-p)}{2} + \cos\alpha(-p,-k)\sin^2\frac{\beta(k,p)}{2}\right) \\
&\quad + \cos\alpha(-k,-p)\left(\cos\alpha(p,k)\sin^2\frac{\beta(p,k)}{2} + \cos\alpha(-p,-k)\sin^2\frac{\beta(-p,k)}{2}\right), \quad (16)
\end{aligned}$$

with E_1 and E_2 as in (9), and where $\cos\alpha(k,p) = \widehat{(k-q)}\cdot\widehat{(k-p)}$ and $\cos\beta(k,p) = \widehat{(q+k)}\cdot\widehat{(q-p)}$. In Ref. [17], we use this gap equation to obtain $\delta\mu_2$, the upper boundary of the crystalline color superconductivity window. This analysis is controlled at asymptotically high densities where the coupling g is weak.

At weak coupling, quark-quark scattering by single-gluon exchange is dominated by forward scattering. In most scatterings, the angular positions of the quarks on their respective Fermi surfaces do not change much. As a consequence, the weaker the coupling the more the physics can be thought of as a sum of many $(1+1)$-dimensional theories, with only rare large-angle scatterings able to connect one direction in momentum space with others [23]. In the LOFF state, small-angle scattering is advantageous because it cannot scatter a pair of quarks out of the region of momentum space in which both members of the pair are in their respective rings, where pairing is favored. This means that it is natural to expect that a forward-scattering-dominated interaction like single-gluon exchange is much more favorable for crystalline color superconductivity than a point-like interaction, which yields s-wave scattering.

Suppose for a moment that we were analyzing a truly $(1+1)$-dimensional theory. The momentum-space geometry of the LOFF state in one spatial dimension is qualitatively different from that in three. Instead of Fermi surfaces, we would have only "Fermi points" at $\pm\mu_u$ and $\pm\mu_d$. The only choice of $|\mathbf{q}|$ which allows pairing between u and d

quarks at their respective Fermi points is $|\mathbf{q}| = \delta\mu$. In $(3+1)$ dimensions, in contrast, $|\mathbf{q}| > \delta\mu$ is favored because it allows LOFF pairing in ring-shaped regions of the Fermi surface, rather than just at antipodal points [12, 13, 14]. Also, in striking contrast to the $(3+1)$-dimensional case, it has long been known that in a true $(1+1)$-dimensional theory with a point-like interaction between fermions, $\delta\mu_2/\Delta_0 \to \infty$ in the weak-interaction limit [24].

We expect that in $(3+1)$-dimensional QCD with the interaction given by single-gluon exchange, as $\mu \to \infty$ and $g(\mu) \to 0$ the $(1+1)$-dimensional results should be approached: the energetically favored value of $|\mathbf{q}|$ should become closer and closer to $\delta\mu$, and $\delta\mu_2/\Delta_0$ should diverge. We derive both these effects in Ref. [17] and furthermore show that both are clearly in evidence already at the rather large coupling $g = 3.43$, corresponding to $\mu = 400$ MeV using the conventions of Refs. [25, 26]. At this coupling, $\delta\mu_2/\Delta_0 \approx 1.2$, meaning that $(\delta\mu_2 - \delta\mu_1) \approx (1.2 - 1/\sqrt{2})\Delta_0$, which is much larger than $(0.754 - 1/\sqrt{2})\Delta_0$. If we go to much higher densities, where the calculation is under quantitative control, we find an even more striking enhancement: when $g = 0.79$ we find $\delta\mu_2/\Delta_0 > 1000$! We see that (relative to expectations based on experience with point-like interactions) the crystalline color superconductivity window is wider by more than four orders of magnitude at this weak coupling, and is about one order of magnitude wider at accessible densities if weak-coupling results are applied there.[2]

We have found that $\delta\mu_2/\Delta_0$ diverges in QCD as the weak-coupling, high-density limit is taken. Applying results valid at asymptotically high densities to those of interest in compact stars, namely $\mu \sim 400$ MeV, we find that even here the crystalline color superconductivity window is an order of magnitude wider than that obtained previously upon approximating the interaction between quarks as point-like. The crystalline color superconductivity window in parameter space may therefore be much wider than previously thought, making this phase a *generic* feature of the phase diagram for cold dense quark matter. The reason for this qualitative increase in $\delta\mu_2$ can be traced back to the fact that gluon exchange at weaker and weaker coupling is more and more dominated by forward-scattering, while point-like interactions describe s-wave scattering. What is perhaps surprising is that even at quite *large* values of g, gluon exchange yields an order of magnitude increase in $\delta\mu_2 - \delta\mu_1$.

This discovery has significant implications for the QCD phase diagram and may have significant implications for compact stars. At high enough baryon density the CFL phase in which all quarks pair to form a spatially uniform BCS condensate is favored. Suppose that as the density is lowered the nonzero strange quark mass induces the formation of some less symmetrically paired quark matter before the density is lowered so much that baryonic matter is obtained. In this less symmetric quark matter, some quarks may yet form a BCS condensate. Those which do not, however, will have

[2] LOFF condensates have also recently been considered in two other contexts. In QCD with $\mu_u < 0$, $\mu_d > 0$ and $\mu_u = -\mu_d$, one has equal Fermi momenta for $\bar{u}$ antiquarks and d quarks, BCS pairing occurs, and consequently a $\langle \bar{u}d \rangle$ condensate forms [27, 28]. If $-\mu_u$ and μ_d differ, and if the difference lies in the appropriate range, a LOFF phase with a spatially varying $\langle \bar{u}d \rangle$ condensate results [27, 28]. The result of Ref. [17] that the LOFF window is much wider than previously thought applies in this context also. Suitably isospin asymmetric nuclear matter may also admit LOFF pairing, as discussed recently in Ref. [29]. Here, the interaction is not forward-scattering dominated.

differing Fermi momenta. These will form a crystalline color superconducting phase if the differences between their Fermi momenta lie within the appropriate window. In QCD, the interaction between quarks is forward-scattering dominated and the crystalline color superconductivity window is consequently wide open. This phase is therefore generic, occurring almost anywhere there are some quarks which cannot form BCS pairs. Evaluating the critical temperature T_c above which the crystalline condensate melts requires solving the nonzero temperature gap equation obtained from (15) as done in Ref. [15] for the case of a point-like interaction. As in that case, we expect that all compact stars which are minutes old or older are much colder than T_c. This suggests that wherever quark matter which is not in the CFL phase occurs within a compact star, crystalline color superconductivity is to be found. As we discuss in the next section, wherever crystalline color superconductivity is found rotational vortices may be pinned resulting in the generation of glitches as the star spins down.

Solidifying the implications of our results requires further work in several directions. First, we must confirm that pushing Fermi surfaces apart via quark mass differences has the same effect as pushing them apart via a $\delta\mu$ introduced by hand. Second, we must extend the analysis to the three flavor theory of interest. And, third, before evaluating the pinning force on a rotational vortex and making predictions for glitch phenomena, we need to understand which crystal structure is favored.

GLITCHES IN COMPACT STARS

We do not yet know whether compact stars feature quark matter cores. And, we do not yet know whether, if they contain quark matter, that quark matter is color-flavor locked, meaning that quarks of all colors and flavors participate in BCS pairing, or whether the BCS condensate leaves some quarks unpaired. The lesson we take from the toy model analysis is that because the interaction between quarks in QCD is dominated by forward scattering, rather than being an s-wave point-like interaction, the difference in Fermi momenta between the unpaired quarks need not fall within a narrow window in order for them to form a crystalline color superconductor.

We wish now to ask whether the presence of a shell of crystalline color superconducting quark matter in a compact star (between the hadronic "mantle" and the CFL "inner core") has observable consequences. A quantitative formulation of this question would allow one either to discover crystalline color superconductivity, or to rule out its presence. (The latter would imply either no quark matter at all, or a single CFL-nuclear interface [10].)

Many pulsars have been observed to glitch. Glitches are sudden jumps in rotation frequency Ω which may be as large as $\Delta\Omega/\Omega \sim 10^{-6}$, but may also be several orders of magnitude smaller. The frequency of observed glitches is statistically consistent with the hypothesis that all radio pulsars experience glitches [30]. Glitches are thought to originate from interactions between the rigid neutron star crust, typically somewhat more than a kilometer thick, and rotational vortices in a neutron superfluid. The inner kilometer of crust consists of a crystal lattice of nuclei immersed in a neutron superfluid [31]. Because the pulsar is spinning, the neutron superfluid (both within the inner crust and deeper inside the star) is threaded with a regular array of rotational vortices. As the pulsar's spin gradually slows, these vortices must gradually move outwards since the rotation frequency of a superfluid is proportional to the density of vortices. Deep within the star, the vortices are free to move outwards. In the crust, however, the vortices are pinned by their interaction with the nuclear lattice. Models [32] differ in important respects as to how the stress associated with pinned vortices is released in a glitch: for example, the vortices may break and rearrange the crust, or a cluster of vortices may suddenly overcome the pinning force and move macroscopically outward, with the sudden decrease in the angular momentum of the superfluid within the crust resulting in a sudden increase in angular momentum of the rigid crust itself and hence a glitch. All the models agree that the fundamental requirements are the presence of rotational vortices in a superfluid and the presence of a rigid structure which impedes the motion of vortices and which encompasses enough of the volume of the pulsar to contribute significantly to the total moment of inertia.

Although it is premature to draw quantitative conclusions, it is interesting to speculate that some glitches may originate deep within a pulsar which features a quark matter core, in a region of that core which is in the crystalline color superconductor phase. A full three-flavor analysis is required, first of all in order to check whether the qualitative conclusions reached in the two-flavor analyses done to date persist, and second of all in order to determine whether the LOFF phase is a superfluid. If the only pairing is between u and d quarks, this 2SC phase is not a superfluid [4], whereas if the LOFF pairing involves the s quarks, as seems likely, a superfluid is obtained [1, 4]. Henceforth, we suppose that the LOFF phase is a superfluid, which means that if it occurs within a pulsar it will be threaded by an array of rotational vortices. It is reasonable to expect that these vortices will be pinned in a LOFF crystal, in which the diquark condensate varies periodically in space. The diquark condensate vanishes at the core of a rotational vortex, and for this reason the vortices will prefer to be located with their cores pinned to the nodes of the LOFF crystal.

A real calculation of the pinning force experienced by a vortex in a crystalline color superconductor must await the determination of the crystal structure of the LOFF phase. We can, however, attempt an order of magnitude estimate along the same lines as that done by Anderson and Itoh [34] for neutron vortices in the inner crust of a neutron star. In that context, this estimate has since been made quantitative [35, 36, 32]. For one specific choice of parameters [14], the LOFF phase is favored over the normal phase by a free energy $F_{\rm LOFF} \sim 5 \times (10\ {\rm MeV})^4$ and the spacing between nodes in the LOFF crystal is $b = \pi/(2|\mathbf{q}|) \sim 9$ fm. The thickness of a rotational vortex is given by the correlation length $\xi \sim 1/\Delta \sim 25$ fm. The pinning energy is the difference between the energy of a section of vortex of length b which is centered on a node of the LOFF crystal vs. one which is centered on a maximum of the LOFF crystal. It is of order $E_p \sim F_{\rm LOFF}\, b^3 \sim 4$ MeV. The resulting pinning force per unit length of vortex is of order $f_p \sim E_p/b^2 \sim (4\ {\rm MeV})/(80\ {\rm fm}^2)$. A complete calculation will be challenging because $b < \xi$, and is likely to yield an f_p which is somewhat less than that we have obtained by dimensional analysis. Note that our estimate of f_p is quite uncertain both because it is only based on dimensional analysis and because the values of Δ, b and $F_{\rm LOFF}$ are uncertain. (We have a reasonable understanding of all the ratios Δ/Δ_0, $\delta\mu/\Delta_0$, q/Δ_0

and consequently $b\Delta_0$ in the LOFF phase. It is of course the value of the BCS gap Δ_0 which is uncertain.) It is premature to compare our crude result to the results of serious calculations of the pinning of crustal neutron vortices as in Refs. [35, 36, 32]. It is nevertheless remarkable that they prove to be similar: the pinning energy of neutron vortices in the inner crust is $E_p \approx 1-3$ MeV and the pinning force per unit length is $f_p \approx (1-3\ \mathrm{MeV})/(200-400\ \mathrm{fm}^2)$.

A quantitative theory of glitches originating within quark matter in a LOFF phase must await further calculations, in particular a three flavor analysis and the determination of the crystal structure of the QCD LOFF phase. However, our rough estimate of the pinning force on rotational vortices in a LOFF region suggests that this force may be comparable to that on vortices in the inner crust of a conventional neutron star. Perhaps, therefore, glitches occurring in a region of crystalline color superconducting quark matter may yield similar phenomenology to that observed. Were this to happen, we can hope that a more detailed analysis would reveal distinctions among observed glitches, with some better understood as conventional glitches, originating in the inner crust, and others better understood as glitches originating deep within the star, in quark matter in the crystalline color superconductor phase. This is surely strong motivation for further investigation.

ACKNOWLEDGMENTS

I am grateful to my collaborators, Mark Alford, Jeff Bowers, Joydip Kundu, Adam Leibovich and Eugene Shuster, with whom I have been exploring crystaline color superconductivity. Let me also thank the organizers of QCD@WORK for bringing together theorists and experimentalists who are putting QCD to work in a variety of different arenas together for a stimulating conference in a very congenial setting. This research was supported in part by the DOE under cooperative research agreement #DF-FC02-94ER40818 and through an OJI Award, and by the A. P. Sloan Foundation.

REFERENCES

1. M. Alford, K. Rajagopal and F. Wilczek, Nucl. Phys. **B537**, 443 (1999) [hep-ph/9804403]
2. R. Rapp, T. Schäfer, E. V. Shuryak and M. Velkovsky, Annals Phys. **280**, 35 (2000) [hep-ph/9904353]; T. Schäfer, Nucl. Phys. **B575**, 269 (2000); [hep-ph/9909574]; I. A. Shovkovy and L. C. Wijewardhana, Phys. Lett. **B470**, 189 (1999) [hep-ph/9910225]; N. Evans, J. Hormuzdiar, S. D. Hsu and M. Schwetz, Nucl. Phys. **B581**, 391 (2000) [hep-ph/9910313].
3. For reviews, see K. Rajagopal and F. Wilczek, hep-ph/0011333; M. Alford, hep-ph/0102047.
4. M. Alford, J. Berges and K. Rajagopal, Nucl. Phys. **B558**, 219 (1999) [hep-ph/9903502].
5. T. Schäfer and F. Wilczek, Phys. Rev. **D60**, 074014 (1999) [hep-ph/9903503].
6. For a review of the effective theory for the mesons of the CFL phase and references to the original literature, see the contribution of R. Casalbuoni to these proceedings, hep-th/0108195.
7. P. F. Bedaque and T. Schäfer, hep-ph/0105150. See also D. B. Kaplan and S. Reddy, hep-ph/0107265. Note that when $m_s \neq 0$, Δ is replaced by five different gap parameters which vary with m_s, at order m_s^2 [4]. Since the Δ's occur quadratically in the free energy, there are contributions to the effective theory of the CFL mesons arising at order m_s^2 and m_s^4. Their effects (along with those of the

electromagnetic and instanton-induced contributions to the meson masses) must be evaluated before firm conclusions about the effects of m_s^4 terms on K^0 condensation can be drawn.
8. V. A. Miransky and I. A. Shovkovy, hep-ph/0108178; T. Schafer, D. T. Son, M. A. Stephanov, D. Toublan and J. J. Verbaarschot, hep-ph/0108210; D. T. Son, hep-ph/0108260.
9. K. Rajagopal and F. Wilczek, Phys. Rev. Lett. **86**, 3492 (2001) [hep-ph/0012039].
10. M. G. Alford, K. Rajagopal, S. Reddy and F. Wilczek, hep-ph/0105009.
11. B. Barrois, Nucl. Phys. **B129**, 390 (1977); D. Bailin and A. Love, Phys. Rept. **107**, 325 (1984); M. Alford, K. Rajagopal and F. Wilczek, Phys. Lett. **B422**, 247 (1998); R. Rapp, T. Schäfer, E. V. Shuryak and M. Velkovsky, Phys. Rev. Lett. **81**, 53 (1998).
12. A. I. Larkin and Yu. N. Ovchinnikov, Zh. Eksp. Teor. Fiz. **47**, 1136 (1964) [Sov. Phys. JETP **20**, 762 (1965)].
13. P. Fulde and R. A. Ferrell, Phys. Rev. **135**, A550 (1964).
14. M. Alford, J. Bowers and K. Rajagopal, Phys. Rev. D **63**, 074016 (2001) [hep-ph/0008208].
15. J. A. Bowers, J. Kundu, K. Rajagopal and E. Shuster, Phys. Rev. D **64**, 014024 (2001) [hep-ph/0101067].
16. R. Casalbuoni, R. Gatto, M. Mannarelli and G. Nardulli, Phys. Lett. B **511**, 218 (2001) [hep-ph/0101326]; R. Casalbuoni, these proceedings, hep-th/0108195.
17. A. K. Leibovich, K. Rajagopal and E. Shuster, hep-ph/0104073.
18. A. M. Clogston, Phys. Rev. Lett. **9**, 266 (1962); B. S. Chandrasekhar, App. Phys. Lett. **1**, 7 (1962).
19. P. F. Bedaque, hep-ph/9910247.
20. S. Takada and T. Izuyama, Prog. Theor. Phys. **41**, 635 (1969).
21. M. LeBellac, *Thermal Field Theory*, Cambridge University Press, (Cambridge, 1996).
22. D. T. Son and M. A. Stephanov, Phys. Rev. **D61**, 074012 (2000) [hep-ph/9910491]; erratum, *ibid.* **D62**, 059902 (2000) [hep-ph/0004095]; D. H. Rischke, Phys. Rev. **D62**, 034007 (2000) [nucl-th/0001040]; G. Carter and D. Diakonov, Nucl. Phys. **B582**, 571 (2000) [hep-ph/0001318]; D. H. Rischke, Phys. Rev. **D62**, 054017 (2000) [nucl-th/0003063]; D. H. Rischke, nucl-th/0103050.
23. D. K. Hong, Phys. Lett. **B473**, 118 (2000) [hep-ph/9812510]; Nucl. Phys. **B582**, 451 (2000) [hep-ph/9905523].
24. A. I. Buzdin and V. V. Tugushev Zh. Eksp. Teor. Fiz. **85**, 735 (1983) [Sov. Phys. JETP **58**, 428 (1983)]; A. I. Buzdin and S. V. Polonskii, Zh. Eksp. Teor. Fiz. **93**, 747 (1987) [Sov. Phys. JETP **66**, 422 (1987)].
25. T. Schäfer and F. Wilczek, Phys. Rev. **D60**, 114033 (1999) [hep-ph/9906512].
26. K. Rajagopal and E. Shuster, Phys. Rev. D **62**, 085007 (2000) [hep-ph/0004074].
27. D. T. Son and M. A. Stephanov, hep-ph/0005225; K. Splittorff, D. T. Son and M. A. Stephanov, hep-ph/0012274.
28. K. Splittorff, D. T. Son and M. A. Stephanov, hep-ph/0012274.
29. A. Sedrakian, nucl-th/0008052. The related unpairing transition was discussed in the absence of LOFF pairing in A. Sedrakian and U. Lombardo, Phys. Rev. Lett. **84**, 602 (2000).
30. M. A. Alpar and C. Ho, Mon. Not. R. Astron. Soc. **204**, 655 (1983). For a recent review, see A.G. Lyne in *Pulsars: Problems and Progress*, S. Johnston, M. A. Walker and M. Bailes, eds., 73 (ASP, 1996).
31. J. Negele and D. Vautherin, Nucl. Phys. **A207**, 298 (1973).
32. For reviews, see D. Pines and A. Alpar, Nature **316**, 27 (1985); D. Pines, in *Neutron Stars: Theory and Observation*, J. Ventura and D. Pines, eds., 57 (Kluwer, 1991); M. A. Alpar, in *The Lives of Neutron Stars*, M. A. Alpar et al., eds., 185 (Kluwer, 1995). For more recent developments and references to further work, see M. Ruderman, Astrophys. J. **382**, 587 (1991); R. I. Epstein and G. Baym, Astrophys. J. **387**, 276 (1992); M. A. Alpar, H. F. Chau, K. S. Cheng and D. Pines, Astrophys. J. **409**, 345 (1993); B. Link and R. I. Epstein, Astrophys. J. **457**, 844 (1996); M. Ruderman, T. Zhu, and K. Chen, Astrophys. J. **492**, 267 (1998); A. Sedrakian and J. M. Cordes, Mon. Not. R. Astron. Soc. **307**, 365 (1999).
33. N. K. Glendenning, Phys. Rev. **D46**, 1274 (1992); N. K. Glendenning, Compact Stars (Springer-Verlag, 1997); F. Weber, J. Phys. G. Nucl. Part. Phys. **25**, R195 (1999).
34. P. W. Anderson and N. Itoh, Nature **256**, 25 (1975).
35. M. A. Alpar, Astrophys. J. **213**, 527 (1977).
36. M. A. Alpar, P. W. Anderson, D. Pines and J. Shaham, Astrophys. J. **278**, 791 (1984).

Color Superconductivity: Symmetries and Effective Lagrangians

Francesco Sannino

NORDITA
Blegdamsvej 17,Copenhagen Ø, DK-2100, Denmark.

Abstract. I briefly review the symmetries and the associated low energy effective Lagrangian for two light flavor Color Superconductivity (2SC).

2SC SYMMETRIES AND EFFECTIVE LAGRANGIAN

Quark matter at very high density is expected to behave as a color superconductor [1]. Possible phenomenological applications include the description of quark stars, neutron star interiors, the physics near the core of collapsing stars and supernova explosions [1, 2, 3]. The color superconductive phase is characterized by its gap energy (Δ) associated to quark-quark pairing which leads to the spontaneous breaking of the color symmetry.

To describe low energy physical processes, where perturbation theory is not applicable, effective Lagrangians based on the global symmetries of the underlying theory are known to play a relevant role. In the case of Color superconductivity effective Lagrangians describe the interactions among the excitations near the fermi surface. The three flavor case (CFL) has been developed in [4]. The low-energy effective Lagrangian for the in medium fermions and the broken sector of the $SU_c(3)$ color group for 2SC has been constructed in Ref. [5]. The effective theories encoding also the electroweak interactions for the low-energy excitations in the 2SC and CFL case can be found in [6]. The light glueball Lagrangian of the unbroken $SU_c(2)$ Yang-Mills sector of the 2SC phase has been constructed in [7].

Here I summarize the effective low energy Lagrangian for two flavors which contains all of the relevant degrees of freedom. First I review the low-energy effective Lagrangian for the 2SC phase of QCD [5, 6]. The latter describes the, in medium, fermions and the broken $SU_c(3)$ gluon sector. I then show how to build the effective Lagrangian describing the light glueballs associated with the unbroken $SU_c(2)$ color subgroup by using the information inherent to the trace anomaly and the medium effects related to a non-vanishing dielectric constant first presented in [8] and confirmed within a different formalism in [9]. Finally the, in medium, glueball to two photon decay process is estimated. The present talk is based on the papers [5, 6, 7, 10, 11].

Quantum Chromo Dynamics with two flavors has gauge symmetry $SU_c(3)$ and global symmetry

$$SU_L(2) \times SU_R(2) \times U_V(1) \ . \tag{1}$$

CP602, *QCD@Work: International Workshop on Quantum Chromodynamics*
edited by P. Colangelo and G. Nardulli

At high matter density a color superconductive phase sets in and the associated diquark condensate leaves invariant the following symmetry group:

$$[SU_c(2)] \times SU_L(2) \times SU_R(2) \times \widetilde{U}_V(1) , \tag{2}$$

where $[SU_c(2)]$ is the unbroken part of the gauge group. The $\widetilde{U}_V(1)$ generator $\widetilde{B}$ is the following linear combination of the previous $U_V(1)$ generator $B = \frac{1}{3}\mathrm{diag}(1,1,1)$ and the broken diagonal generator of the $SU_c(3)$ gauge group $T^8 = \frac{1}{2\sqrt{3}}\mathrm{diag}(1,1,-2)$: $\widetilde{B} = B - \frac{2\sqrt{3}}{3}T^8$. The quarks with color 1 and 2 are neutral under $\widetilde{B}$ and consequently the condensate too ($\widetilde{B}$ is $\sqrt{2}\widetilde{S}$ of Ref. [5]). The superconductive phase for $N_f = 2$ possesses the same global symmetry group of the confined Wigner-Weyl phase [10]. In Reference [10], it was shown that the low-energy spectrum, at finite density, displays the correct quantum numbers to saturate the 't Hooft global anomalies [12]. It was also observed that QCD at finite density can be envisioned, from a global symmetry and anomaly point of view, as a chiral gauge theory. In Reference [11] it was then seen, by using a variety of field theoretical tools, that global anomaly matching conditions hold for any cold but dense gauge theory.

The lowest lying excitations are protected from acquiring a mass by the aforementioned constraints and dominate the low-energy physical processes. The low-energy theorems governing their interactions can be usefully encoded in effective Lagrangians. The dynamics of the Goldstone bosons is efficiently encoded in a non-linear realization framework. Here, see [5], the relevant coset space is G/H with $G = SU_c(3) \times U_V(1)$ and $H = SU_c(2) \times \widetilde{U}_V(1)$ is parameterized by

$$V = \exp(i\xi^i X^i) , \tag{3}$$

where $\{X^i\}$ $i = 1,\cdots,5$ belong to the coset space G/H and are taken to be $X^i = T^{i+3}$ for $i = 1,\cdots,4$ while $X^5 = B + \frac{\sqrt{3}}{3}T^8 = \mathrm{diag}(\frac{1}{2},\frac{1}{2},0)$. T^a are the standard generators of $SU(3)$. The coordinates

$$\xi^i = \frac{\Pi^i}{f} \quad i = 1,2,3,4 , \qquad \xi^5 = \frac{\Pi^5}{\widetilde{f}} , \tag{4}$$

via Π describe the Goldstone bosons.

V transforms non linearly

$$V(\xi) \to u_V\, g\, V(\xi) h^{\dagger}(\xi,g,u) h^{\dagger}_{\widetilde{V}}(\xi,g,u) , \tag{5}$$

with $u_V \in U_V(1)$, $g \in SU_c(3)$, $h(\xi,g,u) \in SU_c(2)$ and $h_{\widetilde{V}}(\xi,g,u) \in \widetilde{U}_V(1)$. It is, also, convenient to define:

$$\omega_\mu = iV^{\dagger}D_\mu V \quad \text{with} \quad D_\mu V = (\partial_\mu - ig_s G_\mu)V , \tag{6}$$

with gluon fields $G_\mu = G_\mu^m T^m$. Following [5] we decompose ω_μ into

$$\omega_\mu^{\parallel} = 2S^a \mathrm{Tr}\left[S^a \omega_\mu\right] \quad \text{and} \quad \omega_\mu^{\perp} = 2X^i \mathrm{Tr}\left[X^i \omega_\mu\right] , \tag{7}$$

where S^a are the unbroken generators of H with $S^{1,2,3} = T^{1,2,3}$, $S^4 = \widetilde{B}/\sqrt{2}$. Summation over repeated indices is assumed.

To be able to include the in medium fermions in the picture we define:

$$\widetilde{\psi} = V^{\dagger}\psi \,, \tag{8}$$

transforming as $\widetilde{\psi} \to h_{\widetilde{V}}(\xi, g, u)h(\xi, g, u)\widetilde{\psi}$ and ψ possesses an ordinary quark transformations (as Dirac spinor).

The simplest non-linearly realized effective Lagrangian describing in medium fermions, the five gluons and their self interactions, up to two derivatives and quadratic in the fermion fields is:

$$\begin{aligned} L = \quad & f^2 a_1 \mathrm{Tr}\left[\omega_0^{\perp}\omega_0^{\perp} - \alpha_1 \vec{\omega}^{\perp}\vec{\omega}^{\perp}\right] + f^2 a_2 \left[\mathrm{Tr}\left[\omega_0^{\perp}\right]\mathrm{Tr}\left[\omega_0^{\perp}\right] - \alpha_2 \mathrm{Tr}\left[\vec{\omega}^{\perp}\right]\mathrm{Tr}\left[\vec{\omega}^{\perp}\right]\right] \\ + \; & b_1 \overline{\widetilde{\psi}} i \left[\gamma^0(\partial_0 - i\omega_0^{\parallel}) + \beta_1 \vec{\gamma}\cdot\left(\vec{\nabla} - i\vec{\omega}^{\parallel}\right)\right]\widetilde{\psi} + b_2 \overline{\widetilde{\psi}}\left[\gamma^0\omega_0^{\perp} + \beta_2\vec{\gamma}\cdot\vec{\omega}^{\perp}\right]\widetilde{\psi} \\ + \; & m_M \overline{\widetilde{\psi}^C}\gamma^5(iT^2)\widetilde{\psi} + \mathrm{h.c.} \,, \end{aligned} \tag{9}$$

where $\widetilde{\psi}^C = i\gamma^2\widetilde{\psi}^*$, $i, j = 1, 2$ are flavor indices and

$$T^2 = S^2 = \frac{1}{2}\begin{pmatrix} \sigma^2 & 0 \\ 0 & 0 \end{pmatrix}, \tag{10}$$

a_1, a_2, b_1 and b_2 are real coefficients while m_M is complex. The breaking of Lorentz invariance to the $O(3)$ subgroup, following [4], has been taken into account by providing different coefficients to the temporal and spatial indices of the Lagrangian, and it is encoded in the coefficients αs and βs. For simplicity, the flavor indices are omitted. From the last two terms, representing a Majorana mass term for the quarks, we deduce that the massless degrees of freedom are the $\psi_{a=3,i}$ which possess the correct quantum numbers to match the 't Hooft anomaly conditions [10]. The generalization to the electroweak processes relevant for the cooling history of compact stars has been investigated in [6].

THE $SU_C(2)$ GLUEBALL EFFECTIVE LAGRANGIAN

The $SU_c(2)$ gauge symmetry does not break spontaneously and it is expected to confine. If the new confining scale is lighter than the superconductive quark-quark gap the associated confined degrees of freedom (light glueballs) [7] can play, together with the true massless quarks a relevant role for the physics of Quark Stars featuring a 2SC superconductive surface layer [3].
Indeed, according to the findings in [8], the medium does lead to partial $SU_c(2)$ screening. In other words the medium is polarizable, i.e., acquires a dielectric constant ε different from unity (in fact $\varepsilon \gg 1$ in the 2SC case [8]) leading to an effectively reduced gauge coupling constant. By assuming locality the $SU_c(2)$ effective action takes the form [8]:

$$S_{eff} = \int d^4x \left[\frac{\varepsilon}{2}\vec{E}^a \cdot \vec{E}^a - \frac{1}{2\lambda}\vec{B}^a \cdot \vec{B}^a\right] \tag{11}$$

with $a = 1,2,3$ and $E_i^a \equiv F_{0i}^a$ and $B_i^a \equiv \frac{1}{2}\varepsilon_{ijk}F_{jk}^a$. Here one assumes an expansion in powers of the fields and derivatives. The gluon speed in this regime is $v = 1/\sqrt{\varepsilon\lambda}$. In Reference [8] the ε and λ were obtained:

$$\varepsilon = 1 + \frac{g_s^2\mu^2}{18\pi^2\Delta^2}\,, \qquad \lambda = 1\,. \tag{12}$$

Equation (12) than suggests that a 2SC color superconductor can have a large positive dielectric constant. This implies that the Coulomb potential between $SU_c(2)$ color charges is reduced in the 2SC medium. $SU_c(2)$ glueballs like particles are expected to emerge. These particles are light with respect to Δ. So, the low-energy $SU_c(2)$ theory should be well represented by the effective Lagrangian describing its hadronic low lying states. This Lagrangian has to be added to the one of Eq. (9) [5] and it has been constructed in [7].

We first rescale the coordinates and the $SU_c(2)$ fields as follows:

$$\hat{x}^0 = \frac{x^0}{\sqrt{\lambda\varepsilon}}\,, \qquad \hat{g} = g_s\left(\frac{\lambda}{\varepsilon}\right)^{\frac{1}{4}} \qquad \hat{A}_0^a = \lambda^{\frac{1}{4}}\varepsilon^{\frac{3}{4}}A_0^a\,, \qquad \hat{A}_i^a = \lambda^{-\frac{1}{4}}\varepsilon^{\frac{1}{4}}A_i^a\,. \tag{13}$$

The $SU_c(2)$ action now becomes:

$$S_{SU(2)} = -\frac{1}{2}\int d^4\hat{x}\,\mathrm{Tr}\left[\hat{F}_{\mu\nu}\hat{F}^{\mu\nu}\right]\,, \tag{14}$$

and $\hat{F}_{\mu,\nu} = \hat{\partial}_\mu\hat{A}_\nu - \hat{\partial}_\nu\hat{A}_\mu + i\hat{g}\left[\hat{A}_\mu,\hat{A}_\nu\right]$ with $\hat{A}_\mu = \hat{A}_\mu^a T^a$ and $a = 1,2,3$. The low-energy effective 3 gluon dynamics in the color superconductor medium (with non-vanishing dielectric constant and magnetic permeability) is similar to the in vacuum theory. The expansion parameter is: $\hat{\alpha} = \frac{\hat{g}^2}{4\pi} = \frac{g_s^2}{4\pi}\sqrt{\frac{\lambda}{\varepsilon}}$. Notice that g_s is the $SU_c(3)$ coupling constant evaluated at the scale μ while we now, following Ref. [8], interpret $\hat{g}$ as the $SU_c(2)$ coupling at Δ. The matching of the scales is encoded in $\sqrt{\lambda/\varepsilon}$.

The, in medium, anomaly-induced effective Lagrangian is based on the trace anomaly arising from the rescaled $SU_c(2)$ [13]:

$$\hat{\theta}_\mu^\mu = -\frac{\beta(\hat{g})}{2\hat{g}}\hat{F}_a^{\mu\nu}\hat{F}_{\mu\nu;a} \equiv \frac{2b}{v}H\,, \tag{15}$$

with $a = 1,2,3$ and we have defined $\beta(\hat{g}) = -b\hat{g}^3/16\pi^2$. At one loop $b = \frac{11}{3}N_c$ with $N_c = 2$ the color number. H is the composite field describing, upon quantization, the scalar glueball [14] in medium and possesses mass-scale dimensions 4. The specific velocity dependence is introduced to properly account for the velocity factors.

The complete simplest light glueball action in the unrescaled coordinates for the, in medium, Yang-Mill theory is:

$$S_{G-ball} = \int d^4x \left\{\frac{c}{2}\sqrt{b}H^{-\frac{3}{2}}\left[\partial^0 H\partial^0 H - v^2\partial^i H\partial^i H\right] - \frac{b}{2}H\log\left[\frac{H}{\hat{\Lambda}^4}\right]\right\}\,. \tag{16}$$

The glueballs move with the same velocity v as the underlying gluons in the 2SC color superconductor. $\hat{\Lambda}$ is the intrinsic scale associated with the theory and can be less than or of the order of few MeVs [8, 7] while c is a constant of order unity.
The glueballs are light (with respect to the gap) and might barely interact with the ungapped fermions. They are stable with respect to the strong interactions unlike ordinary glueballs. We define the mass-dimension one glueball field h via

$$H = \langle H \rangle e^{\frac{h}{F_h}} . \tag{17}$$

By requiring a canonically normalized kinetic term for h one finds $F_h^2 = \frac{c}{\sqrt{2}}\sqrt{2b\langle H \rangle}$, while the glueball mass term is $M_h^2 = \frac{\sqrt{b}}{2c}\sqrt{\langle H \rangle} = \frac{\sqrt{b}}{2c\sqrt{e}}\hat{\Lambda}^2$, which is clearly of the order of $\hat{\Lambda}$ since c is a positive constant of order unity.

Once created, the light $SU_c(2)$ glueballs are stable against strong interactions but not with respect to electromagnetic processes. Indeed, the glueballs couple to two photons via virtual quark loops.

The relevant Lagrangian term, at non zero baryon density, obtained by saturating the electromagnetic trace anomaly is [7]:

$$L_{h\gamma\gamma} = \frac{\widetilde{e}^2}{48\pi^2} \frac{M_h}{\sqrt{2b\langle H \rangle}} \left[\sum_{quarks} \widetilde{Q}^2_{quarks} \right] h \widetilde{F}_{\mu\nu} \widetilde{F}^{\mu\nu} , \tag{18}$$

with $\widetilde{F}_{\mu\nu} = \partial_\mu \widetilde{A}_\nu - \partial_\nu \widetilde{A}_\mu$. Here $\widetilde{A}_\mu$ is the in medium photon field corresponding to the following massless linear combination of the old photon and the eighth gluon [15, 6]:

$$\widetilde{A}_\mu = \cos\theta_Q A_\mu - \sin\theta_Q G^8_\mu , \tag{19}$$

with $\tan\theta_Q = e/(\sqrt{3}g_s)$. The new electric constant is related to the in vacuum one via $\widetilde{e} = e\cos\theta_Q$. $\widetilde{Q}$ is the new electric charge operator associated with the field $\widetilde{A}_\mu$ with $\widetilde{Q} = \tau^3 \times \mathbf{1} + \frac{\widetilde{B}-L}{2} = Q \times \mathbf{1} - \frac{1}{\sqrt{3}}\mathbf{1} \times T^8$, where $L = 0$ is the lepton number, τ^3 the standard Pauli's matrix, Q the quark matrix, while the new baryon number is $\widetilde{B}$, and following the notation of Ref. [6] we have flavor$_{2\times 2}$ $\times$ color$_{3\times 3}$. This leads to the following decay width of the glueballs into two photons in medium:

$$\Gamma[h \to \gamma\gamma] \approx 1.2 \times 10^{-2} \cos\theta_Q^4 \left[\frac{M_h}{1\ \mathrm{MeV}} \right]^5 \mathrm{eV} , \tag{20}$$

where $\alpha = e^2/4\pi \simeq 1/137$. For illustration purposes we consider a glueball mass of the order of 1 MeV which leads to a decay time $\tau \sim 5.5 \times 10^{-14} s$. We used $\cos\theta_Q \sim 1$ since $\theta_Q \sim 2.5°$ [7]. While we are aware of the possible contribution from other hadrons to the saturation of the electromagnetic trace anomaly [16, 17], here we assume it to be dominated by the $SU_c(2)$ glueballs. In any case, it is hard to imagine the photon decay process to be completely switched-off. This shows that a consistent portion of the glue (3/8 or 37.5%) filling the 2SC medium is very rapidly and efficiently converted into electromagnetic radiation.

CONCLUSION

I reviewed the symmetries and the low energy effective Lagrangian for two flavor Color Superconductivity. The effective Lagrangian describes the, in medium, fermions and the broken $SU_c(3)$ gluon sector. The theory has then been extended to incorporate the relevant confining and light (with respect to the gap), $SU_c(2)$ degrees of freedom, i.e. glueballs. It is shown that the light glueballs are unstable to photon decay and estimated the, in medium, two photon decay rate. The present analysis is limited to the zero temperature and high matter density case. However it might be relevant to investigate the role played by a non zero temperature [18].

ACKNOWLEDGMENTS

I thank Roberto Casalbuoni, Zhiyong Duan, Stephen D. Hsu, Rachid Ouyed and Myck Schwetz for sharing part of the work on which this talk is based. For discussions and careful reading of the manuscript I thank Nils Marchal while I am indebted to Joseph Schechter for enlightening discussions and continuous encouragement.

REFERENCES

1. For reviews see K. Rajagopal, F. Wilczek hep-ph/0011333; M. Alford, hep-ph/0102047 ; S. D. Hsu, hep-ph/0003140 , and references therein.
2. D. K. Hong, S. D. Hsu and F. Sannino, hep-ph/0107017.
3. R. Ouyed and F. Sannino, astro-ph/0103022.
4. R. Casalbuoni and R. Gatto, Phys. Lett. B**464**, 11 (1999); Phys. Lett. B**469**, 213 (1999).
5. R. Casalbuoni, Z. Duan and F. Sannino, Phys. Rev. D**62**, 094004, (2000).
6. R. Casalbuoni, Z. Duan and F. Sannino, hep-ph/0011394. Phys. Rev. D**63**, 114026, (2001).
7. R. Ouyed and F. Sannino, Phys. Lett. B**511**, 66, (2001).
8. D.H. Rischke, D.T. Son, M.A. Stephanov, hep-ph/0011379.
9. R. Casalbuoni, R. Gatto, M. Mannarelli and G. Nardulli, hep-ph/0107024.
10. F. Sannino, Phys. Lett. B**480**, 280, (2000).
11. S. Hsu, F. Sannino and M. Schwetz, hep-ph/0006059.
12. G. 't Hooft, in: Recent Developments in Gauge Theories, eds., G. 't Hooft (Plenum Press, New York, 1980).
13. F. Sannino and J. Schechter, Phys. Rev. D**60**, 056004, (1999).
14. J. Schechter, Phys. Rev. **D21**, 3393 (1980).
15. M. Alford, J. Berges and K. Rajagopal, Nucl. Phys. B**571**, 269 (2000).
16. H. Gomm, P. Jain, R. Johnson and J. Schechter, Phys. Rev. **D33**, 801 (1986).
17. H. Gomm, P. Jain, R. Johnson and J. Schechter, Phys. Rev. **D33**, 3476 (1986).
18. N. Marchal and F. Sannino, work in progress.

Effective lagrangians for QCD at high density

Roberto Casalbuoni

Dipartimento di Fisica dell' Universita' di Firenze and Sezione INFN, L.go E. Fermi 2, 50125 Firenze, Italy. E-mail: casalbuoni@fi.infn.it

Abstract. We describe low energy physics in the CFL and LOFF phases by means of effective lagrangians. In the CFL case we present also how to derive expressions for the parameters appearing in the lagrangian via weak coupling calculations taking advantage of the dimensional reduction of fermion physics around the Fermi surface. The Goldstone boson of the LOFF phase turns out to be a phonon satisfying an anisotropic dispersion relation.

INTRODUCTION

Ideas about color superconductivity go back to almost 25 years ago [1], but only recently this phenomenon has received a lot of attention (for recent reviews see ref. [2]). The naive expectation is that at very high density, due to the asymptotic freedom, quarks would form a Fermi sphere of almost free fermions. However, Bardeen, Cooper and Schrieffer proved that the Fermi surface of free fermions is unstable in presence of an attractive, arbitrary small, interaction. Since in QCD the gluon exchange in the $\bar{3}$ channel is attractive one expects the formation of a coherent state of particle/hole pairs (Cooper pairs). For a careful description of the formation of the condensates and of the approximations involved in going from asymptotic densities to finite ones it is useful to see the contribution of K. Rajagopal at this meeting [3]. The phase structure of QCD at high density depends on the number of flavors and there are two very interesting cases, corresponding to two massless flavors (2SC) [1, 4] and to three massless flavors (CFL) [5, 6] respectively. In this talk we will be mainly concerned with the latter case. The two cases correspond to very different patterns of symmetry breaking. If we denote left- and right-handed quark fields by $\psi^{\alpha}_{iL(R)}$ with $\alpha = 1,2,3$, the $SU(3)_c$ color index, and $i = 1,\cdots,N_f$ the flavor index (N_f is the number of massless flavors), in the 2SC phase we have the following structure for the condensate ($C = i\gamma^2\gamma^0$ is the charge-conjugation matrix)

$$\langle q^{\alpha}_{iL(R)} C q^{\beta}_{jL(R)} \rangle \propto \epsilon_{ij}\epsilon^{\alpha\beta 3}. \tag{1}$$

The condensate breaks the color group $SU(3)_c$ down to the subgroup $SU(2)_c$ but it does not break any flavor symmetry. Although the baryon number, B, is broken, there is a combination of B and of the broken color generator, T_8, which is unbroken in the 2SC phase. Therefore no massless Goldstone bosons are present in this phase. On the other hand, five gluon fields acquire mass whereas three are left massless. It is worth to notice that for the electric charge the situation is very similar to the one for the baryon number. Again a linear combination of the broken electric charge and of the broken generator

CP602, *QCD@Work: International Workshop on Quantum Chromodynamics*
edited by P. Colangelo and G. Nardulli

T_8 is unbroken in the 2SC phase. The condensate (1) gives rise to a gap, Δ, for quarks of color 1 and 2, whereas the two quarks of color 3 remain un-gapped (massless). The resulting effective low-energy theory has been described in [7]. In this contribution we will be mainly interested in the formulation of the effective theory for the three massless quarks case. At high density it has been shown that the following condensate is formed [5, 6]

$$\langle q^{\alpha}_{iL(R)} C q^{\beta}_{jL(R)} \rangle \propto \varepsilon^{ijX} \varepsilon_{\alpha\beta X} + \kappa(\delta^i_\alpha \delta^j_\beta + \delta^i_\beta \delta^j_\alpha). \tag{2}$$

Due to the Fermi statistics, the condensate must be symmetric in color and flavor. As a consequence the two terms appearing in eq. (2) correspond to the $(\bar{\mathbf{3}}, \bar{\mathbf{3}})$ and $(\mathbf{6}, \mathbf{6})$ channels of $SU(3)_c \otimes SU(3)_{L(R)}$. It turns out that κ is small [5, 8, 9] and therefore the condensation occurs mainly in the $(\bar{\mathbf{3}}, \bar{\mathbf{3}})$ channel. The expression (2) shows that the ground state is left invariant by a simultaneous transformation of $SU(3)_c$ and $SU(3)_{L(R)}$. This is called Color Flavor Locking (CFL). The symmetry breaking pattern is

$$\begin{array}{c} SU(3)_c \otimes SU(3)_L \otimes SU(3)_R \otimes U(1)_B \otimes U(1)_A \\ \downarrow \\ SU(3)_{c+L+R} \otimes Z_2 \otimes Z_2 \end{array} \tag{3}$$

The $U(1)_A$ symmetry is broken at the quantum level by the anomaly, but it gets restored at very high density since the instanton contribution is suppressed [10, 8, 11]. The Z_2 symmetries arise since the condensate is left invariant by a change of sign of the left- and/or right-handed fields. As for the 2SC case the electric charge is broken but a linear combination with the broken color generator T_8 annihilates the ground state. On the contrary the baryon number is broken. Therefore there are $8+2$ broken global symmetries giving rise to 10 Goldstone bosons. The one associated to $U(1)_A$ gets massless only at very high density. The color group is completely broken and all the gauge particles acquire mass. Also all the fermions are gapped. We will show in the following how to construct an effective lagrangian describing the Goldstone bosons, and how to compute their couplings in the high density limit where the QCD coupling gets weaker. A final problem we will discuss has to do with the fact that when quarks (in particular the strange quark) are massive, their chemical potentials cannot be all equal. This situation has been modeled out in [12]. If the Fermi surfaces of different flavors are too far apart, BCS pairing does not occur. However it might be favorable for different quarks to pair each of one lying at its own Fermi surface and originating a pair of non-zero total momentum. This is the LOFF state first studied by the authors of ref. [13] in the context of electron superconductivity in the presence of magnetic impurities. Since the Cooper pair has non-zero momentum the condensate breaks space symmetries and we will show that in the low-energy spectrum a massless particle, a phonon, the Goldstone boson of the broken translational symmetry, is present. We will construct the effective lagrangian also for this case.

EFFECTIVE THEORY FOR THE CFL PHASE

We start introducing the Goldstone fields as the phases of the condensates in the $(\bar{\mathbf{3}},\bar{\mathbf{3}})$ channel [14, 15]

$$X^i_\alpha \approx \varepsilon^{ijk}\varepsilon_{\alpha\beta\gamma}\langle q^j_{\beta L} q^k_{\gamma L}\rangle^*, \quad Y^i_\alpha \approx \varepsilon^{ijk}\varepsilon_{\alpha\beta\gamma}\langle q^j_{\beta R} q^k_{\gamma R}\rangle^*. \tag{4}$$

Since quarks belong to the representation $(\mathbf{3},\mathbf{3})$ of $SU(3)_c \otimes SU(3)_{L(R)}$ and transform under $U(1)_B \otimes U(1)_A$ according to

$$q_L \to e^{i(\alpha+\beta)} q_L, \quad q_R \to e^{i(\alpha-\beta)} q_R, \quad e^{i\alpha} \in U(1)_B, \quad e^{i\beta} \in U(1)_A, \tag{5}$$

the transformation properties of the fields X and Y under the total symmetry group $G = SU(3)_c \otimes SU(3)_L \otimes SU(3)_R \otimes U(1)_B \otimes U(1)_A$ are ($g_c \in SU(3)_c$, $g_{L(R)} \in SU(3)_{L(R)}$)

$$X \to g_c X g_L^T e^{-2i(\alpha+\beta)}, \quad Y \to g_c Y g_R^T e^{-2i(\alpha-\beta)}. \tag{6}$$

The fields X and Y are $U(3)$ matrices and as such they describe $9+9=18$ fields. Eight of these fields are eaten up by the gauge bosons, producing eight massive gauge particles. Therefore we get the right number of Goldstone bosons, $10 = 18-10$. These fields correspond to the breaking of the global symmetries in G (18 generators) to the symmetry group of the ground state $H = SU(3)_{c+L+R} \otimes Z_2 \otimes Z_2$ (8 generators). For the following it is convenient to separate the $U(1)$ factors in X and Y defining fields belonging to $SU(3)$

$$X = \hat{X} e^{2i(\phi+\theta)}, \quad Y = \hat{Y} e^{2i(\phi-\theta)}, \quad \hat{X},\hat{Y} \in SU(3). \tag{7}$$

The fields ϕ and θ can also be described through the determinants of X and Y

$$d_X = \det(X) = e^{6i(\phi+\theta)}, \quad d_Y = \det(Y) = e^{6i(\phi-\theta)}, \tag{8}$$

The transformation properties under G are

$$\hat{X} \to g_c \hat{X} g_L^T, \quad \hat{Y} \to g_c \hat{Y} g_R^T, \quad \phi \to \phi - \alpha, \quad \theta \to \theta - \beta. \tag{9}$$

The breaking of the global symmetry can be discussed in terms of gauge invariant fields given by d_X, d_Y and

$$\Sigma^i_j = \sum_\alpha (\hat{Y}^j_\alpha)^* \hat{X}^i_\alpha \to \Sigma = \hat{Y}^\dagger \hat{X}. \tag{10}$$

The Σ field describes the 8 Goldstone bosons corresponding to the breaking of the chiral symmetry $SU(3)_L \otimes SU(3)_R$, as it is made clear by the transformation properties of Σ^T, $\Sigma^T \to g_L \Sigma^T g_R^\dagger$. That is Σ^T transforms exactly as the usual chiral field. The other two fields d_X and d_Y provide the remaining two Goldstone bosons related to the breaking of the $U(1)$ factors.

In order to build up an invariant lagrangian, it is convenient to define the following currents

$$J^\mu_X = \hat{X} D^\mu \hat{X}^\dagger = \hat{X}(\partial^\mu \hat{X}^\dagger + \hat{X}^\dagger g^\mu), \quad J^\mu_Y = \hat{Y} D^\mu \hat{Y}^\dagger = \hat{Y}(\partial^\mu \hat{Y}^\dagger + \hat{Y}^\dagger g^\mu), \tag{11}$$

with $g_\mu = ig_s g_\mu^a T^a/2$ the gluon field and $T^a = \lambda_a/2$ the $SU(3)_c$ generators. These currents have simple transformation properties under the full symmetry group G, $J^\mu_{X,Y} \to g_c J^\mu_{X,Y} g_c^\dagger$. The most general lagrangian, up to two derivative terms, invariant under G, the rotation group $O(3)$ (Lorentz invariance is broken by the chemical potential term) and the parity transformation defined as $\hat{X} \leftrightarrow \hat{Y}$, $\phi \to \phi$, $\theta \to -\theta$, is [14]

$$\begin{aligned} L &= -\frac{F_T^2}{4}\mathrm{Tr}\left[(J_X^0 - J_Y^0)^2)\right] - \alpha_T \frac{F_T^2}{4}\mathrm{Tr}\left[(J_X^0 + J_Y^0)^2)\right] + \frac{1}{2}(\partial_0\phi)^2 + \frac{1}{2}(\partial_0\theta)^2 \\ &\quad + \frac{F_S^2}{4}\mathrm{Tr}\left[\left(\vec{J}_X - \vec{J}_Y)^2\right)\right] + \alpha_S \frac{F_S^2}{4}\mathrm{Tr}\left[\left(\vec{J}_X + \vec{J}_Y)^2\right)\right] - \frac{v_\phi^2}{2}|\vec{\nabla}\phi|^2 - \frac{v_\theta^2}{2}|\vec{\nabla}\theta|^2. \end{aligned} \tag{12}$$

Using $SU(3)_c$ color gauge invariance we can choose $\hat{X} = \hat{Y}^\dagger$, making 8 of the Goldstone bosons disappear and giving mass to the gluons. The properly normalized Goldstone bosons, Π^a, are given in this gauge by

$$\hat{X} = \hat{Y}^\dagger = e^{i\Pi^a T^a/F_T}, \tag{13}$$

and expanding eq. (12) at the lowest order in the fields we get

$$L \approx \frac{1}{2}(\partial_0 \Pi^a)^2 + \frac{1}{2}(\partial_0\phi)^2 + \frac{1}{2}(\partial_0\theta)^2 - \frac{v^2}{2}|\vec{\nabla}\Pi^a|^2 - \frac{v_\phi^2}{2}|\vec{\nabla}\phi|^2 - \frac{v_\theta^2}{2}|\vec{\nabla}\theta|^2, \tag{14}$$

with $v = F_s/F_T$. The gluons g_0^a and g_i^a acquire Debye and Meissner masses given by

$$m_D^2 = \alpha_T g_s^2 F_T^2, \quad m_M^2 = \alpha_S v^2 g_s^2 F_T^2. \tag{15}$$

It should be stressed that these are not the true rest masses of the gluons, since there is a large wave function renormalization effect making the gluon masses of order of the gap Δ, rather than μ (see later) [16]. Since this description is supposed to be valid at low energies (we expect much below the gap Δ), we could also decouple the gluons solving their classical equations of motion neglecting the kinetic term. The result from eq. (12) is

$$g_\mu = -\frac{1}{2}\left(\hat{X}\partial_\mu \hat{X}^\dagger + \hat{Y}\partial_\mu \hat{Y}^\dagger\right). \tag{16}$$

It is easy to show that substituting this expression in eq. (12) one gets [16]

$$L = \frac{F_T^2}{4}\left(\mathrm{Tr}[\dot{\Sigma}\dot{\Sigma}^\dagger] - v^2\mathrm{Tr}[\vec{\nabla}\Sigma \cdot \vec{\nabla}\Sigma^\dagger]\right) + \frac{1}{2}\left(\dot{\phi}^2 - v_\phi^2|\vec{\nabla}\phi|^2\right) + \frac{1}{2}\left(\dot{\theta}^2 - v_\phi^2|\vec{\nabla}\theta|^2\right). \tag{17}$$

Notice that the first term is nothing but the chiral lagrangian except for the breaking of the Lorentz invariance. This is a way of seeing the quark-hadron continuity, that is the continuity between the CFL and the nuclear matter phases in three flavor QCD. The identification is perfect if one realizes that in nuclear matter the pairing may occur in such a way to give rise to a superfluid due to the breaking of the baryon number as it happens in the CFL phase [17].

FERMIONS NEAR THE FERMI SURFACE

We will introduce now the formalism described in ref. [18] in order to evaluate the parameters appearing in the effective lagrangian. This formulation is based on the observation that, at very high-density, the energy spectrum of a massless fermion is described by states $|\pm\rangle$ with energies $E_{\pm} = -\mu \pm |\vec{p}$ where μ is the quark number chemical potential. For energies much lower than the Fermi energy μ, only the states $|+\rangle$ close to the Fermi surface. i.e. with $|\vec{p}| \approx \mu$, can be excited. On the contrary, the states $|-\rangle$ have $E_- \approx -2\mu$ and therefore decouple. This can be seen more formally by writing the four-momentum of the fermion as

$$p^\mu = \mu v^\mu + \ell^\mu, \tag{18}$$

where $v^\mu = (0, \vec{v}_F)$, and $\vec{v}_F$ is the Fermi velocity defined as $\vec{v}_F = \partial E/\partial \vec{p}|_{\vec{p}=\vec{p}_F}$. For massless fermions $|\vec{v}_F| = 1$. Since the hamiltonian for a massless Dirac fermion in a chemical potential μ is

$$H = -\mu + \vec{\alpha} \cdot \vec{p}, \quad \vec{\alpha} = \gamma_0 \vec{\gamma}, \tag{19}$$

one has

$$H = -\mu(1 - \vec{\alpha} \cdot \vec{v}_F) + \vec{\alpha} \cdot \vec{\ell}. \tag{20}$$

Then, it is convenient to introduce the projection operators

$$P_{\pm} = \frac{1 \pm \vec{\alpha} \cdot \vec{v}_F}{2}, \tag{21}$$

such that

$$H|+\rangle = \vec{\alpha} \cdot \vec{\ell}|+\rangle, \quad H|-\rangle = (-2\mu + \vec{\alpha} \cdot \vec{\ell})|-\rangle. \tag{22}$$

We can define fields corresponding to the states $|\pm\rangle$ through the decomposition

$$\psi(x) = \sum_{\vec{v}_F} e^{-i\mu v \cdot x} \left[\psi_+(x) + \psi_-(x)\right], \tag{23}$$

where an average over the Fermi velocity $\vec{v}_F$ is performed. The velocity-dependent fields $\psi_{\pm}(x)$ are given by ($v^\mu = (0, \vec{v}_F)$)

$$\psi_{\pm}(x) = e^{i\mu v \cdot x} \left(\frac{1 \pm \vec{\alpha} \cdot \vec{v}_F}{2} \right) \psi(x) = \int_{|\ell| < \delta} \frac{d^4 \ell}{(2\pi)^4} e^{-i\ell \cdot x} \psi_{\pm}(\ell). \tag{24}$$

Since we are interested at physics near the Fermi surface we integrate out all the modes with $|\ell| > \delta$, with δ a cut-off such that $\Delta < \delta \ll \mu$. Substituting inside the Dirac part of the QCD lagrangian density one obtains ($V^\mu = (1, \vec{v}_f)$, $\tilde{V}^\mu = (1, -\vec{v}_F)$)

$$L = \sum_{\vec{v}_F} \left[\psi_+^\dagger iV \cdot D \psi_+ + \psi_-^\dagger (2\mu + i\tilde{V} \cdot D) \psi_- + (\bar{\psi}_+ i \not{D}_\perp \psi_- + \text{h.c.}) \right], \tag{25}$$

where $\not{D}_\perp = D_\mu \gamma_\perp^\mu$ and

$$\gamma_\perp^\mu = \frac{1}{2} \gamma_\nu \left(2g^{\mu\nu} - V^\mu \tilde{V}^\nu - \tilde{V}^\mu V^\nu \right). \tag{26}$$

We notice that the fields appearing in this expression are evaluated at the same Fermi velocity because off-diagonal terms are canceled by the rapid oscillations of the exponential factor in the $\mu \to \infty$ limit. This behavior can be referred to as the Fermi velocity super-selection rule.

At the leading order in $1/\mu$ one has

$$iV \cdot D\psi_+ = 0, \quad \psi_- = -\frac{i}{2\mu}\gamma_0 \not{D}_\perp \psi_+, \tag{27}$$

showing the decoupling of ψ_- for $\mu \to \infty$. The equation for ψ_+ shows also that only the energy and the momentum parallel to the Fermi velocity are relevant variables in the problem. We have an effective two-dimensional theory.

At the next to leading order the effective action for the field ψ_+ is

$$L = \sum_{\vec{v}_F} \left[\psi_+^\dagger iV \cdot D\psi_+ - \frac{1}{2\mu}\psi_+^\dagger (\not{D}_\perp)^2 \psi_+ \right]. \tag{28}$$

The previous remarks apply to any theory describing massless fermions at high density. The next step will be to couple this theory in a $SU(3)_L \otimes SU(3)_R \otimes SU(3)_c$ invariant way to Nambu-Goldstone bosons (NGB) describing the appropriate breaking for the CFL phase (we will not discuss here the determination of the parameters relevant for the $U(1)_{A,B}$ fields, see [11]). Using a gradient expansion we get an explicit expression for the decay coupling constant of the Nambu-Goldstone boson as well for their velocity.

The invariant coupling between fermions and Goldstone fields reproducing the symmetry breaking pattern of eq. (2) is proportional to

$$\gamma_1 Tr[\psi_L^T \hat{X}^\dagger] C Tr[\psi_L \hat{X}^\dagger] + \gamma_2 Tr[\psi_L^T C \hat{X}^\dagger \psi_L \hat{X}^\dagger] + \text{h.c.}, \tag{29}$$

with an analogous expression for the right-handed fields. Here the spinors are meant to be Dirac spinors. The trace is operating over the group indices of the spinors and of the Goldstone fields. Since the vacuum expectation value of the Goldstone fields is $\langle \hat{X} \rangle = \langle \hat{Y} \rangle = 1$, we see that this coupling induces the right breaking of the symmetry. In the following we will consider only the case $\gamma_2 = -\gamma_1 \propto \Delta/2$, where Δ is the gap parameter.

Since the transformation properties under the symmetry group of the fields at fixed Fermi velocity do not differ from those of the quark fields, for both left-handed and right-handed fields we get the effective lagrangian density

$$\begin{aligned} L &= \sum_{\vec{v}_F} \frac{1}{2} \Big[\sum_{A=1}^{9} \left(\psi_+^{A\dagger} iV \cdot D\psi_+^A + \psi_-^{A\dagger} i\tilde{V} \cdot D\psi_-^A - \Delta_A \left(\psi_-^{A\,T} C \psi_+^A + \text{h.c.} \right) \right) \\ &- \Delta \sum_{I=1,3} \left(Tr[(\psi_- X_1^\dagger)^T C \epsilon_I (\psi_+ X_1^\dagger) \epsilon_I] + \text{h.c.} \right) \Big], \end{aligned} \tag{30}$$

where we have introduced the fields $\psi_\pm^A$:

$$\psi_\pm = \frac{1}{\sqrt{2}} \sum_{A=1}^{9} \lambda_A \psi_\pm^A. \tag{31}$$

Here λ_a $(a = 1,...,8)$ are the Gell-Mann matrices normalized as follows: $Tr(\lambda_a\lambda_b) = 2\delta_{ab}$ and $\lambda_9 = \sqrt{2/3}\,\mathbf{1}$. Furthermore $\Delta_1 = \cdots = \Delta_8 = \Delta$, $\Delta_9 = -2\Delta$, and $X_1 = \hat{X} - 1$. Notice that the NGB fields couple to fermionic fields with opposite Fermi velocities. In this expression, as in the following ones, the field ψ_- is defined as ψ_+ with $\vec{v}_F \to -\vec{v}_F$, and therefore it is not the same as the one defined in (24).

The formalism becomes more compact by introducing the Nambu-Gorkov fields

$$\chi = \begin{pmatrix} \psi_+ \\ C\psi_-^* \end{pmatrix}. \tag{32}$$

It is important to realize that the fields χ and $\chi^\dagger$ are not independent variables. In fact, since we integrate over all the Fermi surface, the fields ψ_-^* and ψ_+, appearing in χ, appear also in $\chi^\dagger$ with $\vec{v}_F \to -\vec{v}_F$. In order to avoid this problem we can integrate over half of the Fermi surface, or, taking into account the invariance under $\vec{v}_F \to -\vec{v}_F$, we can simply integrate over all the sphere with a weight $1/8\pi$ instead of $1/4\pi$. Then the first three terms in the lagrangian density (30) become

$$L_0 = \int \frac{d\vec{v}_F}{8\pi}\,\frac{1}{2}\sum_{A=1}^{9} \chi^{A\dagger} \begin{bmatrix} iV\cdot D & \Delta^A \\ \Delta^A & i\tilde{V}\cdot D^* \end{bmatrix} \chi^A, \tag{33}$$

so that, in momentum space the free fermion propagator is

$$S_{AB}(p) = \frac{2\delta_{AB}}{V\cdot p\tilde{V}\cdot p - \Delta_A^2} \begin{bmatrix} \tilde{V}\cdot p & -\Delta_A \\ -\Delta_A & V\cdot p \end{bmatrix}. \tag{34}$$

We are now in position to evaluate the self-energy of the Goldstone bosons through their couplings to the fermions at the Fermi surface. There are two one-loop contributions [16], one from the coupling $\Pi\chi\chi$ and a tadpole from the coupling $\Pi\Pi\chi\chi$ (see eq. (30)). The tadpole diagram contributes only to the mass term and it is essential to cancel the external momentum independent term arising from the other diagram. Therefore, as expected, the mass of the NGB's is zero. The contribution at the second order in the momentum expansion is given by

$$i\frac{21 - 8\ln 2}{72\pi^2 F_T^2} \int \frac{d\vec{v}_F}{4\pi} \sum_{a=1}^{8} \Pi^a V\cdot p\tilde{V}\cdot p\Pi^a. \tag{35}$$

Integrating over the velocities and going back to the coordinate space we get

$$L_{\rm eff}^{\rm kin} = \frac{21 - 8\ln 2}{72\pi^2 F_T^2} \sum_{a=1}^{8} \left(\dot{\Pi}^a\dot{\Pi}^a - \frac{1}{3}|\vec{\nabla}\Pi_a|^2 \right). \tag{36}$$

We can now determine the decay coupling constant F_T through the requirement of getting the canonical normalization for the kinetic term; this implies

$$F_T^2 = \frac{\mu^2(21 - 8\ln 2)}{36\pi^2}, \tag{37}$$

a result obtained by many authors using different methods (for a complete list of the relevant papers see the first reference of [2]). We see also that $v^2 = 1/3$. The other constants appearing in our effective lagrangian can be obtained via a direct calculation of m_D and m_M [16]. This is done evaluating the one-loop contribution to the gluon self-energy. Also in this case there are two contributions, one coming from the gauge coupling to the fermions, whereas the other arises from the next-to leading (in μ) sea-gull contribution to the fermion effective lagrangian in eq. (28). The results we find are [16, 11]

$$m_D^2 = g_s^2 F_T^2, \quad m_M^2 = \frac{1}{3} m_D^2. \tag{38}$$

Comparison with equation (15) shows that

$$\alpha_S = \alpha_T = 1. \tag{39}$$

Performing a gradient expansion of the gluon self-energy one finds that there is a wave function renormalization of order $g_s\mu/\Delta \gg 1$. Extrapolating this result up to momenta of order Δ one gets the result that the physical masses of the gluons are of the order of the gap energy ($\approx 1.70\Delta$) [16]. The origin of the pion velocity $1/\sqrt{3}$ is a direct consequence of the integration over the Fermi velocity. Therefore it is completely general and applies to all the NGB's in the theory, including the ones associated to the breaking of $U(1)_V$ and $U(1)_A$ ($v_\phi^2 = v_\theta^2 = 1/3$); needless to say that higher order terms in the expansion $1/\mu$ could change this result.

The breaking of the Lorentz invariance exhibited by the pion velocity different from one, can be seen also in the matrix element $\langle 0|J_\mu^a|\Pi^b\rangle$. Its evaluation gives [16]

$$\langle 0|J_\mu^a|\Pi^b\rangle = iF_T\delta_{ab}\tilde{p}_\mu, \quad \tilde{p}^\mu = (p^0, \vec{p}/3). \tag{40}$$

The current is conserved, as a consequence of the dispersion relation satisfied by the NGB's.

THE LOFF PHASE

We shall now consider massless quarks of three colors and two different flavors. At finite densities we introduce two chemical potentials, μ_1 and μ_2, for the two species in order to mimic the different mass case. We write

$$\mu_1 - \mu_2 = \delta\mu \ll \mu = \frac{\mu_1 + \mu_2}{2}. \tag{41}$$

The BCS condensation takes place also for $\delta\mu \neq 0$ provided $\delta\mu \ll \Delta$. On the other hand, for $\delta\mu \approx \Delta$, the picture changes significantly. The analysis in [12] shows that there exist two values of $\delta\mu$, $\delta\mu_1$ and $\delta\mu_2$, such that, for $\delta\mu \in (\delta\mu_1, \delta\mu_2)$ the high density quark-gluon matter is in a phase characterized by the breaking of translational and rotational invariance, due to the presence of a scalar and a vector condensate. This phenomenon is called crystalline color superconductivity of QCD and the relative phase is named LOFF phase. The authors of ref. [12] find $\delta\mu_1 = 0.71\Delta$ and $\delta\mu_2 = 0.744\Delta$ for $\mu = 0.4$ GeV and

$\Delta = 40\ MeV$ for a point-like four-fermi coupling. More recently it has been found that in the one-gluon exchange approximation the window opens up considerably [19].

The condensation in the LOFF phase gives rise to two breaking terms in the fermion lagrangian characterized by two gap parameters $\Delta^{(s)}$ and $\Delta^{(v)}$

$$-\frac{1}{2}e^{2i\vec{q}\cdot\vec{x}}\sum_{\vec{v}_F} e^{i\delta\mu\vec{v}_F\cdot\vec{x}}\left[\Delta^{(s)}\varepsilon_{ij}+\vec{v}\cdot\vec{n}\Delta^{(v)}\sigma^1_{ij}\right]\varepsilon^{\alpha\beta 3}\psi_{+\vec{v};i\alpha}C\psi_{-\vec{v};j\beta}-(L\to R), \tag{42}$$

with $\vec{n}=\vec{q}/|\vec{q}|$. The condensates break the space symmetry group. However the discussion of the number of NGB's in the case of space symmetries is a subtle one due to the particular group structure. In fact rotations and translations cannot be considered transformations breaking the symmetries of the theory in an independent way. This is because a translation plus a rotation is physically equivalent to a translation. Let us discuss the consequences of this situation more closely.

We first consider spatial rotations. We define a vector field $\vec{R}(x)$ such that $|\vec{R}|^2=1$ and $\langle\vec{R}\rangle_0=\vec{n}$ [20]. The rotational symmetry is restored by substituting $\vec{v}\cdot\vec{n}\to\vec{v}\cdot\vec{R}$ in the term proportional to $\Delta^{(v)}$. Let us now consider the exponential factor $\exp(2i\vec{q}\cdot\vec{x})$ in (42), which breaks both rotational and translational invariance. By introducing a field $\Phi(x)$ behaving as a scalar under the space group [20], we restore translational and rotational invariance via the substitution $2\vec{q}\cdot\vec{x}\to\Phi(x)$. We assume $\langle\Phi(x)\rangle_0=2\vec{q}\cdot\vec{x}$. We then introduce a field $\phi(x)$ through

$$\Phi(x)=2\vec{q}\cdot\vec{x}+\phi(x),\quad \langle\phi(x)\rangle_0=0, \tag{43}$$

and convenient transformation properties such to compensate the variation of the term $\vec{q}\cdot\vec{x}$ under the space group [20]. The field ϕ acts as the phonon (Nambu-Goldstone boson) field associated to the breaking of the space symmetry. We can now construct the field $\vec{R}$ in terms of $\Phi(x)$ as $\vec{R}=\vec{\nabla}\Phi/|\vec{\nabla}\Phi|$. This expression satisfies the required properties for $\vec{R}(x)$.

Through a bosonization procedure similar to the one employed in the previous Section, one can derive an effective lagrangian for the NGB field. The effective lagrangian must contain only derivative terms. Polynomial terms are indeed forbidden by translation invariance, since ϕ is not a scalar field under space transformations. In order to write the kinetic terms is better to use the field Φ which behaves as a scalar under both rotations and translation. However since the expectation value of the gradient of Φ is given by $\langle\vec{\nabla}\Phi\rangle_0=2\vec{q}\approx\Delta$, we cannot limit the expansion in the spatial derivatives of Φ to any finite order. A real spatial derivative expansion can be made only for the phonon field ϕ. With this in mind the most general invariant lagrangian will contain a tower of space-derivative terms [20]:

$$L(\phi,\partial_\mu\phi)=\frac{f^2}{2}\left[\dot{\Phi}^2-\sum_{n=1}^{\infty}c_n(|\vec{\nabla}\Phi|^2)^n\right]. \tag{44}$$

Here Φ must be thought as a function of the phonon field ϕ. Using

$$|\vec{\nabla}\Phi(x)|^2=4q^2+\frac{4q}{f}\vec{n}\cdot\vec{\nabla}\phi(x)+\frac{1}{f^2}|\vec{\nabla}\phi(x)|^2, \tag{45}$$

at the lowest order in the derivatives of the phonon field ϕ we get (neglecting a constant term):

$$L(\phi,\partial_\mu\phi)=\frac{1}{2}\left[\dot{\phi}^2-v_{\parallel}^2|\vec{\nabla}_{\parallel}\phi|^2-v^2\left(4qf\vec{\nabla}_{\parallel}\phi+|\vec{\nabla}\phi|^2\right)\right], \tag{46}$$

where $\vec{\nabla}_{\parallel}\phi=\vec{n}\cdot\vec{\nabla}\phi$ and $v_{\parallel}^2$, v^2 are constants. Notice that the linear term gives rise to a surface contribution. The lack of rotational invariance in (46) follows from the gradient expansion due to the non-linear transformations undergone by the field $\phi(x)$. This happens also in the analogous expansion for the chiral field. Therefore the physical consequence of the extraction of the expectation value of Φ is an anisotropy in the dispersion relation for the phonon field $\phi(x)$.

ACKNOWLEDGMENTS

I would like to thank R. Gatto, M. Mannarelli and G. Nardulli for their precious collaboration to the papers originating this contribution.

REFERENCES

1. B. Barrois, *Nuclear Physics* **B129**, 390 (1977); S. Frautschi, *Proceedings of workshop on hadronic matter at extreme density*, Erice 1978; D. Bailin and A. Love, *Physics Report* **107**, 325 (1984).
2. K. Rajagopal and F. Wilczek, hep-ph/0011333; S.D.H. Hsu, hep-ph/0003140; D.K. Hong, hep-ph/0101025; M. Alford, hep-ph/0102047.
3. K. Rajagopal, contributed paper to this meeting.
4. M. Alford, K. Rajagopal and F. Wilczek, *Physics Letters* **B422**, 247 (1998), hep-ph/9711395; R. Rapp, T. Schäfer, E.V. Shuryak and M. Velkovsky, *Physical Review Letters* **81**, 53 (1998), hep-ph/9711396.
5. M. Alford, K. Rajagopal and F. Wilczek, *Nuclear Physics* **B537**, 443 (1999), hep-ph/9804403.
6. T. Schäfer and F. Wilczek, *Physical Review Letters* **82**, 3956 (1999), hep-ph/9811473.
7. R. Casalbuoni, Z. Duan and F. Sannino, *Physical Review* **D62**, 094004 (2000), hep-ph/0004207; *ibidem* **D63**, 114026 (2001), hep-ph/0011394.
8. T. Schäfer, *Nuclear Physics* **B575**, 269 (2000), hep-ph/9909574.
9. I.A. Shovkovy and L.C. Wijewardhana, *Physics Letters* **B470**, 189 (1999), hep-ph/9910225.
10. R. Rapp, T. Schäfer, E.V. Shuryak and M. Velkovsky, *Annals of Physics* **280**, 35 (2000), hep-ph/9904353.
11. D.T. Son and M.A. Stephanov, *Physical Review* **D61**, 074012 (2000), hep-ph/9910491; *ibidem* Erratum **D62**, 059902 (2000), hep-ph/0004095.
12. M. Alford, J.A. Bowers and K. Rajagopal, *Physical Review* **D63**, 074016 (2001), hep-ph/0008208.
13. A. Larkin and Y.N. Ovchinnikov, *Soviet Physics JETP* **20**, 762 (1965); P. Fulde and R.A. Ferrel, *Physical Review* **135A**, 550 (1964).
14. R. Casalbuoni and R. Gatto, *Physics Letters* **B464**, 111 (1999), hep-ph/9908227.
15. D.K. Hong, M. Rho and I. Zahed, *Physics Letters* **B468**, 261 (1999), hep-ph/9906551.
16. R. Casalbuoni, R. Gatto and G. Nardulli, *Physics Letters* **B498**, 179 (2001), hep-ph/0010321.
17. T.Schäfer and F. Wilczek, *Physical Review Letters* **82**, 3956 (1999), hep-ph/9811473.
18. D.K. Hong, *Physics Letters* **B473**, 118 (2000), hep-ph/9812510; D.K. Hong *Nuclear Physics* **B582**, 451 (2000), hep-ph/9905523; S.R. Beane, P.F. Bedaque and M.J. Savage, *Physics Letters* **B483**, 131 (2000), hep-ph/0002209.
19. A.K. Leibovich, K. Rajagopal and E. Shuster, hep-ph/0104073.
20. R. Casalbuoni, R. Gatto, M. Mannarelli and G. Nardulli, *Physics Letters* **B511**, 218 (2001), hep-ph/0101326.

Participants

Arunagiri Somasundaram	University of Bari
Aschenauer Elke-Caroline	DESY
Becattini Francesco	INFN and University of Florence
Böttcher Helmut	DESY Zeuthen
Bombaci Ignazio	University of Pisa
Buccella Franco	University of Naples
Caliandro Rocco	INFN and University of Bari
Casalbuoni Roberto	University of Florence
Colangelo Pietro	INFN, Bari
Corianò Claudio	INFN and University of Lecce
Cozza Daniela	INFN and University of Bari
De Carlo Francesco	INFN and University of Bari
Deandrea Aldo	IPN - LYON
De Fazio Fulvia	INFN, Bari
de Rafael Eduardo	CPT - CNRS Luminy - Marseille
De Roeck Albert	CERN
Di Bari Domenico	INFN and University of Bari
Durr Stephan	Paul Scherrer Institut-Villigen
Eeg Jan O.	University of Oslo
Elia Domenico	INFN, Bari
Fajfer Svjetlana	Institute J. Stefan and University of Ljubljana
Fazio Angelo Raffaele	University of Milano
Ferroni Fernando	University La Sapienza, Rome
Fini Rosa Anna	INFN, Bari
Forte Stefano	INFN, Roma 3
Fuster Juan	University of Valencia
Ghidini Bruno	INFN and University of Bari
Grozin Andrey G.	Budker INP, Novosibirsk
Heinz Ulrich	Ohio State University
Heppelmann Steven	Penn State University
Hiorth Aksel	University of Oslo
Hurth Tobias	CERN
Isola Claudia	Ecole Polytechnique - Palaiseau
Jeremie Hannes	University of Montreal
Jones Roger	CERN
Karsch Frithjof	University of Bielefeld
Khodjamirian Alexander	University of Yerevan and Munchen
Kurepin Alexei	Institute for Nuclear Research, Moscow
Lanceri Livio	University of Trieste
Leutwyler Heinrich	University of Bern
Madigojine Dmitri	JINR, Dubna
Mannarelli Massimo	INFN and University of Bari

Manzari Vito	INFN, Bari
Marchesini Giuseppe	University of Milano Bicocca
Martinelli Guido	University La Sapienza, Rome
Nappi Eugenio	INFN, Bari
Nason Paolo	INFN, Milano
Nardulli Giuseppe	INFN and University of Bari
Neubert Matthias	Cornell University
Palano Antimo	INFN and University of Bari
Pancheri Giulia	INFN, Frascati
Paver Nello	INFN and University of Trieste
Pelaez Jose	Complutense University of Madrid
Pettini Giulio	INFN and University of Florence
Pham Tri-nang	CNRS - Ecole Polytechnique , Palaiseau
Pineda Antonio	University of Karlsruhe
Polosa Antonio D.	University of Helsinki
Rajagopal Krishna	MIT, Boston
Ratti Sergio P.	University of Pavia
Safarik Karel	CERN
Sannino Francesco	NORDITA
Santorelli Pietro	University of Naples
Stramaglia Sebastiano	INFN and University of Bari
Ullrich Thomas S.	BNL, BROOKHAVEN

Author Index

A

Agostino, L., 220
Alimonti, G., 220
Anjos, J., 220
Arena, V., 220
Aschenauer, E. C., 83

B

Banfi, A., 108
Becattini, F., 333
Bediaga, I., 220
Bianco, S., 220
Blümlein, J., 94
Boca, G., 220
Bonomi, G., 220
Boschini, M., 220
Böttcher, H., 94
Buccella, F., 248
Butler, J. N., 220

C

Caccianiga, B., 220
Carrillo, S., 220
Casalbuoni, R., 358
Casimiro, E., 220
Cawlfield, C., 220
Cheung, H. W. K., 220
Cho, K., 220
Chung, Y. S., 220
Cinquini, L., 220
Ciuchini, M., 180
Coriano', C., 145
Costa, M. J., 259
Cozza, D., 319
Cumalat, J. P., 220

D

D'Angelo, P., 220
Davenport III, T. F., 220
Deandrea, A., 253
de Miranda, J. M., 220
de Rafael, E., 14
De Roeck, A., 69
DiCorato, M., 220
Dini, P., 220
dos Reis, A. C., 220
Dürr, S., 40

E

Eeg, J. O., 242
Engh, D., 220

F

Fabbri, F. L., 220
Fajfer, S., 242
Faraggi, A. E., 145
Fazio, A. R., 139
Ferroni, F., 153
Forte, S., 60
Franco, E., 180
Fuster, J., 259

G

Gaines, I., 220
Garbincius, P. H., 220
Gardner, R., 220
Garren, L. A., 220
Giammarchi, M., 220
Gianini, G., 220
Gobel, C., 220
Godbole, R. M., 127
Gómez Nicola, A., 34
Gottschalk, E., 220
Grau, A., 127
Grozin, A. G., 271

H

Handler, T., 220
Heinz, U., 271

Heppelmann, S., 121
Hernandez, H., 220
Hosack, M., 220
Hurth, T., 212

I

Inzani, P., 220

J

Jeremie, H., 114
Johns, W. E., 220
Jones, R. W. L., 100

K

Kang, J. S., 220
Karsch, F., 323
Kasper, P. H., 220
Khodjamirian, A., 194
Kim, D. Y., 220
Ko, B. R., 220
Kreymer, A. E., 220
Kryemadhi, A., 220
Kurepin, A. B., 293
Kuschke, R., 220
Kwak, J. W., 220

L

Lanceri, L., 165
Lee, K. B., 220
Leutwyler, H., 3
Leveraro, F., 220
Liguori, G., 220
Link, J., 220
Lopez, A. M., 220

M

Madigojine, D., 28
Magnin, J., 220
Malvezzi, S., 220
Mannel, T., 212
Manzari, V., 299
Marchesini, G., 108
Martinelli, G., 180
Massafferri, A., 220
Menasce, D., 220
Mendez, H., 220
Mendez, L., 220
Merlo, M. M., 220
Mezzadri, M., 220
Milazzo, L., 220
Mitchell, R., 220
Montiel, E., 220
Moroni, L., 220

N

Nason, P., 51
Nehring, M., 220
Neubert, M., 168

O

Olaya, D., 220
O'Reilly, B., 220

P

Palano, A., 229
Pancheri, G., 127
Pantea, D., 220
Paris, A., 220
Park, H., 220
Pedrini, D., 220
Peláez, J. R., 34
Pepe, I. M., 220
Pettini, G., 333
Pham, T. N., 206
Pierini, M., 180
Pineda, A., 265
Pontoglio, C., 220
Prelz, F., 220

Q

Quinones, J., 220

R

Rahimi, A., 220
Rajagopal, K., 339
Ramirez, J. E., 220
Ratti, S. P., 220
Reyes, M., 220
Riccardi, C., 220
Rivera, C., 220
Rovere, M., 220

S

Sala, S., 220
Sánchez-Hernández, A., 220
Sannino, F., 352
Sarwar, S., 220
Segoni, I., 220
Sheaff, M., 220
Sheldon, P. D., 220
Silvestrini, L., 180
Smye, G., 108
Srivastava, Y. N., 127
Stenson, K., 220

T

Tortosa, P., 259

U

Ullrich, T. S., 307
Uribe, C., 220

V

Vaandering, E. W., 220
Vasquez, F., 220
Vitulo, P., 220

W

Webster, M., 220
Wilson, J. R., 220
Wiss, J., 220

X

Xiong, W., 220

Y

Yager, P. M., 220

Z

Zallo, A., 220
Zanderighi, G., 108
Zhang, Y., 220
Zupan, J., 242